OPERATIONAL AMPLIFIERS: Theory and Practice

OPERATIONAL AMPLIFIERS

Theory and Practice

JAMES K. ROBERGE

Massachusetts Institute of Technology

JOHN WILEY & SONS, Inc.

New York · London · Sydney · Toronto

Library of Congress Cataloging in Publication Data:

Roberge, James K   1938–
Operational amplifiers.

Includes bibliographical references and index.
1. Operational amplifiers.   2.   Feedback (Electronics)
I.   Title.
TK7871.58.06R62   629.8'315   75-2309
ISBN 0-471-72585-4

Printed in the United States of America

10 9 8 7 6 5 4 3 2 1

To Nancy

PREFACE

The operational amplifier is responsible for a dramatic and continuing revolution in our approach to analog system design. The availability of high performance, inexpensive devices influences the entire spectrum of circuits and systems, ranging from simple, mass-produced circuits to highly sophisticated equipment designed for complex data collection or processing operations. At one end of this spectrum, modern operational amplifiers have lowered cost and improved performance; at the other end, they allow us to design and implement systems that were previously too complex for consideration.

An appreciation of the importance of this component, gained primarily through research rather than academic experience, prompted me in 1969 to start a course at M.I.T. focusing on the operational amplifier. Initially the course, structured as part of an elective sequence in active devices, concentrated on the circuit techniques needed to realize operational amplifiers and on the application of these versatile elements.

As the course evolved, it became apparent that the operational amplifier had a value beyond that of a circuit component; it was also an excellent instructional vehicle. This device supplied a reason for studying a collection of analytic and design techniques that were necessary for a thorough understanding of operational amplifiers and were also important to the general area of active-circuit design. For example, if we study direct-coupled amplifiers in detail, with proper attention given to transistor-parameter variation with temperature, to loading, and to passive-component peculiarities, we can improve our approach to the design of a large class of circuits dependent on these concepts and also better appreciate operational amplifiers. Similarly, the use of an active load to increase dramatically the voltage gain of a stage is a design technique that has widespread applicability. The integrated-circuit fabrication and design methods responsible for the economical realization of modern operational amplifiers are the same as those used for other linear integrated circuits and also influence the design of many modern discrete-component circuits.

Chapters 7 to 10 reflect the dual role of the operational-amplifier circuit. The presentation is in greater detail than necessary if our only objective is to understand how an operational amplifier functions. However, the depth of the presentation encourages the transfer of this information to other circuit-design problems.

A course based on circuit-design techniques and some applications material was taught for two years. During this period, it became clear that in order to provide the background necessary for the optimum use of operational amplifiers in challenging applications, it was necessary to teach material on classical feedback concepts. These concepts explain the evolution of the topology used for modern amplifiers, suggest configurations that should be used to obtain specific closed-loop transfer functions, and indicate the way to improve the dynamics of operational-amplifier connections.

The linear-system theory course that has become an important part of most engineering educational programs, while providing valuable background, usually does not develop the necessary facility with techniques for the analysis and synthesis of feedback systems. When courses are offered in feedback, they normally use servomechanisms for their examples. Although this material can be transferred to a circuits context, the initial assimilation of these ideas is simplified when instruction is specifically tailored to the intended field of application.

Chapters 2 to 6 and Chapter 13 present the techniques necessary to model, analyze, and design electronic feedback systems. As with the circuit-related material, the detail is greater than the minimum necessary for a background in the design of connections that use operational amplifiers. This detail is justifiable because I use the operational amplifier as a vehicle for presenting concepts valuable for the general area of electronic circuit and system design.

The material included here has been used as the basis for two rather different versions of the M.I.T. course mentioned earlier. One of these concentrates on circuits and applications, using material from Chapters 7 to 10. Some application material is included in the examples in these chapters, and further applications from Chapters 11 and 12 are included as time permits. Some of the elementary feedback concepts necessary to appreciate modern operational-amplifier topologies are also discussed in this version.

The second variation uses the feedback material in Chapters 2 to 6 and Chapter 13 as its central theme. A brief discussion of the topology used for modern operational amplifiers, such as that presented in portions of Chapters 8 and 10, is included in this option. The applications introduced as examples of feedback connections are augmented with topics selected from Chapters 11 and 12.

A laboratory has been included as an integral part of both options. In the circuits variation, students investigate specific circuits such as direct-coupled amplifiers and high-gain stages, and conclude their laboratory experience by designing, building, and testing a simple operational amplifier. In the feedback version, connections of operational amplifiers are used to verify the behavior of linear and nonlinear feedback systems, to compare time-domain and frequency-domain performance indices, and to investigate stability.

We have found it helpful to have ready access to some kind of computational facilities, particularly when teaching the feedback material. The programs made available to the students reduce the manual effort required to draw the various plots and to factor polynomials when exact singularity locations are important.

Both versions of the course have been taught at least twice from notes essentially identical to the book. The student population consisted primarily of juniors and seniors, with occasional graduate students. The necessary background includes an appreciation of active-circuit concepts such as that provided in *Electronic Principles* by P. E. Gray and C. L. Searle (Wiley, New York, 1969), Chapters 1 to 14. An abbreviated circuits preparation is acceptable for the feedback version of the course. Although a detailed linear-systems background stressing formal operational calculus and related topics is not essential, familiarity with concepts such as pole-zero diagrams and elementary relationships between the time and the frequency domain is necessary.

Some of the more advanced applications in Chapters 11 and 12 have been included in a graduate course in analog and analog/digital instrumentation. The success with this material suggests a third possible variation of the course that stresses applications, with feedback and circuit concepts added as necessary to clarify the applications. I have not yet had the opportunity to structure an entire course in this way.

It is a pleasure to acknowledge several of the many individuals who contributed directly or indirectly to this book. High on the list are three teachers and colleagues, Dr. F. Williams Sarles, Jr., Professor Campbell L. Searle, and Professor Leonard A. Gould, who are largely responsible for my own understanding and appreciation of the presented material.

Two students, Jeffrey T. Millman and Samuel H. Maslak, devoted substantial effort to reviewing and improving the book.

Most of the original manuscript and its many revisions were typed and illustrated by Mrs. Janet Lague and Mrs. Rosalind Wood. Miss Susan Garland carefully proofread the final copy.

James K. Roberge

Cambridge, Massachusetts
February, 1975

CONTENTS

CHAPTER I

BACKGROUND AND OBJECTIVES

1.1 INTRODUCTION

An operational amplifier is a high-gain direct-coupled amplifier that is normally used in feedback connections. If the amplifier characteristics are satisfactory, the transfer function of the amplifier with feedback can often be controlled primarily by the stable and well-known values of passive feedback elements.

The term operational amplifier evolved from original applications in analog computation where these circuits were used to perform various mathematical operations such as summation and integration. Because of the performance and economic advantages of available units, present applications extend far beyond the original ones, and modern operational amplifiers are used as general purpose analog data-processing elements.

High-quality operational amplifiers[1] were available in the early 1950s. These amplifiers were generally committed to use with analog computers and were not used with the flexibility of modern units. The range of operational-amplifier usage began to expand toward the present spectrum of applications in the early 1960s as various manufacturers developed modular, solid-state circuits. These amplifiers were smaller, much more rugged, less expensive, and had less demanding power-supply requirements than their predecessors. A variety of these discrete-component circuits are currently available, and their performance characteristics are spectacular when compared with older units.

A quantum jump in usage occurred in the late 1960s, as monolithic integrated-circuit amplifiers with respectable performance characteristics evolved. While certain performance characteristics of these units still do not compare with those of the better discrete-component circuits, the integrated types have an undeniable cost advantage, with several designs available at prices of approximately $0.50. This availability frequently justifies the replacement of two- or three-transistor circuits with operational

---

[1] An excellent description of the technology of this era is available in G. A. Korn and T. M. Korn, *Electronic Analog Computers*, 2nd Ed., McGraw-Hill, New York, 1956.

amplifiers on economic grounds alone, independent of associated perform-ance advantages. As processing and designs improve, the integrated circuit will invade more areas once considered exclusively the domain of the discrete design, and it is probable that the days of the discrete-component circuit, except for specials with limited production requirements, are numbered.

There are several reasons for pursuing a detailed study of operational amplifiers. We must discuss both the theoretical and the practical aspects of these versatile devices rather than simply listing a representative sample of their applications. Since virtually all operational-amplifier connections involve some form of feedback, a thorough understanding of this process is central to the intelligent application of the devices. While partially under-stood rules of thumb may suffice for routine requirements, this design method fails as performance objectives approach the maximum possible use from the amplifier in question.

Similarly, an appreciation of the internal structure and function of opera-tional amplifiers is imperative for the serious user, since such information is necessary to determine various limitations and to indicate how a unit may be modified (via, for example, appropriate connections to its com-pensation terminals) or connected for optimum performance in a given application. The modern analog circuit designer thus needs to understand the internal function of an operational amplifier (even though he may never design one) for much the same reason that his counterpart of 10 years ago required a knowledge of semiconductor physics. Furthermore, this is an area where good design practice has evolved to a remarkable degree, and many of the circuit techniques that are described in following chapters can be applied to other types of electronic circuit and system design.

1.2   THE CLOSED-LOOP GAIN OF AN OPERATIONAL AMPLIFIER

As mentioned in the introduction, most operational-amplifier connec-tions involve feedback. Therefore the user is normally interested in deter-mining the *closed-loop gain* or *closed-loop transfer function* of the amplifier, which results when feedback is included. As we shall see, this quantity can be made primarily dependent on the characteristics of the feedback ele-ments in many cases of interest.

A prerequisite for the material presented in the remainder of this book is the ability to determine the gain of the amplifier-feedback network com-bination in simple connections. The techniques used to evaluate closed-loop gain are outlined in this section.

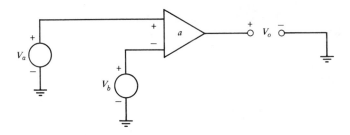

Figure 1.1 Symbol for an operational amplifier.

1.2.1 Closed-Loop Gain Calculation

The symbol used to designate an operational amplifier is shown in Fig. 1.1. The amplifier shown has a differential input and a single output. The input terminals marked $-$ and $+$ are called the *inverting* and the *non-inverting* input terminals respectively. The implied linear-region relationship among input and output variables[2] is

$$V_o = a(V_a - V_b) \qquad (1.1)$$

The quantity a in this equation is the *open-loop gain* or *open-loop transfer function* of the amplifier. (Note that a gain of a is assumed, even if it is not explicitly indicated inside the amplifier symbol.) The dynamics normally associated with this transfer function are frequently emphasized by writing $a(s)$.

It is also necessary to provide operating power to the operational amplifier via power-supply terminals. Many operational amplifiers use balanced (equal positive and negative) supply voltages. The various signals are usually referenced to the common ground connection of these power sup-

---

[2] The notation used to designate system variables consists of a symbol and a subscript. This combination serves not only as a label, but also to identify the nature of the quantity as follows:

Total instantaneous variables:
   lower-case symbols with upper-case subscripts.
Quiescent or operating-point variables:
   upper-case symbols with upper-case subscripts.
Incremental instantaneous variables:
   lower-case symbols with lower-case subscripts.
Complex amplitudes or Laplace transforms of incremental variables:
   upper-case symbols with lower-case subscripts.

Using this notation we would write $v_I = V_I + v_i$, indicating that the instantaneous value of v_I consists of a quiescent plus an incremental component. The transform of v_i is V_i. The notation $V_i(s)$ is often used to reinforce the fact that V_i is a function of the complex variable s.

plies. The power connections are normally not included in diagrams intended only to indicate relationships among signal variables, since eliminating these connections simplifies the diagram.

Although operational amplifiers are used in a myriad of configurations,
many applications are variations of either the inverting connection (Fig.
1.2a) or the noninverting connection (Fig. 1.2b). These connections combine the amplifier with impedances that provide feedback.

The closed-loop transfer function is calculated as follows for the inverting connection. Because of the reference polarity chosen for the intermediate variable V_a,

$$V_o = -aV_a \qquad (1.2)$$

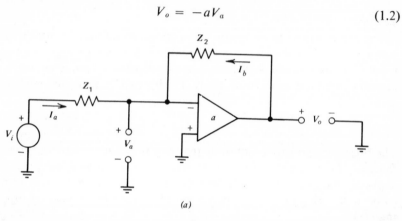

(a)

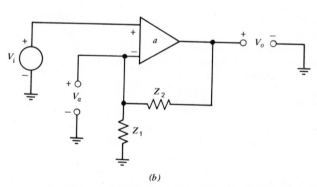

(b)

Figure 1.2 Operational-amplifier connections. (a) Inverting. (b) Noninverting.

where it has been assumed that the output voltage of the amplifier is not modified by the loading of the Z_1-Z_2 network. If the input impedance of the amplifier itself is high enough so that the Z_1-Z_2 network is not loaded significantly, the voltage V_a is

$$V_a = \left(\frac{Z_2}{Z_1 + Z_2}\right) V_i + \left(\frac{Z_1}{Z_1 + Z_2}\right) V_o \tag{1.3}$$

Combining Eqns. 1.2 and 1.3 yields

$$V_o = -\left(\frac{aZ_2}{Z_1 + Z_2}\right) V_i - \left(\frac{aZ_1}{Z_1 + Z_2}\right) V_o \tag{1.4}$$

or, solving for the closed-loop gain,

$$\frac{V_o}{V_i} = \frac{-aZ_2/(Z_1 + Z_2)}{1 + [aZ_1/(Z_1 + Z_2)]} \tag{1.5}$$

The condition that is necessary to have the closed-loop gain depend primarily on the characteristics of the Z_1-Z_2 network rather than on the performance of the amplifier itself is easily determined from Eqn. 1.5. At any frequency ω where the inequality $|a(j\omega)Z_1(j\omega)/[Z_1(j\omega) + Z_2(j\omega)]| \gg 1$ is satisfied, Eqn. 1.5 reduces to

$$\frac{V_o(j\omega)}{V_i(j\omega)} \simeq -\frac{Z_2(j\omega)}{Z_1(j\omega)} \tag{1.6}$$

The closed-loop gain calculation for the noninverting connection is similar. If we assume negligible loading at the amplifier input and output,

$$V_o = a(V_i - V_a) = aV_i - \left(\frac{aZ_1}{Z_1 + Z_2}\right) V_o \tag{1.7}$$

or

$$\frac{V_o}{V_i} = \frac{a}{1 + [aZ_1/(Z_1 + Z_2)]} \tag{1.8}$$

This expression reduces to

$$\frac{V_o(j\omega)}{V_i(j\omega)} \simeq \frac{Z_1(j\omega) + Z_2(j\omega)}{Z_1(j\omega)} \tag{1.9}$$

when $|a(j\omega)Z_1(j\omega)/[Z_1(j\omega) + Z_2(j\omega)]| \gg 1$.

The quantity

$$L = \frac{-aZ_1}{Z_1 + Z_2} \tag{1.10}$$

is the *loop transmission* for either of the connections of Fig. 1.2. The loop transmission is of fundamental importance in any feedback system because it influences virtually all closed-loop parameters of the system. For example, the preceding discussion shows that if the magnitude of loop transmission is large, the closed-loop gain of either the inverting or the non-inverting amplifier connection becomes virtually independent of a. This relationship is valuable, since the passive feedback components that determine closed-loop gain for large loop-transmission magnitude are normally considerably more stable with time and environmental changes than is the open-loop gain a.

The loop transmission can be determined by setting the inputs of a feedback system to zero and breaking the signal path at any point inside the feedback loop.[3] The loop transmission is the ratio of the signal returned by the loop to a test applied at the point where the loop is opened. Figure 1.3 indicates one way to determine the loop transmission for the connections of Fig. 1.2. Note that the topology shown is common to both the inverting and the noninverting connection when input points are grounded.

It is important to emphasize the difference between the loop transmission, which is dependent on properties of both the feedback elements and the operational amplifier, and the open-loop gain of the operational amplifier itself.

1.2.2   The Ideal Closed-Loop Gain

Detailed gain calculations similar to those of the last section are always possible for operational-amplifier connections. However, operational amplifiers are frequently used in feedback connections where loop characteristics are such that the closed-loop gain is determined primarily by the feedback elements. Therefore, approximations that indicate the *ideal closed-loop gain* or the gain that results with perfect amplifier characteristics simplify the analysis or design of many practical connections.

It is possible to calculate the ideal closed-loop gain assuming only two conditions (in addition to the implied condition that the amplifier-feedback network combination is stable[4]) are satisfied.

1. A negligibly small differential voltage applied between the two input terminals of the amplifier is sufficient to produce any desired output voltage.

---

[3] There are practical difficulties, such as insuring that the various elements in the loop remain in their linear operating regions and that loading is maintained. These difficulties complicate the determination of the loop transmission in physical systems. Therefore, the technique described here should be considered a conceptual experiment. Methods that are useful for actual hardware are introduced in later sections.

[4] Stability is discussed in detail in Chapter 4.

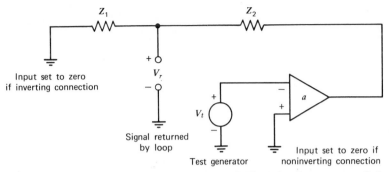

Figure 1.3 Loop transmission for connections of Fig. 1.2. Loop transmission is $V_r/V_t = -a\,Z_1/(Z_1 + Z_2)$.

2. The current required at either amplifier terminal is negligibly small.

The use of these assumptions to calculate the ideal closed-loop gain is first illustrated for the inverting amplifier connection (Fig. 1.2a). Since the noninverting amplifier input terminal is grounded in this connection, condition 1 implies that

$$V_a \simeq 0 \qquad\qquad (1.11)$$

Kirchhoff's current law combined with condition 2 shows that

$$I_a + I_b \simeq 0 \qquad\qquad (1.12)$$

With Eqn. 1.11 satisfied, the currents I_a and I_b are readily determined in terms of the input and output voltages.

$$I_a \simeq \frac{V_i}{Z_1} \qquad\qquad (1.13)$$

$$I_b \simeq \frac{V_o}{Z_2} \qquad\qquad (1.14)$$

Combining Eqns. 1.12, 1.13, and 1.14 and solving for the ratio of V_o to V_i yields the ideal closed-loop gain

$$\frac{V_o}{V_i} = -\frac{Z_2}{Z_1} \qquad\qquad (1.15)$$

The technique used to determine the ideal closed-loop gain is called the *virtual-ground* method when applied to the inverting connection, since in this case the inverting input terminal of the operational amplifier is assumed to be at ground potential.

The noninverting amplifier (Fig. 1.2b) provides a second example of ideal-gain determination. Condition 2 insures that the voltage V_a is not influenced by current at the inverting input. Thus,

$$V_a \simeq \frac{Z_1}{Z_1 + Z_2} V_o \qquad (1.16)$$

Since condition 1 requires equality between V_a and V_i, the ideal closed-loop gain is

$$\frac{V_o}{V_i} = \frac{Z_1 + Z_2}{Z_1} \qquad (1.17)$$

The conditions can be used to determine ideal values for characteristics other than gain. Consider, for example, the input impedance of the two amplifier connections shown in Fig. 1.2. In Fig. 1.2a, the inverting input terminal and, consequently, the right-hand end of impedance Z_1, is at ground potential if the amplifier characteristics are ideal. Thus the input impedance seen by the driving source is simply Z_1. The input source is connected directly to the noninverting input of the operational amplifier in the topology of Fig. 1.2b. If the amplifier satisfies condition 2 and has negligible input current required at this terminal, the impedance loading the signal source will be very high. The noninverting connection is often used as a buffer amplifier for this reason.

The two conditions used to determine the ideal closed-loop gain are deceptively simple in that a complex combination of amplifier characteristics are required to insure satisfaction of these conditions. Consider the first condition. High open-loop voltage gain at anticipated operating frequencies is necessary but not sufficient to guarantee this condition. Note that gain at the frequency of interest is necessary, while the high open-loop gain specified by the manufacturer is normally measured at d-c. This specification is somewhat misleading, since the gain may start to decrease at a frequency on the order of one hertz or less.

In addition to high open-loop gain, the amplifier must have low voltage offset[5] referred to the input to satisfy the first condition. This quantity, defined as the voltage that must be applied between the amplifier input terminals to make the output voltage zero, usually arises because of mismatches between various amplifier components.

Surprisingly, the incremental input impedance of an operational amplifier often has relatively little effect on its input current, since the voltage that appears across this impedance is very low if condition 1 is satisfied.

[5] Offset and other problems with d-c amplifiers are discussed in Chapter 7.

A more important contribution to input current often results from the bias current that must be supplied to the amplifier input transistors.

Many of the design techniques that are used in an attempt to combine the two conditions necessary to approach the ideal gain are described in subsequent sections.

The reason that the satisfaction of the two conditions introduced earlier guarantees that the actual closed-loop gain of the amplifier approaches the ideal value is because of the negative feedback associated with operational-amplifier connections. Assume, for example, that the actual voltage out of the inverting-amplifier connection shown in Fig. 1.2a is more positive than the value predicted by the ideal-gain relationship for a particular input signal level. In this case, the voltage V_a will be positive, and this positive voltage applied to the inverting input terminal of the amplifier drives the output voltage negative until equilibrium is reached. This reasoning shows that it is actually the negative feedback that forces the voltage between the two input terminals to be very small.

Alternatively, consider the situation that results if positive feedback is used by interchanging the connections to the two input terminals of the

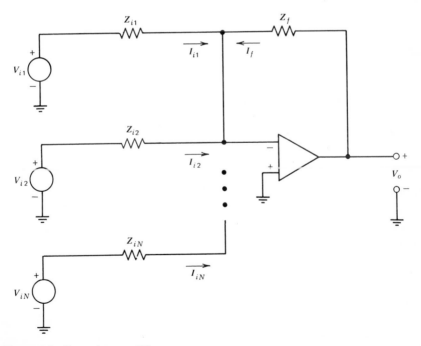

Figure 1.4   Summing amplifier.

amplifier. In this case, the voltage V_a is again zero when V_o and V_i are related by the ideal closed-loop gain expression. However, the resulting equilibrium is unstable, and a small perturbation from the ideal output voltage results in this voltage being driven further from the ideal value until the amplifier saturates. The ideal gain is not achieved in this case in spite of perfect amplifier characteristics because the connection is unstable. As we shall see, negative feedback connections can also be unstable. The ideal gain of these unstable systems is meaningless because they oscillate, producing an output signal that is often nearly independent of the input signal.

1.2.3   Examples

The technique introduced in the last section can be used to determine the ideal closed-loop transfer function of any operational-amplifier connection. The summing amplifier shown in Fig. 1.4 illustrates the use of this technique for a connection slightly more complex than the two basic amplifiers discussed earlier.

Since the inverting input terminal of the amplifier is a virtual ground, the currents can be determined as

$$I_{i1} = \frac{V_{i1}}{Z_{i1}}$$

$$I_{i2} = \frac{V_{i2}}{Z_{i2}}$$

$$.$$
$$.$$
$$.$$

$$I_{iN} = \frac{V_{iN}}{Z_{iN}}$$

$$I_f = \frac{V_o}{Z_f} \qquad (1.18)$$

These currents must sum to zero in the absence of significant current at the inverting input terminal of the amplifier. Thus

$$I_{i1} + I_{i2} + \cdots + I_{iN} + I_f = 0 \qquad (1.19)$$

Combining Eqns. 1.18 and 1.19 shows that

$$V_o = -\frac{Z_f}{Z_{i1}} V_{i1} - \frac{Z_f}{Z_{i2}} V_{i2} - \cdots - \frac{Z_f}{Z_{iN}} V_{iN} \qquad (1.20)$$

We see that this amplifier, which is an extension of the basic inverting-amplifier connection, provides an output that is the weighted sum of several input voltages.

Summation is one of the "operations" that operational amplifiers perform in analog computation. A subsequent development (Section 12.3) will show that if the operations of gain, summation, and integration are combined, an electrical network that satisfies any linear, ordinary differential equation can be constructed. This technique is the basis for analog computation.

Integrators required for analog computation or for any other application can be constructed by using an operational amplifier in the inverting connection (Fig. 1.2a) and making impedance Z_2 a capacitor C and impedance Z_1 a resistor R. In this case, Eqn. 1.15 shows that the ideal closed-loop transfer function is

$$\frac{V_o(s)}{V_i(s)} = - \frac{Z_2(s)}{Z_1(s)} = - \frac{1}{RCs} \tag{1.21}$$

so that the connection functions as an inverting integrator.

It is also possible to construct noninverting integrators using an operational amplifier connected as shown in Fig. 1.5. This topology precedes a noninverting amplifier with a low-pass filter. The ideal transfer function from the noninverting input of the amplifier to its output is (see Eqn. 1.17)

$$\frac{V_o(s)}{V_a(s)} = \frac{RCs + 1}{RCs} \tag{1.22}$$

Since the conditions for an ideal operational amplifier preclude input cur-

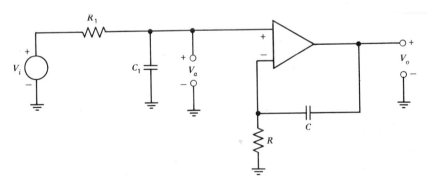

Figure 1.5   Noninverting integrator.

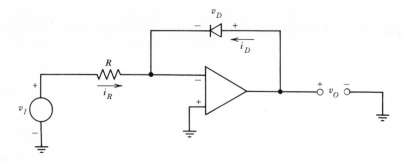

Figure 1.6 Log circuit.

rent, the transfer function from V_i to V_a can be calculated with no loading, and in this case

$$\frac{V_a(s)}{V_i(s)} = \frac{1}{R_1C_1s + 1} \tag{1.23}$$

Combining Eqns. 1.22 and 1.23 shows that the ideal closed-loop gain is

$$\frac{V_o(s)}{V_i(s)} = \left[\frac{1}{R_1C_1s + 1}\right]\left[\frac{RCs + 1}{RCs}\right] \tag{1.24}$$

If the two time constants in Eqn. 1.24 are made equal, noninverting integration results.

The comparison between the two integrator connections hints at the possibility of realizing most functions via either an inverting or a noninverting connection. Practical considerations often recommend one approach in preference to the other. For example, the noninverting integrator requires more external components than does the inverting version. This difference is important because the high-quality capacitors required for accurate integration are often larger and more expensive than the operational amplifier that is used.

The examples considered up to now have involved only linear elements, at least if it is assumed that the operational amplifier remains in its linear operating region. Operational amplifiers are also frequently used in intentionally nonlinear connections. One possibility is the circuit shown in Fig. 1.6.[6] It is assumed that the diode current-voltage relationship is

$$i_D = I_S(e^{qv_D/kT} - 1) \tag{1.25}$$

[6] Note that the notation for the variables used in this case combines lower-case variables with upper-case subscripts, indicating the total instantaneous signals necessary to describe the anticipated nonlinear relationships.

where I_S is a constant dependent on diode construction, q is the charge of an electron, k is Boltzmann's constant, and T is the absolute temperature.

If the voltage at the inverting input of the amplifier is negligibly small, the diode voltage is equal to the output voltage. If the input current is negligibly small, the diode current and the current i_R sum to zero. Thus, if these two conditions are satisfied,

$$- \frac{v_I}{R} = I_S(e^{qv_O/kT} - 1) \tag{1.26}$$

Consider operation with a positive input voltage. The maximum negative value of the diode current is limited to $-I_S$. If $v_I/R > I_S$, the current through the reverse-biased diode cannot balance the current I_R. Accordingly, the amplifier output voltage is driven negative until the amplifier saturates. In this case, the feedback loop cannot keep the voltage at the inverting amplifier input near ground because of the limited current that the diode can conduct in the reverse direction. The problem is clearly not with the amplifier, since no solution exists to Eqn. 1.26 for sufficiently positive values of v_I.

This problem does not exist with negative values for v_I. If the magnitude of i_R is considerably larger than I_S (typical values for I_S are less than 10^{-9} A), Eqn. 1.26 reduces to

$$- \frac{v_I}{R} \simeq I_S e^{qv_O/kT} \tag{1.27}$$

or

$$v_O \simeq \frac{kT}{q} \ln \left(\frac{-v_I}{RI_S} \right) \tag{1.28}$$

Thus the circuit provides an output voltage proportional to the log of the magnitude of the input voltage for negative inputs.

1.3 OVERVIEW

The operational amplifier is a powerful, multifaceted analog data-processing element, and the optimum exploitation of this versatile building block requires a background in several different areas. The primary objective of this book is to help the reader apply operational amplifiers to his own problems. While the use of a "handbook" approach that basically tabulates a number of configurations that others have found useful is attractive because of its simplicity, this approach has definite limitations. Superior results are invariably obtained when the designer tailors the circuit he uses to his own specific, detailed requirements, and to the particular operational amplifier he chooses.

A balanced presentation that combines practical circuit and system design concepts with applicable theory is essential background for the type of creative approach that results in optimum operational-amplifier systems. The following chapters provide the necessary concepts. A second advantage of this presentation is that many of the techniques are readily applied to a wide spectrum of circuit and system design problems, and the material is structured to encourage this type of transfer.

Feedback is central to virtually all operational-amplifier applications, and a thorough understanding of this important topic is necessary in any challenging design situation. Chapters 2 through 6 are devoted to feedback concepts, with emphasis placed on examples drawn from operational-amplifier connections. However, the presentation in these chapters is kept general enough to allow its application to a wide variety of feedback systems. Topics covered include modeling, a detailed study of the advantages and limitations of feedback, determination of responses, stability, and compensation techniques intended to improve stability. Simple methods for the analysis of certain types of nonlinear systems are also included. This indepth approach is included at least in part because I am convinced that a detailed understanding of feedback is the single most important prerequisite to successful electronic circuit and system design.

Several interesting and widely applicable circuit-design techniques are used to realize operational amplifiers. The design of operational-amplifier circuits is complicated by the requirement of obtaining gain at zero frequency with low drift and input current. Chapter 7 discusses the design of the necessary d-c amplifiers. The implications of topology on the dynamics of operational-amplifier circuits are discussed in Chapter 8. The design of the high-gain stages used in most modern operational amplifiers and the factors which influence output-stage performance are also included. Chapter 9 illustrates how circuit design techniques and feedback-system concepts are combined in an illustrative operational-amplifier circuit.

The factors influencing the design of the modern integrated-circuit operational amplifiers that have dramatically increased amplifier usage are discussed in Chapter 10. Several examples of representative present-day designs are included.

A variety of operational-amplifier applications are sprinkled throughout the first 10 chapters to illustrate important concepts. Chapters 11 and 12 focus on further applications, with major emphasis given to clarifying important techniques and topologies rather than concentrating on minor details that are highly dependent on the specifics of a given application and the amplifier used.

Chapter 13 is devoted to the problem of compensating operational amplifiers for optimum dynamic performance in a variety of applications. Discussion of this material is deferred until the final chapter because only then is the feedback, circuit, and application background necessary to fully appreciate the subtleties of compensating modern operational amplifiers available. Compensation is probably the single most important aspect of effectively applying operational amplifiers, and often represents the difference between inadequate and superlative performance. Several examples of the way in which compensation influences the performance of a representative integrated-circuit operational amplifier are used to reinforce the theoretical discussion included in this chapter.

PROBLEMS

P1.1
Design a circuit using a single operational amplifier that provides an ideal input-output relationship

$$V_o = -V_{i1} - 2V_{i2} - 3V_{i3}$$

Keep the values of all resistors used between 10 and 100 kΩ.

Determine the loop transmission (assuming no loading) for your design.

P1.2
Note that it is possible to provide an ideal input-output relationship

$$V_o = V_{i1} + 2V_{i2} + 3V_{i3}$$

by following the design for Problem 1.1 with a unity-gain inverter. Find a more efficient design that produces this relationship using only a single operational amplifier.

P1.3
An operational amplifier is connected to provide an inverting gain with an ideal value of 10. At low frequencies, the open-loop gain of the amplifier is frequency independent and equal to a_0. Assuming that the only source of error is the finite value of open-loop gain, how large should a_0 be so that the actual closed-loop gain of the amplifier differs from its ideal value by less than 0.1%?

P1.4
Design a single-amplifier connection that provides the ideal input-output relationship

$$v_o = -100 \int (v_{i1} + v_{i2}) \, dt$$

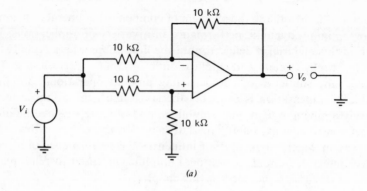

(a)

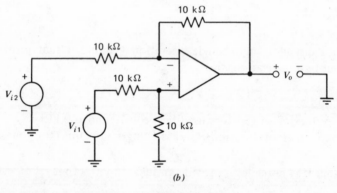

(b)

Figure 1.7 Differential-amplifier connections.

Keep the values of all resistors you use between 10 and 100 kΩ.

P1.5

Design a single-amplifier connection that provides the ideal input-output relationship

$$v_o = +100 \int (v_{i1} + v_{i2}) \, dt$$

using only resistor values between 10 and 100 kΩ. Determine the loop transmission of your configuration, assuming negligible loading.

P1.6

Determine the ideal input-output relationships for the two connections shown in Fig. 1.7.

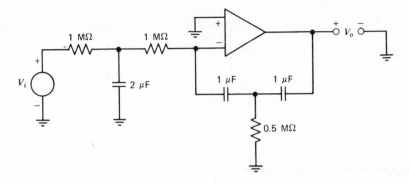

Figure 1.8 Two-pole system.

P1.7

Determine the ideal input-output transfer function for the operational-amplifier connection shown in Fig. 1.8. Estimate the value of open-loop gain required such that the actual closed-loop gain of the circuit approaches its ideal value at an input frequency of 0.01 radian per second. You may neglect loading.

P1.8

Assume that the operational-amplifier connection shown in Fig. 1.9 satisfies the two conditions stated in Section 1.2.2. Use these conditions to determine the output resistance of the connection (i.e., the resistance seen by the load).

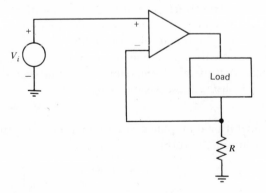

Figure 1.9 Circuit with controlled output resistance.

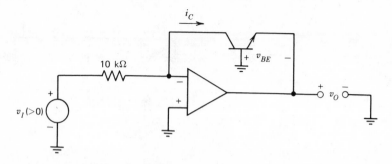

Figure 1.10   Log circuit.

P1.9

Determine the ideal input-output transfer relationship for the circuit shown in Fig. 1.10. Assume that transistor terminal variables are related as

$$i_C = 10^{-13}e^{40\,v_{BE}}$$

where i_C is expressed in amperes and v_{BE} is expressed in volts.

P1.10

Plot the ideal input-output characteristics for the two circuits shown in Fig. 1.11. In part a, assume that the diode variables are related by $i_D = 10^{-13}e^{40\,v_D}$, where i_D is expressed in amperes and v_D is expressed in volts. In part b, assume that $i_D = 0$, $v_D < 0$, and $v_D = 0$, $i_D > 0$.

P1.11

We have concentrated on operational-amplifier connections involving negative feedback. However, several useful connections, such as that shown in Fig. 1.12, use positive feedback around an amplifier. Assume that the linear-region open-loop gain of the amplifier is very high, but that its output voltage is limited to ±10 volts because of saturation of the amplifier output stage. Approximate and plot the output signal for the circuit shown in Fig. 1.12 using these assumptions.

P1.12

Design an operational-amplifier circuit that provides an ideal input-output relationship of the form

$$v_O = K_1 e^{v_I/K_2}$$

where K_1 and K_2 are constants dependent on parameter values used in your design.

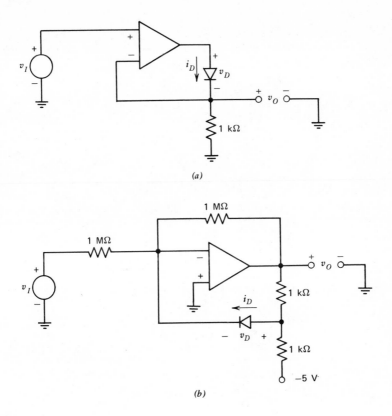

(a)

(b)

Figure 1.11 Nonlinear circuits.

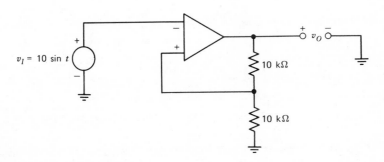

$v_I = 10 \sin t$

Figure 1.12 Schmitt trigger.

19

CHAPTER II

PROPERTIES AND MODELING OF FEEDBACK SYSTEMS

2.1 INTRODUCTION

A *control system* is a system that regulates an output variable with the objective of producing a given relationship between it and an input variable or of maintaining the output at a fixed value. In a feedback control system, at least part of the information used to change the output variable is derived from measurements performed on the output variable itself. This type of *closed-loop* control is often used in preference to *open-loop* control (where the system does not use output-variable information to influence its output) since feedback can reduce the sensitivity of the system to externally applied disturbances and to changes in system parameters. Familiar examples of feedback control systems include residential heating systems, most high-fidelity audio amplifiers, and the iris-retina combination that regulates light entering the eye.

There are a variety of textbooks[1] available that provide detailed treatment on servomechanisms, or feedback control systems where at least one of the variables is a mechanical quantity. The emphasis in this presentation is on feedback amplifiers in general, with particular attention given to feedback connections which include operational amplifiers.

The operational amplifier is a component that is used almost exclusively in feedback connections; therefore a detailed knowledge of the behavior of feedback systems is necessary to obtain maximum performance from these amplifiers. For example, the open-loop transfer function of many operational amplifiers can be easily and predictably modified by means of external

---

[1] G. S. Brown and D. P. Cambell, *Principles of Servomechanisms*, Wiley, New York, 1948; J. G. Truxal, *Automatic Feedback Control System Synthesis*, McGraw-Hill, New York, 1955; H. Chestnut and R. W. Mayer, *Servomechanisms and Regulating System Design*, Vol. 1, 2nd Ed., Wiley, New York, 1959; R. N. Clark, *Introduction to Automatic Control Systems*, Wiley, New York, 1962; J. J. D'Azzo and C. H. Houpis, *Feedback Control System Analysis and Synthesis*, 2nd Ed., McGraw-Hill, New York, 1966; B. C. Kuo, *Automatic Control Systems*, 2nd Ed., Prentice-Hall, Englewood Cliffs, New Jersey, 1967; K. Ogata, *Modern Control Engineering*, Prentice-Hall, Englewood Cliffs, New Jersey, 1970.

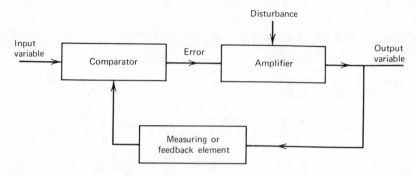

Figure 2.1   A typical feedback system.

components. The choice of the open-loop transfer function used for a particular application must be based on feedback principles.

2.2 SYMBOLOGY

Elements common to many electronic feedback systems are shown in Fig. 2.1. The input signal is applied directly to a comparator. The output signal is determined and possibly operated upon by a feedback element. The difference between the input signal and the modified output signal is determined by the comparator and is a measure of the error or amount by which the output differs from its desired value. An amplifier drives the output in such a way as to reduce the magnitude of the error signal. The system output may also be influenced by disturbances that affect the amplifier or other elements.

We shall find it convenient to illustrate the relationships among variables in a feedback connection, such as that shown in Fig. 2.1, by means of *block diagrams*. A block diagram includes three types of elements.

1. A *line* represents a variable, with an arrow on the line indicating the direction of information flow. A line may split, indicating that a single variable is supplied to two or more portions of the system.

2. A *block* operates on an input supplied to it to provide an output.

3. Variables are added algebraically at a summation point drawn as follows:

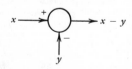

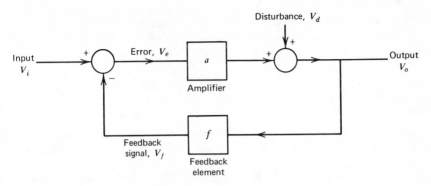

Figure 2.2 Block diagram for the system of Fig. 2.1.

One possible representation for the system of Fig. 2.1, assuming that the input, output, and disturbance are voltages, is shown in block-diagram form in Fig. 2.2. (The voltages are all assumed to be measured with respect to references or grounds that are not shown.) The block diagram implies a specific set of relationships among system variables, including:

1. The error is the difference between the input signal and the feedback signal, or $V_e = V_i - V_f$.
2. The output is the sum of the disturbance and the amplified error signal, or $V_o = V_d + aV_e$.
3. The feedback signal is obtained by operating on the output signal with the feedback element, or $V_f = fV_o$.

The three relationships can be combined and solved for the output in terms of the input and the disturbance, yielding

$$V_o = \frac{aV_i}{1 + af} + \frac{V_d}{1 + af} \qquad (2.1)$$

2.3 ADVANTAGES OF FEEDBACK

There is a frequent tendency on the part of the uninitiated to associate almost magical properties to feedback. Closer examination shows that many assumed benefits of feedback are illusory. The principal advantage is that feedback enables us to reduce the sensitivity of a system to changes in gain of certain elements. This reduction in sensitivity is obtained only in exchange for an increase in the magnitude of the gain of one or more of the elements in the system.

In some cases it is also possible to reduce the effects of disturbances applied to the system. We shall see that this moderation can always, at least conceptually, be accomplished without feedback, although the feedback approach is frequently a more practical solution. The limitations of this technique preclude reduction of such quantities as noise or drift at the input of an amplifier; thus feedback does not provide a method for detecting signals that cannot be detected by other means.

Feedback provides a convenient method of modifying the input and output impedance of amplifiers, although as with disturbance reduction, it is at least conceptually possible to obtain similar results without feedback.

2.3.1 Effect of Feedback on Changes in Open-Loop Gain

As mentioned above, the principal advantage of feedback systems compared with open-loop systems is that feedback provides a method for reducing the sensitivity of the system to changes in the gain of certain elements. This advantage can be illustrated using the block diagram of Fig. 2.2. If the disturbance is assumed to be zero, the closed-loop gain for the system is

$$\frac{V_o}{V_i} = \frac{a}{1 + af} \stackrel{\Delta}{=} A \tag{2.2}$$

(We will frequently use the capital letter A to denote closed-loop gain, while the lower-case a is normally reserved for a forward-path gain.)

The quantity af is the negative of the *loop transmission* for this system. The loop transmission is determined by setting all external inputs (and disturbances) to zero, breaking the system at any point inside the loop, and determining the ratio of the signal returned by the system to an applied test input.[2] If the system is a *negative feedback system*, the loop transmission is negative. The negative sign on the summing point input that is included in the loop shown in Fig. 2.2 indicates that the feedback is negative for this system if a and f have the same sign. Alternatively, the inversion necessary for negative feedback might be supplied by either the amplifier or the feedback element.

Equation 2.2 shows that negative feedback lowers the magnitude of the gain of an amplifier since as f is increased from zero, the magnitude of the closed-loop gain decreases if a and f have this same sign. The result is general and can be used as a test for negative feedback.

It is also possible to design systems with positive feedback. Such systems are not as useful for our purposes and are not considered in detail.

The closed-loop gain expression shows that as the loop-transmission magnitude becomes large compared to unity, the closed-loop gain ap-

---

[2] An example of this type of calculation is given in Section 2.4.1.

proaches the value $1/f$. The significance of this relationship is as follows. The amplifier will normally include active elements whose characteristics vary as a function of age and operating conditions. This uncertainty may be unavoidable in that active elements are not available with the stability required for a given application, or it may be introduced as a compromise in return for economic or other advantages.

Conversely, the feedback network normally attenuates signals, and thus can frequently be constructed using only passive components. Fortunately, passive components with stable, precisely known values are readily available. If the magnitude of the loop transmission is sufficiently high, the closed-loop gain becomes dependent primarily on the characteristics of the feedback network.

This feature can be emphasized by calculating the fractional change in closed-loop gain $d(V_o/V_i)/(V_o/V_i)$ caused by a given fractional change in amplifier forward-path gain da/a, with the result

$$\frac{d(V_o/V_i)}{(V_o/V_i)} = \frac{da}{a}\left(\frac{1}{1 + af}\right) \tag{2.3}$$

Equation 2.3 shows that changes in the magnitude of a can be attenuated to insignificant levels if af is sufficiently large. The quantity $1 + af$ that relates changes in forward-path gain to changes in closed-loop gain is frequently called the *desensitivity* of a feedback system. Figure 2.3 illustrates this desensitization process by comparing two amplifier connections intended to give an input-output gain of 10. Clearly the input-output gain is identically equal to a in Fig. 2.3a, and thus has the same fractional change in gain as does a. Equations 2.2 and 2.3 show that the closed-loop gain for the system of Fig. 2.3b is approximately 9.9, and that the fractional change in closed-loop gain is less than 1% of the fractional change in the forward-path gain of this system.

The desensitivity characteristic of the feedback process is obtained only in exchange for excess gain provided in the system. Returning to the example involving Fig. 2.2, we see that the closed-loop gain for the system is $a/(1 + af)$, while the forward-path gain provided by the amplifier is a. The desensitivity is identically equal to the ratio of the forward-path gain to closed-loop gain. Feedback connections are unique in their ability to automatically trade excess gain for desensitivity.

It is important to underline the fact that changes in the gain of the feedback element have direct influence on the closed-loop gain of the system, and we therefore conclude that it is necessary to observe or measure the output variable of a feedback system accurately in order to realize the advantages of feedback.

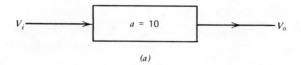

(a)

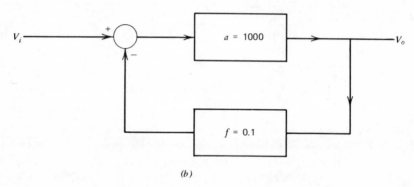

(b)

Figure 2.3 Amplifier connections for a gain of ten. (a) Open loop. (b) Closed loop.

2.3.2 Effect of Feedback on Nonlinearities

Because feedback reduces the sensitivity of a system to changes in open-loop gain, it can often moderate the effects of nonlinearities. Figure 2.4 illustrates this process. The forward path in this connection consists of an amplifier with a gain of 1000 followed by a nonlinear element that might be an idealized representation of the transfer characteristics of a power output stage. The transfer characteristics of the nonlinear element show these four distinct regions:

1. A deadzone, where the output remains zero until the input magnitude exceeds 1 volt. This region models the crossover distortion associated with many types of power amplifiers.
2. A linear region, where the incremental gain of the element is one.
3. A region of soft limiting, where the incremental gain of the element is lowered to 0.1.

4. A region of hard limiting or saturation where the incremental gain of the element is zero.

The performance of the system can be determined by recognizing that, since the nonlinear element is piecewise linear, all transfer relationships must be piecewise linear. The values of all the variables at a breakpoint can be found by an iterative process. Assume, for example, that the variables associated with the nonlinear element are such that this element is at its breakpoint connecting a slope of zero to a slope of $+1$. This condition only occurs for $v_A = 1$ and $v_B = 0$. If $v_B = v_O = 0$, the signal v_F must be zero, since $v_F = 0.1\, v_O$. Similarly, with $v_A = 1$, $v_E = 10^{-3}v_A = 10^{-3}$. Since the relationships at the summing point imply $v_E = v_I - v_F$, or $v_I = v_E + v_F$, v_I must equal 10^{-3}. The values of variables at all other breakpoints can be found by similar reasoning. Results are summarized in Table 2.1.

Table 2.1   Values of Variables at Breakpoints for System of Fig. 2.4

v_I	$v_E = v_I - v_F$	$v_A = 10^3 v_E$	$v_B = v_O$	$v_F = 0.1 v_O$
< -0.258	$v_I + 0.250$	$10^3 v_I + 250$	-2.5	-0.25
-0.258	-0.008	-8	-2.5	-0.25
-0.203	-0.003	-3	-2	-0.2
-10^{-3}	-10^{-3}	-1	0	0
10^{-3}	10^{-3}	1	0	0
0.203	0.003	3	2	0.2
0.258	0.008	8	2.5	0.25
$>\ \ 0.258$	$v_I - 0.250$	$10^3 v_I - 250$	2.5	0.25

The input-output transfer relationship for the system shown in Fig. 2.4c is generated from values included in Table 2.1. The transfer relationship can also be found by using the incremental forward gain, or 1000 times the incremental gain of the nonlinear element, as the value for a in Eqn. 2.2. If the magnitude of signal v_A is less than 1 volt, a is zero, and the incremental closed-loop gain of the system is also zero. If v_A is between 1 and 3 volts, a is 10^3, so the incremental closed-loop gain is 9.9. Similarly, the incremental closed-loop gain is 9.1 for $3 < v_A < 8$.

Note from Fig. 2.4c that feedback dramatically reduces the width of the deadzone and the change in gain as the output stage soft limits. Once the amplifier saturates, the incremental loop transmission becomes zero, and as a result feedback cannot improve performance in this region.

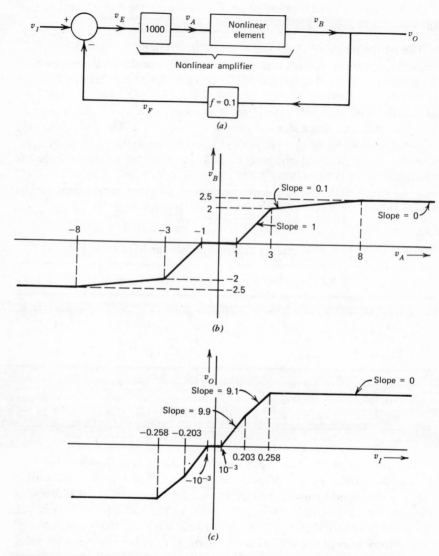

Figure 2.4 The effects of feedback on a nonlinearity. (*a*) System. (*b*) Transfer characteristics of the nonlinear element. (*c*) System transfer characteristics (closed loop). (Not to scale.) (*d*) Waveforms for $v_I(t)$ a unit ramp. (Not to scale.)

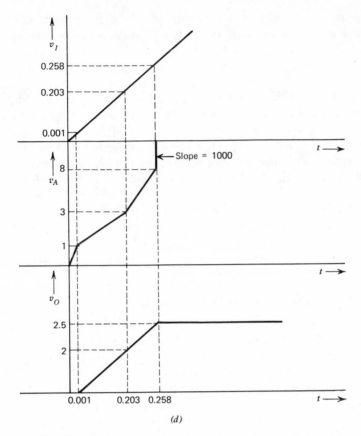

(d)

Figure 2.4—Continued

Figure 2.4d provides insight into the operation of the circuit by comparing the output of the system and the voltage v_A for a unit ramp input. The output remains a good approximation to the input until saturation is reached. The signal into the nonlinear element is "predistorted" by feedback in such a way as to force the output from this element to be nearly linear.

The technique of employing feedback to reduce the effects of nonlinear elements on system performance is a powerful and widely used method that evolves directly from the desensitivity to gain changes provided by feedback. In some applications, feedback is used to counteract the unavoidable nonlinearities associated with active elements. In other applications, feedback is used to maintain performance when nonlinearities result from economic compromises. Consider the power amplifier that provided the motivation for the previous example. The designs for linear power-handling stages are complex and expensive because compensation for the base-to-emitter voltages of the transistors and variations of gain with operating point must be included. Economic advantages normally result if linearity of the power-handling stage is reduced and low-power voltage-gain stages (possibly in the form of an operational amplifier) are added prior to the output stage so that feedback can be used to restore system linearity.

While this section has highlighted the use of feedback to reduce the effects of nonlinearities associated with the forward-gain element of a system, feedback can also be used to produce nonlinearities with well-controlled characteristics. If the feedback element in a system with large loop transmission is nonlinear, the output of the system becomes approximately $v_O = f^{-1}(v_I)$. Here f^{-1} is the inverse of the feedback-element transfer relationship, in the sense that $f^{-1}[f(V)] = V$. For example, transistors or diodes with exponential characteristics can be used as feedback elements around an operational amplifier to provide a logarithmic closed-loop transfer relationship.

2.3.3   Disturbances in Feedback Systems

Feedback provides a method for reducing the sensitivity of a system to certain kinds of disturbances. This advantage is illustrated in Fig. 2.5. Three different sources of disturbances are applied to this system. The disturbance V_{d1} enters the system at the same point as the system input, and might represent the noise associated with the input stage of an amplifier. Disturbance V_{d2} enters the system at an intermediate point, and might represent a disturbance from the hum associated with the poorly filtered voltage often used to power an amplifier output stage. Disturbance V_{d3} enters at the amplifier output and might represent changing load characteristics.

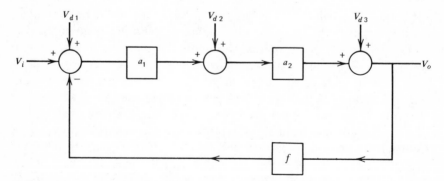

Figure 2.5   Feedback system illustrating effects of disturbances.

The reader should convince himself that the block diagram of Fig. 2.5 implies that the output voltage is related to input and disturbances as

$$V_o = \frac{a_1 a_2 [(V_i + V_{d1}) + (V_{d2}/a_1) + (V_{d3}/a_1 a_2)]}{1 + a_1 a_2 f} \qquad (2.4)$$

Equation 2.4 shows that the disturbance V_{d1} is not attenuated relative to the input signal. This result is expected since V_i and V_{d1} enter the system at the same point, and reflects the fact that feedback cannot improve quantities such as the noise figure of an amplifier. The disturbances that enter the amplifier at other points are attenuated relative to the input signal by amounts equal to the forward-path gains between the input and the points where the disturbances are applied.

It is important to emphasize that the forward-path gain preceding the disturbance, rather than the feedback, results in the relative attenuation of the disturbance. This feature is illustrated in Fig. 2.6. This open-loop system, which follows the forward path of Fig. 2.5 with an attenuator, yields the same output as the feedback system of Fig. 2.5. The feedback system is nearly always the more practical approach, since the open-loop system requires large signals, with attendant problems of saturation and power dissipation, at the input to the attenuator. Conversely, the feedback realization constrains system variables to more realistic levels.

2.3.4 Summary

This section has shown how feedback can be used to desensitize a system to changes in component values or to externally applied disturbances. This desensitivity can only be obtained in return for increases in the gains of various components of the system. There are numerous situations where this type of trade is advantageous. For example, it may be possible to replace a costly, linear output stage in a high-fidelity audio amplifier with a cheaper unit and compensate for this change by adding an inexpensive stage of low-level amplification.

The input and output impedances of amplifiers are also modified by feedback. For example, if the output variable that is fed back is a voltage, the

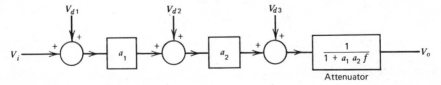

Figure 2.6 Open-loop system illustrating effects of disturbances.

feedback tends to stabilize the value of this voltage and reduce its dependence on disturbing load currents, implying that the feedback results in
lower output impedance. Alternatively, if the information fed back is proportional to output current, the feedback raises the output impedance.
Similarly, feedback can limit input voltage or current applied to an amplifier, resulting in low or high input impedance respectively. A quantitative
discussion of this effect is reserved for Section 2.5.

A word of caution is in order to moderate the impression that performance improvements always accompany increases in loop-transmission
magnitude. Unfortunately, the loop transmission of a system cannot be
increased without limit, since sufficiently high gain invariably causes a system to become *unstable*. A *stable* system is defined as one for which a
bounded output is produced in response to a bounded input. Conversely,
an unstable system exhibits runaway or oscillatory behavior in response to
a bounded input. Instability occurs in high-gain systems because small
errors give rise to large corrective action. The propagation of signals around
the loop is delayed by the dynamics of the elements in the loop, and as a
consequence high-gain systems tend to overcorrect. When this overcorrection produces an error larger than the initiating error, the system is unstable.

This important aspect of the feedback problem did not appear in this
section since the dynamics associated with various elements have been ignored. The problem of stability will be investigated in detail in Chapter 4.

2.4 BLOCK DIAGRAMS

A block diagram is a graphical method of representing the relationships
among variables in a system. The symbols used to form a block diagram
were introduced in Section 2.2. Advantages of this representation include
the insight into system operation that it often provides, its clear indication
of various feedback loops, and the simplification it affords to determining
the transfer functions that relate input and .output variables of the system.
The discussion in this section is limited to linear, time-invariant systems,
with the enumeration of certain techniques useful for the analysis of nonlinear systems reserved for Chapter 6.

2.4.1 Forming the Block Diagram

Just as there are many complete sets of equations that can be written
to describe the relationships among variables in a system, so there are many
possible block diagrams that can be used to represent a particular system.
The choice of block diagram should be made on the basis of the insight it
lends to operation and the ease with which required transfer functions can

be evaluated. The following systematic method is useful for circuits where all variables of interest are node voltages.

1. Determine the node voltages of interest. The selected number of voltages does not have to be equal to the total number of nodes in the circuit, but it must be possible to write a complete, independent set of equations using the selected voltages. One line (which may split into two or more branches in the final block diagram) will represent each of these variables, and these lines may be drawn as isolated segments.

2. Determine each of the selected node voltages as a weighted sum of the other selected voltages and any inputs or disturbances that may be applied to the circuit. This determination requires a set of equations of the form

$$V_j = \sum_{n \neq j} a_{nj} V_n + \sum_m b_{mj} E_m \tag{2.5}$$

where V_k is the kth node voltage and E_k is the kth input or disturbance.

3. The variable V_j is generated as the output of a summing point in the block diagram. The inputs to the summing point come from all other variables, inputs, and disturbances via blocks with transmissions that are the a's and b's in Eqn. 2.5. Some of the blocks may have transmissions of zero, and these blocks and corresponding summing-point inputs can be eliminated.

The set of equations required in Step 2 can be determined by writing node equations for the complete circuit and solving the equation written about the jth node for V_j in terms of all other variables. If a certain node voltage V_k is not required in the final block diagram, the equation relating V_k to other system voltages is used to eliminate V_k from all other members of the set of equations. While this degree of formality is often unnecessary, it always yields a correct block diagram, and should be used if the desired diagram cannot easily be obtained by other methods.

As an example of block diagram construction by this formal approach, consider the common-emitter amplifier shown in Fig. 2.7a. (Elements used for bias have been eliminated for simplicity.) The corresponding small-signal equivalent circuit is obtained by substituting a hybrid-pi[3] model for the transistor and is shown in Fig. 2.7b. Node equations are[4]

---

[3] The hybrid-pi model will be used exclusively for the analysis of bipolar transistors operating in the linear region. The reader who is unfamiliar with the development or use of this model is referred to P. E. Gray and C. L. Searle, *Electronic Principles: Physics, Models, and Circuits*, Wiley, New York, 1969.

[4] G's and R's (or g's and r's) are used to identify corresponding conductances and resistances, while Y's and Z's (or y's and z's) are used to identify corresponding admittances and impedances. Thus for example, $G_A = 1/R_A$ and $z_b = 1/y_b$.

$$G_S V_i = (G_S + g_x)\, V_a \qquad\qquad\qquad -\, g_x\, V_b \qquad\qquad (2.6)$$

$$0 = \qquad -\, g_x\ V_a + [(g_x + g_\pi) + (C_\mu + C_\pi)s]\, V_b \qquad -\, C_\mu s\, V_o$$

$$0 = \qquad\qquad\qquad (g_m - C_\mu s)\ V_b + (G_L + C_\mu s)V_o$$

If the desired block diagram includes all three node voltages, Eqn. 2.6 is arranged so that each member of the set is solved for the voltage at the node about which the member was written. Thus,

$$V_a = \qquad\qquad\qquad \frac{g_x}{g_a}\, V_b \qquad\qquad + \frac{G_S}{g_a}\, V_i \qquad (2.7)$$

$$V_b = \frac{g_x}{y_b}\, V_a \qquad\qquad + \frac{C_\mu s}{y_b}\, V_o$$

$$V_o = \qquad \frac{(C_\mu s - g_m)}{y_o}\, V_b$$

Where

$$g_a = G_S + g_x$$

$$y_b = [(g_x + g_\pi) + (C_\mu + C_\pi)s]$$

$$y_o = G_L + C_\mu s$$

The block diagram shown in Fig. 2.7c follows directly from this set of equations.

Figure 2.8 is the basis for an example that is more typical of our intended use of block diagrams. A simple operational-amplifier model is shown connected as a noninverting amplifier. It is assumed that the variables of interest are the voltages V_b and V_o. The voltage V_o can be related to the other selected voltage, V_b, and the input voltage, V_i, by superposition.

with $V_i = 0$,

$$V_o = -aV_b \qquad (2.8)$$

while with $V_b = 0$,

$$V_o = aV_i \qquad (2.9)$$

The equation relating V_o to other selected voltages and inputs is simply the superposition of the responses represented by Eqns. 2.8 and 2.9, or

$$V_o = aV_i - aV_b \qquad (2.10)$$

The voltage V_b is independent of V_i and is related to V_o as

$$V_b = \frac{Z_1}{Z_1 + Z_2}\, V_o \qquad (2.11)$$

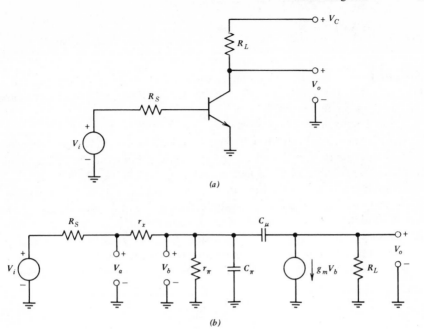

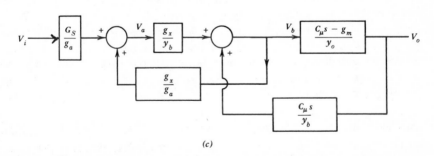

Figure 2.7 Common-emitter amplifier. (*a*) Circuit. (*b*) Incremental equivalent circuit. (*c*) Block diagram.

Equations 2.10 and 2.11 are readily combined to form the block diagram shown in Fig. 2.8*b*.

It is possible to form a block diagram that provides somewhat greater insight into the operation of the circuit by replacing Eqn. 2.10 by the pair of equations

$$V_a = V_i - V_b \tag{2.12}$$

and

$$V_o = aV_a \tag{2.13}$$

Note that the original set of equations were not written including V_a, since V_a, V_b, and V_i form a Kirchhoff loop and thus cannot all be included in an independent set of equations.

The alternate block diagram shown in Fig. 2.8c is obtained from Eqns. 2.11, 2.12, and 2.13. In this block diagram it is clear that the summing point models the function provided by the differential input of the operational amplifier. This same block diagram would have evolved had V_a and V_o been initially selected as the amplifier voltages of interest.

The loop transmission for any system represented as a block diagram can always be determined by setting all inputs and disturbances to zero, breaking the block diagram at any point inside the loop, and finding the signal returned by the loop in response to an applied test signal. One possible point to break the loop is illustrated in Fig. 2.8c. With $V_i = 0$, it is evident that

$$\frac{V_o}{V_t} = \frac{-aZ_1}{Z_1 + Z_2} \tag{2.14}$$

The same result is obtained for the loop transmission if the loop in Fig. 2.8c is broken elsewhere, or if the loop in Fig. 2.8b is broken at any point.

Figure 2.9 is the basis for a slightly more involved example. Here a fairly detailed operational-amplifier model, which includes input and output impedances, is shown connected as an inverting amplifier. A disturbing current generator is included, and this generator can be used to determine the closed-loop output impedance of the amplifier V_o/I_d.

It is assumed that the amplifier voltages of interest are V_a and V_o. The equation relating V_a to the other voltage of interest V_o, the input V_i, and the disturbance I_d, is obtained by superposition (allowing all other signals to be nonzero one at a time and superposing results) as in the preceding example. The reader should verify the results

$$V_a = \frac{Z_i \parallel Z_2}{Z_1 + Z_i \parallel Z_2} V_i + \frac{Z_i \parallel Z_1}{Z_2 + Z_i \parallel Z_1} V_o \tag{2.15}$$

and

$$V_o = \frac{-aZ_2 + Z_o}{Z_2 + Z_o} V_a + (Z_o \parallel Z_2)I_d \tag{2.16}$$

The block diagram of Fig. 2.9b follows directly from Eqns. 2.15 and 2.16.

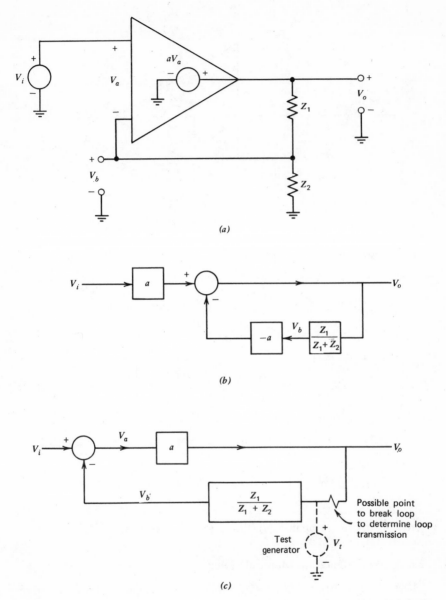

Figure 2.8 Noninverting amplifier. (*a*) Circuit. (*b*) Block diagram. (*c*) Alternative block diagram.

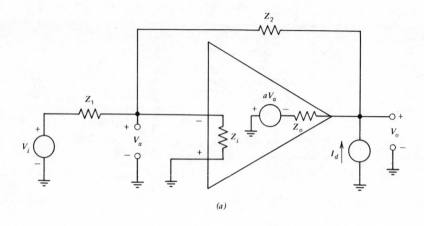

(a)

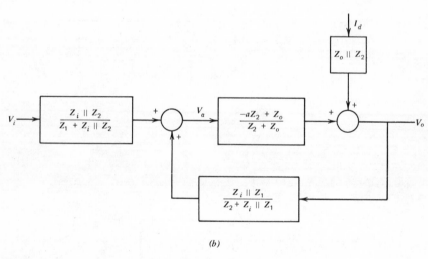

(b)

Figure 2.9 Inverting amplifier. (a) Circuit. (b) Block diagram.

2.4.2 Block-Diagram Manipulations

There are a number of ways that block diagrams can be restructured or reordered while maintaining the correct gain expression between an input or disturbance and an output. These modified block diagrams could be obtained directly by rearranging the equations used to form the block diagram or by using other system variables in the equations. Equivalences that can be used to modify block diagrams are shown in Fig. 2.10.

It is necessary to be able to find the transfer functions relating outputs to inputs and disturbances or the relations among other system variables from the block diagram of the system. These transfer functions can always be found by appropriately applying various equivalences of Fig. 2.10 until a single-loop system is obtained. The transfer function can then be determined by loop reduction (Fig. 2.10h). Alternatively, once the block diagram has been reduced to a single loop, important system quantities are evident. The loop transmission as well as the closed-loop gain approached for large loop-transmission magnitude can both be found by inspection.

Figure 2.11 illustrates the use of equivalences to reduce the block diagram of the common-emitter amplifier previously shown as Fig. 2.7c. Figure 2.11a is identical to Fig. 2.7c, with the exceptions that a line has been replaced with a unity-gain block (see Fig. 2.10a) and an intermediate variable V_c has been defined. These changes clarify the transformation from Fig. 2.11a to 2.11b, which is made as follows. The transfer function from V_c to V_b is determined using the equivalance of Fig. 2.10h, recognizing that the feedback path for this loop is the product of the transfer functions of blocks 1 and 2. The transfer function V_b/V_c is included in the remaining loop, and the transfer function of block 1 links V_o to V_b.

The equivalences of Figs. 2.10b and 2.10h using the identification of transfer functions shown in Fig. 2.11b (unfortunately, as a diagram is reduced, the complexities of the transfer functions of residual blocks increase) are used to determine the overall transfer function indicated in Fig. 2.11c.

The inverting-amplifier connection (Fig. 2.9) is used as another example of block-diagram reduction. The transfer function relating V_o to V_i in Fig. 2.9b can be reduced to single-loop form by absorbing the left-hand block in this diagram (equivalence in Fig. 2.10d). Figure 2.12 shows the result of this absorption after simplifying the feedback path algebraically, eliminating the disturbing input, and using the equivalence of Fig. 2.10e to introduce an inversion at the summing point. The gain of this system approaches the reciprocal of the feedback path for large loop transmission; thus the ideal closed-loop gain is

$$\frac{V_o}{V_i} = -\frac{Z_2}{Z_1} \tag{2.17}$$

The forward gain for this system is

$$\frac{V_o}{V_e} = \left[\frac{Z_i \parallel Z_2}{Z_1 + Z_i \parallel Z_2}\right]\left[\frac{-aZ_2 + Z_o}{Z_2 + Z_o}\right]$$

$$= \left[\frac{Z_i \parallel Z_2}{Z_1 + Z_i \parallel Z_2}\right]\left[\frac{-aZ_2}{Z_2 + Z_o}\right] + \left[\frac{Z_i \parallel Z_2}{Z_1 + Z_i \parallel Z_2}\right]\left[\frac{Z_o}{Z_2 + Z_o}\right] \tag{2.18}$$

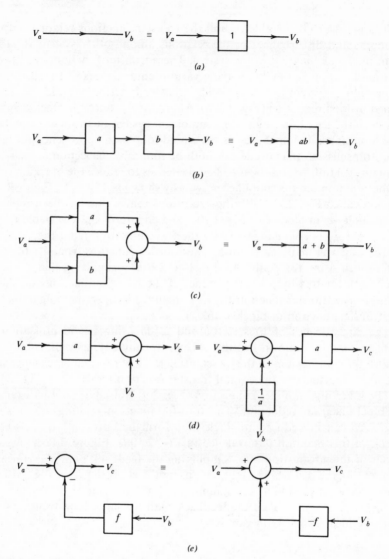

Figure 2.10 Block-diagram equivalences. (*a*) Unity gain of line. (*b*) Cascading. (*c*) Summation. (*d*) Absorption. (*e*) Negation. (*f*) Branching. (*g*) Factoring. (*h*) Loop reduction.

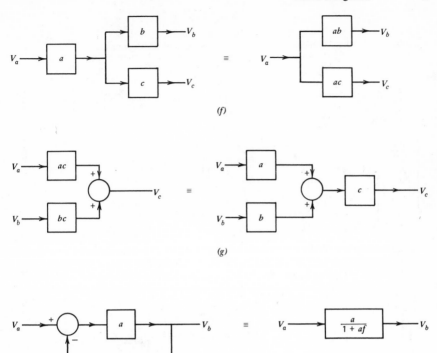

(f)

(g)

(h)

Figure 2.10—Continued

The final term on the right-hand side of Eqn. 2.18 reflects the fact that some fraction of the input signal is coupled directly to the output via the feedback network, even if the amplifier voltage gain a is zero. Since the impedances included in this term are generally resistive or capacitive, the magnitude of this coupling term will be less than one at all frequencies. Similarly, the component of loop transmission attributable to this direct path, determined by setting $a = 0$ and opening the loop is

$$
\left. \frac{V_f}{V_e} \right|_{a=0} = \left[\frac{Z_1}{Z_2} \right] \left[\frac{Z_i \parallel Z_2}{Z_1 + Z_i \parallel Z_2} \right] \left[\frac{Z_o}{Z_2 + Z_o} \right]
$$

$$
= \left[\frac{Z_i Z_1}{Z_i Z_1 + Z_i Z_2 + Z_1 Z_2} \right] \left[\frac{Z_o}{Z_2 + Z_o} \right] \quad (2.19)
$$

and will be less than one in magnitude at all frequencies when the impedances involved are resistive or capacitive.

42

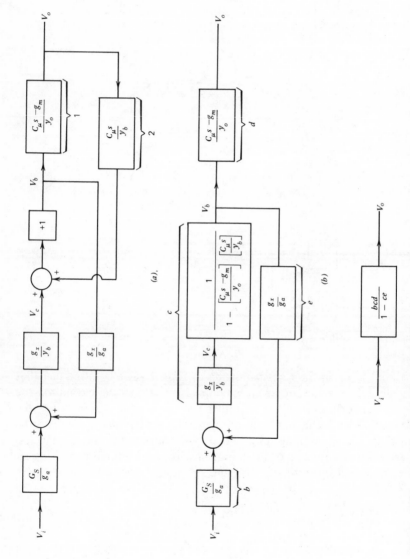

Figure 2.11 Simplification of common-emitter block diagram. (*a*) Original block diagram. (*b*) After eliminating loop generating V_b. (*c*) Reduction to single block.

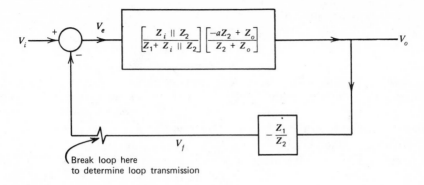

Figure 2.12   Reduced diagram for inverting amplifier.

If the loop-transmission magnitude of the operational-amplifier connection is large compared to one, the component attributable to direct coupling through the feedback network (Eqn. 2.19) must be insignificant. Consequently, the forward-path gain of the system can be approximated as

$$\frac{V_o}{V_e} \simeq \left[\frac{-aZ_2}{Z_2 + Z_o}\right]\left[\frac{Z_i \parallel Z_2}{Z_1 + Z_i \parallel Z_2}\right] \tag{2.20}$$

in this case. The corresponding loop transmission becomes

$$\frac{V_f}{V_e} \simeq \left[\frac{-aZ_1}{Z_2 + Z_o}\right]\left[\frac{Z_i \parallel Z_2}{Z_1 + Z_i \parallel Z_2}\right] \tag{2.21}$$

It is frequently found that the loop-transmission term involving direct coupling through the feedback network can be neglected in practical operational-amplifier connections, reflecting the reasonable hypothesis that the dominant gain mechanism is the amplifier rather than the passive network. While this approximation normally yields excellent results at frequencies where the amplifier gain is large, there are systems where stability calculations are incorrect when the approximation is used. The reason is that stability depends largely on the behavior of the loop transmission at frequencies where its magnitude is close to one, and the gain of the amplifier may not dominate at these frequencies.

2.4.3   The Closed-Loop Gain

It is always possible to determine the gain that relates any signal in a block diagram to an input or a disturbance by manipulating the block diagram until a single path connects the two quantities of interest. Alternatively, it is possible to use a method developed by Mason[5] to calculate gains directly from an unreduced block diagram.

In order to determine the gain between an input or disturbance and any other points in the diagram, it is necessary to identify two topological features of a block diagram. A *path* is a continuous succession of blocks, lines, and summation points that connect the input and signal of interest and along which no element is encountered more than once. Lines may be traversed only in the direction of information flow (with the arrow). It is possible in general to have more than one path connecting an input to an output or other signal of interest. The *path gain* is a product of the gains of all elements in a path. A *loop* is a closed succession of blocks, lines, and summation points traversed with the arrows, along which no element is encountered more than once per cycle. The *loop gain* is the product of gains of all elements in a loop. It is necessary to include the inversions indicated by negative signs at summation points when calculating path or loop gains.

The general expression for the gain or transmission of a block diagram is

$$T = \frac{\sum_a P_a \left(1 - \sum_b L_b + \sum_{c,d} L_c L_d - \sum_{e,f,g} L_e L_f L_g + \cdots - \right)}{1 - \sum_h L_h + \sum_{i,j} L_i L_j - \sum_{k,l,m} L_k L_l L_m + \cdots -} \quad (2.22)$$

The numerator of the gain expression is the sum of the gains of all paths connecting the input and the signal of interest, with each path gain scaled by a *cofactor*. The first sum in a cofactor includes the gains of all loops that do not touch (share a common block or summation point with) the path; the second sum includes all possible products of loop gains for loops that do not touch the path or each other taken two at a time; the third sum includes all possible triple products of loop gains for loops that do not touch the path or each other; etc.

The denominator of the gain expression is called the *determinant* or *characteristic equation* of the block diagram, and is identically equal to one minus the loop transmission of the complete block diagram. The first sum in the characteristic equation includes all loop gains; the second all possible products of the gains of nontouching loops taken two at a time; etc.

Two examples will serve to clarify the evaluation of the gain expression. Figure 2.13 provides the first example. In order to apply Mason's gain formula for the transmission V_o/V_i, the paths and loops are identified and their gains are evaluated. The results are:

$$P_1 = ace$$

[5] S. J. Mason and H. J. Zimmermann, *Electronic Circuits, Signals, and Systems*, Wiley, New York, 1960, Chapter 4, "Linear Signal-Flow Graphs."

$$P_2 = ag$$
$$P_3 = -h$$
$$L_1 = -ab$$
$$L_2 = cd$$
$$L_3 = -ef$$
$$L_4 = -acei$$

The topology of Fig. 2.13 shows that path P_1 shares common blocks with and therefore touches all loops. Path P_2 does not touch loops L_2 or L_3, while path P_3 does not touch any loops. Similarly, loops L_1, L_2, and L_3 do not touch each other, but all touch loop L_4. Equation 2.22 evaluated for this system becomes

$$\frac{V_o}{V_i} = \frac{\begin{aligned}P_1 + P_2\,(1 - L_2 - L_3 + L_2L_3)\\ + P_3(1 - L_1 - L_2 - L_3 - L_4 + L_1L_2 + L_2L_3 + L_1L_3 - L_1L_2L_3)\end{aligned}}{1 - L_1 - L_2 - L_3 - L_4 + L_1L_2 + L_2L_3 + L_1L_3 - L_1L_2L_3}$$

$$(2.23)$$

A second example of block-diagram reduction and some reinforcement of the techniques used to describe a system in block-diagram form is provided by the set of algebraic equations

$$X + Y + Z = 6 \qquad\qquad (2.24)$$
$$X + Y - Z = 0$$
$$2X + 3Y + Z = 11$$

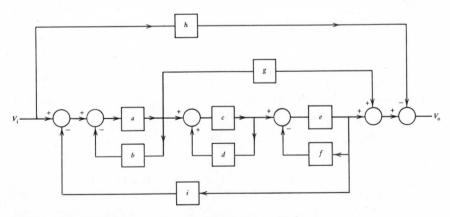

Figure 2.13 Block diagram for gain-expression example.

In order to represent this set of equations in block-diagram form, the three equations are rewritten

$$X = \qquad -Y - Z + 6 \qquad\qquad (2.25)$$
$$Y = \ -X \qquad\quad + Z$$
$$Z = -2X - 3Y \qquad\quad + 11$$

This set of equations is shown in block-diagram form in Fig. 2.14. If we use the identification of loops in this figure, loop gains are

$$L_1 = 1$$
$$L_2 = -3$$
$$L_3 = -3$$
$$L_4 = 2$$
$$L_5 = 2$$

Since all loops touch, the determinant of any gain expression for this system is

$$1 - L_1 - L_2 - L_3 - L_4 - L_5 = 2 \qquad\qquad (2.26)$$

(This value is of course identically equal to the determinant of the coefficients of Eqn. 2.24.)

Assume that the value of X is required. The block diagram shows one path with a transmission of $+1$ connecting the excitation with a value of 6 to X. This path does not touch L_2. There are also two paths (roughly paralleling L_3 and L_5) with transmissions of -1 connecting the excitation with a value of 11 to X. These paths touch all loops. Linearity allows us to combine the X responses related to the two excitations, with the result that

$$X = \frac{6[1 - (-3)] - 11 - 11}{2} = 1 \qquad\qquad (2.27)$$

The reader should verify that this method yields the values $Y = 2$ and $Z = 3$ for the other two dependent variables.

2.5 EFFECTS OF FEEDBACK ON INPUT AND OUTPUT IMPEDANCE

The gain-stabilizing and linearizing effects of feedback have been described earlier in this chapter. Feedback also has important effects on the input and output impedances of an amplifier, with the type of modification dependent on the topology of the amplifier-feedback network combination.

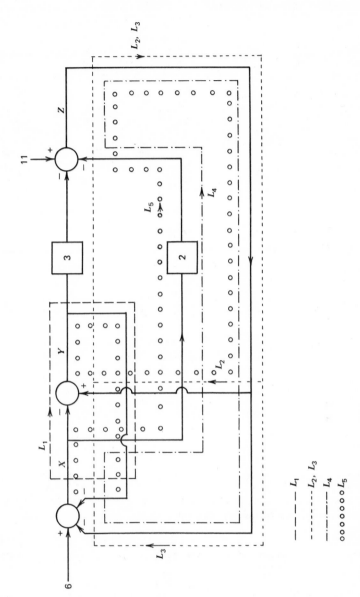

Figure 2.14 Block diagram of Eqn. 2.25.

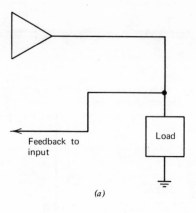

(a)

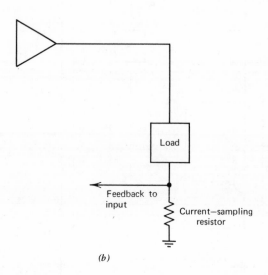

(b)

Figure 2.15   Two possible output topologies. (a) Feedback of load-voltage information. (b) Feedback of load-current information.

Figure 2.15 shows how feedback might be arranged to return information about either the voltage applied to the load or the current flow through it. It is clear from physical arguments that these two output topologies must alter the impedance facing the load in different ways. If the information fed back to the input concerns the output voltage, the feedback tends to reduce changes in output voltage caused by disturbances (changes in load current), thus implying that the output impedance of the amplifier shown in Fig. 2.15*a* is reduced by feedback. Alternatively, if information about load current is fed back, changes in output current caused by disturbances (changes in load voltage) are reduced, showing that this type of feedback raises output impedance.

Two possible input topologies are shown in Fig. 2.16. In Fig. 2.16*a*, the input signal is applied in series with the differential input of the amplifier. If the amplifier characteristics are satisfactory, we are assured that any required output signal level can be achieved with a small amplifier input current. Thus the current required from the input-signal source will be small, implying high input impedance. The topology shown in Fig. 2.16*b* reduces input impedance, since only a small voltage appears across the parallel input-signal and amplifier-input connection.

The amount by which feedback scales input and output impedances is directly related to the loop transmission, as shown by the following example. An operational amplifier connected for high input and high output resistances is shown in Fig. 2.17. The input resistance for this topology is simply the ratio V_i/I_i. Output resistance is determined by including a voltage source in series with the load resistor and calculating the ratio of the change in the voltage of this source to the resulting change in load current, V_l/I_l. If it is assumed that the components of I_l and the current through the sampling resistor R_S attributable to I_i are negligible (implying that the

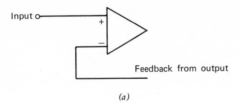

(*a*)

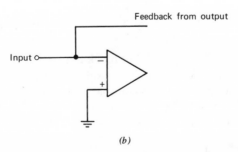

(*b*)

Figure 2.16 Two possible input topologies. (*a*) Input signal applied in series with amplifier input. (*b*) Input signal applied in parallel with amplifier input.

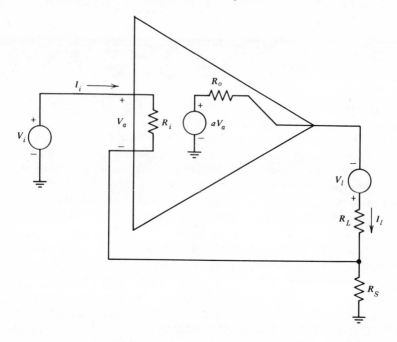

Figure 2.17 Amplifier with high input and output resistances.

amplifier, rather than a passive network, provides system gain) and that $R_i \gg R_S$, the following equations apply.

$$V_a = V_i - R_S I_l \tag{2.28}$$

$$I_l = \frac{aV_a + V_l}{R_o + R_L + R_S} \tag{2.29}$$

$$I_i = \frac{V_a}{R_i} \tag{2.30}$$

These equations are represented in block-diagram form in Fig. 2.18. This block diagram verifies the anticipated result that, since the input voltage is compared with the output current sampled via resistor R_S, the ideal trans-conductance (ratio of I_l to V_i) is simply equal to G_S. The input resistance is evaluated by noting that

$$\frac{I_i}{V_i} = \frac{1}{R_{\text{in}}} = \frac{1}{R_i\{1 + [aR_S/(R_o + R_L + R_S)]\}} \tag{2.31}$$

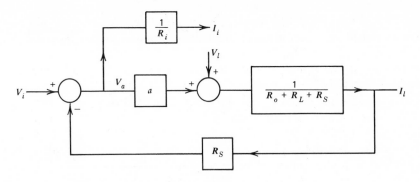

Figure 2.18   Block diagram for amplifier of Fig. 2.17.

or

$$R_{\text{in}} = R_i \left(1 + \frac{aR_S}{R_o + R_L + R_S} \right) \tag{2.32}$$

The output resistance is determined from[6]

$$\frac{I_l}{V_l} = \frac{1}{R_{\text{out}}} = \frac{1}{(R_o + R_L + R_S)\{1 + [aR_S/(R_o + R_L + R_S)]\}} \tag{2.33}$$

yielding

$$R_{\text{out}} = (R_o + R_L + R_S) \left(1 + \frac{aR_S}{R_o + R_L + R_S} \right) \tag{2.34}$$

The essential features of Eqns. 2.32 and 2.34 are the following. If the system has no feedback (e.g., if $a = 0$), the input and output resistances become

$$R'_{\text{in}} = R_i \tag{2.35}$$

and

$$R'_{\text{out}} = R_o + R_L + R_S \tag{2.36}$$

Feedback increases both of these quantities by a factor of $1 + [aR_S/(R_o + R_L + R_S)]$, where $-aR_S/(R_o + R_L + R_S)$ is recognized as the loop transmission. Thus we see that the resistances in this example are increased by the same factor (one minus the loop transmission) as the desensitivity

---

[6] Note that the output resistance in this example is calculated by including a voltage source in series with the load resistor. This approach is used to emphasize that the loop transmission that determines output resistance is influenced by R_L. An alternative development might evaluate the resistance facing the load by replacing R_L with a test generator.

increase attributable to feedback. The result is general, so that input or output impedances can always be calculated for the topologies shown in Figs. 2.15 or 2.16 by finding the impedance of interest with no feedback and scaling it (up or down according to topology) by a factor of one minus the loop transmission.

While feedback offers a convenient method for controlling amplifier input or output impedances, comparable (and in certain cases, superior) results are at least conceptually possible without the use of feedback. Consider, for example, Fig. 2.19, which shows three ways to connect an operational amplifier for high input impedance and unity voltage gain.

The follower connection of Fig. 2.19a provides a voltage gain

$$\frac{V_o}{V_i} = \frac{a}{1 + a} \tag{2.37}$$

or approximately unity for large values of a. The relationship between input impedance and loop transmission discussed earlier in this section shows that the input impedance for this connection is

$$\frac{V_i}{I_i'} = Z_i(1 + a) \tag{2.38}$$

The connection shown in Fig. 2.19b precedes the amplifier with an impedance that, in conjunction with the input impedance of the amplifier, attenuates the input signal by a factor of $1/(1 + a)$. This attenuation combines with the voltage gain of the amplifier itself to provide a composite voltage gain identical to that of the follower connection. Similarly, the series impedance of the attenuator input element adds to the input impedance of the amplifier itself so that the input impedance of the combination is identical to that of the follower.

The use of an ideal transformer as impedance-modifying element can lead to improved input impedance compared to the feedback approach. With a transformer turns ratio of $(a + 1):1$, the overall voltage gain of the transformer-amplifier combination is the same as that of the follower connection, while the input impedance is

$$\frac{V_i}{I_i''} = Z_i(1 + a)^2 \tag{2.39}$$

This value greatly exceeds the value obtained with the follower for large amplifier voltage gain.

The purpose of the above example is certainly not to imply that attenuators or transformers should be used in preference to feedback to modify impedance levels. The practical disadvantages associated with the two

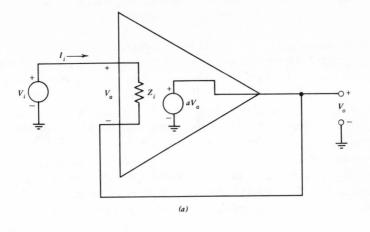

(a)

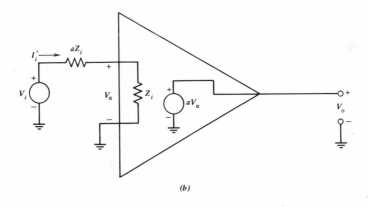

(b)

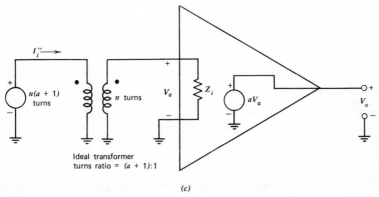

Ideal transformer
turns ratio = $(a + 1)$:1

(c)

Figure 2.19 Unity-gain amplifiers. (*a*) Follower connection. (*b*) Amplifier with input attenuator. (*c*) Amplifier with input transformer.

former approaches, such as the noise accentuation that accompanies large input-signal attenuation and the limited frequency response characteristic of transformers, often preclude their use. The example does, however, serve to illustrate that it is really the power gain of the amplifier, rather than the use of feedback, that leads to the impedance scaling. We can further emphasize this point by noting that the input impedance of the amplifier connection can be increased without limit by following it with a step-up transformer and increasing the voltage attenuation of either the network or the transformer that precedes the amplifier so that the overall gain is one. This observation is a reflection of the fact that the amplifier alone provides infinite power gain since it has zero output impedance.

One rather philosophical way to accept this reality concerning impedance scaling is to realize that feedback is most frequently used because of its fundamental advantage of reducing the sensitivity of a system to changes in the gain of its forward-path element. The advantages of impedance scaling can be obtained *in addition* to desensitivity simply by choosing an appropriate topology.

PROBLEMS

P2.1

Figure 2.20 shows a block diagram for a linear feedback system. Write a complete, independent set of equations for the relationships implied by this diagram. Solve your set of equations to determine the input-to-output gain of the system.

P2.2

Determine how the fractional change in closed-loop gain

$$\frac{d(V_o/V_i)}{V_o/V_i}$$

is related to fractional changes in a_1, a_2, and f for the system shown in Fig. 2.21.

P2.3

Plot the closed-loop transfer characteristics for the nonlinear system shown in Fig. 2.22.

P2.4

The complementary emitter-follower connection shown in Fig. 2.23 is a simple unity-voltage-gain stage that has a power gain approximately equal to the current gain of the transistors used. It has nonlinear transfer charac-

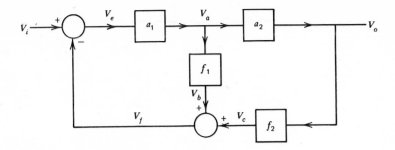

Figure 2.20 Two-loop feedback system.

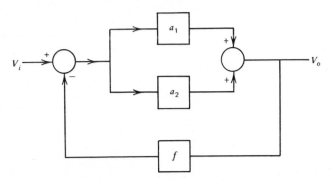

Figure 2.21 Feedback system with parallel forward paths.

teristics, since it is necessary to apply approximately 0.6 volts to the base-to-emitter junction of a silicon transistor in order to initiate conduction.

(a) Approximate the input-output transfer characteristics for the emitter-follower stage.
(b) Design a circuit that combines this power stage with an operational amplifier and any necessary passive components in order to provide a closed-loop gain with an ideal value of $+5$.
(c) Approximate the actual input-output characteristics of your feedback circuit assuming that the open-loop gain of the operational amplifier is 10^5.

P2.5

(a) Determine the incremental gain v_o/v_i for $V_I = 0.5$ and 1.25 for the system shown in Fig. 2.24.

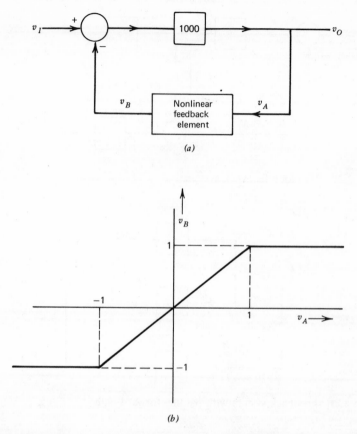

(a)

(b)

Figure 2.22 Nonlinear feedback system. (a) System. (b) Transfer characteristics for nonlinear element.

(b) Estimate the signal v_A for v_I, a unit ramp $[v_I(t) = 0, t < 0, = t, t > 0]$.

(c) For $v_I = 0$, determine the amplitude of the sinusoidal component of v_O.

P2.6

Determine V_o as a function of V_{i1} and V_{i2} for the feedback system shown in Fig. 2.25.

P2.7

Draw a block diagram that relates output voltage to input voltage for an emitter follower. You may assume that the transistor remains linear, and

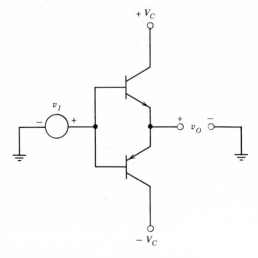

Figure 2.23 Complementary emitter follower.

use a hybrid-pi model for the device. Include elements r_π, r_x, C_π, and C_μ, in addition to the dependent generator, in your model. Reduce the block diagram to a single input-output transfer function.

P2.8

Draw a block diagram that relates V_o to V_i for the noninverting connection shown in Fig. 2.26. Also use block-diagram techniques to determine the impedance at the output, assuming that Z_i is very large.

P2.9

A negative-feedback system used to rotate a roof-top antenna is shown in Fig. 2.27a.

The total inertia of the output member (antenna, motor armature, and pot wiper) is 2 kg·m². The motor can be modeled as a resistor in series with a speed-dependent voltage generator (Fig. 2.27b).

The torque provided by the motor that accelerates the total output-member inertia is 10 N·m per ampere of I_a. The polarity of the motor dependent generator is such that it tends to reduce the value of I_a as the motor accelerates so that I_a becomes zero for a motor shaft velocity equal to $V_m/10$ radians per second.

Draw a block diagram that relates θ_o to θ_i. You may include as many intermediate variables as you wish, but be sure to include V_m and I_a in your diagram. Find the transfer function θ_o/θ_i.

Modify your diagram to include an output disturbance applied to the

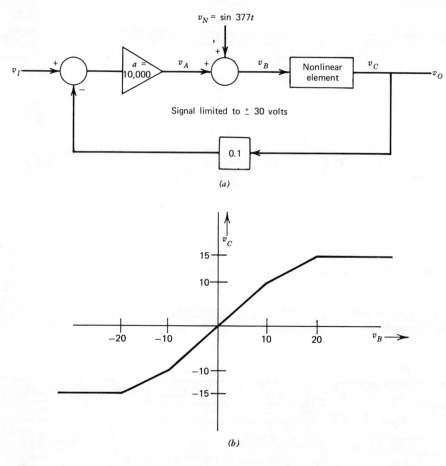

Figure 2.24 Nonlinear system. (*a*) System. (*b*) Transfer characteristics for nonlinear element.

antenna by wind. Calculate the angular error that results from a 1 N·m disturbance.

P2.10

Draw a block diagram for this set of equations:

$$
\begin{aligned}
W + X \qquad\qquad &= 3 \\
X + Y \quad &= 5 \\
Y + Z &= 7 \\
2W + X + Y + Z &= 11
\end{aligned}
$$

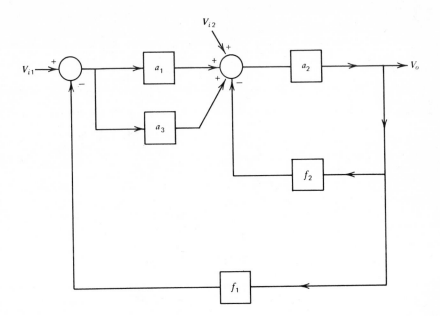

Figure 2.25 Linear block diagram.

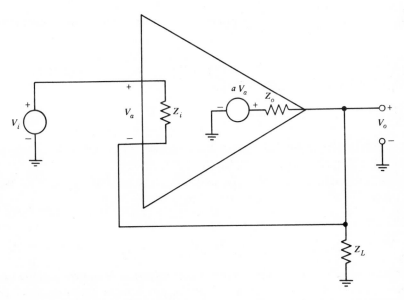

Figure 2.26 Noninverting amplifier.

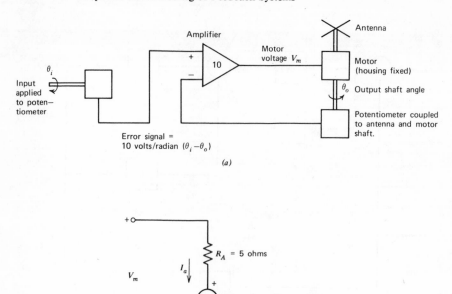

Figure 2.27 Antenna rotator System. (*a*) System configuration. (*b*) Model for motor.

Use the block-diagram reduction equation (Eqn. 2.22) to determine the values of the four dependent variables.

P2.11

The connection shown in Fig. 2.28 feeds back information about both load current and load voltage to the amplifier input. Draw a block diagram that allows you to calculate the output resistance V_o/I_d.

You may assume that $R \gg R_S$ and that the load can be modeled as a resistor R_L. What is the output resistance for very large a?

P2.12

An operational amplifier connected to provide an adjustable output resistance is shown in Fig. 2.29. Find a Thévenin-equivalent circuit facing the load as a function of the potentiometer setting α. You may assume that the resistance R is very large and that the operational amplifier has ideal characteristics.

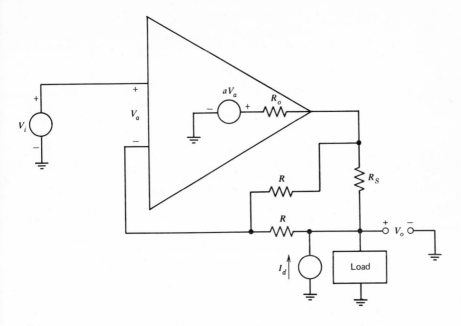

Figure 2.28 Operational-amplifier connection with controlled output resistance.

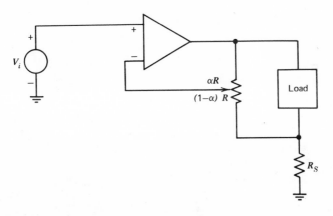

Figure 2.29 Circuit with adjustable output resistance.

CHAPTER III

LINEAR SYSTEM RESPONSE

3.1 OBJECTIVES

The output produced by an operational amplifier (or any other dynamic system) in response to a particular type or class of inputs normally provides the most important characterization of the system. The purpose of this chapter is to develop the analytic tools necessary to determine the response of a system to a specified input.

While it is always possible to determine the response of a linear system to a given input exactly, we shall frequently find that greater insight into the design process results when a system response is approximated by the known response of a simpler configuration. For example, when designing a low-level preamplifier intended for audio signals, we might be interested in keeping the frequency response of the amplifier within ±5% of its midband value over a particular bandwidth. If it is possible to approximate the amplifier as a two- or three-pole system, the necessary constraints on pole location are relatively straightforward. Similarly, if an oscilloscope vertical amplifier is to be designed, a required specification might be that the overshoot of the amplifier output in response to a step input be less than 3% of its final value. Again, simple constraints result if the system transfer function can be approximated by three or fewer poles.

The advantages of approximating the transfer functions of linear systems can only be appreciated with the aid of examples. The LM301A integrated-circuit operational amplifier[1] has 13 transistors included in its signal-transmission path. Since each transistor can be modeled as having two capacitors, the transfer function of the amplifier must include 26 poles. Even this estimate is optimistic, since there is distributed capacitance, comparable to transistor capacitances, associated with all of the other components in the signal path.

Fortunately, experimental measurements of performance can save us from the conclusion that this amplifier is analytically intractable. Figure 3.1a shows the LM301A connected as a unity-gain inverter. Figures 3.1b and 3.1c show the output of this amplifier with the input a −50-mV step

[1] This amplifier is described in Section 10.4.1.

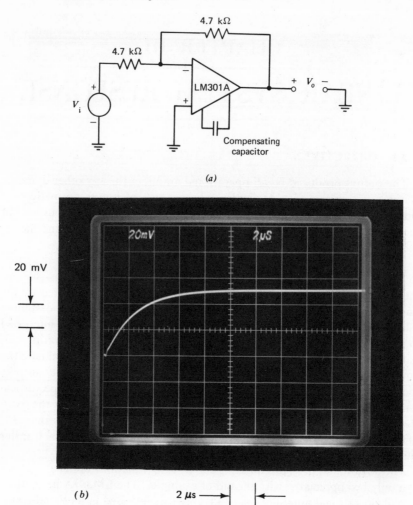

(a)

(b)   2 μs ⟶

Figure 3.1 Step responses of inverting amplifier. (a) Connection. (b) Step response with 220-pF compensating capacitor. (c) Step response with 12-pF compensating capacitor.

for two different values of compensating capacitor.[2] The responses of an R-C network and an R-L-C network when excited with +50-mV steps supplied from the same generator used to obtain the previous transients are shown in Figs. 3.2a and 3.2b, respectively. The network transfer functions are

$$\frac{V_o(s)}{V_i(s)} = \frac{1}{2.5 \times 10^{-6}s + 1} \qquad (3.1)$$

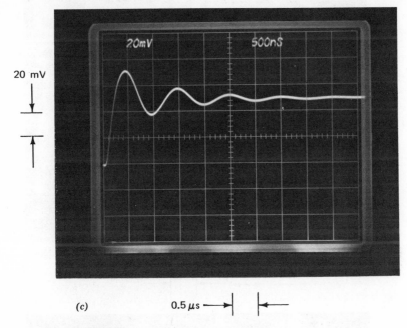

20 mV

(c)   0.5 μs →| |←

Figure 3.1—Continued

for the response shown in Fig. 3.2a and

$$\frac{V_o(s)}{V_i(s)} = \frac{1}{2.5 \times 10^{-14}s^2 + 7 \times 10^{-8}s + 1} \qquad (3.2)$$

for that shown in Fig. 3.2b. We conclude that there are many applications where the first- and second-order transfer functions of Eqns. 3.1 and 3.2 adequately model the closed-loop transfer function of the LM301A when connected and compensated as shown in Fig. 3.1.

This same type of modeling process can also be used to approximate the open-loop transfer function of the operational amplifier itself. Assume that the input impedance of the LM301A is large compared to 4.7 kΩ and that its output impedance is small compared to this value at frequencies of interest. The closed-loop transfer function for the connection shown in Fig. 3.1 is then

$$\frac{V_o(s)}{V_i(s)} = \frac{-a(s)}{2 + a(s)} \qquad (3.3)$$

---

[2] Compensation is a process by which the response of a system can be modified advantageously, and is described in detail in subsequent sections.

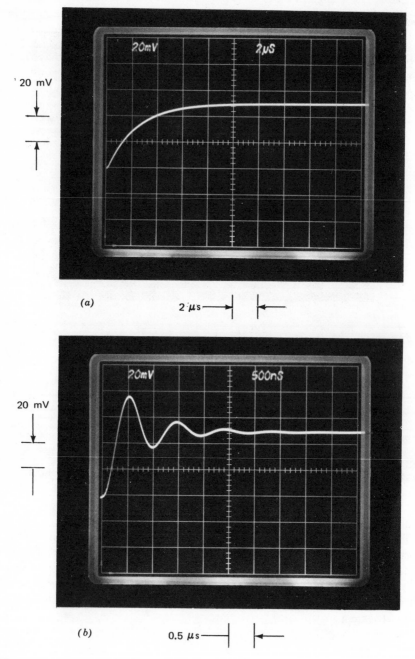

Figure 3.2 Step responses for first- and second-order networks. (a) Step response for $V_o(s)/V_i(s) = 1/(2.5 \times 10^{-6}s + 1)$. (b) Step response for $V_o(s)/V_i(s) = 1/(2.5 \times 10^{-14}s^2 + 7 \times 10^{-8}s + 1)$.

where $a(s)$ is the unloaded open-loop transfer function of the amplifier. Substituting approximate values for closed-loop gain (the negatives of Eqns. 3.1 and 3.2) into Eqn. 3.3 and solving for $a(s)$ yields

$$a(s) \simeq \frac{8 \times 10^5}{s} \tag{3.4}$$

and

$$a(s) \simeq \frac{2.8 \times 10^7}{s(3.5 \times 10^{-7}s + 1)} \tag{3.5}$$

as approximate open-loop gains for the amplifier when compensated with 220-pF and 12-pF capacitors, respectively. We shall see that these approximate values are quite accurate at frequencies where the magnitude of the loop transmission is near unity.

3.2 LAPLACE TRANSFORMS[3]

Laplace Transforms offer a method for solving any linear, time-invariant differential equation, and thus can be used to evaluate the response of a linear system to an arbitrary input. Since it is assumed that most readers have had some contact with this subject, and since we do not intend to use this method as our primary analytic tool, the exposure presented here is brief and directed mainly toward introducing notation and definitions that will be used later.

3.2.1 Definitions and Properties

The Laplace transform of a time function $f(t)$ is defined as

$$\mathcal{L}[f(t)] \triangleq F(s) \triangleq \int_0^\infty f(t)e^{-st}\,dt \tag{3.6}$$

where s is a complex variable $\sigma + j\omega$. The inverse Laplace transform of the complex function $F(s)$ is

$$\mathcal{L}^{-1}[F(s)] \triangleq f(t) \triangleq \frac{1}{2\pi j} \int_{\sigma_1 - j\infty}^{\sigma_1 + j\infty} F(s)e^{st}\,ds \tag{3.7}$$

---

[3] A complete discussion is presented in M. F. Gardner and J. L. Barnes, *Transients in Linear Systems*, Wiley, New York, 1942.

In this section we temporarily suspend the variable and subscript notation used elsewhere and conform to tradition by using a lower-case variable to signify a time function and the corresponding capital for its transform.

The direct-inverse transform pair is unique[4] so that

$$\mathcal{L}^{-1}\mathcal{L}[f(t)] = f(t) \qquad (3.8)$$

if $f(t) = 0$, $t < 0$, and if $\int_0^\infty |f(t)| e^{-\sigma_1 t} \, dt$ is finite for some real value of σ_1.

A number of theorems useful for the analysis of dynamic systems can be developed from the definitions of the direct and inverse transforms for functions that satisfy the conditions of Eqn. 3.8. The more important of these theorems include the following.

1. *Linearity*

$$\mathcal{L}[af(t) + bg(t)] = [aF(s) + bG(s)]$$

where a and b are constants.

2. *Differentiation*

$$\mathcal{L}\left[\frac{df(t)}{dt}\right] = sF(s) - \lim_{t \to 0^+} f(t)$$

(The limit is taken by approaching $t = 0$ from positive t.)

3. *Integration*

$$\mathcal{L}\left[\int_0^t f(\tau) \, d\tau\right] = \frac{F(s)}{s}$$

4. *Convolution*

$$\mathcal{L}\left[\int_0^t f(\tau)g(t - \tau) \, d\tau\right] = \mathcal{L}\left[\int_0^t f(t - \tau)g(\tau) \, d\tau\right] = F(s)G(s)$$

5. *Time shift*

$$\mathcal{L}[f(t - \tau)] = F(s)e^{-s\tau}$$

if $f(t - \tau) = 0$ for $(t - \tau) < 0$, where τ is a positive constant.

6. *Time scale*

$$\mathcal{L}[f(at)] = \frac{1}{a} F\left[\frac{s}{a}\right]$$

where a is a positive constant.

7. *Initial value*

$$\lim_{t \to 0^+} f(t) = \lim_{s \to \infty} sF(s)$$

---

[4] There are three additional constraints called the Direchlet conditions that are satisfied for all signals of physical origin. The interested reader is referred to Gardner and Barnes.

8. *Final value*

$$\lim_{t\to\infty} f(t) = \lim_{s\to 0} sF(s)$$

Theorem 4 is particularly valuable for the analysis of linear systems, since it shows that the Laplace transform of a system output is the product of the transform of the input signal and the transform of the impulse response of the system.

3.2.2   Transforms of Common Functions

The defining integrals can always be used to convert from a time function to its transform or vice versa. In practice, tabulated values are frequently used for convenience, and many mathematical or engineering references[5] contain extensive lists of time functions and corresponding Laplace transforms. A short list of Laplace transforms is presented in Table 3.1.

The time functions corresponding to ratios of polynomials in s that are not listed in the table can be evaluated by means of a *partial fraction* expansion. The function of interest is written in the form

$$F(s) = \frac{p(s)}{q(s)} = \frac{p(s)}{(s + s_1)(s + s_2) \cdots (s + s_n)} \tag{3.9}$$

It is assumed that the order of the numerator polynomial is less than that of the denominator. If all of the roots of the denominator polynomial are *first order* (i.e., $s_i \neq s_j$, $i \neq j$),

$$F(s) = \sum_{k=1}^{n} \frac{A_k}{s + s_k} \tag{3.10}$$

where

$$A_k = \lim_{s\to -s_k} [(s + s_k)F(s)] \tag{3.11}$$

If one or more roots of the denominator polynomial are *multiple roots*, they contribute terms of the form

$$\sum_{k=1}^{m} \frac{B_k}{(s + s_i)^k} \tag{3.12}$$

[5] See, for example, A. Erdeyli (Editor) *Tables of Integral Transforms*, Vol. 1, Bateman Manuscript Project, McGraw-Hill, New York, 1954 and R. E. Boly and G. L. Tuve, (Editors), *Handbook of Tables for Applied Engineering Science*, The Chemical Rubber Company, Cleveland, 1970.

Table 3.1   Laplace Transform Pairs

$F(s)$	$f(t), t \geq 0$ $[f(t) = 0, t < 0]$
1	Unit impulse $u_0(t)$
$\dfrac{1}{s}$	Unit step $u_{-1}(t)$ $[f(t) = 1, t \geq 0]$
$\dfrac{1}{s^2}$	Unit ramp $u_{-2}(t)$ $[f(t) = t, t \geq 0]$
$\dfrac{1}{s^{n+1}}$	$\dfrac{t^n}{n!}$
$\dfrac{1}{s + a}$	e^{-at}
$\dfrac{1}{(s + a)^{n+1}}$	$\dfrac{t^n}{(n)!} e^{-at}$
$\dfrac{1}{s(\tau s + 1)}$	$1 - e^{-t/\tau}$
$\dfrac{\omega}{(s + a)^2 + \omega^2}$	$e^{-at} \sin \omega t$
$\dfrac{s + a}{(s + a)^2 + \omega^2}$	$e^{-at} \cos \omega t$
$\dfrac{1}{s^2/\omega_n^2 + 2\zeta s/\omega_n + 1}$	$\dfrac{\omega_n}{\sqrt{1 - \zeta^2}} e^{-\zeta\omega_n t}(\sin \omega_n \sqrt{1 - \zeta^2}\, t), \zeta < 1$
$\dfrac{1}{s(s^2/\omega_n^2 + 2\zeta s/\omega_n + 1)}$	$1 - \dfrac{e^{-\zeta\omega_n t}}{\sqrt{1 - \zeta^2}} \sin \left[\omega_n\sqrt{1 - \zeta^2}\, t + \tan^{-1}\left(\dfrac{\sqrt{1 - \zeta^2}}{\zeta} \right) \right]$, $\zeta < 1$

where m is the order of the multiple root located at $s = -s_i$. The B's are determined from the relationship

$$B_k = \lim_{s \to -s_i} \left\{ \frac{1}{(m - k)!} \frac{d^{m-k}}{ds^{m-k}} [(s + s_i)^m F(s)] \right\} \qquad (3.13)$$

Because of the linearity property of Laplace transforms, it is possible to find the time function $f(t)$ by summing the contributions of all components of $F(s)$.

The properties of Laplace transforms listed earlier can often be used to determine the transform of time functions not listed in the table. The rectangular pulse shown in Fig. 3.3 provides one example of this technique. The pulse (Fig. 3.3a) can be decomposed into two steps, one with an amplitude of $+A$ starting at $t = t_1$, summed with a second step of ampli-

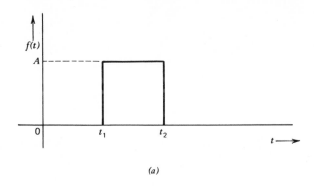

(a)

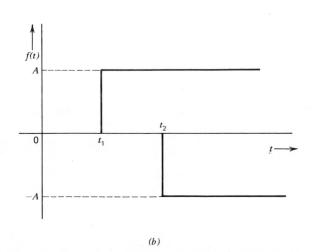

(b)

Figure 3.3 Rectangular pulse. (a) Signal. (b) Signal decomposed in two steps.

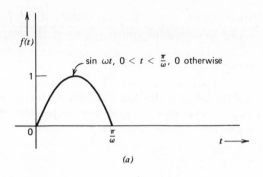

(a)

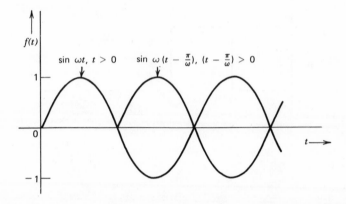

(b)

Figure 3.4 Sinusoidal pulse. (*a*) Signal. (*b*) Signal decomposed into two sinusoids. (*c*) First derivative of signal. (*d*) Second derivative of signal.

tude $-A$ starting at $t = t_2$. Theorems 1 and 5 combined with the transform of a unit step from Table 3.1 show that the transform of a step with amplitude A that starts at $t = t_1$ is $(A/s)e^{-st_1}$. Similarly, the transform of the second component is $-(A/s)e^{-st_2}$. Superposition insures that the transform of $f(t)$ is the sum of these two functions, or

$$F(s) = \frac{A}{s} (e^{-st_1} - e^{-st_2}) \tag{3.14}$$

The sinusoidal pulse shown in Fig. 3.4 is used as a second example. One approach is to represent the single pulse as the sum of two sinusoids

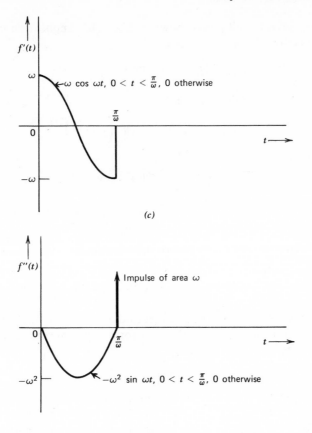

(c)

Impulse of area ω

$-\omega^2 \sin \omega t, 0 < t < \frac{\pi}{\omega}$, 0 otherwise

(d)

Figure 3.4—Continued

exactly as was done for the rectangular pulse. Table 3.1 shows that the transform of a unit-amplitude sinusoid starting at time $t = 0$ is $\omega/(s^2 + \omega^2)$. Summing transforms of the components shown in Fig. 3. 4b yields

$$F(s) = \frac{\omega}{s^2 + \omega^2} [1 + e^{-s(\pi/\omega)}] \qquad (3.15)$$

An alternative approach involves differentiating $f(t)$ twice. The derivative of $f(t)$, $f'(t)$, is shown in Fig. 3.4c. Since $f(0) = 0$, theorem 2 shows that

$$\mathcal{L}[f'(t)] = sF(s) \qquad (3.16)$$

The second derivative of $f(t)$ is shown in Fig. 3.4d. Application of theorem 2 to this function[6] leads to

$$\mathcal{L}[f''(t)] = s\mathcal{L}[f'(t)] - \lim_{t \to 0+} f'(t) = s^2 F(s) - \omega \qquad (3.17)$$

However, Fig. 3.4d indicates that

$$f''(t) = -\omega^2 f(t) + \omega u_0 \left(t - \frac{\pi}{\omega} \right) \qquad (3.18)$$

Thus

$$\mathcal{L}[f''(t)] = -\omega^2 F(s) + \omega e^{-s(\pi/\omega)} \qquad (3.19)$$

Combining Eqns. 3.17 and 3.19 yields

$$s^2 F(s) - \omega = -\omega^2 F(s) + \omega e^{-s(\pi/\omega)} \qquad (3.20)$$

Equation 3.20 is solved for $F(s)$ with the result that

$$F(s) = \frac{\omega}{s^2 + \omega^2} [1 + e^{-s(\pi/\omega)}] \qquad (3.21)$$

Note that this development, in contrast to the one involving superposition, does not rely on knowledge of the transform of a sinusoid, and can even be used to determine this transform.

3.2.3 Examples of the Use of Transforms

Laplace transforms offer a convenient method for the solution of linear, time-invariant differential equations, since they replace the integration and differentiation required to solve these equations in the time domain by algebraic manipulation. As an example, consider the differential equation

$$\frac{d^2 x}{dt^2} + 3 \frac{dx}{dt} + 2x = e^{-t} \qquad t > 0 \qquad (3.22)$$

subject to the initial conditions

$$x(0^+) = 2 \qquad \frac{dx}{dt}(0^+) = 0$$

The transform of both sides of Eqn. 3.22 is taken using theorem 2 (applied twice in the case of the second derivative) and Table 3.1 to determine the Laplace transform of e^{-t}.

$$s^2 X(s) - sx(0^+) - \frac{dx}{dt}(0^+) + 3sX(s) - 3x(0^+) + 2X(s) = \frac{1}{s+1} \qquad (3.23)$$

---

[6] The portion of this expression involving $\lim t \to 0^+$ could be eliminated if a second impulse $\omega u_0(t)$ were included in $f''(t)$.

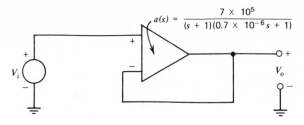

Figure 3.5 Unity-gain follower.

Collecting terms and solving for $X(s)$ yields

$$X(s) = \frac{2s^2 + 8s + 7}{(s + 1)^2(s + 2)} \tag{3.24}$$

Equations 3.10 and 3.12 show that since there is one first-order root and one second-order root,

$$X(s) = \frac{A_1}{(s + 2)} + \frac{B_1}{(s + 1)} + \frac{B_2}{(s + 1)^2} \tag{3.25}$$

The coefficients are evaluated with the aid of Eqns. 3.11 and 3.13, with the result that

$$X(s) = \frac{-1}{s + 2} + \frac{3}{s + 1} + \frac{1}{(s + 1)^2} \tag{3.26}$$

The inverse transform of $X(s)$, evaluated with the aid of Table 3.1, is

$$x(t) = -e^{-2t} + 3e^{-t} + te^{-t} \tag{3.27}$$

The operational amplifier connected as a unity-gain noninverting amplifier (Fig. 3.5) is used as a second example illustrating Laplace techniques. If we assume loading is negligible,

$$\frac{V_o(s)}{V_i(s)} = \frac{a(s)}{1 + a(s)} = \frac{7 \times 10^5}{(s + 1)(0.7 \times 10^{-6}s + 1) + 7 \times 10^5}$$

$$\simeq \frac{1}{10^{-12}s^2 + 1.4 \times 10^{-6}s + 1} \tag{3.28}$$

If the input signal is a unit step so that $V_i(s)$ is $1/s$,

$$V_o(s) = \frac{1}{s(10^{-12}s^2 + 1.4 \times 10^{-6}s + 1)}$$

$$= \frac{1}{s[s^2/(10^6)^2 + 2(0.7)s/10^6 + 1]} \tag{3.29}$$

The final term in Eqn. 3.29 shows that the quadratic portion of the expression has a natural frequency $\omega_n = 10^6$ and a damping ratio $\zeta = 0.7$. The corresponding time function is determined from Table 3.1, with the result

$$f(t) = 1 - \frac{e^{-0.7 \times 10^6 t}}{0.7} \sin (0.7 \times 10^6 t + 45°) \qquad (3.30)$$

3.3   TRANSIENT RESPONSE

The *transient response* of an element or system is its output as a function of time following the application of a specified input. The test signal chosen to excite the transient response of the system may be either an input that is anticipated in normal operation, or it may be a mathematical abstraction selected because of the insight it lends to system behavior. Commonly used test signals include the impulse and time integrals of this function.

3.3.1   Selection of Test Inputs

The mathematics of linear systems insures that the same system information is obtainable independent of the test input used, since the transfer function of a system is clearly independent of inputs applied to the system. In practice, however, we frequently find that certain aspects of system performance are most easily evaluated by selecting the test input to accentuate features of interest.

For example, we might attempt to evaluate the d-c gain of an operational amplifier with feedback by exciting it with an impulse and measuring the net area under the impulse response of the amplifier. This approach is mathematically sound, as shown by the following development. Assume that the closed-loop transfer function of the amplifier is $G(s)$ and that the corresponding impulse response [the inverse transform of $G(s)$] is $g(t)$. The properties of Laplace transforms show that

$$\int_0^t g(t) \, dt = \frac{1}{s} G(s) \qquad (3.31)$$

The final value theorem applied to this function indicates that the net area under impulse response is

$$\lim_{t \to \infty} \int_0^t g(t) \, dt = \lim_{s \to 0} s \frac{1}{s} G(s) = G(0) \qquad (3.32)$$

Unfortunately, this technique involves experimental pitfalls. The first of these is the choice of the time function used to approximate an impulse.

In order for a finite-duration pulse to approximate an impulse satisfactorily, it is necessary to have[7]

$$t_p \ll \frac{1}{|s_m|} \qquad (3.33)$$

where t_p is the width of the pulse and s_m is the frequency of the pole of $G(s)$ that is located furthest from the origin.

It may be difficult to find a pulse generator that produces pulses narrow enough to test high-frequency amplifiers. Furthermore, the narrow pulse frequently leads to a small-amplitude output with attendant measurement problems. Even if a satisfactory impulse response is obtained, the tedious task of integrating this response (possibly by counting boxes under the output display on an oscilloscope) remains. It should be evident that a far more accurate and direct measurement of d-c gain is possible if a constant input is applied to the amplifier.

Alternatively, high-frequency components of the system response are not excited significantly if slowly time-varying inputs are applied as test inputs. In fact, systems may have high-frequency poles close to the imaginary axis in the s-plane, and thus border on instability; yet they exhibit well-behaved outputs when tested with slowly-varying inputs.

For systems that have neither a zero-frequency pole nor a zero in their transfer function, the step response often provides the most meaningful evaluation of performance. The d-c gain can be obtained directly by measuring the final value of the response to a unit step, while the initial discontinuity characteristic of a step excites high-frequency poles in the system transfer function. Adequate approximations to an ideal step are provided by rectangular pulses with risetimes

$$t_r \ll \frac{1}{|s_m|} \qquad (3.34)$$

(s_m as defined earlier) and widths

$$t_w \gg \frac{1}{|s_n|} \qquad (3.35)$$

where s_n is the frequency of the pole in the transfer function located closest to the origin. Pulse generators with risetimes under 1 ns are available, and these generators can provide useful information about amplifiers with bandwidths on the order of 100 MHz.

---

[7] While this statement is true in general, if only the d-c gain of the system is required, any pulse can be used. An extension of the above development shows that the area under the response to any unit-area input is identical to the area under the impulse response.

3.3.2 Approximating Transient Responses

Examples in Section 3.1 indicated that in some cases it is possible to approximate the transient response of a complex system by using that of a much simpler system. This type of approximation is possible whenever the transfer function of interest is dominated by one or two poles.

Consider an amplifier with a transfer function

$$\frac{V_o(s)}{V_i(s)} = \frac{a_0 \prod_{i=1}^{m} (\tau_{zi}s + 1)}{\prod_{j=1}^{n} (\tau_{pj}s + 1)} \qquad n > m, \quad \text{all } \tau > 0 \qquad (3.36)$$

The response of this system to a unit-step input is

$$v_o(t) = \mathcal{L}^{-1}\left[\frac{1}{s}\frac{V_o(s)}{V_i(s)}\right] = a_o + \sum_{k=1}^{n} A_k e^{-t/\tau_{pk}} \qquad (3.37)$$

The A's obtained from Eqn. 3.11 after slight rearrangement are

$$A_k = -a_0 \frac{\prod_{i=1}^{m}\left(-\dfrac{\tau_{zi}}{\tau_{pk}} + 1\right)}{\prod_{\substack{j=1 \\ j \neq k}}^{n}\left(-\dfrac{\tau_{pj}}{\tau_{pk}} + 1\right)} \qquad (3.38)$$

Assume that $\tau_{p1} \gg$ all other τ's. In this case, which corresponds to one pole in the system transfer function being much closer to the origin than all other singularities, Eqn. 3.38 can be used to show that $A_1 \simeq a_0$ and all other A's $\simeq 0$ so that

$$v_o(t) \simeq a_0(1 - e^{-t/\tau_{p1}}) \qquad (3.39)$$

This single-exponential transient response is shown in Fig. 3.6. Experience shows that the single-pole response is a good approximation to the actual response if remote singularities are a factor of five further from the origin than the dominant pole.

The approximate result given above holds even if some of the remote singularities occur in complex conjugate pairs, providing that the pairs are located at much greater distances from the origin in the s plane than the dominant pole. However, if the real part of the complex pair is not more negative than the location of the dominant pole, small-amplitude, high-frequency damped sinusoids may persist after the dominant transient is completed.

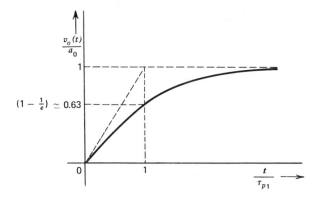

Figure 3.6 Step response of first-order system.

Another common singularity pattern includes a complex pair of poles much closer to the origin in the s plane than all other poles and zeros. An argument similar to that given above shows that the transfer function of an amplifier with this type of singularity pattern can be approximated by the complex pair alone, and can be written in the standard form

$$\frac{V_o(s)}{V_i(s)} = \frac{a_o}{s^2/\omega_n^2 + 2\zeta s/\omega_n + 1} \tag{3.40}$$

The equation parameters ω_n and ζ are called the *natural frequency* (expressed in radians per second) and the *damping ratio*, respectively. The physical significance of these parameters is indicated in the s-plane plot shown as Fig. 3.7. The relative pole locations shown in this diagram correspond to the *underdamped* case ($\zeta < 1$). Two other possibilities are the *critically damped* pair ($\zeta = 1$) where the two poles coincide on the real axis and the *overdamped* case ($\zeta > 1$) where the two poles are separated on the real axis. The denominator polynomial can be factored into two roots with real coefficients for the later two cases and, as a result, the form shown in Eqn. 3.40 is normally not used. The output provided by the amplifier described by Eqn. 3.40 in response to a unit step is (from Table 3.1).

$$v_o(t) = a_o \left[1 - \frac{1}{\sqrt{1 - \zeta^2}} e^{-\zeta \omega_n t} \sin \left(\sqrt{1 - \zeta^2} \, \omega_n t + \Phi \right) \right] \tag{3.41}$$

where

$$\Phi = \tan^{-1} \left[\frac{\sqrt{1 - \zeta^2}}{\zeta} \right]$$

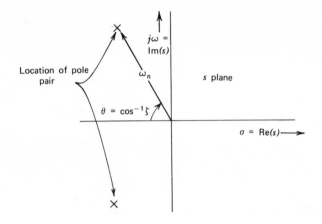

Figure 3.7 *s*-plane plot of complex pole pair.

Figure 3.8 is a plot of $v_o(t)$ as a function of normalized time $\omega_n t$ for various values of damping ratio. Smaller damping ratios, corresponding to complex pole pairs with the poles nearer the imaginary axis, are associated with step responses having a greater degree of overshoot.

The transient responses of third- and higher-order systems are not as easily categorized as those of first- and second-order systems since more parameters are required to differentiate among the various possibilities. The situation is simplified if the relative pole positions fall into certain patterns. One class of transfer functions of interest are the Butterworth filters. These transfer functions are also called *maximally flat* because of properties of their frequency responses (see Section 3.4). The step responses of Butterworth filters also exhibit fairly low overshoot, and because of these properties feedback amplifiers are at times compensated so that their closed-loop poles form a Butterworth configuration.

The poles of an *n*th-order Butterworth filter are located on a circle centered at the origin of the *s*-plane. For *n* even, the poles make angles $\pm$ $(2k + 1)$ $90°/n$ with the negative real axis, where k takes all possible integral values from 0 to $(n/2) - 1$. For *n* odd, one pole is located on the negative real axis, while others make angles of $\pm k$ $(180°/n)$ with the negative real axis where k takes integral values from 1 to $(n/2) - (1/2)$. Thus, for example, a first-order Butterworth filter has a single pole located at $s = -\omega_n$. The second-order Butterworth filter has its poles located $\pm 45°$ from the negative real axis, corresponding to a damping ratio of 0.707.

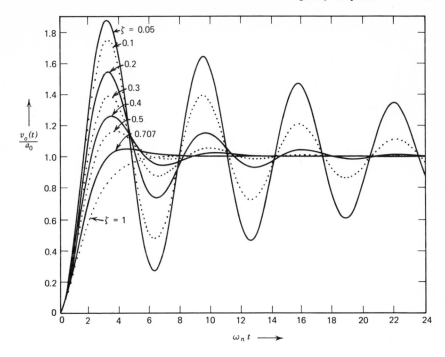

Figure 3.8   Step responses of second-order system.

The transfer functions for third- and fourth-order Butterworth filters are

$$B_3(s) = \frac{1}{s^3/\omega_n^3 + 2s^2/\omega_n^2 + 2s/\omega_n + 1} \tag{3.42}$$

and

$$B_4(s) = \frac{1}{s^4/\omega_n^4 + 2.61s^3/\omega_n^3 + 3.42s^2/\omega_n^2 + 2.61s/\omega_n + 1} \tag{3.43}$$

respectively. Plots of the pole locations of these functions are shown in Fig. 3.9. The transient outputs of these filters in response to unit steps are shown in Fig. 3.10.

3.4   FREQUENCY RESPONSE

The frequency response of an element or system is a measure of its steady-state performance under conditions of sinusoidal excitation. In

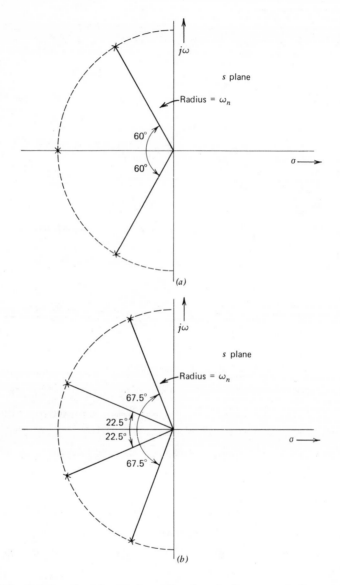

Figure 3.9 Pole locations for third- and fourth-order Butterworth filters. (*a*) Third-Order. (*b*) Fourth-order.

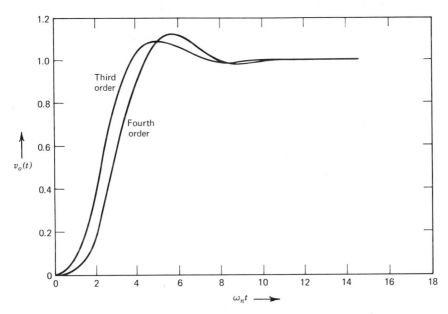

Figure 3.10 Step responses for third- and fourth-order Butterworth filters.

steady state, the output of a linear element excited with a sinusoid at a frequency ω (expressed in radians per second) is purely sinusoidal at frequency ω. The frequency response is expressed as a gain or magnitude $M(\omega)$ that is the ratio of the amplitude of the output to the input sinusoid and a phase angle $\phi(\omega)$ that is the relative angle between the output and input sinusoids. The phase angle is positive if the output leads the input. The two components that comprise the frequency response of a system with a transfer function $G(s)$ are given by

$$M(\omega) = |G(j\omega)| \qquad (3.44a)$$

$$\phi(\omega) = \measuredangle G(j\omega) = \tan^{-1} \frac{\mathrm{Im}[G(j\omega)]}{\mathrm{Re}[G(j\omega)]} \qquad (3.44b)$$

It is frequently necessary to determine the frequency response of a system with a transfer function that is a ratio of polynomials in s. One possible method is to evaluate the frequency response by substituting $j\omega$ for s at all frequencies of interest, but this method is cumbersome, particularly for high-order polynomials. An alternative approach is to present the information concerning the frequency response graphically, as described below.

The transfer function is first factored so that both the numerator and denominator consist of products of first- and second-order terms with real coefficients. The function can then be written in the general form

$$
G(s) = \frac{a_0}{s^n} \left[\prod_{\substack{\text{first-}\\\text{order}\\\text{zeros}}} (\tau_h s + 1) \right] \left[\prod_{\substack{\text{complex}\\\text{zero}\\\text{pairs}}} \left(\frac{s^2}{\omega_{ni}{}^2} + \frac{2\zeta_i s}{\omega_{ni}} + 1 \right) \right]
$$

$$
\times \left[\prod_{\substack{\text{first-}\\\text{order}\\\text{poles}}} \frac{1}{(\tau_j s + 1)} \right] \left[\prod_{\substack{\text{complex}\\\text{pole}\\\text{pairs}}} \frac{1}{(s^2/\omega_{nk}{}^2 + 2\zeta_k s/\omega_{nk} + 1)} \right] \qquad (3.45)
$$

While several methods such as Lin's method[8] are available for factoring polynomials, this operation can be tedious unless machine computation is employed, particularly when the order of the polynomial is large. Fortunately, in many cases of interest the polynomials are either of low order or are available from the system equations in factored form.

Since $G(j\omega)$ is a function of a complex variable, its angle $\phi(\omega)$ is the sum of the angles of the constituent terms. Similarly, its magnitude $M(\omega)$ is the product of the magnitudes of the components. Furthermore, if the magnitudes of the components are plotted on a logarithmic scale, the log of M is given by the sum of the logs corresponding to the individual components.[9]

Plotting is simplified by recognizing that only four types of terms are possible in the representation of Eqn. 3.45:

1. Constants, a_0.
2. Single- or multiple-order differentiations or integrations, s^n, where n can be positive (differentiations) or negative (integrations).
3. First-order terms $(\tau s + 1)$, or its reciprocal.
4. Complex conjugate pairs $s^2/\omega_n{}^2 + 2\zeta s/\omega_n + 1$, or its reciprocal.

---

[8] S. N. Lin, "A Method of Successive Approximations of Evaluating the Real and Complex Roots of Cubic and Higher-Order Equations," *J. Math. Phys.*, Vol. 20, No. 3, August, 1941, pp. 231–242.

[9] The decibel, equal to 20 $\log_{10}$ [magnitude] is often used for these manipulations. This usage is technically correct only if voltage gains or current gains between portions of a circuit with identical impedance levels are considered. The issue is further confused when the decibel is used indiscriminately to express dimensioned quantities such as transconductances. We shall normally reserve this type of presentation for loop-transmission manipulations (the loop transmission of any feedback system must be dimensionless), and simply plot signal ratios on logarithmic coordinates.

It is particularly convenient to represent each of these possible terms as a plot of M (on a logarithmic magnitude scale) and ϕ (expressed in degrees) as a function of ω (expressed in radians per second) plotted on a logarithmic frequency axis. A logarithmic frequency axis is used because it provides adequate resolution in cases where the frequency range of interest is wide and because the relative shape of a particular response curve on the log axis does not change as it is frequency scaled. The magnitude and angle of any rational function can then be determined by adding the magnitudes and angles of its components. This representation of the frequency response of a system or element is called a *Bode plot*.

The magnitude of a term a_0 is simply a frequency-independent constant, with an angle equal to 0° or 180° depending on whether the sign of a_0 is positive or negative, respectively.

Both differentiations and integrations are possible in feedback systems. For example, a first-order high-pass filter has a single zero at the origin and, thus, its voltage transfer ratio includes a factor s. A motor (frequently used in mechanical feedback systems) includes a factor $1/s$ in the transfer function that relates mechanical shaft angle to applied motor voltage, since a constant input voltage causes unlimited shaft rotation. Similarly, various types of phase detectors are examples of purely electronic elements that have a pole at the origin in their transfer functions. This pole results because the voltage out of such a circuit is proportional to the phase-angle difference between two input signals, and this angle is equal to the integral of the frequency difference between the two signals. We shall also see that it is often convenient to approximate the transfer function of an amplifier with high d-c gain and a single low-frequency pole as an integration.

The magnitude of a term s^n is equal to ω^n, a function that passes through 1 at $\omega = 1$ and has a slope of n on logarithmic coordinates. The angle of this function is $n \times 90°$ at all frequencies.

The magnitude of a first order pole $1/(\tau s + 1)$ is

$$M = \frac{1}{\sqrt{\tau^2\omega^2 + 1}} \tag{3.46}$$

while the angle of this function is

$$\phi = -\tan^{-1}\tau\omega \tag{3.47}$$

The magnitude and angle for the first-order pole are plotted as a function of normalized frequency in Fig. 3.11. An essential feature of the magnitude function is that it can be approximated by two straight lines, one lying along the $M = 1$ line and the other with a slope of -1, which intersect at $\omega = 1/\tau$. (This frequency is called the *corner frequency*.) The maximum

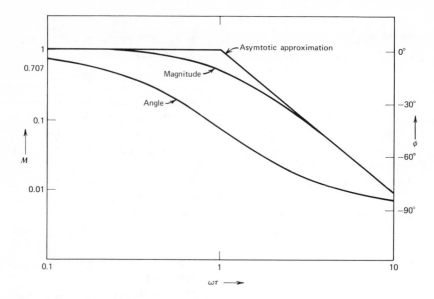

Figure 3.11   Frequency response of first-order system.

departure of the actual curves from the asymptotic representation is a factor of 0.707 and occurs at the corner frequency. The magnitude and angle for a first-order zero are obtained by inverting the curves shown for the pole, so that the magnitude approaches an asymptotic slope of $+1$ beyond the corner frequency, while the angle changes from 0 to $+90°$.

The magnitude for a complex-conjugate pole pair

$$\frac{1}{s^2/\omega_n{}^2 + 2\zeta s/\omega_n + 1}$$

is

$$M = \frac{1}{\sqrt{\dfrac{4\zeta^2\omega^2}{\omega_n{}^2} + \left(1 - \dfrac{\omega^2}{\omega_n{}^2}\right)^2}} \qquad (3.48)$$

with the corresponding angle

$$\phi = -\tan^{-1}\frac{2\zeta\omega/\omega_n}{1 - \omega^2/\omega_n{}^2} \qquad (3.49)$$

These functions are shown in Bode-plot form as a parametric family of curves plotted against normalized frequency ω/ω_n in Fig. 3.12. Note that

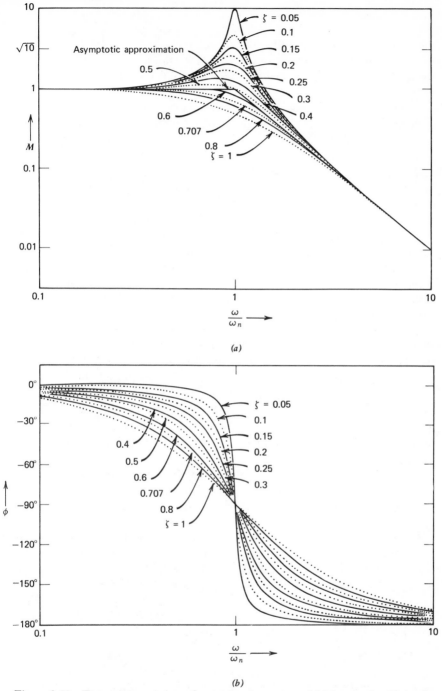

Figure 3.12 Frequency response of second-order system. (*a*) Magnitude. (*b*) Angle.

the asymptotic approximation to the magnitude is reasonably accurate providing that the damping ratio exceeds 0.25. The corresponding curves for a complex-conjugate zero are obtained by inverting the curves shown in Fig. 3.12.

It was stated in Section 3.3.2 that feedback amplifiers are occasionally adjusted to have Butterworth responses. The frequency responses for third- and fourth-order Butterworth filters are shown in Bode-plot form in Fig. 3.13. Note that there is no peaking in the frequency response of these maximally-flat transfer functions. We also see from Fig. 3.12 that the damping ratio of 0.707, corresponding to the two-pole Butterworth configuration, divides the second-order responses that peak from those which do not. The reader should recall that the flatness of the Butterworth response refers to its *frequency response*, and that the step responses of all Butterworth filters exhibit overshoot.

The value associated with Bode plots stems in large part from the ease with which the plot for a complex system can be obtained. The overall system transfer function can be obtained by the following procedure. First, the magnitude and phase curves corresponding to all the terms included in the transfer function of interest are plotted. When the first- and second-order curves (Figs. 3.11 and 3.12) are used, they are located along the frequency axis so that their corner frequencies correspond to those of the represented factors. Once these curves have been plotted, the magnitude of the complete transfer function at any frequency is obtained by adding the linear distances from unity magnitude of all components at the frequency of interest. The same type of graphical addition can be used to obtain the complete phase curve. Dividers, or similar aids, can be used to perform the graphical addition.

In practice, the asymptotic magnitude curve is usually sketched by drawing a series of intersecting straight lines with appropriate slope changes at intersections. Corrections to the asymptotic curve can be added in the vicinity of singularities if necessary.

The information contained in a Bode plot can also be presented as a *gain-phase* plot, which is a more convenient representation for some operations. Rectangular coordinates are used, with the ordinate representing the magnitude (on a logarithmic scale) and the abscissa representing the phase angle in degrees. Frequency expressed in radians per second is a parameter along the gain-phase curve. Gain-phase plots are frequently drawn by transferring data from a Bode plot.

The transfer function

$$G(s) = \frac{10^7(10^{-4}s + 1)}{s(0.01s + 1)(s^2/10^{12} + 2(0.2)s/10^6 + 1)} \qquad (3.50)$$

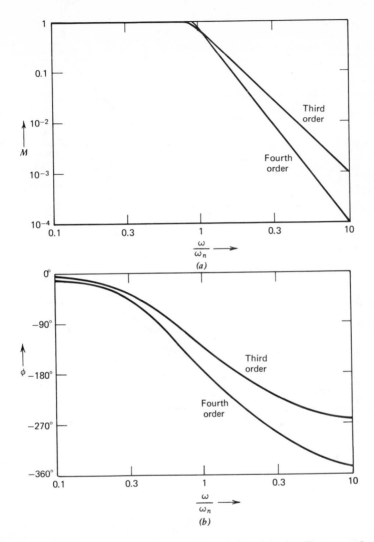

Figure 3.13 Frequency response of third- and fourth-order Butterworth filters. (*a*) Magnitude. (*b*) Angle.

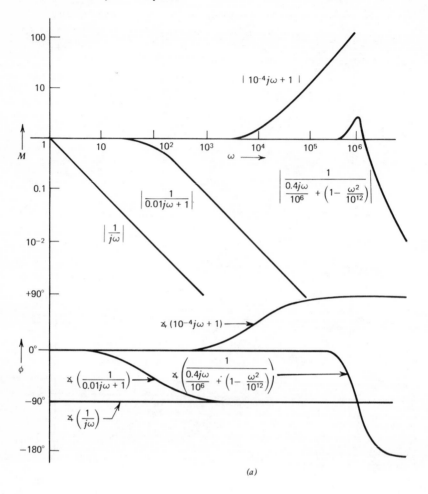

(a)

Figure 3.14 Bode plot of $\dfrac{10^7(10^{-4}s + 1)}{s(0.01s + 1)(s^2/10^{12} + 2(0.2)s/10^6 + 1)}.$ (a) Individual factors. (b) Bode plot.

is used to illustrate construction of Bode and gain-phase plots. This function includes these five factors:

1. A constant 10^7.
2. A single integration.
3. A first-order pole with a time constant of 0.01 second, corresponding to a corner frequency of 100 radians per second.

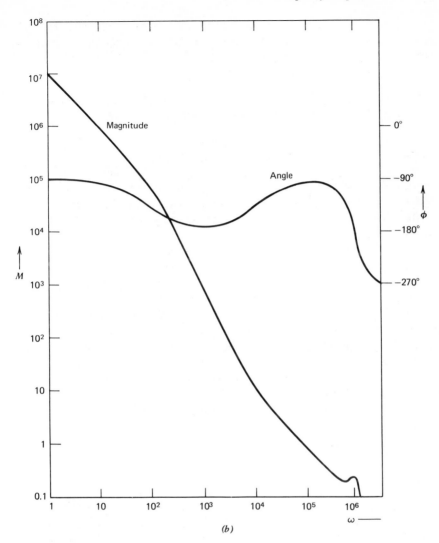

(b)

Figure 3.14—Continued

4. A first-order zero with a time constant of 10^{-4} seconds, corresponding to a corner frequency of 10^4 radians per second.

5. A complex-conjugate pole pair with a natural frequency of 10^6 radians per second and a damping ratio of 0.2.

The individual factors are shown in Bode-plot form on a common frequency scale in Fig. 3.14a. These factors are combined to yield the Bode plot for the complete transfer function in Fig. 3.14b. The same information is presented in gain-phase form in Fig. 3.15.

3.5 RELATIONSHIPS BETWEEN TRANSIENT RESPONSE AND FREQUENCY RESPONSE

It is clear that either the impulse response (or the response to any other transient input) of a linear system or its frequency response completely characterize the system. In many cases experimental measurements on a closed-loop system are most easily made by applying a transient input. We may, however, be interested in certain aspects of the frequency response of the system such as its *bandwidth* defined as the frequency where its gain drops to 0.707 of the midfrequency value.

Since either the transient response or the frequency response completely characterize the system, it should be possible to determine performance in one domain from measurements made in the other. Unfortunately, since the measured transient response does not provide an equation for this response, Laplace techniques cannot be used directly unless the time response is first approximated analytically as a function of time. This section lists several approximate relationships between transient response and frequency response that can be used to estimate one performance measure from the other. The approximations are based on the properties of first- and second-order systems.

It is assumed that the feedback path for the system under study is frequency independent and has a magnitude of unity. A system with a frequency-independent feedback path f_0 can be manipulated as shown in Fig. 3.16 to yield a scaled, unity-feedback system. The approximations given are valid for the transfer function V_a/V_i, and V_o can be determined by scaling values for V_a by $1/f_0$.

It is also assumed that the magnitude of the d-c loop transmission is very large so that the closed-loop gain is nearly one at d-c. It is further assumed that the singularity closest to the origin in the s plane is either a pole or a complex pair of poles, and that the number of poles of the function exceeds the number of zeros. If these assumptions are satisfied, many practical systems have time domain-frequency domain relationships similar to those of first- or second-order systems.

The parameters we shall use to describe the transient response and the frequency response of a system include the following.

(a) Rise time t_r. The time required for the step response to go from 10 to 90% of final value.

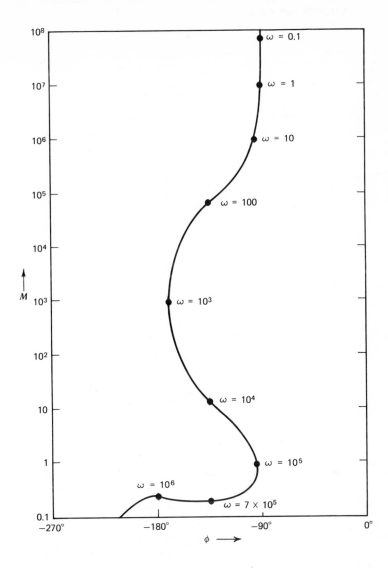

Figure 3.15 Gain phase plot of $\dfrac{10^7(10^{-4}s + 1)}{s(0.1s + 1)(s^2/10^{12} + 2(0.2)s/10^6 + 1)}$.

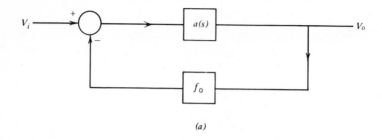

(a)

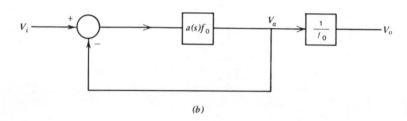

(b)

Figure 3.16 System topology for approximate relationships. (a) System with frequency-independent feedback path. (b) System represented in scaled, unity-feedback form.

(b) The maximum value of the step response P_0.

(c) The time at which P_0 occurs t_p.

(d) Settling time t_s. The time after which the system step response remains within 2% of final value.

(e) The error coefficient e_1. (See Section 3.6.) This coefficient is equal to the time delay between the output and the input when the system has reached steady-state conditions with a ramp as its input.

(f) The bandwidth in radians per second ω_h or hertz f_h ($f_h = \omega_h/2\pi$). The frequency at which the response of the system is 0.707 of its low-frequency value.

(g) The maximum magnitude of the frequency response M_p.

(h) The frequency at which M_p occurs ω_p.

These definitions are illustrated in Fig. 3.17.

For a first-order system with $V_o(s)/V_i(s) = 1/(\tau s + 1)$, the relationships are

$$t_r = 2.2\tau = \frac{2.2}{\omega_h} = \frac{0.35}{f_h} \tag{3.51}$$

$$P_0 = M_p = 1 \tag{3.52}$$

$$t_p = \infty \tag{3.53}$$

$$t_s = 4\tau \tag{3.54}$$

$$e_1 = \tau \tag{3.55}$$

$$\omega_p = 0 \tag{3.56}$$

For a second-order system with $V_o(s)/V_i(s) = 1/(s^2/\omega_n^2 + 2\zeta s/\omega_n + 1)$ and $\theta \triangleq \cos^{-1}\zeta$ (see Fig. 3.7) the relationships are

$$t_r \simeq \frac{2.2}{\omega_h} = \frac{0.35}{f_h} \tag{3.57}$$

$$P_0 = 1 + \exp\frac{-\pi\zeta}{\sqrt{1 - \zeta^2}} = 1 + e^{-\pi/\tan\theta} \tag{3.58}$$

$$t_p = \frac{\pi}{\omega_n \sqrt{1 - \zeta^2}} = \frac{\pi}{\omega_n \sin\theta} \tag{3.59}$$

$$t_s \simeq \frac{4}{\zeta\omega_n} = \frac{4}{\omega_n \cos\theta} \tag{3.60}$$

$$e_1 = \frac{2\zeta}{\omega_n} = \frac{2\cos\theta}{\omega_n} \tag{3.61}$$

$$M_p = \frac{1}{2\zeta \sqrt{1 - \zeta^2}} = \frac{1}{\sin 2\theta} \qquad \zeta < 0.707, \theta > 45° \tag{3.62}$$

$$\omega_p = \omega_n \sqrt{1 - 2\zeta^2} = \omega_n \sqrt{-\cos 2\theta} \qquad \zeta < 0.707, \theta > 45° \tag{3.63}$$

$$\omega_h = \omega_n \left(1 - 2\zeta^2 + \sqrt{2 - 4\zeta^2 + 4\zeta^4}\right)^{1/2} \tag{3.64}$$

If a system step response or frequency response is similar to that of an approximating system (see Figs. 3.6, 3.8, 3.11, and 3.12) measurements of t_r, P_0, and t_p permit estimation of ω_h, ω_p, and M_p or vice versa. The steady-state error in response to a unit ramp can be estimated from either set of measurements.

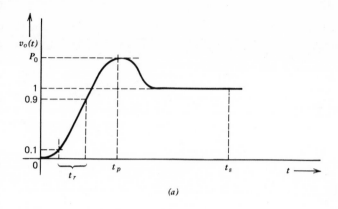

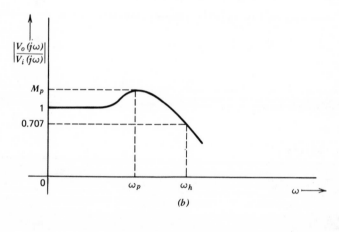

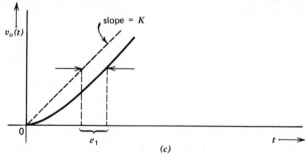

Figure 3.17 Parameters used to describe transient and frequency responses. (*a*) Unit-step response. (*b*) Frequency response. (*c*) Ramp response.

One final comment concerning the quality of the relationship between 0.707 bandwidth and 10 to 90% step risetime (Eqns. 3.51 and 3.57) is in order. For virtually any system that satisfies the original assumptions, independent of the order or relative stability of the system, the product $t_r f_h$ is within a few percent of 0.35. This relationship is so accurate that it really isn't worth measuring f_h if the step response can be more easily determined.

3.6   ERROR COEFFICIENTS

The response of a linear system to certain types of transient inputs may be difficult or impossible to determine by Laplace techniques, either because the transform of the transient is cumbersome to evaluate or because the transient violates the conditions necessary for its transform to exist. For example, consider the angle that a radar antenna makes with a fixed reference while tracking an aircraft, as shown in Fig. 3.18. The pointing angle determined from the geometry is

$$\theta = \tan^{-1}\left[\frac{v}{l}t\right] \tag{3.65}$$

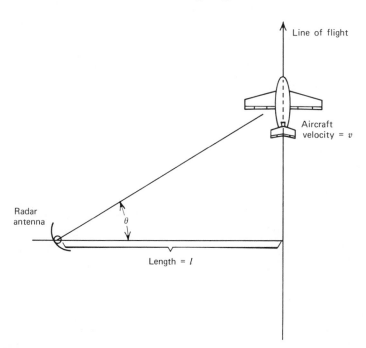

Figure 3.18   Radar antenna tracking an airplane.

assuming that $\theta = 0$ at $t = 0$. This function is not transformable using our form of the Laplace transform, since it is nonzero for negative time and since no amount of time shift makes it zero for negative time. The expansion introduced in this section provides a convenient method for evaluating the performance of systems excited by transient inputs, such as Eqn. 3.65, for which all derivatives exist at all times.

3.6.1   The Error Series

Consider a system, initially at rest and driven by a single input, with a transfer function $G(s)$. Furthermore, assume that $G(s)$ can be expanded in a power series in s, or that

$$G(s) = g_0 + g_1 s + g_2 s^2 + \cdots + \qquad (3.66)$$

If the system is excited by an input $v_i(t)$, the output signal as a function of time is

$$v_o(t) = \mathcal{L}^{-1}[G(s)V_i(s)]$$
$$= \mathcal{L}^{-1}[g_0 V_i(s) + g_1 s V_i(s) + g_2 s^2 V_i(s) + \cdots +] \quad (3.67)$$

If Eqn. 3.66 is inverse transformed term by term, and the differentiation property of Laplace transforms is used to simplify the result, we see that[8]

$$v_o(t) = g_0 v_i(t) + g_1 \frac{dv_i(t)}{dt} + g_2 \frac{d^2 v_i(t)}{dt^2} + \cdots + \qquad (3.68)$$

The complete series yields the correct value for $v_o(t)$ in cases where the function $v_i(t)$ and all its derivatives exist at all times.

In practice, the method is normally used to evaluate the error (or difference between ideal and actual output) that results for a specified input. If Eqn. 3.68 is rewritten using the error $e(t)$ as the dependent parameter, the resultant series

$$e(t) = e_0 v_i(t) + e_1 \frac{dv_i(t)}{dt} + e_2 \frac{d^2 v_i(t)}{dt^2} + \cdots + \qquad (3.69)$$

is called an *error series*, and the e's on the right-hand side of this equation are called *error coefficients*.

The error coefficients can be obtained by two equivalent expansion methods. A formal mathematical approach shows that

$$e_k = \frac{1}{k!} \frac{d^k}{ds^k} \left[\frac{V_e(s)}{V_i(s)} \right]_{s=0} \qquad (3.70)$$

---

[8] A mathematically satisfying development is given in G. C. Newton, Jr., L. A. Gould, and J. F. Kaiser, *Analytical Design of Linear Feedback Controls*, Wiley, New York, 1957, Appendix C. An expression that bounds the error when the series is truncated is also given in this reference.

where $V_e(s)/V_i(s)$ is the input-to-error transfer function for the system. Alternatively, synthetic division can be used to write the input-to-error transfer function as a series in ascending powers of s. The coefficient of the s^k term in this series is e_k.

While the formal mathematics require that the complete series be used to determine the error, the series converges rapidly in cases of practical interest where the error is small compared to the input signal. (Note that if the error is the same order of magnitude as the input signal in a unity-feedback system, comparable results can be obtained by turning off the system.) Thus in reasonable applications, a few terms of the error series normally suffice. Furthermore, the requirement that all derivatives of the input signal exist can be usually relaxed if we are interested in errors at times separated from the times of discontinuities by at least the settling time of the system. (See Section 3.5 for a definition of settling time.)

3.6.2 Examples

Some important properties of feedback amplifiers can be illustrated by applying error-coefficient analysis methods to the inverting-amplifier connection shown in Fig. 3.19a. A block diagram obtained by assuming negligible loading at the input and output of the amplifier is shown in Fig. 3.19b. An error signal is generated in this diagram by comparing the actual output of the amplifier with the ideal value, $-V_i$. The input-to-error transfer function from this block diagram is

$$\frac{V_e(s)}{V_i(s)} = \frac{-1}{1 + a(s)/2} \tag{3.71}$$

Operational amplifiers are frequently designed to have an approximately single-pole open-loop transfer function, implying

$$a(s) \simeq \frac{a_0}{\tau s + 1} \tag{3.72}$$

The error coefficients assuming this value for $a(s)$ are easily evaluated by means of synthetic division since

$$\frac{V_e(s)}{V_i(s)} = \frac{-1}{1 + a_0/2(\tau s + 1)} = \frac{-2 - 2\tau s}{a_0 + 2 + 2\tau s}$$

$$= -\frac{2}{a_0 + 2} - \frac{2\tau}{a_0 + 2}\left(1 - \frac{2}{a_0 + 2}\right)s$$

$$+ \frac{4\tau^2}{(a_0 + 2)^2}\left(1 - \frac{2}{a_0 + 2}\right)s^2 + \cdots + \tag{3.73}$$

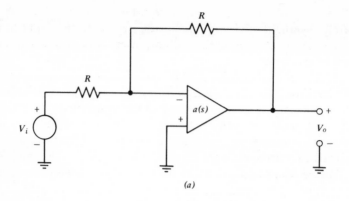

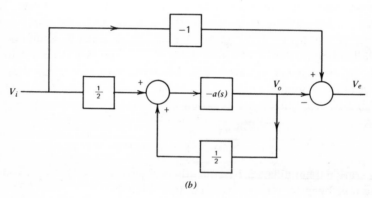

Figure 3.19 Unity-gain inverter. (*a*) Connection. (*b*) Block diagram including error signal.

If a_0, the amplifier d-c gain, is large, the error coefficients are

$$e_0 \simeq -\frac{2}{a_0}$$

$$e_1 \simeq -\frac{2\tau}{a_0}$$

$$e_2 \simeq \frac{4\tau^2}{a_0^2}$$

.
.
.

$$e_n = \frac{(-2)^n \tau^n}{a_0{}^n} \qquad n \geq 1 \qquad (3.74)$$

The error coefficients are easily interpreted in terms of the loop transmission of the amplifier-feedback network combination in this example. The magnitude of the zero-order error coefficient is equal to the reciprocal of the d-c loop transmission. The first-order error-coefficient magnitude is equal to the reciprocal of the frequency (in radians per second) at which the loop transmission is unity, while the magnitude of each subsequent higher-order error coefficient is attenuated by a factor equal to this frequency. These results reinforce the conclusion that feedback-amplifier errors are reduced by large loop transmissions and unity-gain frequencies.

If this amplifier is excited with a ramp $v_i(t) = Rt$, the error after any start-up transient has died out is

$$v_e(t) = e_0 v_i(t) + e_1 \frac{dv_i(t)}{dt} + \cdots + = -\frac{2Rt}{a_0} - \frac{2R\tau}{a_0} \qquad (3.75)$$

Because the maximum input-signal level is limited by linearity considerations, (the voltage Rt must be less than the voltage at which the amplifier saturates) the second term in the error series frequently dominates, and in these cases the error is

$$v_e(t) \simeq -\frac{2R\tau}{a_0} \qquad (3.76)$$

implying the actual ramp response of the amplifier lags behind the ideal output by an amount equal to the slope of the ramp divided by the unity-loop-transmission frequency. The ramp response of the amplifier, assuming that the error series is dominated by the e_1 term, is compared with the ramp response of a system using an infinite-gain amplifier in Fig. 3.20. The steady-state ramp error, introduced earlier in Eqns. 3.55 and 3.61 and illustrated in Fig. 3.17c, is evident in this figure.

One further observation lends insight into the operation of this type of system. If the relative magnitudes of the input signal and its derivatives are constrained so that the first-order (or higher) terms in the error series dominate, the open-loop transfer function of the amplifier can be approximated as an integration.

$$a(s) \simeq \frac{a_0}{\tau s} \qquad (3.77)$$

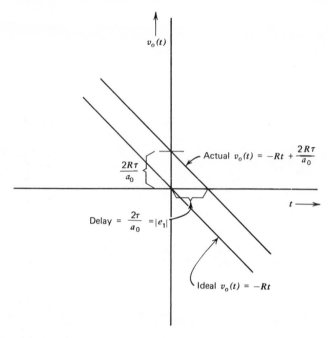

Figure 3.20 Ideal and actual ramp responses.

In order for the output of an amplifier with this type of open-loop gain to be a ramp, it is necessary to have a constant error signal applied to the amplifier input.

Pursuing this line of reasoning further shows how the open-loop transfer function of the amplifier should be chosen to reduce ramp error. Error is clearly reduced if the quantity a_0/τ is increased, but such an increase requires a corresponding increase in the unity-loop-gain frequency. Unfortunately oscillations result for sufficiently high unity-gain frequencies. Alternatively, consider the result if the amplifier open-loop transfer function approximates a double integration

$$a(s) \simeq \frac{a_0(\tau s + 1)}{s^2} \tag{3.78}$$

(The zero is necessary to insure stability. See Chapter 4.) The reader should verify that both e_0 and e_1 are zero for an amplifier with this open-loop transfer function, implying that the steady-state ramp error is zero. Further manipulation shows that if the amplifier open-loop transfer function includes an nth order integration, the error coefficients e_0 through e_{n-1} are zero.

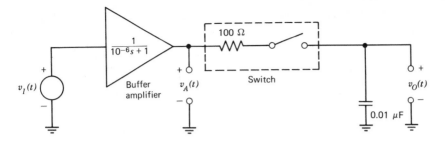

Figure 3.21 Sample-and-hold circuit.

The use of error coefficients to analyze systems excited by pulse signals is illustrated with the aid of the sample-and-hold circuit shown in Fig. 3.21. This circuit consists of a buffer amplifier followed by a switch and capacitor. In practice the switch is frequently realized with a field-effect transistor, and the 100-Ω resistor models the on resistance of the transistor. When the switch is closed, the capacitor is charged toward the voltage v_I through the switch resistance. If the switch is opened at a time t_A, the voltage $v_O(t)$ should ideally maintain the value $v_I(t_A)$ for all time greater than t_A. The buffer amplifier is included so that the capacitor charging current is supplied by the amplifier rather than the signal source. A second buffer amplifier is often included following the capacitor to isolate it from loads, but this second amplifier is not required for the present example.

There are a variety of effects that degrade the performance of a sample-and-hold circuit. One important source of error stems from the fact that $v_O(t)$ is generally not equal to $v_I(t)$ unless $v_I(t)$ is time invariant because of the dynamics of the buffer amplifier and the switch-capacitor combination. Thus an incorrect value is held when the switch is opened.

Error coefficients can be used to predict the magnitude of this tracking error as a function of the input signal and the system dynamics. For purposes of illustration, it is assumed that the buffer amplifier has a single-pole transfer function such that

$$\frac{V_a(s)}{V_i(s)} = \frac{1}{10^{-6}s + 1} \tag{3.79}$$

Since the time constant associated with the switch-capacitor combination is also 1 μs, the input-to-output transfer function with the switch closed (in which case the system is linear, time-invariant) is

$$\frac{V_o(s)}{V_i(s)} = \frac{1}{(10^{-6}s + 1)^2} \tag{3.80}$$

With the switch closed the output is ideally equal to the input, and thus the input-to-error transfer function is

$$\frac{V_e(s)}{V_i(s)} = 1 - \frac{V_o(s)}{V_i(s)} = \frac{10^{-12}s^2 + 2 \times 10^{-6}s}{(10^{-6}s + 1)^2} \tag{3.81}$$

The first three error coefficients associated with Eqn. 3.81, obtained by means of synthetic division, are

$$e_0 = 0$$

$$e_1 = 2 \times 10^{-6} \text{ sec} \tag{3.82}$$

$$e_2 = -3 \times 10^{-12} \text{ sec}^2$$

Sample-and-hold circuits are frequently used to process pulses such as radar echos after these signals have passed through several amplifier stages. In many cases the pulse following amplification can be well approximated by a Gaussian signal, and for this reason a signal

$$v_i(t) = e^{-(10^{10}t^2/2)} \tag{3.83}$$

is used as a test input.

The first two derivatives of $v_i(t)$ are

$$\frac{dv_i(t)}{dt} = -10^{10}te^{-(10^{10}t^2/2)} \tag{3.84}$$

and

$$\frac{d^2v_i(t)}{dt^2} = -10^{10}e^{-(10^{10}t^2/2)} + 10^{20}t^2e^{-(10^{10}t^2/2)} \tag{3.85}$$

The maximum magnitude of dv_i/dt is 6.07×10^4 volts per second occurring at $t = \pm 10^{-5}$ seconds, and the maximum magnitude of d^2v_i/dt^2 is 10^{10} volts per second squared at $t = 0$. If the first error coefficient is used to estimate error, we find that a tracking error of approximately 0.12 volt (12% of the peak-signal amplitude) is predicted if the switch is opened at $t = \pm 10^{-5}$ seconds. The error series converges rapidly in this case, with its second term contributing a maximum error of 0.03 volt at $t = 0$.

PROBLEMS

P3.1
An operational amplifier is connected to provide a noninverting gain of 10. The small-signal step response of the connection is approximately first order with a 0 to 63% risetime of 1 μs. Estimate the quantity $a(s)$ for the

amplifier, assuming that loading at the amplifier input and output is insignificant.

P3.2

The transfer function of a linear system is

$$A(s) = \frac{1}{(s^2 + 0.5s + 1)(0.1s + 1)}$$

Determine the step response of this system. Estimate (do not calculate exactly) the percentage overshoot of this system in response to step excitation.

P3.3

Use the properties of Laplace transforms to evaluate the transform of the triangular pulse signal shown in Fig. 3.22.

P3.4

Use the properties of Laplace transforms to evaluate the transform of the pulse signal shown in Fig. 3.23.

P3.5

The response of a certain linear system is approximately second order, with a d-c gain of one. Measured performance shows that the peak value of the response to a unit step is 1.38 and that the time for the step response to first pass through one is 0.5 μs. Determine second-order parameters that can be used to model the system. Also estimate the peak value of the output that results when a unit impulse is applied to the input of the system and the time required for the system impulse response to first return to zero. Estimate the quantities M_p and f_h for this system.

P3.6

A high-fidelity audio amplifier has a transfer function

$$A(s) = \frac{100s}{(0.05s + 1)(s^2/4 \times 10^{10} + s/2 \times 10^5 + 1)(0.5 \times 10^{-6}s + 1)}$$

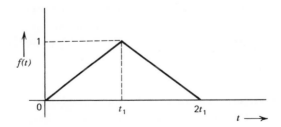

Figure 3.22   Triangular pulse.

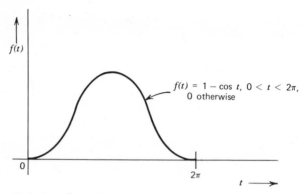

Figure 3.23 Raised cosine pulse.

Plot this transfer function in both Bode and gain-phase form. Recognize that the high- and low-frequency singularities of this amplifier are widely spaced and use this fact to estimate the following quantities when the amplifier is excited with a 10-mV step.

(a) The peak value of the output signal.
(b) The time at which the peak value occurs.
(c) The time required for the output to go from 2 to 18 V.
(d) The time until the output droops to 7.4 V.

P3.7

An oscilloscope vertical amplifier can be modeled as having a transfer function equal to $A_0/(10^{-9}s + 1)^5$. Estimate the 10 to 90% rise time of the output voltage when the amplifier is excited with a step-input signal.

P3.8

An asymptotic plot of the measured open-loop frequency response of an operational amplifier is shown in Fig. 3.24a. The amplifier is connected as shown in Fig. 3.24b. (You may neglect loading.) Show that lower values of α result in more heavily damped responses. Determine the value of α that results in the closed-loop step response of the amplifier having an overshoot of 20% of final value. What is the 10 to 90% rise time in response to a step for this value of α?

P3.9

A feedback system has a forward gain $a(s) = K/s(\tau s + 1)$ and a feedback gain $f = 1$. Determine conditions on K and τ so that e_0 and e_2 are

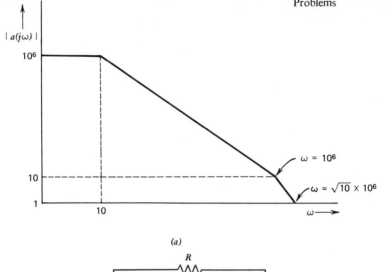

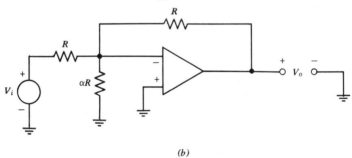

Figure 3.24 Inverting amplifier. (*a*) Amplifier open-loop response. (*b*) Connection.

both zero. What is the steady-state error in response to a unit ramp for this system?

P3.10

An operational amplifier connected as a unity-gain noninverting amplifier is excited with an input signal

$$v_i(t) = 5 \tan^{-1} 10^5 t$$

Estimate the error between the actual and ideal outputs assuming that the open-loop transfer function can be approximated as indicated below. (Note that these transfer functions all have identical values for unity-gain frequency.)

(a) $a(s) = 10^7/s$
(b) $a(s) = 10^{13}(10^{-6}s + 1)/s^2$
(c) $a(s) = 10^{19}(10^{-6}s + 1)^2/s^3$

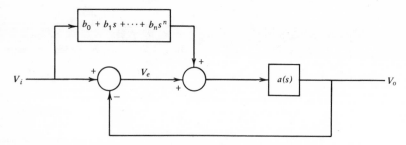

Figure 3.25   System with feedforward path.

P3.11

The system shown in Fig. 3.25 uses a feedforward path to reduce errors. How should the b's be chosen to reduce error coefficients e_0 through e_n to zero? Can you think of any practical disadvantages to this scheme?

CHAPTER IV
STABILITY

4.1 THE STABILITY PROBLEM

The discussion of feedback systems presented up to this point has tacitly assumed that the systems under study were *stable*. A stable system is defined in general as one which produces a bounded output in response to any bounded input. Thus stability implies that

$$\int_{-\infty}^{\infty} |v_O(t)| \, dt \leq M < \infty \tag{4.1}$$

for *any* input such that

$$\int_{-\infty}^{\infty} |v_I(t)| \, dt \leq N < \infty \tag{4.2}$$

If we limit our consideration to linear systems, stability is independent of the input signal, and the sufficient and necessary condition for stability is that all poles of system transfer function lie in the left half of the s plane. This condition follows directly from Eqn. 4.1, since any right-half-plane poles contribute terms to the output that grow exponentially with time and thus are unbounded. Note that this definition implies that a system with poles on the imaginary axis is unstable, since its output is not bounded unless its input is rather carefully chosen.

The origin of the stability problem can be described in intuitively appealing through nonrigorous terms as follows. If a feedback system detects an error between the actual and desired outputs, it attempts to reduce this error to zero. However, changes in the error signal that result from corrective action do not occur instantaneously because of time delays around the loop. In a high-gain system, these delays can cause a tendency to overcorrect. If the magnitude of the overcorrection exceeds the magnitude of the initial error, instability results. Signal amplitudes grow exponentially until some nonlinearity limits further growth, at which time the system either saturates or oscillates in a constant-amplitude fashion called a *limit cycle*.[1] The feedback system designer must always temper his desire to

---

[1] The effect of nonlinearities on the steady-state amplitude reached by an unstable system is investigated in Chapter 6.

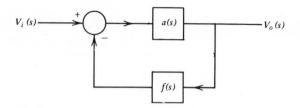

Figure 4.1   Block diagram of single-loop amplifier.

provide a large magnitude and a high unity-gain frequency for the loop transmission with the certain knowledge that sufficiently high values for these quantities invariably lead to instability.

As a specific example of a system with potentially unstable behavior, consider a simple single-loop system of the type shown in Fig. 4.1, with

$$a(s) = \frac{a_0}{(s + 1)^3} \tag{4.3}$$

and

$$f(s) = 1 \tag{4.4}$$

The loop transmission for this system is

$$-a(s)f(s) = \frac{-a_0}{(s + 1)^3} \tag{4.5}$$

or for sinusoidal excitation,

$$-a(j\omega)f(j\omega) = \frac{-a_0}{(j\omega + 1)^3} = \frac{-a_0}{-j\omega^3 - 3\omega^2 + 3j\omega + 1} \tag{4.6}$$

If we evaluate Eqn. 4.6 at $\omega = \sqrt{3}$, we find that

$$-a(j\sqrt{3})f(j\sqrt{3}) = \frac{a_0}{8} \tag{4.7}$$

If the quantity a_0 is chosen equal to 8, the system has a real, *positive* loop transmission with a magnitude of one for sinusoidal excitation at three radians per second.

We might suspect that a system with a loop transmission of $+1$ is capable of oscillation, and this suspician can be confirmed by examining the closed-loop transfer function of the system with $a_0 = 8$. In this case,

$$A(s) = \frac{a(s)}{1 + a(s)f(s)} = \frac{8}{s^3 + 3s^2 + 3s + 9}$$

$$= \frac{8}{(s + 3)(s + j\sqrt{3})(s - j\sqrt{3})} \tag{4.8}$$

This transfer function has a negative, real-axis pole and a pair of poles located on the imaginary axis at $s = \pm j\sqrt{3}$. An argument based on the properties of partial-fraction expansions (see Section 3.2.2) shows that the response of this system to many common (bounded) transient signals includes a constant-amplitude sinusoidal component.

Further increases in low-frequency loop-transmission magnitude move the pole pair into the right-half plane. For example, if we combine the forward-path transfer function

$$a(s) = \frac{64}{(s + 1)^3} \tag{4.9}$$

with unity feedback, the resultant closed-loop transfer function is

$$
\begin{aligned}
A(s) &= \frac{64}{s^3 + 3s^2 + 3s + 65} \\
&= \frac{64}{(s + 5)(s - 1 + j2\sqrt{3})(s - 1 - j2\sqrt{3})} \tag{4.10}
\end{aligned}
$$

With this value for a_0, the system transient response will include a sinusoidal component with an exponentially growing envelope.

If the dynamics associated with the loop transmission remain fixed, the system will be stable only for values of a_0 less than 8. This stability is achieved at the expense of desensitivity. If a value of $a_0 = 1$ is used so that

$$a(s)f(s) = \frac{1}{(s + 1)^3} \tag{4.11}$$

we find all closed-loop poles are in the left-half plane, since

$$
\begin{aligned}
A(s) &= \frac{1}{s^3 + 3s^2 + 3s + 2} \\
&= \frac{1}{(s + 2)(s + 0.5 + j\sqrt{3}/2)(s + 0.5 - j\sqrt{3}/2)} \tag{4.12}
\end{aligned}
$$

in this case.

In certain limited cases, a binary answer to the stability question is sufficient. Normally, however, we shall be interested in more quantitative information concerning the "degree" of stability of a feedback system. Frequently used measures of relative stability include the peak magnitude of the frequency response, the fractional overshoot in response to a step input, the damping ratio associated with the dominant pole pair, or the variation of a certain parameter that can be tolerated without causing absolute instability. Any of the measures of relative stability mentioned above can be found by direct calculations involving the system transfer function. While such determinations are practical with the aid of machine computation, insight into system operation is frequently obscured if this process is used. The techniques described in this chapter are intended not only to provide answers to questions concerning stability, but also (and more important) to indicate how to improve the performance of unsatisfactory systems.

4.2 THE ROUTH CRITERION

The Routh test is a mathematical method that can be used to determine the number of zeros of a polynomial with positive real parts. If the test is applied to the denominator polynomial of a transfer function (also called the *characteristic equation*) the absence of any right-half-plane zeros of the characteristic equation guarantees system stability. One computational advantage of the Routh test is that it is not necessary to factor the polynomial to apply the test.

4.2.1 Evaluation of Stability

The test is described for a polynomial of the form

$$P(s) = a_0 s^n + a_1 s^{n-1} + \cdots + a_{n-1}s + a_n \qquad (4.13)$$

A necessary but not sufficient condition for all the zeros of Eqn. 4.13 to have negative real parts is that all the a's be present and that they all have the same sign. If this necessary condition is satisfied, an array of numbers is generated from the a's as follows. (This example is for n even. For n odd, a_n terminates the second row.)

a_0	a_2	a_4	.	.	a_{n-2}	a_n
a_1	a_3	a_5	.	.	a_{n-1}	0
$\dfrac{a_1 a_2 - a_0 a_3}{a_1} = b_1$	$\dfrac{a_1 a_4 - a_0 a_5}{a_1} = b_2$	.	.	.	$\dfrac{a_1 a_n - a_0 \cdot 0}{a_1} = b_{n/2}$	0
$\dfrac{b_1 a_3 - a_1 b_2}{b_1} = c_1$	$\dfrac{b_1 a_5 - a_1 b_3}{b_1} = c_2$	.	.	.	0	0
$\dfrac{c_1 b_2 - b_1 c_2}{c_1} = d_1$	.	.	.	.	0	0
.	.	.	.	.	.	.
.	.	.	.	.	.	.
0	0	.	.	.	0	0

$$(4.14)$$

As the array develops, progressively more elements of each row become zero, until only the first element of the $n + 1$ row is nonzero. The total number of sign changes in the first column is then equal to the number of zeros of the original polynomial that lie in the right-half plane.

The use of the Routh criterion is illustrated using the polynomial

$$P(s) = s^4 + 9s^3 + 14s^2 + 266s + 260 \qquad (4.15)$$

Since all coefficients are real and positive, the necessary condition for all roots of Eqn. 4.15 to have negative real parts is satisfied. The array is

$$
\begin{array}{ccc}
1 & 14 & 260 \\
9 & 266 & 0 \\
\dfrac{9 \times 14 - 1 \times 266}{9} = -\dfrac{140}{9} & \dfrac{9 \times 260 - 1 \times 0}{9} = 260 & 0 \\
\text{(sign change)} & & \\
\dfrac{-(140/9) \times 266 - 9 \times 260}{-(140/9)} = +\dfrac{2915}{7} & 0 & 0 \\
\text{(sign change)} & & \\
\dfrac{(2915/7) \times 260 - [-(140/9) \times 0]}{2915/7} = 260 & 0 & 0 \\
\end{array}
$$

$$(4.16)$$

The two sign changes in the first column indicate two right-half-plane zeros. This result can be verified by factoring the original polynomial, showing that

$$s^4 + 9s^3 + 14s^2 + 266s + 260 = (s - 1 + j5)(s - 1 - j5)(s + 1)(s + 10)$$

$$(4.17)$$

A second example is provided by the polynomial

$$P(s) = s^4 + 13s^3 + 58s^2 + 306s + 260 \qquad (4.18)$$

The corresponding array is

$$
\begin{array}{ccc}
1 & 58 & 260 \\
13 & 306 & 0 \\
\dfrac{13 \times 58 - 1 \times 306}{13} = \dfrac{448}{13} & \dfrac{13 \times 260 - 1 \times 0}{13} = 260 & 0 \\
\dfrac{(448/13) \times 306 - 13 \times 260}{448/13} = \dfrac{23287}{112} & 0 & 0 \\
\dfrac{(23287/112) \times 260 - (448/13) \times 0}{23287/112} = 260 & 0 & 0 \\
\end{array}
$$

$$(4.19)$$

Factoring verifies the result that there are no right-half-plane zeros for this polynomial, since

$$s^4 + 13s^3 + 58s^2 + 306s + 260$$

$$= (s + 1 + j5)(s + 1 - j5)(s + 1)(s + 10) \quad (4.20)$$

Two kinds of difficulties can occur when applying the Routh test. It is possible that the first element in one row of the array is zero. In this case, the original polynomial is multiplied by $s + \alpha$, where α is any positive real number, and the test is repeated. This procedure is illustrated using the polynomial

$$P(s) = s^5 + s^4 + 10s^3 + 10s^2 + 20s + 5 \quad (4.21)$$

The first element of the third row of the array is zero.

1	10	20
1	10	5
0	15	0 (4.22)

The difficulty is resolved by multiplying Eqn. 4.21 by $s + 1$, yielding

$$P'(s) = s^6 + 2s^5 + 11s^4 + 20s^3 + 30s^2 + 25s + 5 \quad (4.23)$$

The array for Eqn. 4.23 is

1	11	30	5
2	20	25	0
1	17.5	5	0
-15	15	0	0
-18.5	5	0	0
10.95	0	0	0
5	0	0	0 (4.24)

Since multiplication by $s + 1$ did not add any right-half-plane zeros to Eqn. 4.21, we conclude that the two right-half-plane zeros indicated by the array of Eqn. 4.24 must be contained in the original polynomial.

The second possibility is that an entire row becomes zero. This condition indicates that there is a pair of roots on the imaginary axis, a pair of real roots located symmetrically with respect to the origin, or both kinds of pairs in the original polynomial. The terms in the row above the all-zero row are used as coefficients of an equation in even powers of s called the *auxiliary equation*. The zeros of this equation are the pairs mentioned above. The auxiliary equation can be differentiated with respect to s, and the resultant coefficients are used in place of the all-zero row to continue the array. This type of difficulty is illustrated with the polynomial

$$P(s) = s^4 + 11s^3 + 11s^2 + 11s + 10 = (s + j)(s - j)(s + 1)(s + 10) \tag{4.25}$$

The array is

1	11	10
11	11	0
10	10	0
0	0	0

(4.26)

The auxiliary equation is

$$Q(s) = 10s^2 + 10 \tag{4.27}$$

The roots of the equation are the two imaginary zeros of Eqn. 4.25. Differentiating Eqn. 4.27 and using the nonzero coefficient to replace the first element of row 4 of Eqn. 4.26 yields a new array.

1	11	10
11	11	0
10	10	0
20	0	0
10	0	0

(4.28)

The absence of sign changes in the array verifies that the original polynomial has no zeros in the right-half plane.

Note that, while there are no closed-loop poles in the right-half plane, a system with a characteristic equation given by Eqn. 4.25 is unstable by our definition since it has a pair of poles on the imaginary axis. Examining only the left-hand column of the Routh array only identifies the number of right-half-plane zeros of the tested polynomial. Imaginary-axis zeros can be found by the manipulations involving the auxiliary equation.

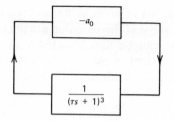

Figure 4.2   Block diagram of phase-shift oscillator.

4.2.2   Use as a Design Aid

The Routh criterion is most frequently used to determine the stability of a feedback system. In certain cases, however, more quantitative design information is obtainable, as illustrated by the following examples.

A phase-shift oscillator can be constructed by applying sufficient negative feedback around a network that has three or more poles. If an amplifier with frequency-independent gain is combined with a network with three coincident poles, the block diagram for the resultant system is as shown in Fig. 4.2. The value of a_0 necessary to sustain oscillations can be determined by Routh analysis.[2]

Stability investigations for Fig. 4.2 are complicated by the fact that the oscillator has no input; thus we cannot use the poles of an input-to-output transfer function to determine stability. We should note that the stability of a linear system is a property of the system itself and is thus independent of input signals that may be applied to it. Any unstable physical system will demonstrate its instability with no input, since runaway behavior will be stimulated by always present noise. Even in a purely mathematical linear system, stability is determined by the location of the closed-loop poles, and these locations are clearly input independent.

The analysis of the oscillator is initiated by recalling that the characteristic equation of any feedback system is one minus its loop transmission. Therefore

$$P(s) = 1 + \frac{a_0}{(\tau s + 1)^3} \tag{4.29}$$

In this and other calculations involving the characteristic equation, it is possible to clear fractions since the location of the zeros are not altered

---

[2] The Routh test applied to this example offers computational advantages compared to the direct factoring used for a similar transfer function in the example of Section 4.1.

by this operation. After clearing fractions and identifying coefficients, the Routh array is

$$
\begin{array}{ll}
\tau^3 & 3\tau \\[6pt]
3\tau^2 & 1 + a_0 \\[6pt]
\dfrac{(8 - a_0)\tau}{3} & 0 \\[6pt]
1 + a_0 & 0
\end{array}
\qquad (4.30)
$$

Assuming τ is positive, roots with positive real parts occur for $a_0 < -1$ (one right-half-plane zero) and for $a_0 > +8$ (two right-half-plane zeros). Laplace analysis indicates that generation of a constant-amplitude sinusoidal oscillation requires a pole pair on the imaginary axis. In practice, a complex pole pair is located slightly to the right of the imaginary axis. An intentionally introduced nonlinearity can then be used to limit the amplitude of the oscillation (see Section 6.3.3). Thus, a practical oscillator circuit is obtained with $a_0 > 8$.

The frequency of oscillation with $a_0 = 8$ can be determined by examining the array with this value for a_0. Under these conditions the third row becomes all zero. The auxiliary equation is

$$
Q(s) = 3\tau^2 s^2 + 9 \qquad (4.31)
$$

and the equation has zeros at $s = \pm j\sqrt{3}/\tau$, indicating oscillation at $\sqrt{3}/\tau$ radians per second for $a_0 = 8$.

As a second example of the type of design information that can be obtained via Routh analysis, consider an operational amplifier with an open-loop transfer function

$$
a(s) = \frac{a_0}{(s + 1)\,(10^{-6}s + 1)\,(10^{-7}s + 1)} \qquad (4.32)
$$

It is assumed that this amplifier is connected as a unity-gain noninverting amplifier, and we wish to determine the range of values of a_0 for which all closed-loop poles have real parts more negative than -2×10^5 sec^{-1}. This condition on closed-loop pole location implies that any pulse response of the system will decay at least as fast as $Ke^{-2\times10^5 t}$ after the exciting pulse returns to zero. The constant K is dependent on conditions at the time the input becomes zero.

The characteristic equation for the amplifier is (after dropping insignificant terms)

$$
P(s) = 10^{-13}s^3 + 1.1 \times 10^{-6}s^2 + s + 1 + a_0 \qquad (4.33)
$$

In order to determine the range of a_0 for which all zeros of this characteristic equation have real parts more negative than $-2 \times 10^5 \text{ sec}^{-1}$, it is only necessary to make a change of variable in Eqn. 4.33 and apply Routh's criterion to the modified equation. In particular, application of the Routh test to a polynomial obtained by substituting

$$\lambda = s + c \qquad (4.34)$$

will determine the number of zeros of the original polynomial with real parts more positive than $-c$, since this substitution shifts singularities in the s plane to the right by an amount c as they are mapped into the λ plane. If the indicated substitution is made with $c = 2 \times 10^5 \text{ sec}^{-1}$, Eqn. 4.33 becomes

$$P(\lambda) = 10^{-13}\lambda^3 + 10^{-6}\lambda^2 + 0.57\lambda - 1.57 \times 10^5 + a_0 \qquad (4.35)$$

The Routh array is

10^{-13}	0.57
10^{-6}	$-1.57 \times 10^5 + a_0$
$0.59 - 10^{-7} a_0$	0
$-1.57 \times 10^5 + a_0$	0

$$(4.36)$$

This array shows that Eqn. 4.33 has one zero with a real part more positive than $-2 \times 10^5 \text{ sec}^{-1}$ for $a_0 < 1.57 \times 10^5$, and has two zeros to the right of the dividing line for $a_0 > 5.9 \times 10^6$. Accordingly, all zeros have real parts more negative than $-2 \times 10^5 \text{ sec}^{-1}$ only for

$$1.57 \times 10^5 < a_0 < 5.9 \times 10^6 \qquad (4.37)$$

4.3   ROOT-LOCUS TECHNIQUES

A single-loop feedback amplifier is shown in the block diagram of Fig. 4.1. The closed-loop transfer function for this amplifier is

$$\frac{V_0(s)}{V_i(s)} = A(s) = \frac{a(s)}{1 + a(s)f(s)} \qquad (4.38)$$

Root-locus techniques provide a method for finding the poles of the closed-loop transfer function $A(s)$ [or equivalently the zeros of $1 + a(s)f(s)$] given the poles and zeros of $a(s)f(s)$ and the d-c loop-transmission magnitude $a_0 f_0$.[3] Notice that since the quantity $a_0 f_0$ must appear in one or more terms

---

[3] If the loop transmission has one or more zeros at the origin so that its d-c magnitude is zero, the closed-loop poles are found from the midband value of af.

of the characteristic equation, the locations of the poles of $A(s)$ must depend on a_0f_0. A *root-locus diagram* consists of a collection of branches or loci in the s plane that indicate how the locations of the poles of $A(s)$ change as a_0f_0 varies.

The root-locus diagram provides useful information concerning the performance of a feedback system since the relative stability of any linear system is uniquely determined by its close-loop pole locations. We shall find that approximate root-locus diagrams are easily and rapidly sketched, and that they provide readily interpreted insight into how the closed-loop performance of a system responds to changes in its loop transmission. We shall also see that root-locus techniques can be combined with simple algebraic methods to yield exact answers in certain cases.

4.3.1   Forming the Diagram

A simple example that illustrates several important features of root-locus techniques is provided by the system shown in Fig. 4.1 with a feedback transfer function f of unity and a forward transfer function

$$a(s) = \frac{a_0}{(\tau_a s + 1)(\tau_b s + 1)} \tag{4.39}$$

The corresponding closed-loop transfer function is

$$A(s) = \frac{a(s)}{1 + a(s)f(s)} = \frac{a_0}{\tau_a \tau_b s^2 + (\tau_a + \tau_b)s + (1 + a_0)} \tag{4.40}$$

The closed-loop poles can be determined by factoring the characteristic equation of $A(s)$, yielding

$$s_1 = \frac{-(\tau_a + \tau_b) + \sqrt{(\tau_a + \tau_b)^2 - 4(1 + a_0)\tau_a \tau_b}}{2\tau_a \tau_b} \tag{4.41a}$$

$$s_2 = \frac{-(\tau_a + \tau_b) - \sqrt{(\tau_a + \tau_b)^2 - 4(1 + a_0)\tau_a \tau_b}}{2\tau_a \tau_b} \tag{4.41b}$$

The root-locus diagram in Fig. 4.3 is drawn with the aid of Eqn. 4.41. The important features of this diagram include the following.

(a) The loop-transmission pole locations are shown. (Loop-transmission zeros are also indicated if they are present.)

(b) The poles of $A(s)$ coincide with loop-transmission poles for $a_0 = 0$.

(c) As a_0 increases, the locations of the poles of $A(s)$ change along the loci as shown. Arrows indicate the direction of changes that result for increasing a_0.

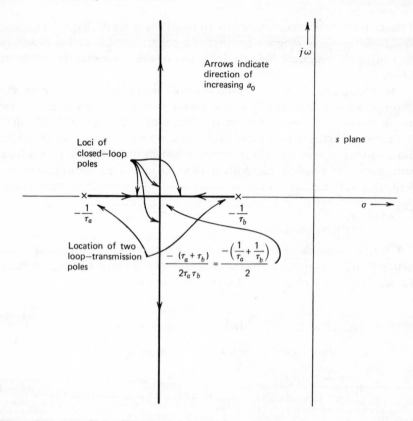

Figure 4.3   Root-locus diagram for second-order system.

(d) The two poles coincide at the arithmetic mean of the loop-transmission pole locations for zero radicand in Eqn. 4.41, or for

$$a_0 = \frac{(\tau_a + \tau_b)^2}{4\tau_a\tau_b} - 1 \qquad (4.42)$$

(e) For increases in a_0 beyond the value of Eqn. 4.42, the closed-loop pole pair is complex with constant real part and a damping ratio that is a monotonic decreasing function of a_0. Consequently, ω_n increases with increasing a_0 in this range.

Certain important features of system behavior are evident from the diagram. For example, the system does not become unstable for any positive value of a_0. However, the relative stability decreases as a_0 increases beyond the value indicated in Eqn. 4.42.

It is always possible to draw a root-locus diagram by directly factoring the characteristic equation of the system under study as in the preceding example. Unfortunately, the effort involved in factoring higher-order polynomials makes machine computation mandatory for all but the simplest systems. We shall see that it is possible to approximate the root-locus diagrams and thus retain the insight often lost with machine computation when absolute accuracy is not required.

The key to developing the rules used to approximate the loci is to realize that closed-loop poles occur only at zeros of the characteristic equation or at frequencies s_1 such that[4]

$$1 + a(s_1)f(s_1) = 0 \tag{4.43a}$$

or

$$a(s_1)f(s_1) = -1 \tag{4.43b}$$

Thus, if the point s_1 is a point on a branch of the root-locus diagram, the two conditions

$$\left| a(s_1)f(s_1) \right| = 1 \tag{4.44a}$$

and

$$\measuredangle\, a(s_1)f(s_1) = (2n + 1)\, 180° \tag{4.44b}$$

where n is any integer, must be satisfied. The angle condition is the more important of these two constraints for purposes of forming a root-locus diagram. The reason is that since we plot the loci as $a_0 f_0$ is varied, it is possible to find a value for a $a_0 f_0$ that satisfies the magnitude condition at any point in the s plane where the angle condition is satisfied.

By concentrating primarily on the angle condition, we are able to formulate a set of rules that greatly simplify root-locus-diagram construction compared with brute-force factoring of the characteristic equation. Here are some of the rules we shall use.

1. The number of branches of the diagram is equal to the number of poles of $a(s)f(s)$. Each branch starts at a pole of $a(s)f(s)$ for small values of $a_0 f_0$ and approaches a zero of $a(s)f(s)$ either in the finite s plane or at infinity for large values of $a_0 f_0$. The starting and ending points are demonstrated by considering

$$a(s)f(s) = a_0 f_0 g(s) \tag{4.45}$$

where $g(s)$ contains the frequency-dependent portion of the loop trans-

---

[4] It is assumed throughout that the system under study is a negative feedback system with the topology shown in Fig. 4.1.

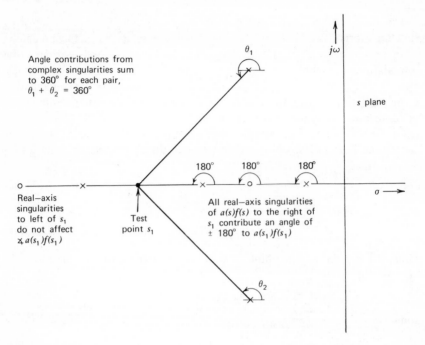

Figure 4.4   Loci on real axis.

mission and the value of $g(0) \triangleq g_0$ is unity. Rearranging Eqn. 4.44 and using this notation yields

$$|g(s_1)| = \frac{1}{a_0 f_0} \qquad (4.46)$$

at any point s_1 on a branch of the root-locus diagram. Thus for small values of $a_0 f_0$, $|g(s_1)|$ must be large, implying that the point s_1 is close to a pole of $g(s)$. Conversely, a large value of $a_0 f_0$ requires proximity to a zero of $g(s)$.

2. Branches of the diagram lie on the real axis to the left of an odd number of real-axis poles and zeros of $a(s)f(s)$.[5] This rule follows directly from Eqn. 4.44b as illustrated in Fig. 4.4. Each real-axis zero of $a(s)f(s)$ to the right of s_1 adds 180° to the angle of $a(s_1)f(s_1)$ while each real-axis pole to the right of s_1 subtracts 180° from the angle. Real-axis singularities to the left of point s_1 do not influence the angle of $a(s_1)f(s_1)$. Similarly, since complex singularities must always occur in conjugate pairs, the net angle con-

[5] Special care is necessary for systems with right-half-plane open-loop singularities. See Section 4.3.3.

tribution from these singularities is zero. This rule is thus sufficient to satisfy Eqn. 4.44b. We are further guaranteed that branches *must* exist on all segments of the real axis to the left of an odd number of singularities of $a(s)f(s)$, since there is some value of $a_0 f_0$ that will exactly satisfy Eqn. 4.44a at every point on these segments, and the satisfaction of Eqns. 4.44a and 4.44b is both necessary and sufficient for the existence of a pole of $A(s)$.

3. The two separate branches of the diagram that must exist between pairs of poles or pairs of zeros on segments of the real axis that satisfy rule 2 must at some point depart from or enter the real axis at right angles to it. Frequently the precise break-away point is not required in order to sketch the loci to acceptable accuracy. If it is necessary to have an exact location, it can be shown that the break-away points are the solutions of the equation

$$\frac{d[g(s)]}{ds} = 0 \tag{4.47}$$

for systems without coincident singularities.

4. If the number of poles of $a(s)f(s)$ exceeds the number of zeros of this function by two or more, the average distance of the poles of $A(s)$ from the imaginary axis is independent of $a_0 f_0$. This rule evolves from a property of algebraic polynomials. Consider a polynomial

$$P(s) = (a_1 s + a_1 s_1)(a_2 s + a_2 s_2)(a_3 s + a_3 s_3) \cdots (a_n s + a_n s_n)$$

$$= (a_1 a_2 \cdots a_n)(s + s_1)(s + s_2)(s + s_3) \cdots (s + s_n)$$

$$= (a_1 a_2 \cdots a_n)[s^n + (s_1 + s_2 + s_3 + \cdots + s_n)s^{n-1}$$

$$+ \cdots + s_1 s_2 s_3 \cdots s_n] \tag{4.48}$$

From the final expression of Eqn. 4.48, we see that the ratio of the coefficients of the s^{n-1} term and the s^n term (denoted as $-n\bar{s}$) is

$$-n\bar{s} = s_1 + s_2 + s_3 + \cdots + s_n \tag{4.49}$$

Since imaginary components of terms on the right-hand side of Eqn. 4.49 must occur in conjugate pairs and thus cancel, the quantity

$$\bar{s} = -\frac{(s_1 + s_2 + s_3 + \cdots + s_n)}{n} \tag{4.50}$$

is the average distance of the roots of $P(s)$ from the imaginary axis. In order to apply Eqn. 4.50 to the characteristic equation of a feedback system, assume that

$$a(s)f(s) = a_0 f_0 \frac{p(s)}{q(s)} \tag{4.51}$$

Then

$$A(s) = \frac{a(s)}{1 + a(s)f(s)} = \frac{a(s)}{1 + a_0 f_0 [p(s)/q(s)]} = \frac{a(s)q(s)}{q(s) + a_0 f_0 p(s)} \qquad (4.52)$$

If the order of $q(s)$ exceeds that of $p(s)$ by two or more, the ratio of the co-efficients of the two highest-order terms of the characteristic equation of $A(s)$ is independent of $a_0 f_0$, and thus the average distance of the poles of $A(s)$ from the imaginary axis is a constant.

5. For large values of $a_0 f_0$, $P - Z$ branches approach infinity, where P and Z are the number of poles and finite-plane zeros of $a(s)f(s)$, respectively. These branches approach asymptotes that make angles with the real axis given by

$$\theta_n = \frac{(2n + 1)\,180°}{P - Z} \qquad (4.53)$$

In Eqn. 4.53, n assumes all integer values from 0 to $P - Z - 1$. The asymptotes all intersect the real axis at a point

$$\frac{\Sigma \text{ real parts of poles of } a(s)f(s) - \Sigma \text{ real parts of zeros of } a(s)f(s)}{P - Z}$$

The proof of this rule is left to Problem P4.4.

6. Near a complex pole of $a(s)f(s)$, the angle of a branch with respect to the pole is

$$\theta_p = 180° + \Sigma \measuredangle z - \Sigma \measuredangle p \qquad (4.54)$$

where $\Sigma \measuredangle z$ is the sum of the angles of vectors drawn from all the zeros of $a(s)f(s)$ to the complex pole in question and $\Sigma \measuredangle p$ is the sum of the angles of vectors drawn from all other poles of $a(s)f(s)$ to the complex pole. Similarly, the angle a branch makes with a loop-transmission zero in the vicinity of the zero is

$$\theta_z = 180° - \Sigma \measuredangle z + \Sigma \measuredangle p \qquad (4.55)$$

These conditions follow directly from Eqn. 4.44b.

7. If the singularities of $a(s)f(s)$ include a group much nearer the origin than all other singularities of $a(s)f(s)$, the higher-frequency singularities can be ignored when determining loci in the vicinity of the origin. Figure 4.5 illustrates this situation. It is assumed that the point s_1 is on a branch if the high-frequency singularities are ignored, and thus the angle of the low-frequency portion of $a(s)f(s)$ evaluated at $s = s_1$ must be $(2n + 1)\,180°$. The geometry shows that the angular contribution attributable to remote singularities such as that indicated as θ_1 is small. (The two angles from a remote complex-conjugate pair also sum to a small angle.) Small changes in the

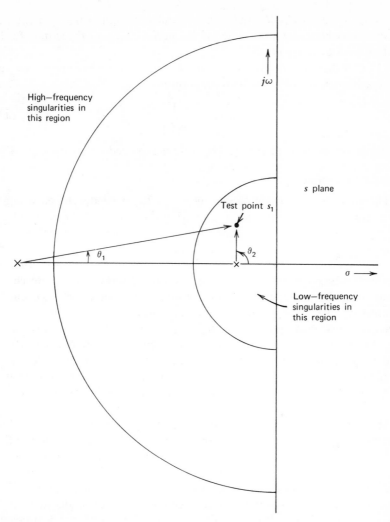

High—frequency
singularities in
this region

s plane

Test point s_1

θ_1

θ_2

σ —→

Low—frequency
singularities in
this region

Figure 4.5 Loci in vicinity of low-frequency singularities.

location of s_1 that can cause relatively large changes in the angle (e.g., θ_2) from low-frequency singularities offset the contribution from remote singularities, implying that ignoring the remote singularities results in insignificant changes in the root-locus diagram in the vicinity of the low-frequency singularities. Furthermore, all closed-loop pole locations will lie relatively close to their starting points for low and moderate values of $a_0 f_0$. Since the discussion of Section 3.3.2 shows that $A(s)$ will be dominated by its lowest-frequency poles, the higher-frequency singularities of $a(s)f(s)$ can be ignored when we are interested in the performance of the system for low and moderate values of a_0f_0.

8. The value of a_0f_0 required to make a closed-loop pole lie at the point s_1 on a branch of the root-locus diagram is

$$a_0f_0 = \frac{1}{|g(s_1)|} \qquad (4.56)$$

where $g(s)$ is defined in rule 1. This rule is required to satisfy Eqn. 4.44a.

4.3.2 Examples

The root-locus diagram shown in Fig. 4.3 can be developed using the rules given above rather than by factoring the denominator of the closed-loop transfer function. The general behavior of the two branches on the real axis is determined using rules 2 and 3. While the break-away point can be found from Eqn. 4.47, it is easier to use either rule 4 or rule 5 to establish off-axis behavior. Since the average distance of the closed-loop poles from the imaginary axis must remain constant for this system [the number of poles of $a(s)f(s)$ is two greater than the number of its zeros], the branches must move parallel to the imaginary axis after they leave the real axis. Furthermore, the average distance must be identical to that for $a_0f_0 = 0$, and thus the segment parallel to the imaginary axis must be located at $-\frac{1}{2}[(1/\tau_a) + (1/\tau_b)]$. Rule 5 gives the same result, since it shows that the two branches must approach vertical asymptotes that intersect the real axis at $-\frac{1}{2}[(1/\tau_a) + (1/\tau_b)]$.

More interesting root-locus diagrams result for systems with more loop-transmission singularities. For example, the transfer function of an amplifier with three common-emitter stages normally has three poles at moderate frequencies and three additional poles at considerably higher frequencies. Rule 7 indicates that the three high-frequency poles can be ignored if this type of amplifier is used in a feedback connection with moderate values of d-c loop transmission. If it is assumed that frequency-independent negative feedback is applied around the three-stage amplifier, a representative *af* product could be[6]

$$a(s)f(s) = \frac{a_0f_0}{(s + 1)(0.5s + 1)(0.1s + 1)} \qquad (4.57)$$

---

[6] The corresponding pole locations at -1, -2, and -10 sec^{-1} are unrealistically low for most amplifiers. These values result, however, if the transfer function for an amplifier with poles at -10^6, -2×10^6, and -10^7 sec^{-1} is normalized using the microsecond rather than the second as the basic time unit. Such frequency scaling will often be used since it eliminates some of the unwieldy powers of 10 from our calculations.

The root-locus diagram for this system is shown in Fig. 4.6. Rule 2 determines the diagram on the real axis, while rule 5 establishes the asymptotes. Rule 4 can be used to estimate the branches off the real axis, since the branches corresponding to the two lower-frequency poles must move to the right to balance the branch going left from the high-frequency pole. The break-away point can be determined from Eqn. 4.47, with

$$\frac{d[g(s)]}{ds} = \frac{-[0.15s^2 + 1.3s + 1.6]}{[(s + 1)(0.5s + 1)(0.1s + 1)]^2} \tag{4.58}$$

Zeros of Eqn. 4.58 are at -7.2 sec^{-1} and -1.47 sec^{-1}. The higher-frequency location is meaningless for this problem, and in fact corresponds to a break-away point which results if positive feedback is applied around the amplifier. Note that the break-away point can be accurately estimated using rule 7. If the relatively higher-frequency pole at 10 sec^{-1} is ignored, a break-away point at -1.5 sec^{-1} results for the remaining two-pole transfer function.

Algebraic manipulations can be used to obtain more quantitative information about the system. Figure 4.6 shows that the system becomes unstable as two poles move into the right-half plane for sufficiently large values of $a_0 f_0$. The value of $a_0 f_0$ that moves the pair of closed-loop poles onto the imaginary axis is found by applying Routh's criterion to the characteristic equation of the system, which is (after clearing fractions)

$$P(s) = (s + 1)(0.5s + 1)(0.1s + 1) + a_0 f_0 \tag{4.59}$$

$$= 0.05s^3 + 0.65s^2 + 1.6s + 1 + a_0 f_0$$

The Routh array is

$$
\begin{array}{cc}
0.05 & 1.6 \\
0.65 & 1 + a_0 f_0 \\
\dfrac{1}{0.65}(0.99 - 0.05a_0 f_0) & 0 \\
1 + a_0 f_0 & 0
\end{array}
\tag{4.60}
$$

Two sign reversals indicating instability occur for $a_0 f_0 > 19.8$. With this value of $a_0 f_0$, the auxiliary equation is

$$Q(s) = 0.65s^2 + 20.8 \tag{4.61}$$

The roots of this equation indicate that the poles cross the imaginary axis at $s = \pm j(5.65)$.

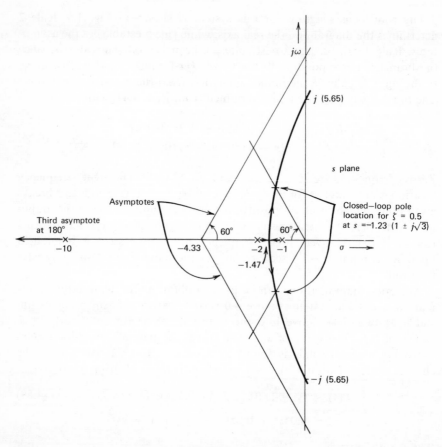

Figure 4.6 Root-locus diagram for third-order system.

It is also possible to determine values for $a_0 f_0$ that result in specified closed-loop pole configurations. This type of calculation is illustrated by finding the value of $a_0 f_0$ required to provide a damping ratio of 0.5, corresponding to complex-pair poles located 60° from the real axis. The magnitude of the ratio of the imaginary part to the real part of the pole location for a pole pair with $\zeta = 0.5$ is $\sqrt{3}$. Thus the characteristic equation for this system, when the damping ratio of the complex pole pair is 0.5, is

$$P'(s) = (s + \gamma)(s + \beta + j\sqrt{3}\beta)(s + \beta - j\sqrt{3}\beta)$$

$$= s^3 + (\gamma + 2\beta)s^2 + 2\beta(\gamma + 2\beta)s + 4\gamma\beta^2 \qquad (4.62)$$

where $-\gamma$ is the location of the real-axis pole.

The parameters are determined by multiplying Eqn. 4.59 by 20 (to make the coefficient of the s^3 term unity) and equating the new equation to $P'(s)$.

$$s^3 + 13s^2 + 32s + 20(1 + a_0 f_0)$$
$$= s^3 + (\gamma + 2\beta)s^2 + 2\beta(\gamma + 2\beta)s + 4\gamma\beta^2 \qquad (4.63)$$

Equation 4.63 is easily solved for γ, β, and $a_0 f_0$, with the results

$$\gamma = 10.54$$
$$\beta = 1.23$$
$$a_0 f_0 = 2.2 \qquad (4.64)$$

Several features of the system are evident from this analysis. Since the complex pair is located at $s = -1.23 \, (1 \pm j\sqrt{3})$ when the real-axis pole is located at $s = -10.54$, a two-pole approximation based on the pair should accurately model the transient or frequency response of the system. The relatively low desensitivity $1 + a_0 f_0 = 3.2$ results if the damping ratio of the complex pair is made 0.5, and any increase in desensitivity will result in poorer damping. The earlier analysis shows that attempts to increase desensitivity beyond 20.8 result in instability.

Note that since there was only one degree of freedom (the value of $a_0 f_0$) existed in our calculations, only one feature of the closed-loop pole pattern could be controlled. It is not possible to force arbitrary values for more than one of the three quantities defining the closed-loop pole locations (ζ and ω_n for the pair and the location of the real pole) unless more degrees of design freedom are allowed.

Another example of root-locus diagram construction is shown in Fig. 4.7, the diagram for

$$a(s)f(s) = \frac{a_0 f_0}{(s + 1)\,(s^2/8 + s/2 + 1)} \qquad (4.65)$$

Rule 5 establishes the asymptotes, while rule 6 is used to determine the loci near the complex poles. The value of $a_0 f_0$ for which the complex pair of poles enters the right-half plane and the frequency at which they cross the imaginary axis are found by Routh's criterion. The reader should verify that these poles cross the imaginary axis at $s = \pm j2\sqrt{3}$ for $a_0 f_0 = 6.5$.

The root-locus diagram for a system with

$$a(s)f(s) = \frac{a_0 f_0(0.5s + 1)}{s(s + 1)} \qquad (4.66)$$

is shown in Fig. 4.8. Rule 2 indicates that branches are on the real axis between the two loop-transmission poles and to the left of the zero. The

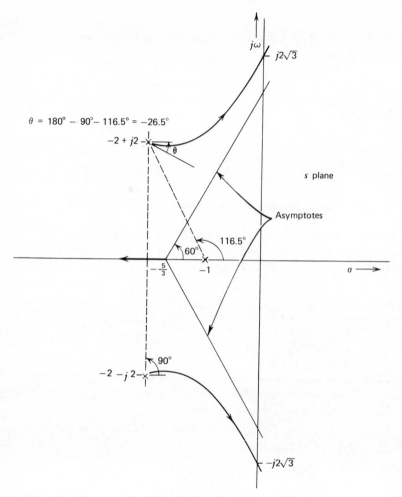

Figure 4.7 Root-locus diagram for $a(s)f(s) = a_0f_0/[(s + 1)(s^2/8 + s/2 + 1)]$.

points of departure from and reentry to the real axis are obtained by solving

$$\frac{d}{ds}\left[\frac{(0.5s + 1)}{s(s + 1)}\right] = 0 \qquad (4.67)$$

yielding $s = -2 \pm \sqrt{2}$.

4.3.3 Systems With Right-Half-Plane Loop-Transmission Singularities

It is necessary to be particularly careful about the sign of the loop transmission when root-locus diagrams are drawn for systems with right-half-

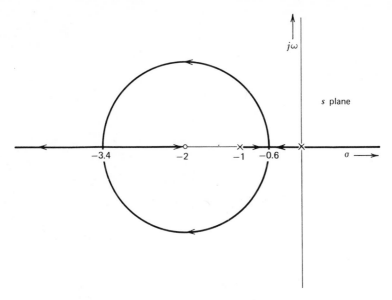

Figure 4.8   Root-locus for diagram $a(s)f(s) = a_0f_0(0.5s + 1)/[s(s + 1)]$.

plane loop-transmission singularities. Some systems that are unstable without feedback have one or more loop-transmission poles in the right-half plane. For example, a large rocket does not become aerodynamically stable until it reaches a certain critical speed, and would tip over shortly after lift off if the thrust were not vectored by means of a feedback system. It can be shown that the transfer function of the rocket alone includes a real-axis right-half-plane pole.

A more familiar example arises from a single-stage common-emitter amplifier. The transfer function of this type of amplifier includes a pole at moderate frequency, a second pole at high frequency, and a high-frequency right-half-plane zero that reflects the signal fed forward from input to output through the collector-to-base capacitance of the transistor. A representative af product for this type of amplifier with frequency-independent feedback applied around it is

$$a(s)f(s) = \frac{a_0f_0(-10^{-3}s + 1)}{(10^{-3}s + 1)(s + 1)} \qquad (4.68)$$

The singularities for this amplifier are shown in Fig. 4.9. If the root-locus rules are applied blindly, we conclude that the low-frequency pole moves to the right, and enters the right-half plane for d-c loop-transmission magnitudes in excess of one. Fortunately, experimental evidence refutes

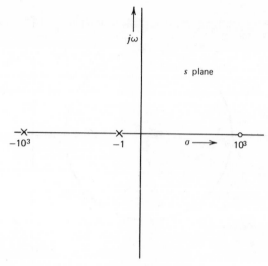

Figure 4.9   Singularities for common-emitter amplifier.

this result. The difficulty stems from the sign of the low-frequency gain. It has been assumed throughout this discussion that loop transmission is negative at low frequency so that the system has negative feedback. The rules were developed assuming the topology shown in Fig. 4.1 where negative feedback results when a_0 and f_0 have the same sign. If we consider positive feedback systems, Eqn. 4.44b must be changed to

$$\sphericalangle \ a(s_1)f(s_1) = n \ 360° \qquad (4.69)$$

where n is any integer, and rules evolved from the angle condition must be appropriately modified. For example, rule 2 is changed to "branches lie on the real axis to the left of an even number of real-axis singularities for positive feedback systems."

The singularity pattern shown in Fig. 4.9 corresponds to a transfer function

$$a'(s)f'(s) = \frac{a_0 f_0 (10^{-3}s - 1)}{(10^{-3}s + 1)(s + 1)} = \frac{-a_0 f_0 (-10^{-3}s + 1)}{(10^{-3}s + 1)(s + 1)} \qquad (4.70)$$

because the vector from the zero to $s = 0$ has an angle of 180°. The sign reversal associated with the zero when plotted in the s plane diagram has changed the sign of the d-c loop transmission compared with that of Eqn. 4.68. One way to reverse the effects of this sign change is to substitute Eqn. 4.69 for Eqn. 4.44b and modify all angle-dependent rules accordingly.

A far simpler technique that works equally well for amplifiers with the right-half plane zeros located at high frequencies is to ignore these zeros when forming the root-locus diagram. Since elimination of these zeros eliminates associated sign reversals, no modification of the rules is necessary. Rule 7 insures that the diagram is not changed for moderate magnitudes of loop transmission by ignoring the high-frequency zeros.

4.3.4 Location of Closed-Loop Zeros

A root-locus diagram indicates the location of the closed-loop poles of a feedback system. In addition to the stability information provided by the pole locations, we may need the locations of the closed-loop zeros to determine some aspects of system performance.

The method used to determine the closed-loop zeros is developed with the aid of Fig. 4.10. Part *a* of this figure shows the block diagram for a single-loop feedback system. The diagram of Fig. 4.10*b* has the same input-output transfer function as that of Fig. 4.10*a*, but has been modified so that

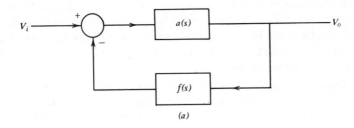

(a)

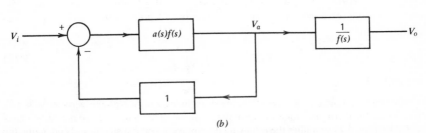

(b)

Figure 4.10 System used to determine closed-loop zeros. (*a*) Single-loop feedback system. (*b*) Modified block diagram.

the feedback path inside the loop has unity gain. We first consider the closed-loop transfer function

$$\frac{V_a(s)}{V_i(s)} = \frac{a(s)f(s)}{1 + a(s)f(s)} \tag{4.71}$$

A root-locus diagram gives the pole locations for this closed-loop transfer function directly, since the diagram indicates the frequencies at which the denominator of Eqn. 4.71 is zero. The zeros of Eqn. 4.71 coincide with the zeros of the transfer function $a(s)f(s)$. However, from Fig. 4.10b,

$$A(s) = \left[\frac{V_o(s)}{V_i(s)}\right] = \left[\frac{V_a(s)}{V_i(s)}\right]\left[\frac{V_o(s)}{V_a(s)}\right] = \left[\frac{V_a(s)}{V_i(s)}\right]\left[\frac{1}{f(s)}\right] \tag{4.72}$$

Thus in addition to the singularities associated with Eqn. 4.71, $A(s)$ has poles at poles of $1/f(s)$, or equivalently at zeros of $f(s)$, and has zeros at poles of $f(s)$. The additional poles of Eqn. 4.72 cancel the zeros of $f(s)$ in Eqn. 4.71, with the net result that $A(s)$ has zeros at zeros of $a(s)$ and at poles of $f(s)$. It is important to recognize that the zeros of $A(s)$ are independent of $a_0 f_0$.

An alternative approach is to recognize that zeros of $A(s)$ occur at zeros of the numerator of this function *and* at frequencies where the denominator becomes infinite while the numerator remains finite. The later condition is satisfied at poles of $f(s)$, since this term is included in the denominator of $A(s)$ but not in its numerator.

Note that the singularities of $A(s)$ are particularly easy to determine if the feedback path is frequency independent. In this case, (as always) closed-loop poles are obtained directly from the root-locus diagram. The zeros of $a(s)$, which are the only zeros plotted in the diagram when $f(s) = f_0$, are also the zeros of $A(s)$.

These concepts are illustrated by means of two examples of frequency-selective feedback amplifiers. Amplifiers of this type can be constructed by combining twin-T networks with operational amplifiers. A twin-T network can have a voltage transfer function that includes complex zeros with positive, negative, or zero real parts. It is assumed that a twin-T with a voltage-transfer ratio[7]

$$T(s) = \frac{s^2 + 1}{s^2 + 2s + 1} \tag{4.73}$$

is available.

[7] The transfer function of a twin-T network includes a third real-axis zero, as well as a third pole. Furthermore, none of the poles coincide. The departure from reality represented by Eqn. 4.73 simplifies the following development without significantly changing the conclusions. The reader who is interested in the transfer function of this type of network is referred to J. E. Gibson and F. B. Tuteur, *Control System Components*, McGraw-Hill, New York, 1958, Section 1.26.

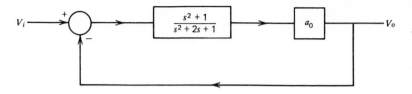

Figure 4.11 Rejection amplifier.

Figures 4.11 and 4.12 show two ways of combining this network with an amplifier that is assumed to have constant gain a_0 at frequencies of interest. Since both of these systems have the same loop transmission, they have identical root-locus diagrams as shown in Fig. 4.13. The closed-loop poles leave the real axis for any finite value of a_0 and approach the j-axis zeros along circular arcs. The closed-loop pole location for one particular value of a_0 is also indicated in this figure.

The rejection amplifier (Fig. 4.11) is considered first. Since the connection has a frequency-independent feedback path, its closed-loop zeros are the two shown in the root-locus diagram. If the signal V_i is a constant-amplitude sinusoid, the effects of the closed-loop poles and zeros very nearly cancel except at frequencies close to one radian per second. The closed-loop frequency response is indicated in Fig. 4.14a. As a_0 is increased, the distance between the closed-loop poles and zeros becomes smaller. Thus the band of frequencies over which the poles and zeros do not cancel becomes narrower, implying a sharper notch, as a_0 is increased.

The bandpass amplifier combines the poles from the root-locus diagram with a second-order closed-loop zero at $s = -1$, corresponding to the pole pair of $f(s)$. The closed-loop transfer function has no other zeros, since $a(s)$ has no zeros in this connection. The frequency response for this amplifier is shown in Fig. 4.14b. In this case the amplifier becomes more selective and provides higher gain at one radian per second as a_0 increases, since the damping ratio of the complex pole pair decreases.

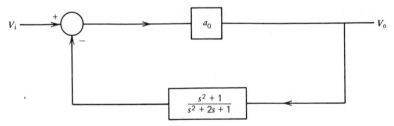

Figure 4.12 Bandpass amplifier.

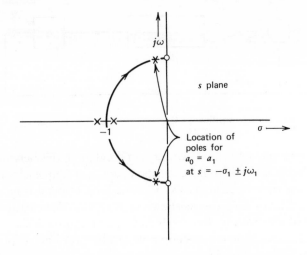

Figure 4.13   Root-locus diagram for systems of Figs. 4.11 and 4.12.

4.3.5   Root Contours

The root-locus method allows us to determine how the locations of the closed-loop poles of a feedback system change as the magnitude of the low-frequency loop transmission is varied. There are many systems where relative stability as a function of some parameter other than gain is required. We shall see, for example, that the location of an open-loop singularity in the transfer function of an operational amplifier is frequently varied to compensate the amplifier and thus improve its performance in a given application. Root-locus techniques could be used to plot a family of root-locus diagrams corresponding to various values for a system parameter other than gain. It is also possible to extend root-locus concepts so that the variation in closed-loop pole location as a function of some single parameter other than gain is determined for a fixed value of $a_0 f_0$. The generalized root-locus diagram that results from this extension is called a *root-contour* diagram.

In order to see how the root contours are constructed, we recall that the characteristic equation for a negative feedback system can be written in the form

$$P(s) = q(s) + a_0 f_0 p(s) \tag{4.74}$$

where it is assumed that

$$a(s)f(s) = a_0 f_0 \frac{p(s)}{q(s)}$$

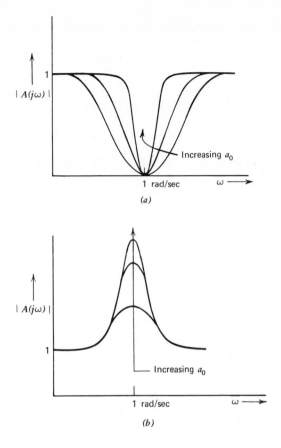

Figure 4.14. Frequency responses for selective amplifiers. (*a*) Rejection amplifier. (*b*) Bandpass amplifier.

If the $a_0 f_0$ product is constant, but some other system parameter τ varies, the characteristic equation can be rewritten

$$P(s) = q'(s) + \tau p'(s) \tag{4.75}$$

All of the terms that multiply τ are included in $p'(s)$ in Eqn. 4.75, so that $q'(s)$ and $p'(s)$ are both independent of τ. The root-contour diagram as a function of τ can then be drawn by applying the construction rules to a singularity pattern that has poles at zeros of $q'(s)$ and zeros at zeros of $p'(s)$.

An operational amplifier connected as a unity-gain follower is used to illustrate the construction of a root-contour diagram. This connection has unity feedback, and it is assumed that the amplifier open-loop transfer function is

$$a(s) = \frac{10^6(\tau s + 1)}{(s + 1)^2} \tag{4.76}$$

The characteristic equation after clearing fractions is

$$P(s) = s^2 + 2s + (10^6 + 1) + \tau 10^6 s \tag{4.77}$$

Identifying terms in accordance with Eqn. 4.75 results in

$$p'(s) = 10^6 s \tag{4.78a}$$

$$q'(s) = s^2 + 2s + 10^6 + 1 \simeq s^2 + 2s + 10^6 \tag{4.78b}$$

Thus the singularity pattern used to form the root contours has a zero at the origin and complex poles at $s = -1 \pm j10^3$. The root-contour diagram is shown in Fig. 4.15. Rule 8 is used to find the value of τ necessary to locate

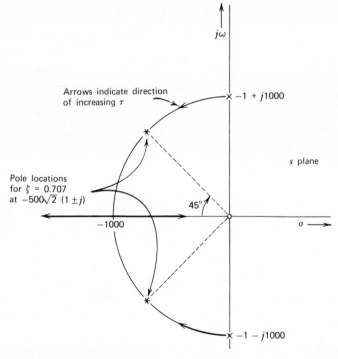

Figure 4.15 Root-contour diagram for $p'(s)/q'(s) = 10^6 s /(s^2 + 2s + 10^6)$.

the complex pole pair 45° from the negative real axis corresponding to a damping ratio of 0.707. From Eqn. 4.56, the required value is

$$\tau = \left| \frac{q'(s)}{p'(s)} \right|_{s \, = \, -500\sqrt{2} \, (1+j)}$$

$$= \left| \frac{s^2 + 2s + 10^6}{10^6 s} \right|_{s \, = \, -500\sqrt{2} \, (1+j)} = \sqrt{2} \times 10^{-3} \qquad (4.79)$$

4.4 STABILITY BASED ON FREQUENCY RESPONSE

The Routh criterion and root-locus methods provide information concerning the stability of a feedback system starting with either the characteristic equation or the loop-transmission singularities of the system. Thus both of these techniques require that the system loop transmission be expressible as a ratio of polynomials in s. There are two possible difficulties. The system may include elements with transfer functions that cannot be expressed as a ratio of finite polynomials. A familiar example of this type of element is the pure time delay of τ seconds with a transfer function $e^{-s\tau}$. A second possibility is that the available information about the system consists of an experimentally determined frequency response. Approximating the measured data in a form suitable for Routh or root-locus analysis may not be practical.

The methods described in this section evaluate the stability of a feedback system starting from its loop transmission as a function of frequency. The only required data are the magnitude and angle of this transmission, and it is not necessary that these data be presented as analytic expressions. As a result, stability can be determined directly from experimental results.

4.4.1 The Nyquist Criterion

It is necessary to develop a method for determining absolute and relative stability information for feedback systems based on the variation of their loop transmissions with frequency. The topology of Fig. 4.1 is assumed. If there is some frequency ω at which

$$a(j\omega)f(j\omega) = -1 \qquad (4.80)$$

the loop transmission is $+1$ at this frequency. It is evident that the system can then oscillate at the frequency ω, since it can in effect supply its own driving signal without an externally applied input. This kind of intuitive argument fails in many cases of practical interest. For example, a system with a loop transmission of $+10$ at some frequency may or may not be stable depending on the loop-transmission values at other frequencies. The Nyquist criterion can be used to resolve this and other stability questions.

The test determines if there are any values of s with positive real parts for which $a(s)f(s) = -1$. If this condition is satisfied, the characteristic equation of the system has a right-half-plane zero implying instability. In order to use the Nyquist criterion, the function $a(s)f(s)$ is evaluated as s takes on values along the contour shown in the s-plane plot of Fig. 4.16. The contour includes a segment of the imaginary axis and is closed with a large semi-circle of radius R that lies in the right half of the s plane. The values of $a(s)f(s)$ as s varies along the indicated contour are plotted in gain-phase form in an af plane. A possible af-plane plot is shown in Fig. 4.17. The symmetry about the 0° line in the af plane is characteristic of all such plots since $\mathrm{Im}[a(j\omega)f(j\omega)] = -\mathrm{Im}[a(-j\omega)f(-j\omega)]$.

Our objective is to determine if there are any values of s that lie in the shaded region of Fig. 4.16 for which $a(s)f(s) = -1$. This determination is simplified by recognizing that the transformation involved maps closed contours in the s plane into closed contours in the af plane. Furthermore,

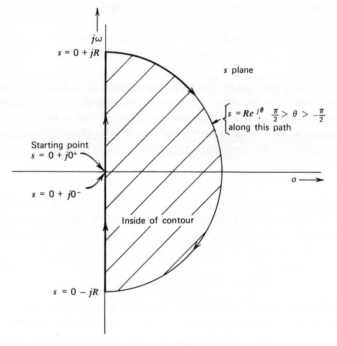

Figure 4.16   Contour Used to evaluate $a(s)f(s)$.

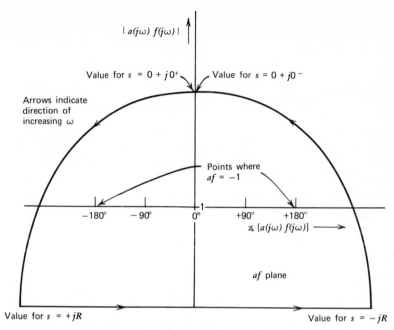

Figure 4.17 Plot of $a(s)f(s)$ as s varies along contour of Fig. 4.16.

all values of s that lie on one side of a contour in the s plane must map to values of af that lie on one side of the corresponding contour in the af plane. The -1 points are clearly indicated in the af-plane plot. Thus the only remaining task is to determine if the shaded region in Fig. 4.16 maps to the inside or to the outside of the contour in Fig. 4.17. If it maps to the inside, there are two values of s in the right-half plane for which $a(s)f(s) = -1$, and the system is unstable.

The form of the af-plane plot and corresponding regions of the two plots are easily determined from $a(s)f(s)$ as illustrated in the following examples. Figure 4.18 indicates the general shape of the s-plane and af-plane plots for

$$a(s)f(s) = \frac{10^3}{(s+1)(0.1s+1)(0.01s+1)} \tag{4.81}$$

Note that the magnitude of af in this example is 10^3 and its angle is zero at $s = 0$. As s takes on values approaching $+jR$, the angle of af changes from $0°$ toward $-270°$, and its magnitude decreases. These relationships are readily obtained from the usual vector manipulations in the s plane. For a sufficiently large value of R, the magnitude of af is arbitrarily small,

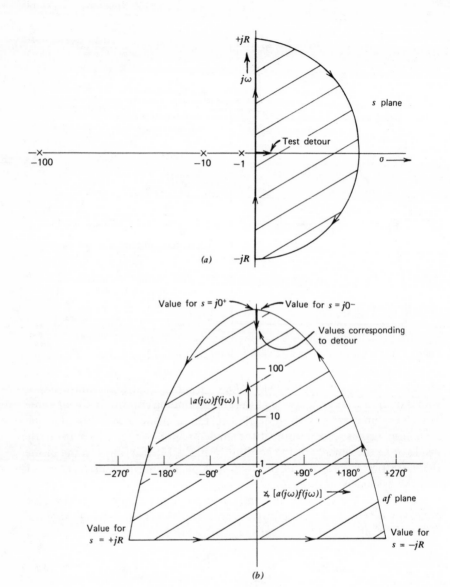

Figure 4.18 Nyquist test for $a(s)f(s) = 10^3/[(s + 1)(0.1s + 1)(0.01s + 1)]$. (*a*) *s*-plane plot. (*b*) *af*-plane plot.

and its angle is nearly $-270°$. As s assumes values in the right-half plane along a semicircle of radius R, the magnitude of af remains constant (for R much greater than the distance of any singularities of af from the origin), and its angle changes from $-270°$ to $0°$ as s goes from $+jR$ to $+R$. The remainder of the af-plane plot must be symmetric about the $0°$ line.

In order to show that the two shaded regions correspond to each other, a small detour from the contour in the s plane is made at $s = 0$ as indicated in Fig. 4.18a. As s assumes real positive values, the magnitude of $a(s)f(s)$ decreases, since the distance from the point on the test detour to each of the poles increases. Thus the detour produces values in the af plane that lie in the shaded region. While we shall normally use a test detour to determine corresponding regions in the two planes, the angular relationships indicated in this example are general ones. Because of the way axes are chosen in the two planes, right-hand turns in one plane map to left-hand turns in the other. A consequence of this reversal is illustrated in Fig. 4.18. Note that if we follow the contour in the s plane in the direction of the arrows, the shaded region is to our right. The angle reversal places the corresponding region in the af plane to the left when its boundary is followed in the direction of the arrows.

Since the two -1 points lie in the shaded region of the af plane, there are two values of s in the right-half plane for which $a(s)f(s) = -1$ and the system is unstable. Note that if a_0f_0 is reduced, the contour in the af plane slides downward and for sufficiently small values of a_0f_0 the system is stable. A geometric development or the Routh criterion shows that the system is stable for positive values of a_0f_0 smaller than 122.21.

Contours with the general shape shown in Fig. 4.19 result if a zero is added at the origin changing $a(s)f(s)$ to

$$a(s)f(s) = \frac{10^3 s}{(s + 1)(0.1s + 1)(0.01s + 1)} \tag{4.82}$$

In order to avoid angle and magnitude uncertainties that result if the s-plane contour passes through a singularity, a small-radius circular arc is used to avoid the zero. Two test detours on the s-plane contour are shown. As the first is followed, the magnitude of af increases since the dominant effect is that of leaving the zero. As the second test detour is followed, the magnitude of af increases since this detour approaches three poles and only one zero. The location of the shaded region in the af plane indicates that the -1 points remain outside this region for all positive values of a_0 and, therefore, the system is stable for any amount of negative feedback.

The Nyquist test can also be used for systems that have one or more loop-transmission poles in the right-half plane and thus are unstable without

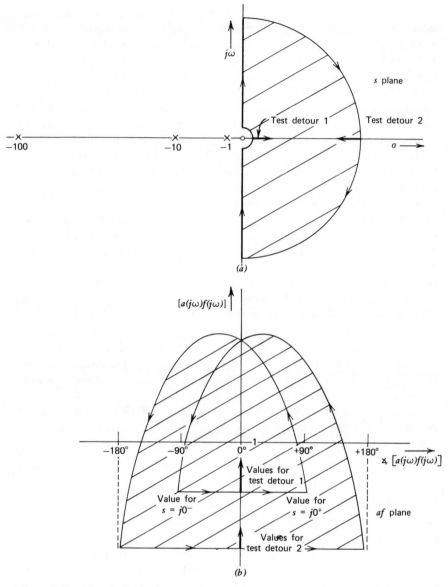

Figure 4.19 Nyquist test for $a(s)f(s) = 10^3s/[(s + 1)(0.1s + 1)(0.01s + 1)]$. (a) s-plane plot. (b) af-plane plot.

feedback. An example of this type of system results for

$$a(s)f(s) = \frac{a_0}{s - 1} \tag{4.83}$$

with s-plane and af-plane plots shown in Figs. 4.20a and 4.20b. The line indicated by $+$ marks in the af-plane plot is an attempt to show that for this system the angle must be continuous as s changes from $j0^-$ to $j0^+$. In order to preserve this necessary continuity, we must realize that $+180°$ and $-180°$ are identical angles, and conceive of the af plane as a cylinder joined at the $\pm180°$ lines. This concept is made somewhat less disturbing by using polar coordinates for the af-plane plot as shown in Fig. 4.20c. Here the -1 point appears only once. The use of the test detour shows that values of s in the right-half plane map outside of a circle that extends from 0 to $-a_0$ as shown in Fig. 4.20c. The location of the -1 point in either af-plane plot shows that the system is stable only for $a_0 > 1$.

Note that the -1 points in the af plane corresponding to angles of $\pm180°$ collapse to one point when the af cylinder necessary for the Nyquist construction for this example is formed. This feature and the nature of the af contour show that when a_0 is less than one, there is only one value of s for which $a(s)f(s) = -1$. Thus this system has a single closed-loop pole on the positive real axis for values of a_0 that result in instability.

This system indicates another type of difficulty that can be encountered with systems that have right-half-plane loop-transmission singularities. The angle of $a(j\omega)f(j\omega)$ is $180°$ at low frequencies, implying that the system actually has positive feedback at these frequencies. (Recall the additional inversion included at the summation point in our standard representation.) The s-plane representation (Fig. 4.20a) is consistent since it indicates an angle of $180°$ for $s = 0$. Thus no procedural modification of the type described in Section 4.3.3 is necessary in this case.

4.4.2 Interpretation of Bode Plots

A Bode plot does not contain the information concerning values of af as the contour in the s plane is closed, which is necessary to apply the Nyquist test. Experience shows that the easiest way to determine stability from a Bode plot of an arbitrary loop transmission is to roughly sketch a complete af-plane plot and apply the Nyquist test as described in Section 4.4.1. For many systems of practical interest, however, it is possible to circumvent this step and use the Bode information directly.

The following two rules evolve from the Nyquist test for systems that have negative feedback at low or mid frequencies and that have no right-half-plane singularities in their loop transmission.

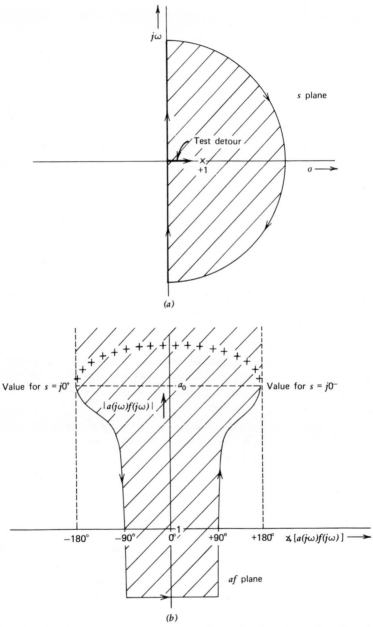

Figure 4.20 Nyquist test for $a(s)f(s) = a_0/(s - 1)$. (a) s-plane plot. (b) af-plane plot. (c) af-plane plot (polar coordinates).

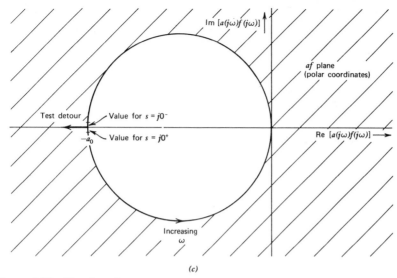

Figure 4.20—Continued

1. If the magnitude of *af* is 1 at only one frequency, the system is stable if the angle of *af* is between $+180°$ and $-180°$ at the unity-gain frequency.

2. If the angle of *af* passes through $+180°$ or $-180°$ at only one frequency, the system is stable if the magnitude of *af* is less than 1 at this frequency.

Information concerning the relative stability of a feedback system can also be determined from a Bode plot for the following reason. The values of *s* for which $af = -1$ are the closed-loop pole locations of a feedback system. The Nyquist test exploits this relationship in order to determine the absolute stability of a system. If the system is stable, but a pair of -1's of *af* occur for values of *s* close to the imaginary axis, the system must have a pair of closed-loop poles with a small damping ratio.

The quantities shown in Fig. 4.21 provide a useful estimation of the proximity of -1's of *af* to the imaginary axis and thus indicate relative stability. The *phase margin* is the difference between the angle of *af* and $-180°$ at the frequency where the magnitude of *af* is 1. A phase margin of $0°$ indicates closed-loop poles on the imaignary axis, and therefore the phase margin is a measure of the additional negative phase shift at the unity-magnitude frequency that will cause instability. Similarly, the *gain margin* is the amount of gain increase required to make the magnitude of *af* unity at the frequency where the angle of *af* is $-180°$, and represents the

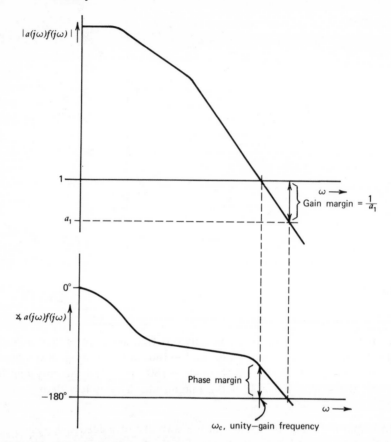

Figure 4.21   Loop-transmission quantities.

amount of increase in $a_0 f_0$ required to cause instability. The frequency at which the magnitude of af is unity is called the *unity-gain frequency* or the *crossover frequency*. This parameter characterizes the relative frequency response or speed of the time response of the system.

A particularly valuable feature of analysis based on the loop-transmission characteristics of a system is that the gain margin and the phase margin, quantities that are quickly and easily determined using Bode techniques, give surprisingly good indications of the relative stability of a feedback system. It is generally found that gain margins of three or more combined with phase margins between 30 and 60° result in desirable trade-offs between bandwidth or rise time and relative stability. The smaller values for gain and phase margin correspond to lower relative stability and are avoided if small overshoot in response to a step or small frequency-response peaking is necessary or if there is the possibility of severe changes in parameter values.

The closed-loop bandwidth and rise time are almost directly related to the unity-gain frequency for systems with equal gain and phase margins. Thus any changes that increase the unity-gain frequency while maintaining constant values for gain and phase margins tend to increase closed-loop bandwidth and decrease closed-loop rise time.

Certain relationships between these three quantities and the corresponding closed-loop performance are given in the following section. Prior to presenting these relationships, it is emphasized that the simplicity and excellence of results associated with frequency-response analysis makes this method a frequently used one, particularly during the initial design phase. Once a tentative design based on these concepts is determined, more detailed information, such as the exact location of closed-loop singularities or the transient response of the system may be investigated, frequently with the aid of machine computation.

4.4.3 Closed-Loop Performance in Terms of Loop-Transmission Parameters

The quantity $a(j\omega)f(j\omega)$ can generally be quickly and accurately obtained in Bode-plot form. The effects of system-parameter changes on the loop transmission are also easily determined. Thus approximate relationships between the loop transmission and closed-loop performance provide a useful and powerful basis for feedback-system design.

The input-output relationship for a system of the type illustrated in Fig. 4.10a is

$$A(s) = \frac{V_o(s)}{V_i(s)} = \frac{a(s)}{1 + a(s)f(s)} \tag{4.84}$$

If the system is stable, the closed-loop transfer function of the system can be approximated for limiting values of loop transmission as

$$A(j\omega) \simeq \frac{1}{f(j\omega)} \qquad |a(j\omega)f(j\omega)| \gg 1 \tag{4.85a}$$

$$A(j\omega) \simeq a(j\omega) \qquad |a(j\omega)f(j\omega)| \ll 1 \tag{4.85b}$$

One objective in the design of feedback systems is to insure that the approximation of Eqn. 4.85a is valid at all frequencies of interest, so that the system closed-loop gain is controlled by the feedback element. The approximation of Eqn. 4.85b is relatively unimportant, since the system is effective operating without feedback in this case. While we normally do not expect to have the system provide precisely controlled closed-loop gain at frequencies where the magnitude of the loop transmission is close to one, the discussion of Section 4.4.2 shows that the relative stability of a system is largely determined by its performance in this frequency range.

The *Nichols chart* shown in Fig. 4.22 provides a convenient method of evaluating the closed-loop gain of a feedback system from its loop trans-

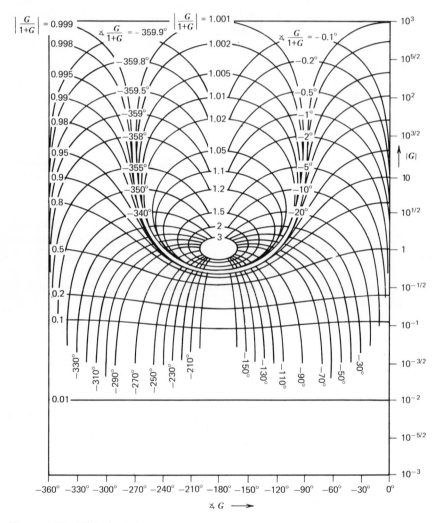

Figure 4.22 Nichols chart.

mission, and is particularly valuable when neither of the limiting approximations of Eqn. 4.85 is valid. This chart relates $G/(1 + G)$ to G where G is any complex number. In order to use the chart, the value of G is located on the rectangular gain-phase coordinates. The angle and magnitude of $G/(1 + G)$ are than read directly from the curved coordinates that intersect the value of G selected.

The gain-phase coordinates shown in Fig. 4.22 cover the complete $0°$ to $-360°$ range in angle and a ratio of 10^6 in magnitude. This magnitude range is unnecessary, since the approximations of Eqn. 4.85 are usually valid when the loop-transmission magnitude exceeds 10 or is less than 0.1. Similarly, the range of angles of greatest interest is that which surrounds the $-180°$ value and which includes anticipated phase margins. The Nichols chart shown in Fig. 4.23 is expanded to provide greater resolution in the region where it will normally be used.

One effective way to view the Nichols chart is as a three-dimensional surface, with the height of the surface proportional to the magnitude of the closed-loop transfer function corresponding to the loop-transmission parameters that define the point of interest. This visualization shows a "mountain" (with a peak of infinite height) where the loop transmission is $+1$.

The Nichols chart can be used directly for any unity-gain feedback system. The transformation indicated in Fig. 4.10b shows that the chart can be used for arbitrary single-loop systems by observing that

$$A(j\omega) = \frac{a(j\omega)}{1 + a(j\omega)f(j\omega)} = \left[\frac{a(j\omega)f(j\omega)}{1 + a(j\omega)f(j\omega)}\right]\left[\frac{1}{f(j\omega)}\right] \qquad (4.86)$$

The closed-loop frequency response is determined by multiplying the factor $a(j\omega)f(j\omega)/[1 + a(j\omega)f(j\omega)]$ obtained via the Nichols chart by $1/f(j\omega)$ using Bode techniques.

One quantity of interest for feedback systems with frequency-independent feedback paths is the peak magnitude M_p, equal to the ratio of the maximum magnitude of $A(j\omega)$ to its low-frequency magnitude (see Section 3.5). A large value for M_p indicates a relatively less stable system, since it shows that there is some frequency for which the characteristic equation approaches zero and thus that there is a pair of closed-loop poles near the imaginary axis at approximately the peaking frequency. Feedback amplifiers are frequently designed to have M_p's between 1.1 and 1.5. Lower values for M_p imply greater relative stability, while higher values indicate that stability has been compromised in order to obtain a larger low-frequency loop transmission and a higher crossover frequency.

The value of M_p for a particular system can be easily determined from the Nichols chart. Furthermore, the chart can be used to evaluate the

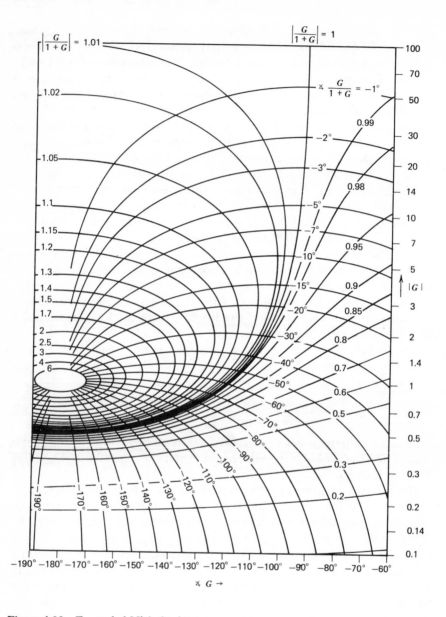

Figure 4.23 Expanded Nichols chart.

effects of variations in loop transmission on M_p. One frequently used manipulation determines the relationship between M_p and a_0f_0 for a system with fixed loop-transmission singularities. The quantity $a(j\omega)f(j\omega)/a_0f_0$ is first plotted on gain-phase coordinates using the same scale as the Nichols chart. If this plot is made on tracing paper, it can be aligned with the Nichols chart and slid up or down to illustrate the effects of different values of a_0f_0. The closed-loop transfer function is obtained directly from the Nichols chart by evaluating $A(j\omega)$ at various frequencies, while the highest magnitude curve of the Nichols chart touched by $a(j\omega)f(j\omega)$ for a particular value of a_0f_0 indicates the corresponding M_p.

Figure 4.24 shows this construction for a system with $f = 1$ and

$$a(s) = \frac{a_0}{(s + 1)(0.1s + 1)} \tag{4.87}$$

The values of a_0 for the three loop transmissions are 8.5, 22, and 50. The corresponding M_p's are 1, 1.4, and 2, respectively.

While the Nichols chart is normally used to determine the closed-loop function from the loop transmission, it is possible to use it to go the other way; that is, to determine $a(j\omega)f(j\omega)$ from $A(j\omega)$. This transformation is occasionally useful for the analysis of systems for which only closed-loop measurements are practical. The transformation yields good results when the magnitude of $a(j\omega)f(j\omega)$ is close to one. Furthermore, the approximation of Eqn. 4.85b shows tha $A(j\omega) \simeq a(j\omega)$ when the magnitude of the loop transmission is small. However, Eqn. 4.85a indicates that $A(j\omega)$ is essentially independent of the loop transmission when the loop-transmission magnitude is large. Examination of the Nichols chart confirms this result since it shows that very small changes in the closed-loop magnitude or angle translate to very large changes in the loop transmission for large loop-transmission magnitudes. Thus even small errors in the measurement of $A(j\omega)$ preclude estimation of large values for $a(j\omega)f(j\omega)$ with any accuracy.

The relative stability of a feedback system and many other important characteristics of its closed-loop response are largely determined by the behavior of its loop transmission at frequencies where the magnitude of this quantity is close to unity. The approximations presented below relate closed-loop quantities defined in Section 3.5 to the loop-transmission properties defined in Section 4.4.2. These approximations are useful for predicting closed-loop response, comparing the performance of various systems, and estimating the effects of changes in loop transmission on closed-loop performance.

The assumptions used in Section 3.5, in particular that f is one at all

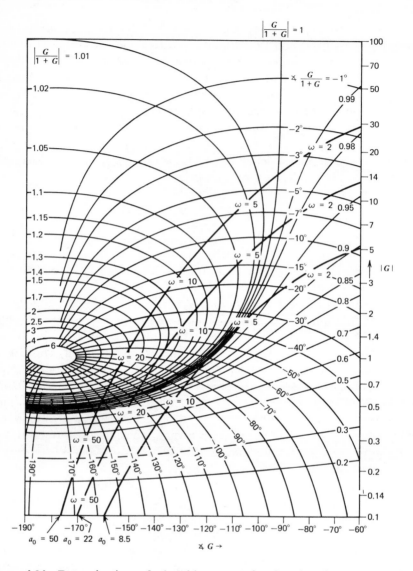

Figure 4.24 Determination of closed-loop transfer function for $a(s) = a_0/[(s + 1)(0.1s + 1)], f = 1$.

frequencies, that a_0 is large, and that the lowest frequency singularity of $a(s)$ is a pole, are assumed here. Under these conditions,

$$M_p \simeq \frac{1}{\sin \phi_m} \qquad (4.88)$$

where ϕ_m is the phase margin. The considerations that lead to this approximation are illustrated in Fig. 4.25. This figure shows several closed-loop-magnitude curves in the vicinity of $M_p = 1.4$ and assumes that the system phase margin is 45°. Since the point $|G| = 1$, $\measuredangle\ G = -135°$ must exist

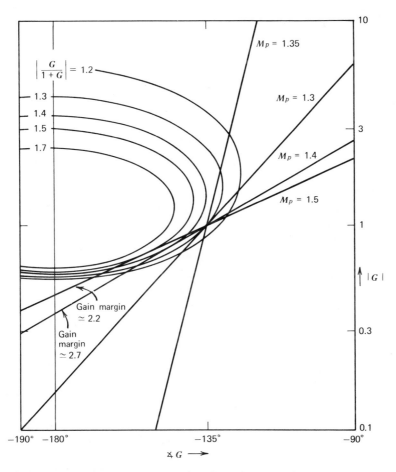

Figure 4.25 M_p for several systems with 45° of phase margin.

for a system with a 45° phase margin, there is no possible way that M_p can be less than approximately 1.3, and the loop-transmission gain-phase curve must be quite specifically constrained for M_p just to equal this value. If it is assumed that the magnitude and angle of G are linearly related, the linear constructions included in Fig. 4.25 show that M_p cannot exceed approximately 1.5 unless the gain margin is very small. Well-behaved systems are actually most likely to have a gain-phase curve that provides an extended region of approximate tangency to the $M_p = 1.4$ curve for a phase margin of 45°. Similar arguments hold for other values of phase margin, and the approximation of Eqn. 4.88 represents a good fit to the relationship between phase margin and corresponding M_p.

Two other approximations relate the system transient response to its crossover frequency ω_c.

$$\frac{0.6}{\omega_c} < t_r < \frac{2.2}{\omega_c} \tag{4.89}$$

The shorter values of rise time correspond to lower values of phase margin.

$$t_s > \frac{4}{\omega_c} \tag{4.90}$$

The limit is approached only for systems with large phase margins.

We shall see that the open-loop transfer function of many operational amplifiers includes one pole at low frequencies and a second pole in the vicinity of the unity-gain frequency of the amplifier. If the system dynamics are dominated by these two poles, the damping ratio and natural frequency of a second-order system that approximates the actual closed-loop system can be obtained from Bode-plot parameters of a system with a frequency-independent feedback path using the curves shown in Fig. 4.26a. The curves shown in Fig. 4.26b relate peak overshoot and M_p for a second-order system to damping ratio and are derived using Eqns. 3.58 and 3.62. While the relationships of Fig. 4.26a are strictly valid only for a system with two widely spaced poles in its loop transmission, they provide an accurate approximation providing two conditions are satisfied.

1. The system loop-transmission magnitude falls off as $1/\omega$ at frequencies between one decade below crossover and the next higher frequency singularity.

2. Additional negative phase shift is provided in the vicinity of the crossover frequency by other components of the loop transmission.

The value of these curves is that they provide a way to determine an approximating second-order system from either phase margin, M_p, or peak

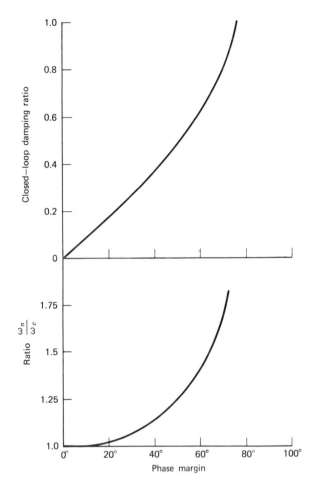

Figure 4.26*a* Closed-loop quantities from loop-transmission parameters for system with two widely spaced poles. Damping ratio and natural frequency as a function of phase margin and crossover frequency.

overshoot of a complex system. The validity of this approach stems from the fact that most systems must be dominated by one or two poles in the vicinity of the crossover frequency in order to yield acceptable performance. Examples illustrating the use of these approximations are included in later sections. We shall see that transient responses based on the approximation are virtually indistinguishable from those of the actual system in many cases of interest.

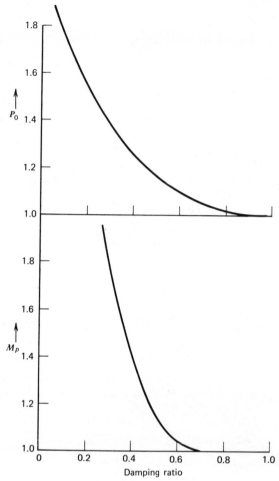

Figure 4.26b P_0 and M_p versus damping ratio for second-order system.

The first significant error coefficient for a system with unity feedback can also be determined directly from its Bode plot. If the loop transmission includes a wide range of frequencies below the crossover frequency where its magnitude is equal to k/ω^n, the error coefficients e_0 through e_{n-1} are negligible and e_n equals $1/k$.

PROBLEMS

P4.1

Find the number of right-half-plane zeros of the polynomial

$$P(s) = s^5 + s^4 + 3s^3 + 4s^2 + s + 2$$

P4.2

A phase-shift oscillator is constructed with a loop transmission

$$L(s) = -\frac{a_0}{(\tau s + 1)^4}$$

Use the Routh condition to determine the value of a_0 that places a pair of closed-loop poles on the imaginary axis. Also determine the location of the poles. Use this information to factor the characteristic equation of the system, thus finding the location of all four closed-loop poles for the critical value of a_0.

P4.3

Describe how the Routh test can be modified to determine the real parts of all singularities in a polynomial. Also explain why this modification is usually of little value as a computational aid to factoring the polynomial.

P4.4

Prove the root-locus construction rule that establishes the angle and intersection of branch asymptotes with the real axis.

P4.5

Sketch root-locus diagrams for the loop-transmission singularity pattern shown in Fig. 4.27. Evaluate part c for moderate values of $a_0 f_0$, and part d for both moderate and very large values of $a_0 f_0$.

P4.6

Consider two systems, both with $f = 1$. One of these systems has a forward-path transfer function

$$a(s) = \frac{a_0(0.5s + 1)}{(s + 1)(0.01s + 1)(0.51s + 1)}$$

while the second system has

$$a'(s) = \frac{a_0(0.51s + 1)}{(s + 1)(0.01s + 1)(0.5s + 1)}$$

Common sense dictates that the closed-loop transfer functions of these systems should be very nearly identical and, furthermore, that both should be similar to a system with

$$a''(s) = \frac{a_0}{(s + 1)(0.01s + 1)}$$

[The closely spaced pole-zero doublets in $a(s)$ and $a'(s)$ should effectively cancel out.] Use root-locus diagrams to show that the closed-loop responses are, in fact, similar.

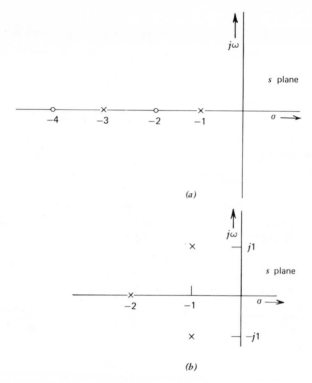

Figure 4.27 Loop-transmission singularity patterns.

P4.7

An operational amplifier has an open-loop transfer function

$$a(s) = \frac{10^6}{(0.1s + 1)(10^{-6}s + 1)^2}$$

This amplifier is combined with two resistors in a noninverting-amplifier configuration. Neglecting loading, determine the value of closed-loop gain that results when the damping ratio of the complex closed-loop pole pair is 0.5.

P4.8

An operational amplifier has an open-loop transfer function

$$a(s) = \frac{10^5}{(\tau s + 1)(10^{-6}s + 1)}$$

The quantity τ can be adjusted by changing the amplifier compensation.

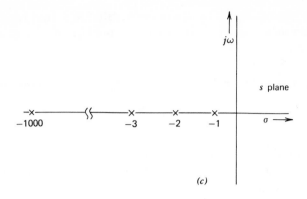

(c)

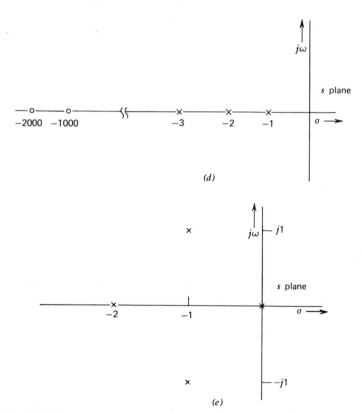

(d)

(e)

Figure 4.27—Continued

161

Use root-contour techniques to determine a value of τ that results in a closed-loop damping ratio of 0.707 when the amplifier is connected as a unity-gain inverter.

P4.9

A feedback system that includes a time delay has a loop transmission

$$L(s) = -\frac{a_0 e^{-0.01s}}{(s + 1)}$$

Use the Nyquist test to determine the maximum value of a_0 for stable operation. What value of a_0 should be selected to limit M_p to a factor of 1.4? (You may assume that the feedback path of the system is frequency independent.)

P4.10

We have been investigating the stability of feedback systems that are generally low pass in nature, since the transfer functions of most operational-amplifier connections fall in this category. However, stability problems also arise in high-pass systems. For example, a-c coupled feedback amplifiers designed for use at audio frequencies sometimes display a low-frequency instability called "motor-boating." Use the Nyquist test to demonstrate the possibility of this type of instability for an amplifier with a loop transmission

$$L(s) = -\frac{a_0 s^3}{(s + 1)(0.1s + 1)^2}$$

Also show the potentially unstable behavior using root-locus methods. For what range of values of a_0 is the amplifier stable?

P4.11

Develop a modification of the Nyquist test that enables you to determine if a feedback system has any closed-loop poles with a damping ratio of less than 0.707. Illustrate your test by forming the modified Nyquist diagram for a system with $a(s) = a_0/(s + 1)^2$, $f(s) = 1$. For what value of a_0 does the damping ratio of the closed-loop pole pair equal 0.707? Verify your answer by factoring the characteristic equation for this value of a_0.

P4.12

The open-loop transfer function of an operational amplifier is

$$a(s) = \frac{10^5}{(0.1s + 1)(10^{-6}s + 1)^2}$$

Determine the gain margin, phase margin, crossover frequency, and M_p for this amplifier when used in a feedback connection with $f = 1$. Also find the value of f that results in an M_p of 1.1. What are the values of phase and gain margin and crossover frequency with this value for f?

P4.13

A feedback system is constructed with

$$a(s) = \frac{10^6(0.01s + 1)^2}{(s + 1)^3}$$

and an adjustable, frequency-independent value for f. As f is increased from zero, it is observed that the system is stable for very small values of f, then becomes unstable, and eventually returns to stable behavior for sufficiently high values of f. Explain this performance using Nyquist and root-locus analysis. Use the Routh criterion to determine the two borderline values for f.

P4.14

An operational amplifier with a frequency-independent feedback path exhibits 40% overshoot and 10 to 90% rise time of 0.5 μs in response to a step input. Estimate the phase margin and crossover frequency of the feedback connection, assuming that its performance is dominated by two widely separated loop-transmission poles.

P4.15

Consider a feedback system with

$$a(s) = \frac{a_0}{s[(s^2/2) + s + 1]}$$

and $f(s) = 1$.

Show that by appropriate choice of a_0, the closed-loop poles of the system can be placed in a third-order Butterworth pattern. Find the crossover frequency and the phase margin of the loop transmission when a_0 is selected for the closed-loop Butterworth response. Use these quantities in conjunction with Fig. 4.26 to find the damping ratio and natural frequency of a second-order system that can be used to approximate the transient response of the third-order Butterworth filter. Compare the peak overshoot and rise time of the approximating system in response to a step with those of the Butterworth response (Fig. 3.10). Note that, even though this system is considerably different from that used to develop Fig. 4.26, the approximation predicts time-domain parameters with fair accuracy.

CHAPTER V

COMPENSATION

5.1 OBJECTIVES

The discussion up to this point has focused on methods used to analyze the performance of a feedback system with a given set of parameters. The results of such analysis frequently show that the performance of the feedback system is unacceptable for a given application because of such deficiencies as low desensitivity, slow speed of response, or poor relative stability. The process of modifying the system to improve performance is called *compensation*.

Compensation usually reduces to a trial-and-error procedure, with the experience of the designer frequently playing a major role in the eventual outcome. One normally assumes a particular form of compensation and then evaluates the performance of the system to see if objectives have been met. If the performance remains inadequate, alternate methods of compensation are tried until either objectives are met, or it becomes evident that they cannot be achieved.

The type of compensation that can be used in a specific application is usually highly dependent on the components that form the system. The general principles that guide the compensation process will be described in this chapter. Most of these ideas will be reviewed and reinforced in later chapters after representative amplifier topologies and applications have been introduced.

5.2 SERIES COMPENSATION

One way to change the performance of a feedback system is to alter the transfer function of either its forward-gain path or its feedback path. This technique of modifying a series element in a single-loop system is called *series compensation*. The changes may involve the d-c gain of an element or its dynamics or both.

5.2.1 Adjusting the D-C Gain

One conceptually straightforward modification that can be made to the loop transmission is to vary its d-c or midband value $a_0 f_0$. This modification has a direct effect on low-frequency desensitivity, since we have seen that the attenuation to changes in forward-path gain provided by feedback is equal to $1 + a_0 f_0$.

The closed-loop dynamics are also dependent on the magnitude of the low-frequency loop transmission. The example involving Fig. 4.6 showed how root-locus methods are used to determine the relationship between $a_0 f_0$ and the damping ratio of a dominant pole pair. A second approach to the control of closed-loop dynamics by adjusting $a_0 f_0$ for a specific value of M_p was used in the example involving Fig. 4.24.

An assumption common to both of these previous examples was that the value of $a_0 f_0$ could be selected without altering the singularities included in the loop transmission. For certain types of feedback systems independence of the d-c magnitude and the dynamics of the loop transmission is realistic. The dynamics of servomechanisms, for example, are generally dominated by mechanical components with bandwidths of less than 100 Hz. A portion of the d-c loop transmission of a servomechanism is often provided by an electronic amplifier, and these amplifiers can provide frequency-independent gain into the high kilohertz or megahertz range. Changing the amplifier gain changes the value of $a_0 f_0$ but leaves the dynamics associated with the loop transmission virtually unaltered.

This type of independence is frequently absent in operational amplifiers. In order to increase gain, stages may have to be added, producing significant changes in dynamics. Lowering the gain of an amplifying stage may also change dynamics because, for example, of a relationship between the input capacitance and voltage gain of a common-emitter amplifier. A further practical difficulty arises in that there is generally no predictable way to change the d-c open-loop gain of available discrete- or integrated-circuit operational amplifiers from the available terminals.

An alternative approach involves modification of the d-c loop transmission by means of the feedback network connected around the amplifier. The connection of Fig. 5.1a illustrates one possibility. The block diagram for this amplifier, assuming negligible loading at either input or output, is shown in part b of this figure, while the block diagram after reduction to unity-feedback form is shown in part c. If the shunt resistance R from the inverting input to ground is an open circuit, the d-c value of the loop transmission is completely determined by a_0 and the ideal closed-loop gain $-R_2/R_1$. However, inclusion of R provides an additional degree of freedom so that the d-c loop transmission and the ideal gain can be changed independently.

This technique is illustrated for a unity-gain inverter ($R_1 = R_2$) and

$$a(s) = \frac{10^6}{(s + 1)(10^{-5}s + 1)} \tag{5.1}$$

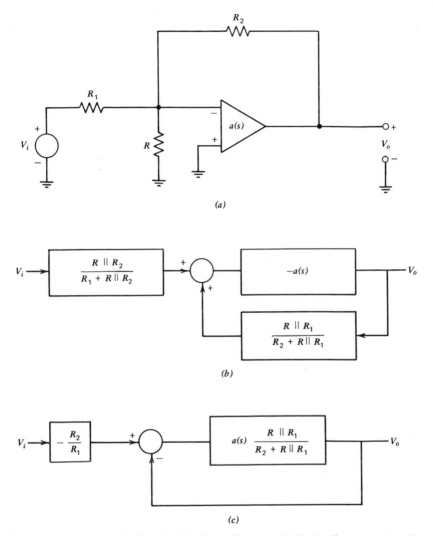

Figure 5.1 Inverter. (*a*) Circuit. (*b*) Block diagram. (*c*) Block diagram reduced to unity-feedback form.

A Bode plot of this transfer function is shown in Fig. 5.2. If R is an open circuit, the magnitude of the loop transmission is one at approximately 2.15×10^5 radians per second, since the magnitude of $a(s)$ at this frequency is equal to the factor of two attenuation provided by the R_1-R_2 network. The phase margin of the system is 25°, and Fig. 4.26*a* shows that the closed-loop damping ratio is 0.22. Since Fig. 4.26 was generated assuming this type of loop transmission, it yields exact results in this case. If the resistor

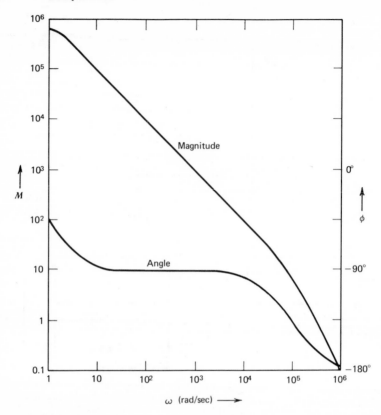

Figure 5.2 Bode plot of $10^6/[(s + 1)(10^{-5}s + 1)]$.

R is made equal to $0.2R_1$, the loop-transmission unity-gain frequency is lowered to 10^5 radians per second by the factor-of-seven attenuation provided by the network, and phase margin and damping ratio are increased to $45°$ and 0.42, respectively. One penalty paid for this type of attenuation at the input terminals of the amplifier is that the voltage offset and noise at the output of the amplifier are increased for a given offset and noise at the amplifier input terminals (see Problem P5.2).

5.2.2 Creating a Dominant Pole

Elementary considerations show that a single-pole loop transmission results in a stable system for any amount of negative feedback, and that the closed-loop bandwidth of such a system increases with increasing a_0f_0. Similarly, if the loop transmission in the vicinity of the unity-gain frequency is dominated by one pole, ample phase margin is easily obtained. Because of the ease of stabilizing approximately single-pole systems, many types of compensation essentially reduce to making one pole dominate the loop transmission.

One brute-force method for making one pole dominate the loop transmission of an amplifier is simply to connect a capacitor from a node in the signal path to ground. If a large enough capacitor is used, the gain of the amplifier will drop below one at a frequency where other amplifier poles can be ignored. The obvious disadvantage of this approach to compensation is that it may drastically reduce the closed-loop bandwidth of the system.

A feedback system designed to hold the value of its output constant independent of disturbances is called a *regulator*. Since the output need not track a rapidly varying input, closed-loop bandwidth is an unimportant parameter. If a dominant pole is included in the output portion of a regulator, the low-pass characteristics of this pole may actually improve system performance by attenuating disturbances even in the absence of feedback.

One possible type of voltage regulator is shown in simplified form in Fig. 5.3. An operational amplifier is used to compare the output voltage with a fixed reference. The operational amplifier drives a series regulator stage that consists of a transistor with an emitter resistor. The series regulator isolates the output of the circuit from an unregulated source of voltage. The load includes a parallel resistor-capacitor combination and a disturbing current source. The current source is included for purposes of analysis and will be used to determine the degree to which the circuit rejects load-current changes. The dominant pole in the system is assumed to occur because of the load, and it is further assumed that the operational amplifier and series transistor contribute no dynamics at frequencies where the loop-transmission magnitude exceeds one.

The block diagram of Fig. 5.3b models the regulator if it is assumed that the common-base current gain of the transistor is one and that the resistor R is large compared to the reciprocal of the transistor transconductance. This diagram verifies the single-pole nature of the system loop transmission.

As mentioned earlier, the objective of the circuitry is to minimize changes in load voltage that result from changes in the disturbing current and the unregulated voltage. The disturbance-to-output closed-loop transfer functions that indicate how well the regulator achieves this objective are

$$\frac{V_l}{I_d} = \frac{R/a_0}{RC_L s/a_0 + (1 + R/a_0 R_L)} \tag{5.2}$$

and

$$\frac{V_l}{V_u} = \frac{1/a_0}{RC_L s/a_0 + (1 + R/a_0 R_L)} \tag{5.3}$$

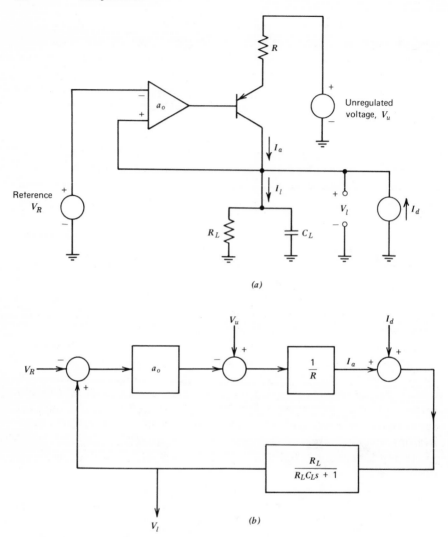

Figure 5.3 Voltage regulator. (*a*) Circuit. (*b*) Block diagram.

If sinusoidal disturbances are considered, the magnitude of either disturbance-to-output transfer function is a maximum at d-c, and decreases with increasing frequency because of the low-pass characteristics of the load. Increasing C_L improves performance, since it lowers the frequency at which the disturbance is attenuated significantly compared to its d-c value. If it is assumed that arbitrary loads can be connected to the regu-

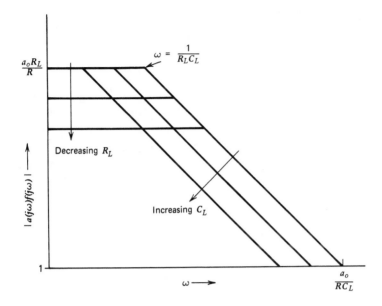

Figure 5.4 Effect of changing load parameters on the Bode plot of a voltage regulator.

lator (which is the usual situation, if, for example, this circuit is used as a laboratory power supply), the values of R_L and C_L must be considered variable. The minimum value of C_L can be constrained by including a capacitor with the regulation circuitry. The load-capacitor value increases as external loads are connected to the regulator because of the decoupling capacitors usually associated with these loads. Similarly, R_L decreases with increasing load to some minimum value determined by loading limitations.

The compensation provided by the pole at the output of the regulator maintains stability as R_L and C_L change, as illustrated in the Bode plot of Fig. 5.4. (The negative of the loop transmission for this plot is $a_0 R_L / R(R_L C_L s + 1)$, determined directly from Fig. 5.3b.) Note that the unity-gain frequency can be limited by constraining the maximum value of the $a_0 / R C_L$ ratio, and thus crossover can be forced before other system elements affect dynamics. The phase margin of the system remains close to 90° as R_L and C_L vary over wide limits.

5.2.3   Lead and Lag Compensation

If the designer is free to modify the dynamics of the loop transmission as well as its low-frequency magnitude, he has considerably more control over the closed-loop performance of the system. The rather simple modification of making a single pole dominate has already been discussed.

The types of changes that can be made to the dynamics of the loop transmission are constrained, even in purely mathematical systems. It is tempting to think that systems could be improved, for example, by adding positive phase shift to the loop transmission without changing its magnitude characteristics. This modification would clearly improve the phase margin of a system. Unfortunately, the magnitude and angle characteristics of physically realizable transfer functions are not independent, and transfer functions that provide positive phase shift also have a magnitude that increases with increasing frequency. The magnitude increase may result in a higher system crossover frequency, and the additional negative phase shift that results from other elements in the loop may negate hoped-for advantages.

The way that series compensation is implemented and the types of compensating transfer functions that can be obtained in practical systems are even further constrained by the hardware realities of the feedback system being compensated. The designer of a servomechanism normally has a wide variety of compensating transfer functions available to him, since the electrical networks and amplifiers usually used to compensate servomechanisms have virtually unlimited bandwidth relative to the mechanical portions of the system. Conversely, we should remember that the choices of the feedback-amplifier designer are more restricted because the ways that the transfer function of an amplifier can be changed, particularly near its unity-gain frequency where transistor bandwidth limitations dominate performance, are often severely constrained.

Two distinct types of transfer functions are normally used for the series compensation of feedback systems, and these types can either be used separately or can be combined in one system. A *lead transfer function* can be realized with the network shown in Fig. 5.5. The transfer function of this network is

$$\frac{V_o(s)}{V_i(s)} = \frac{1}{\alpha}\left[\frac{\alpha\tau s + 1}{\tau s + 1}\right] \tag{5.4}$$

where $\alpha = (R_1 + R_2)/R_2$ and $\tau = (R_1 \parallel R_2)C$. As the name implies, this network provides positive or leading phase shift of the output signal relative to the input signal at all frequencies. Lead-network parameters are usually selected to locate its singularities near the crossover frequency of the system being compensated. The positive phase shift of the network then improves the phase margin of the system. In many cases, the lead network has negligible effect on the magnitude characteristics of the compensated system at or below the crossover frequency, since we shall see that a

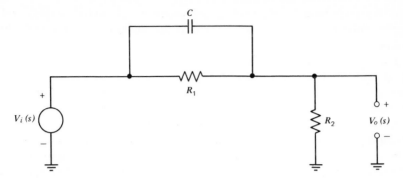

Figure 5.5 Lead network.

lead network provides substantial phase shift before its magnitude increases significantly.

The *lag network* shown in Fig. 5.6 has the transfer function

$$\frac{V_o(s)}{V_i(s)} = \frac{\tau s + 1}{\alpha \tau s + 1} \tag{5.5}$$

where $\alpha = (R_1 + R_2)/R_2$ and $\tau = R_2C$. The singularities of this type of network are usually located well below crossover in order to reduce the crossover frequency of a system so that the negative phase shift associated with other elements in the system is reduced at the unity-gain frequency. This effect is possible because of the attenuation of the lag network at frequencies above both its singularities.

The maximum magnitude of the phase angle associated with either of these transfer functions is

$$\phi_{\max} = \sin^{-1}\left[\frac{\alpha - 1}{\alpha + 1}\right] \tag{5.6}$$

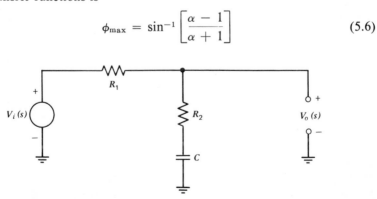

Figure 5.6 Lag network.

and this magnitude occurs at the geometric mean of the frequencies of the two singularities. The gain of either network at its maximum-phase-shift frequency is $1/\sqrt{\alpha}$.

The magnitudes and angles of lead transfer functions for α values of 5, 10, and 20, are shown in Bode-plot form in Fig. 5.7. Figure 5.8 shows corresponding curves for lag transfer functions. The corner frequencies for the poles of the plotted functions are normalized to one in these figures.

As mentioned earlier, an important feature of the lead transfer function is that it provides substantial positive phase shift over a range of frequencies below its zero location without a significant increase in magnitude. The reason stems from a basic property of real-axis singularities. At frequencies below the zero location, this singularity dominates the lead transfer function, so

$$\frac{V_o(s)}{V_i(s)} \simeq \frac{1}{\alpha}(\alpha\tau s + 1) \tag{5.7}$$

The magnitude and angle of this function are

$$M = \frac{1}{\alpha}[\sqrt{1 + (\alpha\tau\omega)^2}] \tag{5.8a}$$

$$\phi = \tan^{-1}\alpha\tau\omega \tag{5.8b}$$

At a small fraction of the zero location, $\alpha\tau\omega \ll 1$, so

$$M \simeq \frac{1}{\alpha}\left[1 + \frac{(\alpha\tau\omega)^2}{2}\right] \tag{5.9a}$$

$$\phi \simeq \alpha\tau\omega \tag{5.9b}$$

Since the angle increases linearly with frequency in this region while the magnitude increases quadratically, the angle change is relatively larger at a given frequency. The same sort of reasoning applies even if the zero is located at or slightly below crossover. Figure 5.7 shows that the positive phase shift of a lead transfer function with a reasonable value of α is approximately 40° at its zero location, while the magnitude increase is only a factor of 1.4. Much of this advantage is lost at frequencies beyond the geometric mean of the singularities, since the positive phase shift decreases beyond this frequency, while the magnitude continues to increase.

We should recognize that an isolated zero can be used in place of a lead transfer function, and that this type of transfer function actually has phase-shift characteristics superior to those of the zero-pole pair. However, the unlimited high-frequency gain implied by an isolated zero is clearly unachievable, at least at sufficiently high frequencies. Thus the form of the

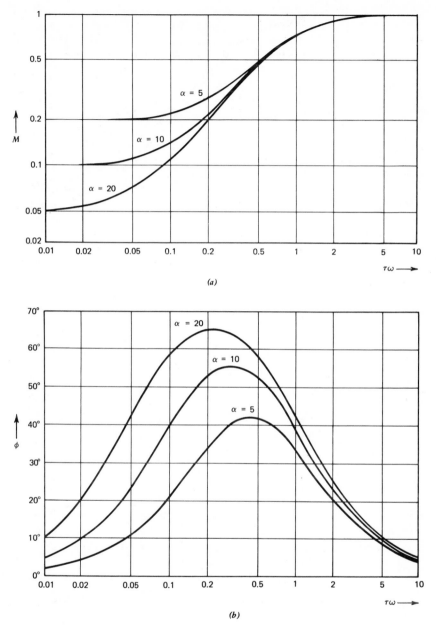

Figure 5.7 Lead network characteristics for $V_o(s)/V_i(s) = (1/\alpha)$ $[(\alpha\tau s + 1)/(\tau s + 1)]$. (*a*) Magnitude. (*b*) Angle.

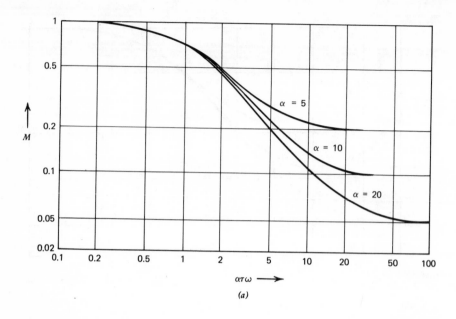

(a)

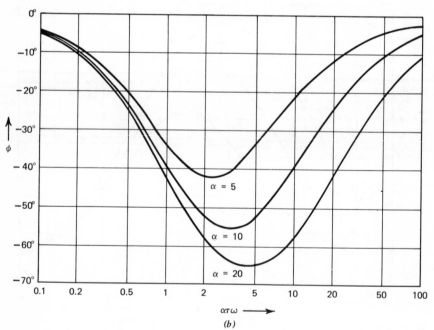

(b)

Figure 5.8 Lag network characteristics for $V_o(s)/V_i(s) = (\tau s + 1)/(\alpha \tau s + 1)$. (a) Magnitude. (b) Angle.

176

lead transfer function introduced earlier reflects the realities of physical systems.

The important feature of the lag transfer function illustrated in Fig. 5.8 is that at frequencies well above the zero location, it provides a magnitude attenuation equal to the ratio of the two singularity locations and negligible phase shift. It can thus be used to reduce the magnitude of the loop transmission without significantly adding to the negative phase shift of this transmission at moderate frequencies.

5.2.4 Example

Lead and lag networks were originally developed for use in servomechanisms, and provide a powerful means for compensation when their singularities can be located arbitrarily with respect to other system poles and when independent adjustment of the low-frequency loop-transmission magnitude is possible. Even without this flexibility, which is usually absent with operational-amplifier circuits, lead or lag compensation can provide effective control of closed-loop performance in certain configurations. As an example, consider the noninverting gain-of-ten amplifier connection shown in Fig. 5.9. It is assumed that the input admittance and output impedance of the operational amplifier are small. The open-loop transfer function of the operational amplifier is[1]

$$a(s) = \frac{5 \times 10^5}{(s + 1)(10^{-4}s + 1)(10^{-5}s + 1)} \tag{5.10}$$

and it is assumed that the user cannot alter this function. When connected as shown in Fig. 5.9 the value of f is 0.1, and thus the negative of the loop transmission is

$$a(s)f(s) = \frac{5 \times 10^4}{(s + 1)(10^{-4}s + 1)(10^{-5}s + 1)} \tag{5.11}$$

[1] While an analytic expression is used for $a(s)$ in this example, the reader should realize that the open-loop transfer function of an operational amplifier will generally not be available in this form. Note, however, that an experimentally determined Bode plot is completely acceptable for all of the required manipulations, and that this information can always be determined.

The general characteristics of the assumed open-loop transfer function are typical of many operational amplifiers, in that this quantity is dominated by a single pole at low frequencies. At frequencies closer to the unity-gain frequency, additional negative phase shift results from effects related to transistor limitations. As we shall see in later sections, these effects constrain the ultimate performance capabilities of the amplifier.

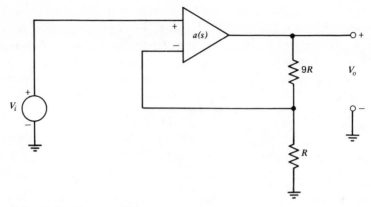

Figure 5.9 Gain-of-ten amplifier.

The closed-loop gain is

$$\frac{V_o(s)}{V_i(s)} = A(s) = \frac{a(s)}{1 + a(s)f(s)}$$

$$\simeq \frac{10}{2 \times 10^{-14}s^3 + 2.2 \times 10^{-9}s^2 + 2 \times 10^{-5}s + 1} \quad (5.12)$$

A Bode plot of Eqn. 5.11 (Fig. 5.10) shows that the system crossover frequency is 2.1×10^4 radians per second, its phase margin is 13°, and the gain margin is 2.

While the problem statement precludes altering $a(s)$, we can introduce a lead transfer function into the loop transmission by including a capacitor across the upper resistor in the feedback network. The topology is shown in Fig. 5.11a, with a block diagram shown in Fig. 5.11b. The negative of the loop transmission for the system is

$$a'(s)f'(s) = \frac{5 \times 10^4(9RCs + 1)}{(s + 1)(10^{-4}s + 1)(10^{-5}s + 1)(0.9RCs + 1)} \quad (5.13)$$

Several considerations influence the selection of the R-C product that locates the singularities of the lead network. As mentioned earlier, the objective of a lead network is to provide positive phase shift in the vicinity of the crossover frequency, and maximum positive phase shift from the network results if crossover occurs at the geometric mean of the zero-pole pair. However, the network singularities and the crossover frequently cannot be adjusted independently for this system, since if the zero of the lead network is located at a frequency below about 3×10^4 radians per second, the crossover frequency increases. An increase in crossover frequency in-

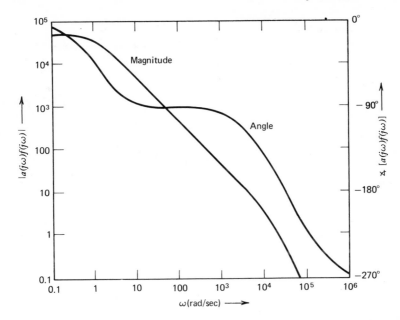

Figure 5.10   Bode plot for uncompensated grain-of-ten amplifier. $af = 5 \times 10^4/$
$[(s + 1)(10^{-4}s + 1)(10^{-5}s + 1)]$.

creases the negative phase shift of the amplifier at this frequency, offsetting
in part the positive phase shift of the network. A related consideration involves the effect of the lead network on the ideal closed-loop gain of the
amplifier since the network is introduced in the feedback path and the ideal
gain is reciprocally related to the feedback transfer function. If the lead-
network zero is located at a low frequency, a low-frequency closed-loop
pole that reduces the closed-loop bandwidth of the system results.

A reasonable compromise in this case is to locate the zero of the lead
network near the unity-gain frequency, in an attempt to obtain positive
phase shift from the network without a significant increase in the crossover
frequency. The choice $RC = 4.44 \times 10^{-6}$ seconds locates the zero at $2.5 \times$
10^4 radians per second. A Bode plot of Eqn. 5.13 for this value of RC is
shown in Fig. 5.12. The unity-gain frequency is increased slightly to $2.5 \times$
10^4 radians per second, while the phase margin is increased to the respectable value of 47°. Gain margin is 14.

A lag transfer function can be introduced into the forward path of the
amplifier by shunting a series resistor-capacitor network between its input
terminals as shown in Fig. 5.13a. Note that the same loop transmission
could be obtained by shunting the R-valued resistor with the R_1-C network,

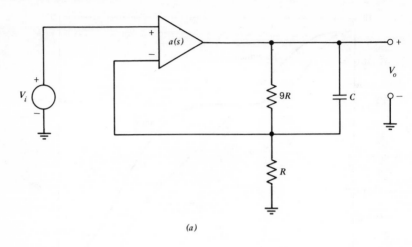

(a)

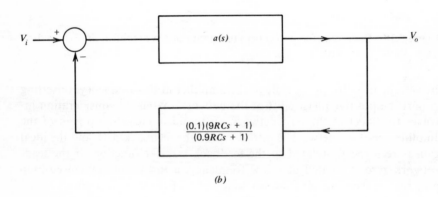

(b)

Figure 5.11 Gain-of-ten amplifier with lead network in feedback path. (a) Circuit. (b) Block diagram.

since both the bottom end of the R-valued resistor and the noninverting input of the amplifier are connected to incrementally grounded points. If this later option were used, the R_1-C network would introduce the lag transfer function into the feedback path of the topology. Consequently, the ideal closed-loop transfer function would include the reciprocal of the lag function. Since the singularities of lag networks are generally located at low frequencies, the closed-loop transfer function could be adversely influenced at frequencies of interest. (See Problem P5.7.)

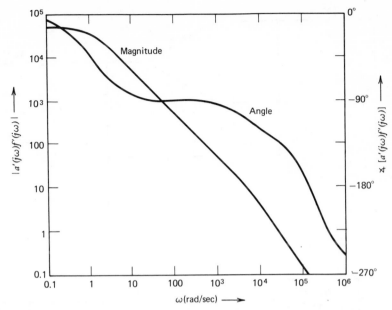

Figure 5.12 Bode plot for lead-compensated gain-of-ten amplifier. $a'f' = 5 \times 10^4(4 \times 10^{-5}s + 1)/[(s + 1)(10^{-4}s + 1)(10^{-5}s + 1)(4 \times 10^{-6}s + 1)]$.

The system block diagram for the topology of Fig. 5.13a is shown in Fig. 5.13b. In this case, the lag transfer function appears in both the feedback path and a forward path outside the loop. The block diagram can be rearranged as shown in Fig. 5.13c, and this final diagram shows that including the R_1-C network between amplifier inputs leaves the ideal closed-loop gain unchanged. The negative of the loop transmission for Fig. 5.13c is

$$a''(s)f''(s) = 0.1 \frac{(\tau s + 1)}{(\alpha \tau s + 1)} a(s) \tag{5.14}$$

where

$$\alpha = \frac{R_1 + 0.9R}{R_1} \quad \text{and} \quad \tau = R_1C$$

As mentioned earlier, the singularities of a lag transfer function are generally located well below the system crossover frequency so that the lag network does not deteriorate phase margin significantly. A frequently used rule of thumb suggests locating the zero of the lag network at one-tenth of the crossover frequency that results following compensation, since this value yields a maximum negative phase contribution of 5.7° from the network at crossover. We also, rather arbitrarily, decide to choose the lag-network parameters to yield a phase margin of approximately 47°, the same

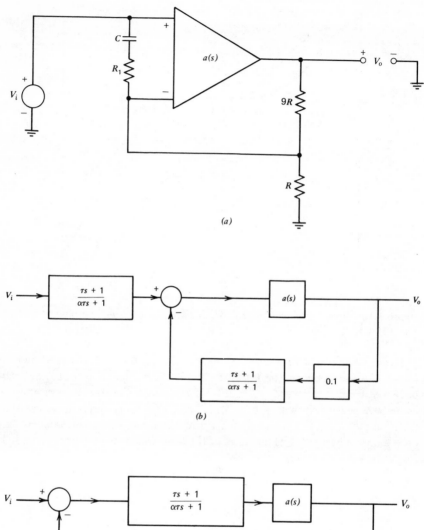

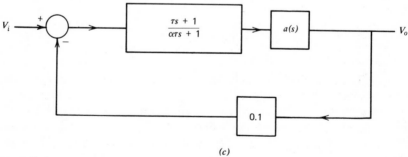

Figure 5.13 Gain-of-ten amplifier with lag compensation. (*a*) Circuit. (*b*) Block diagram. (*c*) Block diagram following rearrangement.

value as that of the system compensated with a lead network. The Bode plot of the system without compensation, Fig. 5.10, aids in selecting lag-network parameters. This plot indicates an uncompensated phase angle of $-128°$ and an uncompensated magnitude of 6.2 at a frequency of 6.7×10^3 radians per second. If the value of 6.2 is the chosen high-frequency attenuation α of the lag network, the compensated crossover frequency will be 6.7×10^3 radians per second. The 5° of negative phase shift anticipated from a properly located lag network combines with the $-128°$ of phase shift of the system prior to compensation to yield a compensated phase margin of 47°. The zero of the lag network is located at 6.7×10^2 radians per second, a factor 10 below crossover. These design objectives are met with $R_1 = 0.173R$ and $R_1C = 1.5 \times 10^{-3}$ seconds. With these values, the negative of the loop transmission is

$$a''(s)f''(s) = \frac{5 \times 10^4(1.5 \times 10^{-3}s + 1)}{(s + 1)(10^{-4}s + 1)(10^{-5}s + 1)(9.3 \times 10^{-3}s + 1)} \quad (5.15)$$

This transfer function, plotted in Fig. 5.14, indicates predicted values for crossover frequency and phase margin. The gain margin is 15.

Two other modifications of the loop transmission result in Bode plots that are similar to that of the lag-compensated system in the vicinity of the crossover frequency. One possibility is to lower the value of a_0f_0 by a factor of 6.2 (see Section 5.2.1). The required reduction can be accomplished by simply using the shunt-resistor value determined for lag compensation directly across the input terminals of the operational amplifier. This modification results in the same crossover frequency as that of the lag-compensated amplifier, and has several degrees more phase margin since it does not have the slight negative phase shift associated with the lag network at crossover. Unfortunately, the lowered a_0f_0 results in a lower value for desensitivity compared with that of the lag-compensated amplifier at all frequencies below the zero of the network.

A second possibility is to move the lowest-frequency pole of the loop transmission back by a factor of 6.2. This modification might be made to the amplifier itself, or could be accomplished by appropriate selection of lag-network components. The effect on parameters in the vicinity of crossover is essentially identical to that of reducing a_0f_0. Desensitivity is retained at d-c with this method, but is lowered at intermediate frequencies compared to that provided by lag compensation. These two approaches to compensating the amplifier described here are investigated in detail in Problem P5.8.

The discussion of series compensation up to this point has focused on the use of the frequency-domain concepts of phase margin, gain margin,

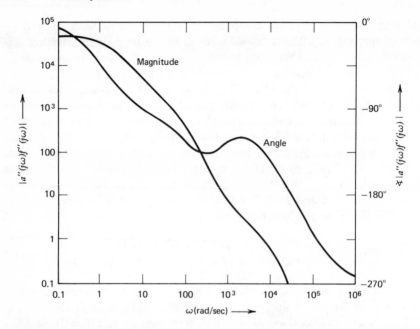

Figure 5.14 Bode plot for lag compensated gain-of-ten amplifier. $a''f'' = 5 \times 10^4(1.5 \times 10^{-3}s + 1)/[(s + 1)(10^{-4}s + 1)(10^{-5}s + 1)(9.3 \times 10^{-3}s + 1)]$.

and crossover frequency to determine compensating-network parameters. Root-locus methods cannot be used directly since the value of a_0f_0 is not varied to effect compensation. However, the root-locus sketches for the uncompensated, lead-compensated, and lag-compensated systems shown in Fig. 5.15 do lend a degree of insight into system behavior. (There is significant distortion in these sketches, since it is not convenient to present sketches accurately where the singularities are located several decades apart.)

The root-locus diagram of Fig. 5.15a illustrates the change in closed-loop pole location as a function of a_0f_0 for the uncompensated system. Adding the lead network (Fig. 5.15b) shifts the dominant branches to the left and, thus, improves the damping ratio of this pair of poles for a given value of a_0f_0.

The effect of lag compensation is somewhat more subtle. The root-locus diagram of Fig. 5.15c is virtually identical to that of Fig. 5.15a except in the immediate vicinity of the lag-network singularity pair. However, a gain calculation using rule 8 (Section 4.3.1) shows that the value of a_0f_0 required to reach a given damping ratio for the dominant pair is higher by approximately a factor of α when the lag network is included.

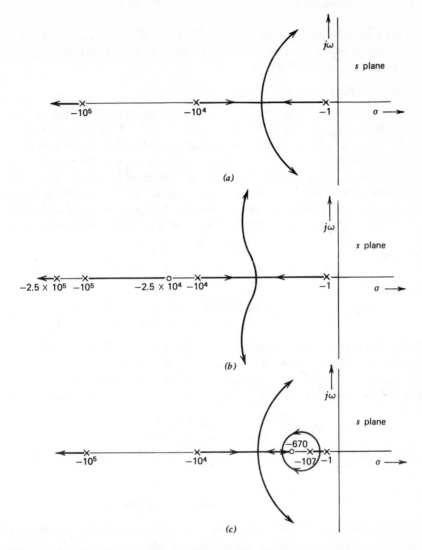

Figure 5.15 Root-locus diagrams illustrating compensation of gain-of-ten amplifier. (*a*) Uncompensated. (*b*) Lead compensated. (*c*) Lag compensated.

Root contours can also be used to show the effects of varying a single parameter of either the lead or the lag network. This design approach is explored in Problems P5.9 and P5.10.

5.2.5 Evaluation of the Effects of Compensation

There are several ways to demonstrate the improvement in performance provided by compensation. Since the parameters of the compensating transfer function are usually determined with the aid of loop-transmission Bode plots, one simple way to evaluate various types of compensation is to compare the desensitivity obtained from them. The considerations used to determine lead- and lag-compensation parameters for an operational amplifier connected to provide a gain of 10 were described in detail in Section 5.2.4. The resulting loop transmissions, repeated here for convenience, are

$$a'(s)f'(s) = \frac{5 \times 10^4(4 \times 10^{-5}s + 1)}{(s + 1)(10^{-4}s + 1)(10^{-5}s + 1)(4 \times 10^{-6}s + 1)} \qquad (5.16)$$

and

$$a''(s)f''(s) = \frac{5 \times 10^4(1.5 \times 10^{-3}s + 1)}{(s + 1)(10^{-4}s + 1)(10^{-5}s + 1)(9.3 \times 10^{-3}s + 1)} \qquad (5.17)$$

for the lead- and lag-compensated cases, respectively. The phase-margin obtained by either method is approximately 47°.

It was mentioned that the stability of the uncompensated amplifier could be improved by either lowering $a_0 f_0$ by a factor of 6.2, resulting in

$$a'''(s)f(s)''' = \frac{8.1 \times 10^3}{(s + 1)(10^{-4} s + 1)(10^{-5} s + 1)} \qquad (5.18)$$

or by lowering the location of the first pole by the same factor, yielding

$$a''''(s)f''''(s) = \frac{5 \times 10^4}{(6.2 s + 1)(10^{-4} s + 1)(10^{-5} s + 1)} \qquad (5.19)$$

Either of these approaches results in a crossover frequency identical to that of the lag-compensated system and a phase margin of approximately 52°.

The magnitude portions of the loop transmissions for these four cases are compared in Fig. 5.16. The relative desensitivities that are achieved at various frequencies, as well as the relative crossover frequencies, are evident in this figure.

An alternative way to evaluate various compensation techniques is to compare the error coefficients that are obtained using them. This approach is explored in Problem P5.11. As expected, systems with greater desensitivity generally also have smaller-magnitude error coefficients.

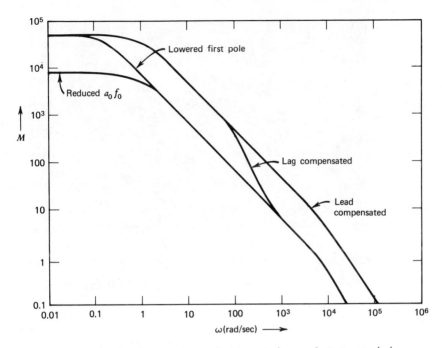

Figure 5.16 Effects of various types of compensation on loop-transmission magnitude.

The discussion of compensation up to now has focused on the use of Bode plots, since this is usually the quickest way to find compensating parameters. However, design objectives are frequently stated in terms of transient response, and the inexperienced designer often feels an act of faith is required to accept the principle that systems with properly chosen values for phase margin, gain margin, and crossover frequency will produce satisfactory transient responses. The step responses shown in Fig. 5.17 are offered as an aid to establishing this necessary faith.

Figure 5.17a shows the step response of the gain-of-ten amplifier without compensation. The large peak overshoot and poor damping of the ringing reflect the low phase margin of the system. The overshoot and damping for the lead compensated, lag compensated, and reduced $a_0 f_0$ cases (Figs. 5.17b, 5.17c, and 5.17d, respectively) are significantly improved, as anticipated in view of the much higher phase margins of these connections. The step response obtained by lowering the frequency of the first pole in the loop is not shown, since it is indistinguishable from Fig. 5.17d.

Certain features of these step responses are evident from the figures. The peak overshoot exhibited by the amplifier with reduced $a_0 f_0$ is slightly

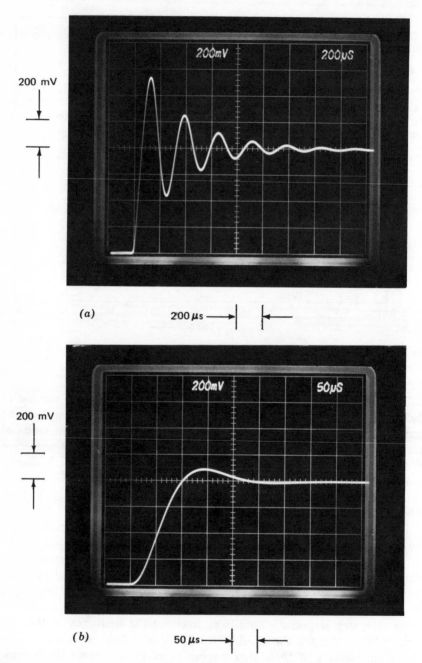

200 mV

200mV   200µS

(a)   200 µs

200 mV

200mV   50µS

(b)   50 µs

Figure 5.17   Response of gain-of-ten amplifier to an 80-mV step. (a) No compensation. (b) Lead compensated. (c) Lag compensated. (d) Lowered $a_0 f_0$. (e) Lead compensation in forward path. (f) Second-order approximation to (c).

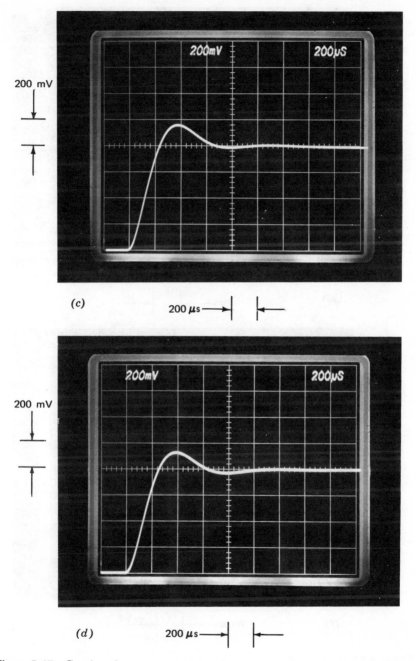

200 mV

200mV   200µS

(c)   200 µs

200 mV

200mV   200µS

(d)   200 µs

Figure 5.17—Continued

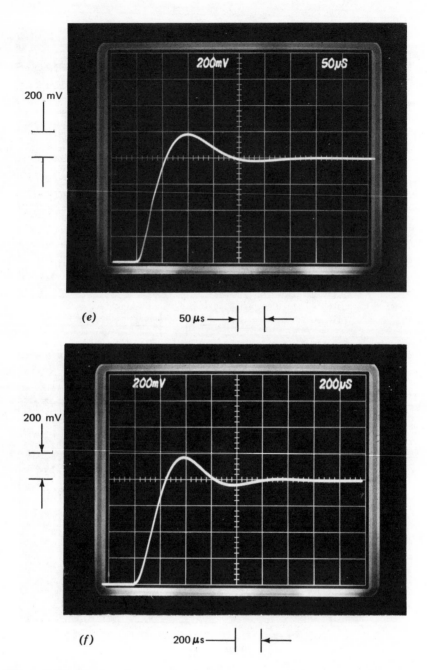

200 mV

(e)

50 µs →

200 mV

(f)

200 µs →

Figure 5.17—Continued

190

less than that of the amplifier with lag compensation, reflecting slightly higher phase margin. Similarly, the rise time of lag-compensated amplifier is very slightly faster, again reflecting the influence of relative phase margin on the performance of these two systems with identical crossover frequencies. The smaller peak overshoot of the lead-compensated system does not imply greater relative stability for this amplifier, but rather occurs because of the influence of the lead network in the feedback path on the ideal closed-loop gain.

Figure 5.17e shows the step response that results if lead compensation is provided in the forward path rather than in the feedback path. Thus the loop transmission for this transient response is identical to that of Fig. 5.17b (Eqn. 5.16), but the feedback path for the system illustrated in Fig. 5.17e is frequency independent. While forward-path lead compensation was prohibited by the problem statement of the earlier examples, Fig. 5.17e provides a more realistic indication of relative stability than does Fig. 5.17b, since Fig. 5.17e is obtained from a system with a frequency-independent ideal gain. The difference between these two systems with identical loop transmissions arises because of differences in the closed-loop zero locations (see Section 4.3.4).

The peak overshoot and relative damping of Figs. 5.17c and 5.17e are virtually identical, demonstrating that, at least for this example, equal values of phase margin result in equal relative stability for the lead- and lag-compensated systems. The rise time of Fig. 5.17e is approximately one-quarter that of Fig. 5.17c, and this ratio is virtually identical to the ratio of the crossover frequencies of the two amplifiers.

The step response of Fig. 5.17f is that of a second-order system with $\zeta = 0.45$ and $\omega_n = 8.5 \times 10^3$ radians per second. These values were obtained using Fig. 4.26a to determine a second-order approximating system to the lag-compensated amplifier. The similarity of Figs. 5.17c and 5.17f is another example of the accuracy that is frequently obtained when complex systems are approximated by first- or second-order ones. The loop transmission for the lag-compensated system (Eqn. 5.17) includes four poles and one zero. However, this quantity has only a single-pole roll off between 6.7×10^2 radians per second and the crossover frequency, with a second pole in the vicinity of crossover. It can thus be well approximated as a system with two widely separated poles, the model from which Fig. 4.26 was developed.

5.2.6 Related Considerations

Several additional comments concerning the relative benefits of different series compensation methods are in order. The evaluation of performance in the previous example seems to imply advantages for lead compensation. The lead-compensated amplifier appears superior if desensitivity at various frequencies, error-coefficient magnitude, or speed of transient response is used as the indicator of performance. Furthermore, if the lead transfer function is included in the feedback path, the amplifier exhibits better-damped transient responses than can be obtained from other types of compensation selected to yield equivalent phase margin. The advantages associated with lead compensation primarily reflect the higher value for cross-over frequency and the correspondingly higher closed-loop bandwidth that is frequently possible with this method. It should be emphasized, however, that bandwidth in excess of requirements usually deteriorates overall performance. Larger bandwidth increases the noise susceptibility of an amplifier and frequently leads to greater stability problems because of stray inductance or capacitance.

Lead compensation usually aggravates the stability problem if the loop also includes elements that provide large negative phase shift over a wide frequency range without a corresponding magnitude attenuation. (While the constraints of physical realizability preclude elements that provide positive phase shift without an amplitude increase, the less useful converse described above occurs with distressing frequency.) For example, consider a system that combines a frequency-independent gain in a loop with a τ-second time delay such as that provided by a delay line. The negative of the loop transmission for this system is

$$a(s)f(s) = a_0 e^{-s\tau} \qquad (5.20)$$

The time delay is an element that has a gain magnitude of one at all frequencies and a negative phase shift that is linearly related to frequency. The Nyquist diagram (Fig. 5.18) for this system shows that it is unstable for $a_0 > 1$. The use of lead compensation compounds the problem, since the positive phase shift of the lead network cannot counteract the unlimited negative phase shift of the time delay, while the magnitude increase of the lead function further lowers the maximum low frequency desensitivity consistent with stable operation.

The correct approach is to use a dominant pole to decrease the magnitude of the loop transmission before the phase shift of time delay becomes excessive. The limiting case of an integrator (pole at the origin) works well, and this modification results in

$$a(s)f(s) = \frac{a_0}{s} e^{-s\tau} \qquad (5.21)$$

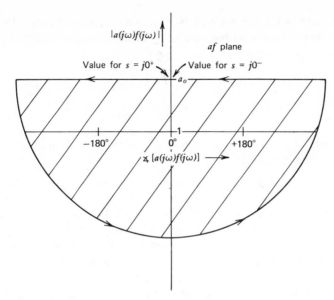

Figure 5.18 Nyquist test for $a(s)f(s) = a_0 e^{-s\tau}$.

The desensitivity of this function is infinite at d-c. The reader should convince himself that the system is absolutely stable for any positive value of $a_0 < \pi/2\tau$, and that at least 45° of phase margin is obtained with positive $a_0 < \pi/4\tau$.

The use of lag compensation introduces a type of error that compromises its value in some applications. If the step response of a lag-compensated amplifier is examined in sufficient detail, it is often found to include a long time-constant, small-amplitude "tail," which may increase inordinately the time required to settle to a small fraction of final value. Similarly, while the error coefficient e_1 may be quite small, the time required for the ramp error to reach its steady-state value may seem incompatible with the amplifier crossover frequency.

As an aid to understanding this problem, consider a system with $f(s) = 1$ and

$$a(s) = \frac{1000(0.1s + 1)}{s(s + 1)} \qquad (5.22)$$

This transfer function is an idealized representation of a system that combines a single dominant pole with lag compensation to improve desensitivity. The zero of the lag network is located a factor of 10 below the cross-over frequency. The closed-loop transfer function is

$$A(s) = \frac{a(s)}{1 + a(s)f(s)} = \frac{(0.1s + 1)}{10^{-3}s^2 + 0.101s + 1}$$

$$= \frac{(0.1s + 1)}{(0.09s + 1)(0.011s + 1)} \quad (5.23)$$

The response of this system to a unit step is easily evaluated via Laplace techniques, with the result

$$v_o(t) = 1 - 1.126e^{-t/0.011} + 0.126e^{-t/0.09} \quad (5.24)$$

This step response reaches 10% of final value in 0.02 second, a reasonable value in view of the 100 radian per second crossover frequency of the system. However, the time required to reach 1% of final value is 0.23 second because of the final term in Eqn. 5.24. Note that if $a(s)$ is changed to $100/s$, a transfer function with the same unity-gain frequency as Eqn. 5.22 and *less* gain magnitude at all frequencies below 10 radians per second, the time required for the system step response to reach 1% of final value is approximately 0.05 second.

The root-locus diagram for the system (Fig. 5.19) clarifies the situation. The system has a closed-loop zero with a corner frequency at 10 radians per second since the zero shown in the diagram is a forward-path singularity. The feedback forces one closed-loop pole close to this zero. The resultant closely spaced pole-zero doublet adds a long-time-constant tail

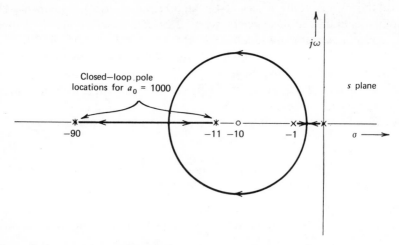

Figure 5.19   Root-locus diagram for $a(s)f(s) = a_0(0.1s + 1)/[s(s + 1)]$.

to the otherwise well-behaved system transient response. The reader should recall that it is precisely this type of doublet that deteriorates the step response of a poorly compensated oscilloscope probe. Since linear system relationships require that the ramp response be the integral of the step response, the time required for the ramp error to reach final value is similarly delayed.

Similar calculations show that as the lag transfer function is moved further below crossover, the amplitude of the tail decreases, but its time constant increases. We conclude that while lag compensation is a powerful technique for improving desensitivity, it must be used with care when the time required for the step response to settle to a small fraction of its final value or the time required for the ramp error to reach final value is constrained.

It should be emphasized that a closed-loop pole will generally be located close to any open-loop zero with a break frequency below the crossover frequency. Thus the type of tail associated with lag compensation can also result with, for example, lead compensation that often includes a zero below crossover. The performance difference results because the zero and the closed-loop pole that approaches it to form a doublet are usually located

Table 5.1   Comparison of Series-Compensating Methods

Type	Special Considerations	Advantages	Disadvantages
Reduced $a_0 f_0$		Simplicity.	Lowest desensitivity.
Create dominant pole	Lower the frequency of the existing dominant pole if possible. Locate at the output of a regulator.	Can improve noise immunity of system. Usually the type of choise for a regulator.	Lowers bandwidth.
Lag	Locate well below crossover frequency.	Better desensitivity than either of above.	May add undesirable "tail" to transient response.
Lead	Locate zero near crossover frequency.	Greatest desensitivity. Lowest error coefficients. Fastest transient response.	Increases sensitivity to noise. Cannot be used with fixed elements that contribute excessive negative phase shift.

close to the crossover frequency for lead compensation. Thus the decay time of the resultant tail, which is determined by the closed-loop pole in question, does not greatly lengthen the settling time of the system.

It is difficult to develop generalized rules concerning compensation, since the proper approach is highly dependent on the fixed elements included in the loop, on the types of inputs anticipated, on the performance criterion chosen, and on numerous other factors. In spite of this reservation, Table 5.1 is an attempt to summarize the most important features of the four types of series compensation described in this section.

5.3 FEEDBACK COMPENSATION

Series compensation is accomplished by adding a cascaded element to a single-loop feedback system. Feedback compensation is implemented by adding a feedback element which creates a two-loop system. One possible topology is illustrated in Fig. 5.20. The closed-loop transfer function for this system is

$$\frac{V_o}{V_i} = \frac{a_1 a_2/(1 + a_2 f_2)}{1 + a_1 a_2 f_1/(1 + a_2 f_2)} \tag{5.25}$$

A series-compensated system with a feedback element identical to the major-loop feedback element of Fig. 5.20 is shown in Fig. 5.21. The two feedback elements are identical since it is assumed that the same ideal closed-loop transfer function is required from the two systems. The closed-loop transfer function for the series-compensated system is

$$\frac{V_o}{V_i} = \frac{a_3 a_4}{1 + a_3 a_4 f_1} \tag{5.26}$$

The closed-loop transfer functions of the feedback- and series-compensated systems will be equal if f_2 is selected so that

$$a_3 = \frac{a_1 a_2}{(1 + a_2 f_2) a_4} \tag{5.27a}$$

or

$$f_2 = \frac{a_1 a_2 - a_3 a_4}{a_2 a_3 a_4} \tag{5.27b}$$

The above analysis suggests that one way to select appropriate feedback compensation is first to determine the series compensation that yields acceptable performance and then convert to equivalent feedback compensation. In practice, this approach is normally *not* used, but rather the series compensation is determined to exploit potential advantages of this method.

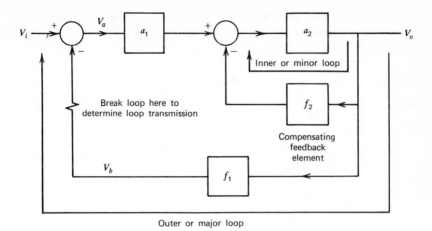

Figure 5.20   Topology for feedback compensation.

We shall see that if an operational amplifier is designed to accept feedback compensation, the use of this technique often results in performance superior to that which can be achieved with series compensation. The frequent advantage of feedback compensation is not a consequence of any error in the mathematics that led to the equivalence of Eqn. 5.27 but instead is a result of practical factors that do not enter into these calculations. For example, the compensating network required to obtain specified closed-loop performance is often easier to determine and implement and may be less sensitive to variations in other amplifier parameters in the case of a feedback-compensated amplifier. Similarly, problems associated with nonlinearities and noise are often accentuated by series compensation, yet may actually be reduced by feedback compensation.

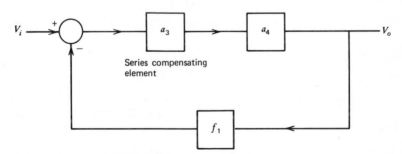

Figure 5.21   Series-compensated system.

The approach to finding the type of feedback compensation that should be used in a given application is to consider the negative of the loop transmission for the system of Fig. 5.20. This quantity is

$$\frac{V_b}{V_a} = a_1 f_1 \frac{a_2}{1 + a_2 f_2} \tag{5.28}$$

If the inner loop is stable (i.e., if $1 + a_2 f_2$ has no zeros in the right half of the s plane), then

$$\frac{V_b(j\omega)}{V_a(j\omega)} \simeq \frac{a_1(j\omega)f_1(j\omega)}{f_2(j\omega)} \qquad |\, a_2(j\omega)f_2(j\omega)\,| \gg 1 \tag{5.29a}$$

and

$$\frac{V_b(j\omega)}{V_a(j\omega)} \simeq a_1(j\omega)f_1(j\omega)a_2(j\omega) \qquad |\, a_2(j\omega)f_2(j\omega)\,| \ll 1 \tag{5.29b}$$

In practice, system parameters are frequently selected so that the magnitude of the transmission of the minor loop is large at frequencies where the magnitude of the major loop transmission is close to one. The approximation of Eqn. 5.29a can then be used to determine a value for f_2 that insures stability for the system.

A simple example of feedback compensation is provided by the operational-amplifier model shown in Fig. 5.22a. The model is an idealization of a common amplifier topology that will be investigated in detail in subsequent sections. The amplifier modeled includes a first stage with wide bandwidth compared to the rest of the circuit driving into a second stage that has relatively low input impedance and that dominates the uncompensated dynamics of the amplifier. The compensation is provided by a two-port network that is connected around the second stage and that forms a minor loop. This network is constrained to be passive. A block diagram for the amplifier is shown in Fig. 5.22b. The quantity Y_c is the short-circuit transfer admittance of the compensating network, I_n/V_n.[2]

If no compensation is used, the open-loop transfer function for the amplifier is

$$\frac{V_o(s)}{V_i(s)} = -\frac{10^6}{(10^{-3}s + 1)^2} \tag{5.30}$$

If a wire is connected from the output of the amplifier back to its input, creating a major loop with $f = 1$, the phase margin of the resultant system is approximately $0.12°$.

---

[2] The convention used to define Y_c is at variance with normal two-port notation, which would change the reference direction for I_n. This form is used since it results in fewer minus signs in subsequent equations.

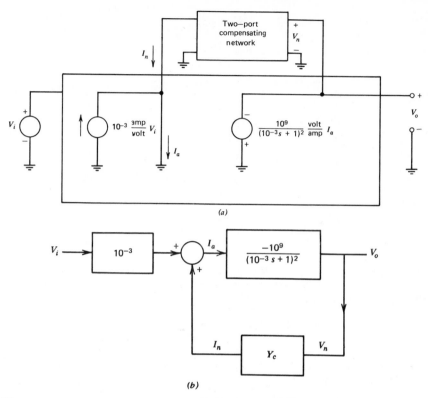

Figure 5.22   Operational amplifier. (*a*) Model. (*b*) Block diagram.

When feedback compensation is included, the block diagram shows that the amplifier transfer function is

$$\frac{V_o(s)}{V_i(s)} = \frac{-10^6/(10^{-3}s + 1)^2}{1 + 10^9 Y_c/(10^{-3}s + 1)^2} \tag{5.31}$$

One way to improve the phase margin of this amplifier when used in a feedback connection is to make $V_o(s)/V_i(s)$ dominated by a single pole. Equation 5.31 shows that

$$\frac{V_o(j\omega)}{V_i(j\omega)} \simeq \frac{-10^{-3}}{Y_c(j\omega)} \quad \text{when} \quad \left| \frac{10^9 Y_c(j\omega)}{(10^{-3}j\omega + 1)^2} \right| \gg 1 \tag{5.32}$$

If a single capacitor C is used for the compensating network, $Y_c = Cs$ and

$$\frac{V_o(j\omega)}{V_i(j\omega)} \simeq \frac{-10^{-3}}{j\omega C} \tag{5.33}$$

for all frequencies such that

$$\left| \frac{10^9 Cj\omega}{(10^{-3}j\omega + 1)^2} \right| \gg 1$$

The exact expression for the amplifier open-loop transfer function with this compensation is

$$\frac{V_o(s)}{V_i(s)} = \frac{-10^6/(10^{-3}s + 1)^2}{1 + 10^9 Cs/(10^{-3}s + 1)^2}$$

$$= \frac{-10^6}{10^{-6}s^2 + (2 \times 10^{-3} + 10^9 C)s + 1} \quad (5.34)$$

If an 840-pF capacitor is used for C, the transfer function becomes

$$\frac{V_o(s)}{V_i(s)} = \frac{-10^6}{(0.84s + 1)(1.19 \times 10^{-6}s + 1)} \quad (5.35)$$

and a phase margin of at least 45° is assured for frequency-independent feedback with any magnitude less than one applied around the amplifier. With this value of compensating feedback element,

$$- \frac{V_o(j\omega)}{V_i(j\omega)} \simeq \frac{1.19 \times 10^6}{j\omega} = \frac{10^{-3}}{Cj\omega} = \frac{10^{-3}}{Y_c(j\omega)} \quad (5.36)$$

at any frequency between 1.19 radians per second and 0.84×10^6 radians per second. The two bounding frequencies are those at which the magnitude of the compensating loop transmission is one. The essential point is that minor-loop feedback controls the transfer function of the amplifier over nearly six decades of frequency. We also note that even though a dominant pole has been created by means of feedback compensation, the unity-gain frequency of the compensated amplifier (approximately 8×10^5 radians per second) remains close to the uncompensated value of 10^6 radians per second.

Feedback compensation is a powerful and frequently used compensating technique for modern operational amplifiers. Several examples of this type of compensation will be provided after the circuit topologies of representative amplifiers have been described.

PROBLEMS

P5.1

An operational amplifier has an open-loop transfer function

$$a(s) = \frac{2 \times 10^5}{(0.1s + 1)(10^{-5}s + 1)^2}$$

Design a connection that uses this amplifier to provide an ideal gain of -10. Include provision to lower the magnitude of the loop transmission so that the overshoot in response to a unit step is 10%. You may use the curves of Fig. 4.26 as an aid to determining the required attenuation.

P5.2

An operational amplifier is connected as shown in Fig. 5.23a. The value of α is adjusted to control the stability of the connection. Assume that noise associated with the amplifier can be modeled as shown in Fig. 5.23b. Evaluate the noise at the amplifier output as a function of α, neglecting loading at the input and the output of the amplifier. Note that an increase in the noise at the amplifier output implies a decrease in signal-to-noise ratio, since the gain from input to output is essentially independent of α.

P5.3

A certain feedback amplifier can be modeled as shown in Fig. 5.24. You may assume that the operational amplifier included in this diagram is ideal. Select a value for the capacitor C that results in a system phase margin of 45°.

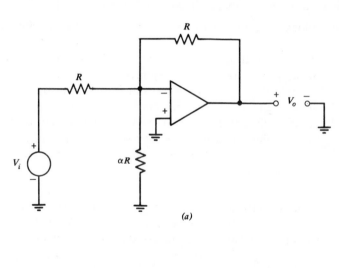

(a)

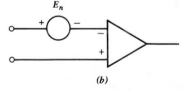

(b)

Figure 5.23 Evaluation of noise at the output of an inverting amplifier. (a) Inverter connection. (b) Method for modeling noise at amplifier input.

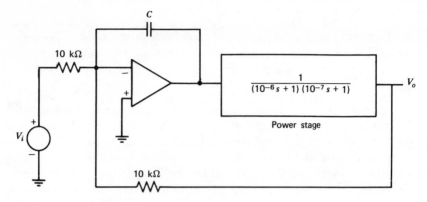

Figure 5.24 Feedback system with dominant pole.

P5.4

A speed-control system combines a high-power operational amplifier in a loop with a motor and a tachometer as shown in Fig. 5.25. The tachometer provides a voltage proportional to output shaft velocity, and this voltage is used as the feedback signal to effect speed control.

(a) Draw a block diagram for this system that includes the effects of the disturbing torque.

(b) Determine compensating component values (R and C) as a function of J_L so that the system loop transmission is $-100/s$.

(c) Show that, with this type of loop transmission, the steady-state output velocity is independent of any constant load torque.

(d) Use an error-coefficient analysis to show that the system is less sensitive to time-varying disturbing torques when larger values of J_L are used. Assume that R and C are changed with J_L to maintain the loop transmission indicated in part b.

P5.5

Show that the network illustrated in Fig. 5.26 can be used to combine a lag transfer function with a lead transfer function located at a higher frequency. Determine network parameters that will result in the transfer function

$$\frac{V_o(s)}{V_i(s)} = \frac{(0.1s + 1)(10^{-2}s + 1)}{(s + 1)(10^{-3}s + 1)}$$

P5.6

The loop transmission of a feedback system can be approximated as

$$L(s) = -\frac{10^6}{s^2}$$

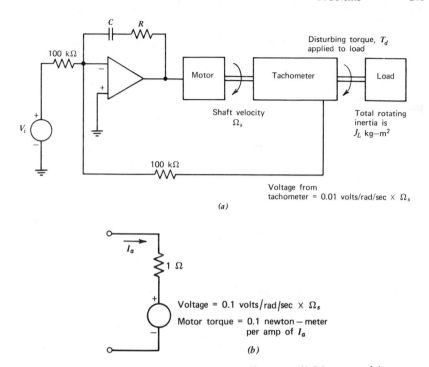

(b)

Figure 5.25 Speed-control system. (*a*) System diagram. (*b*) Motor model.

in the vicinity of the unity-gain frequency. Assume that a lead transfer function (Eqn. 5.4) with a value of $\alpha = 10$ can be added to the loop transmission. How should the transfer function be located to maximize phase margin? What values of phase margin and crossover frequency result?

P5.7
Use a block diagram to show that a lag transfer function can be introduced into the loop transmission of the gain-of-ten amplifier (Fig. 5.9) by shunting the R-valued resistor with an appropriate network.

(a) Choose network parameters so that the system loop transmission is given by Eqn. 5.15.
(b) Find the closed-loop transfer function and plot the closed-loop step response for the gain-of-ten amplifier using values found in part *a*, *assuming that the operational-amplifier characteristics are ideal.*
(c) Estimate the closed-loop step response for this connection assuming that the amplifier open-loop transfer function is as given by Eqn. 5.10.
(d) Compare the performance of the lag-compensated system developed in this problem with that shown in Fig. 5.13 considering both the sta-

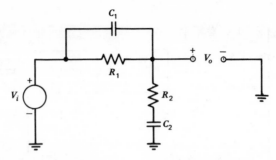

Figure 5.26 Lag-Lead network.

bility and the ideal closed-loop transfer function of the two connections.

P5.8

It was mentioned in Section 5.2.4 that alternative compensation possibilities for the gain-of-ten amplifier include lowering the magnitude of the loop transmission at all frequencies by a factor of 6.2 and lowering the location of the lowest-frequency pole in the loop transfer function by a factor of 6.2 by selecting appropriate lag-network parameters.

(a) Determine topologies and component values to implement both of these compensation schemes.
(b) Draw loop-transmission Bode plots for these two methods of compensation.
(c) Compare the relative stability produced by these methods with that provided by the lag compensation described in Section 5.2.4.

P5.9

The negative of the loop transmission for the lead-compensated gain-of-ten amplifier described in Section 5.2.4 is

$$a(s)f(s) = \frac{5 \times 10^4(10\tau s + 1)}{(s + 1)(10^{-4}s + 1)(10^{-5}s + 1)(\tau s + 1)}$$

where τ is determined by the resistor and capacitor values used in the feedback network (see Eqn. 5.13). Use root contours to evaluate the stability of the gain-of-ten amplifier as a function of the parameter τ. Find the value of τ that maximizes the damping ratio of the dominant pole pair. *Note.* Since it is necessary to factor third- and fourth-order polynomials in order to complete this problem, the use of machine computation is suggested. Numerical calculations are also suggested to evaluate the maximum damping ratio.

P5.10

The negative of the loop transmission for the lag-compensated amplifier is

$$a(s)f(s) = \frac{5 \times 10^4(\tau s + 1)}{(s + 1)(10^{-4}s + 1)(10^{-5}s + 1)(\alpha\tau s + 1)}$$

It was shown in Section 5.2.4 that reasonable stability results for $\alpha = 6.2$ and a value of τ that locates the lag-function zero a factor of 10 below crossover. Use root contours to evaluate stability as a function of the zero location $(1/\tau)$ for $\alpha = 6.2$. The note concerning the advisability of machine computation mentioned in Problem P5.9 applies to this calculation as well.

P5.11

Determine the first three error coefficients for the four loop transmissions of the gain-of-ten amplifier described by Eqns. 5.16 through 5.19. Assume that the lead compensation is obtained in the feedback path (see Section 5.2.4) while all other compensations can be considered to be located in the forward path.

P5.12

A feedback system includes a factor

$$\frac{(s^2/12) - (s/2) + 1}{(s^2/12) + (s/2) + 1}$$

in its loop transmission.

Assume that you have complete freedom in the choice of d-c loop-transmission magnitude and the selection of additional singularities in the loop transmission. Determine the type of compensation that will maximize the desensitivity of this system.

P5.13

Calculate the settling time (to 1% of final value for a step input) for the gain-of-ten amplifier with lag compensation (Eqn. 5.15). Contrast this value with that of a first-order system with an identical crossover frequency.

P5.14

A model for an operational amplifier using minor-loop compensation is shown in block-diagram form in Fig. 5.27.

(a) Assume that the series compensating element has a transfer function $a_c(s) = 1$. Find values for b and τ such that a major loop formed by feeding V_o directly back to V_i will have a crossover frequency of 10^3 radians per second, approximately 55° of phase margin, and maximum desensitivity at frequencies below crossover subject to these constraints.

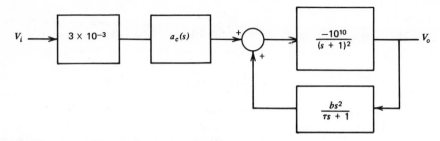

Figure 5.27   Operational-amplifier model.

Draw an open-loop Bode plot for the amplifier with these values for b and τ.

(b) Now assume that $b = 0$. Can you find a value for $a_c(s)$ that results in the same asymptotic open-loop magnitude characteristics as you obtained in part a, subject to the constraint that $|\, a_c(j\omega)\,| \leq 1$ for all ω?

P5.15

This problem includes a laboratory portion that can be performed with commonly available test equipment and that will give you experience compensating a system with well-defined dynamics. The experimental vehicle is the circuit shown in Fig. 5.28, which gives quite repeatable operational-amplifier-like characteristics. The suggested experiments use the configuration at relatively low frequencies, so that the inevitable stray circuit elements have little effect on the measured performance.

The dynamics of the circuit should first be standardized. Connect it as an inverting amplifier as shown in Fig. 5.29.

Select the capacitor C connected between pins 1 and 8 of the LM301A so that the configuration is just on the verge of instability. An estimated value should be around 5000 pF. Please remember that the amplifier reacts very poorly (usually by dying) if pins 1 or 5 are shorted to almost any potential.

Note. The assumptions required for linear analysis are severely compromised if the peak-to-peak magnitude of the input signal exceeds approximately 50 mV. It is also necessary to have the driving source impedance low in this and other connections. A resistive divider attenuating the signal-generator output and located close to the amplifier is suggested.

After this standardization, it is claimed that if the loads applied to the amplifier are much higher than the output impedance of the network involving the 0.15 μF capacitor, etc., we can approximate $a(s)$ as

$$a(s) \simeq \frac{5 \times 10^4}{(s + 1)(10^{-3}s + 1)(10^{-4}s + 1)}$$

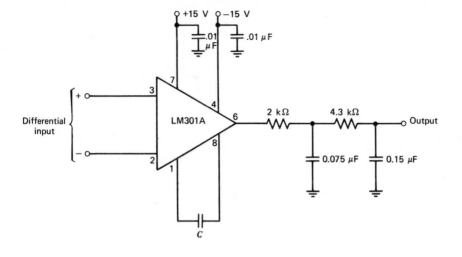

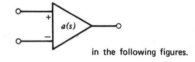

Note. This complete circuit will be denoted as

in the following figures.

Figure 5.28 Amplifier with controlled dynamics. Pin numbers are for TO-99 and minidip packages.

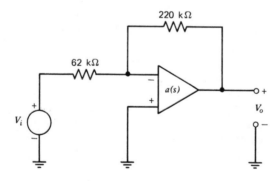

Figure 5.29 Inverting configuration.

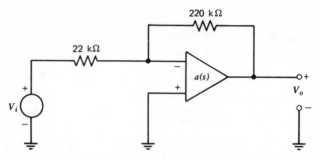

Figure 5.30   Inverting gain-of-ten amplifier.

for purposes of stability analysis. This transfer function is not unique and, in general, functions of the form

$$a(s) = \frac{5 \times 10^4 \tau}{(\tau s + 1)(10^{-3}s + 1)(10^{-4}s + 1)}$$

will yield equivalent results in your analysis providing $\tau \gg 10^{-3}$ seconds.

Supply a convincing argument why the above family of transfer functions properly represents the operational amplifier that you have just brought to the verge of oscillation. Note that simply showing the two given expressions are equivalent is *not* sufficient. You must show why they can be used to analyze the standardized circuit.

Use a Bode plot to determine the phase margin of the connection shown in Fig. 5.30 when the standardized amplifier is used. Predict a value for M_p based on the phase margin, and compare your prediction with measured results.

You are to compensate the system to improve its phase margin to 60° by reducing $a_0 f_0$ and by using lag and lead compensating techniques. You may not change the value of C or elements in the network connected to the output of the LM301A, nor load the network unreasonably to implement compensation.

Analytically determine the topology and element values you will use for each of the three forms of compensation. It may not be possible to meet the phase-margin objective using lead compensation alone; if you find this to be the case, you may reduce $a_0 f_0$ slightly so that the design goal can be achieved.

Compensate the amplifier in the laboratory and convince yourself that the step responses you measure are reasonable for systems with 60° of phase margin. Also correlate the rise times of the responses with your predicted values for crossover frequencies.

CHAPTER VI

NONLINEAR SYSTEMS

6.1 INTRODUCTION

The techniques discussed up to this point have all been developed for the analysis of linear systems. While the computational advantages of the assumption of linearity are legion, this assumption is often unrealistic, since virtually all physical systems are nonlinear when examined in sufficient detail. In addition to systems where the nonlinearity represents an undesired effect, there are many systems that are intentionally designed for or to exploit nonlinear performance characteristics.

Analytic difficulties arise because most of the methods we have learned are dependent on the principle of superposition, and nonlinear systems violate this condition. Time-domain methods such as convolution and frequency-domain methods based on transforms usually cannot be applied directly to nonlinear systems. Similarly, the blocks in a nonlinear block diagram cannot be shuffled with impunity. The absolute stability question may no longer have a binary answer, since nonlinear systems can be stable for certain classes of inputs and unstable for others.

The difficulty of effectively handling nonlinear differential equations is evidenced by the fact that the few equations we know how to solve are often named for the solvers. While considerable present and past research has been devoted to this area, it is clear that much work remains to be done. For many nonlinear systems the only methods that yield useful results involve experimental evaluation or machine computation.

This chapter describes two methods that can be used to determine the response or stability of certain types of nonlinear systems. The methods, while certainly not suited to the analysis of general nonlinear systems, are relatively easy to apply to many physical systems. Since they represent straightforward extensions of previously studied linear techniques, the insight characteristic of linear-system analysis is often retained.

6.2 LINEARIZATION

One direct and powerful method for the analysis of nonlinear systems involves approximation of the actual system by a linear one. If the approximating system is correctly chosen, it accurately predicts the behavior of the actual system over some restricted range of signal levels.

This technique of linearization based on a tangent approximation to a nonlinear relationship is familiar to electrical engineers, since it is used to model many electronic devices. For example, the bipolar transistor is a highly nonlinear element. In order to develop a linear-region model such as the hybrid-pi model to predict the circuit behavior of this device, the relationships between base-to-emitter voltage and collector and base current are linearized. Similarly, if the dynamic performance of the transistor is of interest, linearized capacitances that relate incremental changes in stored charge to incremental changes in terminal voltages are included in the model.

6.2.1   The Approximating Function

The tangent approximation is based on the use of a Taylor's series estimation of the function of interest. In general, it is assumed that the output variable of an element is a function of N input variables

$$v_O = F(v_{I1}, v_{I2}, \ldots, v_{IN}) \tag{6.1}$$

The output variable is expressed for small variation $v_{i1}, v_{i2}, \ldots, v_{iN}$ about input-variable operating points $V_{I1}, V_{I2}, \ldots, V_{IN}$ by noting that

$$v_O = V_O + v_o = F(V_{I1}, V_{I2}, \ldots, V_{IN})$$

$$+ \sum_{j=1}^{N} \left. \frac{\partial V_O}{\partial V_{Ij}} \right|_{V_{I1}, V_{I2}, \ldots, V_{IN}} v_{ij}$$

$$+ \frac{1}{2!} \sum_{k,l=1}^{N} \left. \frac{\partial^2 V_O}{\partial V_{Ik} \partial V_{Il}} \right|_{V_{I1}, V_{I2}, \ldots, V_{IN}} v_{ik} v_{il} + \cdots + \tag{6.2}$$

(Recall that the variable and subscript notation used indicates that v_O is a total variable, V_O is its operating-point value, and v_o its incremental component.)

The expansion of Eqn. 6.2 is valid at any operating point where the derivatives exist.

Since the various derivatives are assumed bounded, the function can be adequately approximated by the first-order terms over some restricted range of inputs. Thus

$$V_O + v_o \simeq F(V_{I1}, V_{I2}, \ldots, V_{IN}) + \sum_{j=1}^{N} \left. \frac{\partial V_O}{\partial V_{Ij}} \right|_{V_{I1}, V_{I2}, \ldots, V_{IN}} v_{ij} \tag{6.3}$$

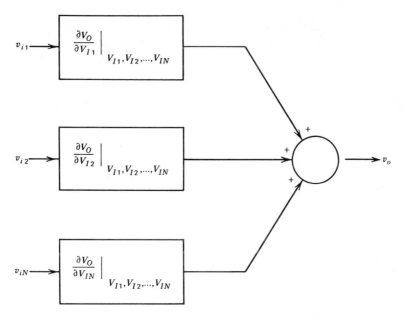

Figure 6.1   Linearized block diagram.

The constant terms in Eqn. 6.3 are substracted out, leaving

$$v_o \simeq \sum_{j=1}^{N} \frac{\partial V_o}{\partial V_{Ij}} \bigg|_{V_{I1}, V_{I2}, \ldots, V_{IN}} v_{ij} \qquad (6.4)$$

Equation 6.4 can be used to develop linear-system equations that relate incremental rather than total variables and that approximate the incremental behavior of the actual system over some restricted range of operation. A block diagram of the relationships implied by Eqn. 6.4 is shown in Fig. 6.1.

6.2.2   Analysis of an Analog Divider

Certain types of signal-processing operations require that the ratio of two analog variables be determined, and this function can be performed by a divider. Division is frequently accomplished by applying feedback around an analog multiplier, and several commercially available multipliers can be converted to dividers by making appropriate jumpered connections to the output amplifier included in these units. A possible divider connection of this type is shown in Fig. 6.2a.

The multiplier scale factor shown in this figure is commonly used since it provides a full-scale output of 10 volts for two 10-volt input signals. It

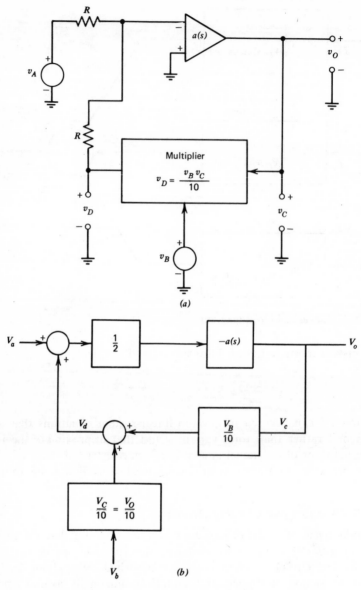

Figure 6.2 Analog divider. (*a*) Circuit. (*b*) Linearized block diagram.

is assumed that the multiplying element itself has no dynamics and thus the speed of response of the system is determined by the operational amplifier.

The ideal relationship between input and output variables can easily be determined using the virtual-ground method. If the current at the inverting input of the amplifier is small and if the magnitude of the loop transmission is high enough so that the voltage at this terminal is negligible, the circuit relationships are

$$v_A + v_D = 0 \tag{6.5}$$

and

$$v_D = \frac{v_B v_C}{10} = \frac{v_B v_O}{10} \tag{6.6}$$

Solving Eqns. 6.5 and 6.6 for v_O in terms of v_A and v_B yields

$$v_O = -\frac{10 v_A}{v_B} \tag{6.7}$$

System dynamics are determined by linearizing the multiplying-element characteristics. Applying Eqn. 6.3 to the variables of Eqn. 6.6 shows that

$$V_D + v_d \simeq \frac{V_B V_C}{10} + \frac{V_B v_c}{10} + \frac{V_C v_b}{10} \tag{6.8}$$

The incremental portion of this equation is

$$v_d = \frac{V_B v_c}{10} + \frac{V_C v_b}{10} \tag{6.9}$$

This relationship combined with other circuit constraints (assuming the operational amplifier has infinite input impedance and zero output impedance) is used to develop the incremental block diagram shown in Fig. 6.2b.

The incremental dependence of V_o on V_a, assuming that v_B is constant, is

$$\frac{V_o(s)}{V_a(s)} = \frac{-a(s)/2}{1 + V_B a(s)/20} \tag{6.10}$$

If the operational-amplifier transfer function is approximately single pole so that

$$a(s) = \frac{a_0}{\tau s + 1} \tag{6.11}$$

and a_0 is very large, Eqn. 6.10 reduces to

$$\frac{V_o(s)}{V_a(s)} \simeq \frac{-10/V_B}{(20\tau/V_B a_0)s + 1} \tag{6.12}$$

Several features are evident from this transfer function. First, if V_B is negative, the system is unstable. Second, the incremental step response of the system is first order, with a time constant of $20\tau/V_B a_0$ seconds. These features indicate two of the many ways that nonlinearities can affect the performance of a system. The stability of the circuit depends on an input-signal level. Furthermore, if V_B is positive, the transient response of the circuit becomes faster with increasing V_B, since the loop transmission depends on the value of this input.

6.2.3   A Magnetic-Suspension System

An electromechanical system that provides a second example of linearized analysis is illustrated in Fig. 6.3. The purpose of the system is to suspend an iron ball in the field of an electromagnet. Only vertical motion of the ball is considered.

In order to suspend the ball it is necessary to cancel the downward gravitational force on the ball with an upward force produced by the magnet. It is clear that stabilization with constant current is impossible, since while a value of x_B for which there is no net force on the ball exists, a small deviation from this position changes the magnetic force in such a way as to accelerate the ball further from equilibrium. This effect can be cancelled by appropriately controlling the magnet current as a function of measured ball position.

For certain geometries and with appropriate choice of the reference position for x_B, the magnetic force f_M exerted on the ball in an upward direction is

$$f_M = \frac{C i_M^2}{x_B^2} \tag{6.13}$$

where C is a constant.

Assuming incremental changes x_b and i_m about operating-point values X_B and I_M, respectively,

$$f_M = F_M + f_m = \frac{C I_M^2}{X_B^2} + \frac{2 C I_M}{X_B^2} i_m - \frac{2 C I_M^2}{X_B^3} x_b$$

$$+ \text{ higher-order terms} \tag{6.14}$$

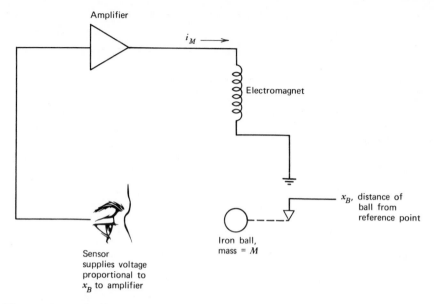

Figure 6.3 Magnetic-suspension system.

The equation of motion of the ball is

$$\frac{Md^2x_B}{dt^2} = Mg - f_M \tag{6.15}$$

where g is the acceleration of gravity. Equilibrium or operating-point values are selected so that

$$Mg = \frac{CI_M^2}{X_B^2} \tag{6.16}$$

When we combine Eqns. 6.14 and 6.15 and assume operation about the equilibrium point, the linearized relationship among incremental variables becomes

$$\frac{M\,d^2x_b}{dt^2} - \frac{2CI_M^2}{X_B^3}\,x_b = -\frac{2CI_M}{X_B^2}\,i_m \tag{6.17}$$

Equation 6.17 is transformed and rearranged as

$$\frac{s^2 X_b(s)}{k^2} - X_b(s) = X_b(s)\left(\frac{s}{k}+1\right)\left(\frac{s}{k}-1\right) = -\frac{X_B}{I_M}I_m(s) \tag{6.18}$$

where $k^2 = 2CI_M^2/MX_B^3$.

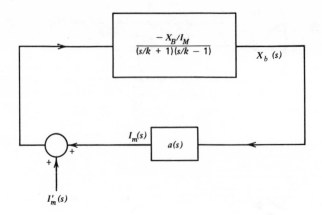

Figure 6.4 Linearized block diagram for system of Fig. 6.3.

Feedback is applied to the system by making i_m a linear function of x_b, or

$$I_m(s) = a(s)X_b(s) \qquad (6.19)$$

Equations 6.18 and 6.19 are used to draw the linearized block diagram shown in Fig. 6.4. [The input $I'_m(s)$ is used as a test input later in the analysis.]

The loop transmission for this system

$$L(s) = -\frac{a(s)\dfrac{X_B}{I_M}}{\left(\dfrac{s}{k}+1\right)\left(\dfrac{s}{k}-1\right)} \qquad (6.20)$$

contains a pole in the right-half plane that reflects the fact that the system is unstable in the absence of feedback. A naive attempt at stabilization for this type of system involves cancellation of the right-half-plane pole with a zero of $a(s)$. While such cancellation works when the singularities in question are in the left-half plane, it is doomed to failure in this case. Although the pole could seemingly be removed from the loop transmission by this method,[1] consider the closed-loop transfer function that relates X_b to a disturbance I'_m.

---

[1] Component tolerances preclude exact cancellation in any but a mathematical system.

If $a(s)$ is selected as $a'(s)(s/k - 1)$, this transfer function is

$$\frac{X_b(s)}{I'_m(s)} = \frac{\dfrac{-X_B/I_M}{(s/k + 1)(s/k - 1)}}{1 + \dfrac{a'(s)\,X_B/I_M}{s/k + 1}} \tag{6.21}$$

Equation 6.21 contains a right-half-plane pole implying exponentially growing responses for x_b even though this growth is not observed as a change in i_m.

A satisfactory method for compensating the system can be determined by considering the root-locus diagrams shown in Fig. 6.5. Figure 6.5a is the diagram for frequency-independent feedback with $a(s) = a_0$. As a_0 is increased, the two poles come together and branch out along the imaginary axis. This diagram shows that it is possible to remove the closed-loop pole from the right-half plane if a_0 is appropriately chosen. However, the poles cannot be moved into the left-half plane, and thus the system exhibits undampened oscillatory responses. The system can be stabilized by including a lead transfer function in $a(s)$. It is possible to move all closed-loop poles to the left-half plane for *any* lead-network parameters coupled with a sufficiently high value of a_0. Figure 6.5b illustrates the root trajectories for one possible choice of lead-network singularities.

6.3   DESCRIBING FUNCTIONS

Describing functions provide a method for the analysis of nonlinear systems that is closely related to the linear-system techniques involving Bode or gain-phase plots. It is possible to use this type of analysis to determine if *limit cycles* (constant-amplitude periodic oscillations) are possible for a given system. It is also possible to use describing functions to predict the response of certain nonlinear systems to purely sinusoidal excitation, although this topic is not covered here.[2] Unfortunately, since the frequency response and transient response of nonlinear systems are not directly related, the determination of transient response is not possible via describing functions.

6.3.1   The Derivation of the Describing Function

A describing function describes the behavior of a nonlinear element for purely sinusoidal excitation. Thus the input signal applied to the nonlinear

[2] G. J. Thaler and M. P. Pastel, *Analysis and Design of Nonlinear Feedback Control Systems*, McGraw-Hill, New York, 1962.

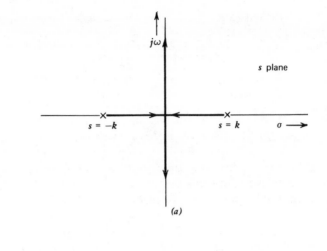

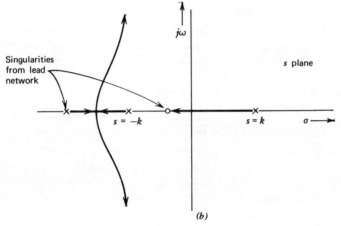

Figure 6.5 Root-locus diagrams for magnetic-suspension system. (*a*) Uncompensated. (*b*) With lead compensation.

element to determine its describing function is

$$v_I = E \sin \omega t \qquad (6.22)$$

If the nonlinearity does not rectify the input (produce a d-c output) and does not introduce subharmonics, the output of the nonlinear element can be expanded in a Fourier series of the form

$$v_O = A_1(E, \omega) \cos \omega t + B_1(E, \omega) \sin \omega t + A_2(E, \omega) \cos 2\omega t$$
$$+ B_2(E, \omega) \sin 2\omega t + \cdots + \qquad (6.23)$$

The describing function for the nonlinear element is defined as

$$G_D(E, \omega) = \frac{\sqrt{A_1^2(E, \omega) + B_1^2(E, \omega)}}{E} \measuredangle \tan^{-1} \frac{A_1(E, \omega)}{B_1(E, \omega)} \qquad (6.24)$$

The describing-function characterization of a nonlinear element parallels the transfer-function characterization of a linear element. If the transfer function of a linear element is evaluated for $s = j\omega$, the magnitude of resulting function of a complex variable is the ratio of the amplitudes of the output and input signals when the element is excited with a sinusoid at a frequency ω. Similarly, the angle of the function is the phase angle between the output and input signals under sinusoidal steady-state conditions. For linear elements these quantities must be independent of the amplitude of excitation.

The describing function indicates the relative amplitude and phase angle of the *fundamental component* of the output of a nonlinear element when the element is excited with a sinusoid. In contrast to the case with linear elements, these quantities can be dependent on the amplitude as well as the frequency of the excitation.

Two examples illustrate the derivation of the describing function for nonlinear elements. Figure 6.6 shows the transfer characteristics of a saturating nonlinearity together with input and output waveforms for sinusoidal excitation. Since the transfer characteristics for this element are not dependent on the dynamics of the input signal, it is clear that the describing function must be frequency independent.

If the input amplitude E is less than E_M,

$$v_O = Kv_I \qquad (6.25)$$

In this case,

$$G_D = K \measuredangle 0° \qquad E < E_M \qquad (6.26)$$

For $E \geq E_M$, the output signal over the interval $0 \leq \omega t \leq \pi$ is

$$v_O = Kv_I \qquad 0 \leq \omega t < \alpha \qquad \text{or} \qquad \pi - \alpha < \omega t \leq \pi \qquad (6.27a)$$

$$v_O = KE_M \qquad \alpha \leq \omega t \leq \pi - \alpha \qquad (6.27b)$$

where

$$\alpha = \sin^{-1} \frac{E_M}{E}$$

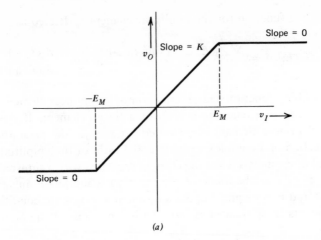

(a)

Figure 6.6 Relationships for a saturating nonlinearity. (a) Transfer characteristics for saturating element. (b) Input and output waveforms for sinusoidal excitation.

The coefficients A_1 and B_1 are in this case,

$$A_1 = \frac{2}{\pi} \int_0^\alpha KE \sin \omega t \cos \omega t \, d\omega t + \frac{2}{\pi} \int_\alpha^{\pi-\alpha} KE_M \cos \omega t \, d\omega t$$

$$+ \frac{2}{\pi} \int_{\pi-\alpha}^\pi KE \sin \omega t \cos \omega t \, d\omega t = 0 \qquad (6.28)$$

$$B_1 = \frac{2}{\pi} \int_0^\alpha KE \sin^2 \omega t \, d\omega t + \frac{2}{\pi} \int_\alpha^{\pi-\alpha} KE_M \sin \omega t \, d\omega t$$

$$+ \frac{2}{\pi} \int_{\pi-\alpha}^\pi KE \sin^2 \omega t \, d\omega t$$

$$= \frac{2KE}{\pi} \left[\sin^{-1} \frac{E_M}{E} + \frac{E_M}{E} \sqrt{1 - \left(\frac{E_M}{E}\right)^2} \right] \qquad (6.29)$$

Using Eqn. 6.24, we obtain

$$G_D(E) = K \angle 0° \qquad E \le E_M \qquad (6.30a)$$

$$G_D(E) = \frac{2K}{\pi} \left(\sin^{-1} R + R \sqrt{1 - R^2} \right) \angle 0° \qquad E > E_M \quad (6.30b)$$

where $R = E_M/E$.

The transfer characteristics of an element with hysteresis, such as a Schmitt trigger or a relay, are shown in Fig. 6.7a. The memory associated

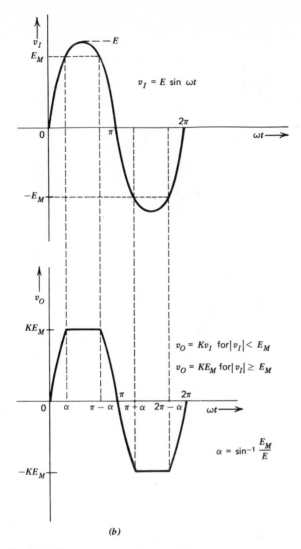

$$v_I = E \sin \omega t$$

$$v_O = Kv_I \text{ for} |v_I| < E_M$$

$$v_O = KE_M \text{ for} |v_I| \geq E_M$$

$$\alpha = \sin^{-1} \frac{E_M}{E}$$

(b)

Figure 6.6—Continued

with this type of element produces a phase shift between the fundamental component of the output and the input sinusoid applied to it as shown in Fig. 6.7b. It is necessary for the peak amplitude of the input signal to exceed E_M in order to have the output signal other than a constant.

Several features of the output signal permit writing the describing function for this element. The relevant relationships include the following.

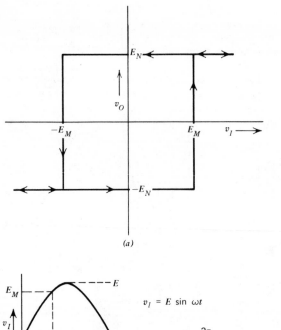

(a)

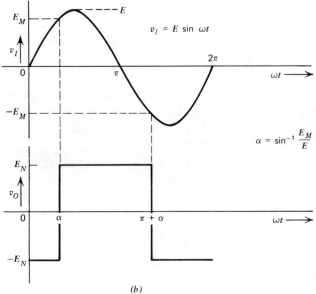

$$v_I = E \sin \omega t$$

$$\alpha = \sin^{-1} \frac{E_M}{E}$$

(b)

Figure 6.7 Relationships for an element with hysteresis. (a) Transfer characteristics. (b) Input and output waveforms for sinusoidal excitation.

(a) While there is phase shift between the input signal and the fundamental component of the output, neither the amount of this phase shift nor the amplitude of the output signal are dependent on the excitation frequency.

(b) The amplitude of the fundamental component of a square wave with a peak amplitude E_N is $4E_N/\pi$.

(c) The relative phase shift between the input signal and the fundamental component of the output is $\sin^{-1}(E_M/E)$, with the output lagging the input.

Table 6.1 Describing Functions

Nonlinearity Input $= v_I = E \sin \omega t$	Describing Function (All are frequency independent.)
	$G_D(E) = K \measuredangle 0° \quad E \leq E_M$ $G_D(E) = \dfrac{2K}{\pi}\left(\sin^{-1}R + R\sqrt{1 - R^2}\right) \measuredangle 0°,$ $E > E_M$ where $R = \dfrac{E_M}{E}$
	$G_D(E) = \dfrac{4E_N}{\pi E} \measuredangle 0°$
	$G_D(E) = 0 \measuredangle 0° \quad E \leq E_M$ $G_D(E) = K\left[1 - \dfrac{2}{\pi}\left(\sin^{-1}R + R\sqrt{1 - R^2}\right)\right]\measuredangle 0°,$ $E > E_M$ where $R = \dfrac{E_M}{E}$

Table 6.1—Continued

Nonlinearity Input $= v_I = E \sin \omega t$	Describing Function (All are frequency independent.)
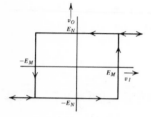	$G_D(E) = 0 \not\!\angle 0° \qquad E \leq E_M$ $G_D(E) = \dfrac{4E_N}{\pi E} \sqrt{1 - R^2} \not\!\angle 0° \qquad E > E_M$ where $R = \dfrac{E_M}{E}$
	E must exceed E_M or a d-c term results. $G_D(E) = \dfrac{4E_N}{\pi E} \not\!\angle - \sin^{-1} R$ where $R = \dfrac{E_M}{E}$

Combining these relationships shows that

$$G_D(E) = \frac{4E_N}{\pi E} \not\!\angle -\sin^{-1} \frac{E_M}{E} \qquad E \geq E_M \tag{6.31}$$

$$G_D(E) \qquad \text{undefined otherwise}$$

Table 6.1 lists the describing functions for several common nonlinearities. Since the transfer characteristics shown are all independent of the frequency of the input signal, the corresponding describing functions are dependent only on input-signal amplitude. While this restriction is not necessary to use describing-function techniques, the complexity associated with describing-function analysis of systems that include frequency-dependent nonlinearities often limits its usefulness.

The linearity of the Fourier series can be exploited to determine the describing function of certain nonlinearities from the known describing functions of other elements. Consider, for example, the soft-saturation characteristics shown in Fig. 6.8a. The input-output characteristics for this element can be duplicated by combining two tabulated elements as shown in

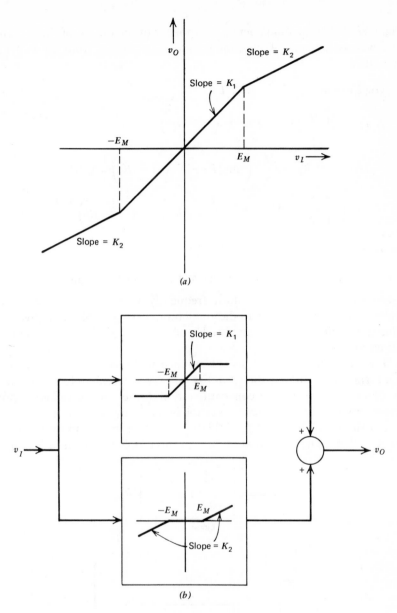

Figure 6.8 Soft saturation as a combination of two nonlinearities. (*a*) Transfer characteristics. (*b*) Decomposition into two nonlinearities.

Fig. 6.8*b*. Since the fundamental component of the output of the system of Fig. 6.8*b* is the sum of the fundamental components from the two non-linearities

$$G_D(E) = K_1 \measuredangle 0° \qquad E \le E_M \qquad (6.32a)$$

$$G_D(E) = \left[\frac{2K_1}{\pi} \left(\sin^{-1}R + R \sqrt{1 - R^2} \right) \right.$$

$$+ K_2 - \frac{2K_2}{\pi} \left(\sin^{-1}R + R \sqrt{1 - R^2} \right) \Big] \measuredangle 0°$$

$$= \left[K_2 + \frac{2(K_1 - K_2)}{\pi} \left(\sin^{-1}R + R \sqrt{1 - R^2} \right) \right] \measuredangle 0° \quad (6.32b)$$

for $E > E_M$, where $R = \sin^{-1}(E_M/E)$.

6.3.2   Stability Analysis with the Aid of Describing Functions

Describing functions are most frequently used to determine if limit cycles (stable-amplitude periodic oscillations) are possible for a given system, and to determine the amplitudes of various signals when these oscillations are present.

Describing-function analysis is simplified if the system can be arranged in a form similar to that shown in Fig. 6.9. The inverting block is included to represent the inversion conventionally indicated at the summing point in a negative-feedback system. Since the intent of the analysis is to examine the possibility of steady-state oscillations, system input and output points are irrevelant. The important feature of the topology shown in Fig. 6.9 is

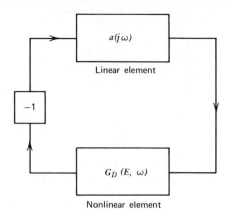

Figure 6.9   System arranged for describing-function analysis.

that a single nonlinear element appears in a loop with a single linear element. The linear element shown can of course represent the reduction of a complex interconnection of linear elements in the original system to a single transfer function. The techniques described in Sections 2.4.2 and 2.4.3 are often useful for these reductions.

The system shown in Fig. 6.10 illustrates a type of manipulation that simplifies the use of describing functions in certain cases. A limiter consisting of back-to-back Zener diodes is included in a circuit that also contains an amplifier and a resistor-capacitor network. The Zener limiter is assumed to have the piecewise-linear characteristics shown in Fig. 6.10b.

The describing function for the nonlinear network that includes R_1, R_2, C, and the limiter could be calculated by assuming a sinusoidal signal for v_B and finding the amplitude and relative phase angle of the fundamental component of v_A. The resulting describing function would be frequency

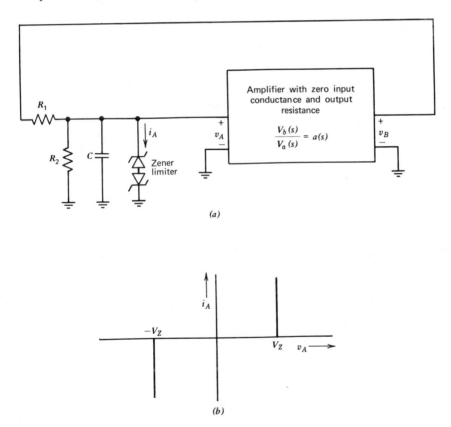

(a)

(b)

Figure 6.10   Nonlinear system. (a) Circuit. (b) Zener-limiter characteristics.

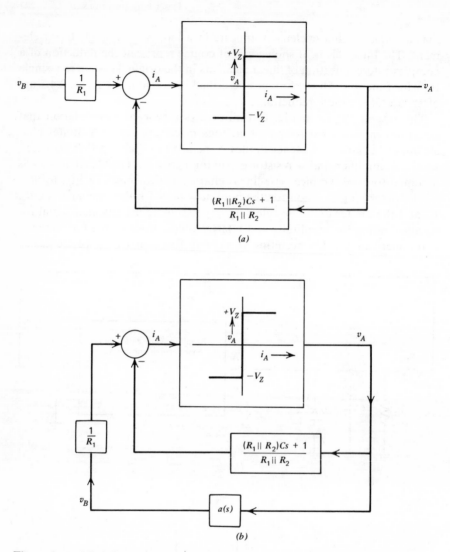

Figure 6.11 Modeling system of Fig. 6.10 as a single loop. (*a*) Block-diagram representation of nonlinear network. (*b*) Block diagram representation of complete system. (*c*) Reduced to form of Fig. 6.9.

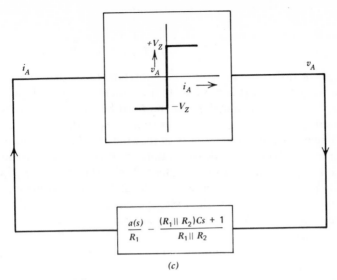

$$\frac{a(s)}{R_1} - \frac{(R_1 \| R_2)Cs + 1}{R_1 \| R_2}$$

(c)

Figure 6.11—Continued

dependent. A more satisfactory representation results if the value of the Zener current i_A is determined as a function of the voltage applied to the network.

$$i_A = \frac{v_B}{R_1} - \frac{v_A}{R_1 \| R_2} - C\frac{dv_A}{dt} \qquad (6.33)$$

The Zener limiter forces the additional constraints

$$v_A = +V_Z \qquad i_A > 0 \qquad (6.34a)$$

$$v_A = -V_Z \qquad i_A < 0 \qquad (6.34b)$$

Equations 6.33 and 6.34 imply that the block diagram shown in Fig. 6.11a can be used to relate the variables in the nonlinear network. The pleasing feature of this representation is that the remaining nonlinearity can be characterized by a frequency-independent describing function. Figure 6.11b illustrates the block diagram that results when the network is combined with the amplifier. The two linear paths in this diagram are combined in Fig. 6.11c, which is the form suggested for analysis.

Once a system has been reduced to the form shown in Fig. 6.9, it can be analyzed by means of describing functions. The describing-function approximation states that oscillations may be possible if particular values of E_1 and ω_1 exist such that

$$a(j\omega_1)G_D(E_1, \omega_1) = -1 \qquad (6.35a)$$

or

$$a(j\omega_1) = \frac{-1}{G_D(E_1, \omega_1)} \qquad (6.35b)$$

The satisfaction of Eqn. 6.35 does not guarantee that the system in question will oscillate. It is possible that a system satisfying Eqn. 6.35 will be stable for a range of signal levels and must be triggered into oscillation by, for example, exceeding a particular signal level at the input to the non-linear element. A second possibility is that the equality of Eqn. 6.35 does not describe a stable-amplitude oscillation. In this case, if it is assumed that the system is oscillating with parameter values given in Eqn. 6.35, a small amplitude perturbation is divergent and leads to either an increasing or a decreasing amplitude. As we shall see, the method can be used to resolve these questions. The describing-function analysis also predicts that if stable-amplitude oscillations exist, the frequency of the oscillations will be ω_1 and the amplitude of the fundamental component of the signal applied to the nonlinearity will be E_1.

The above discussion shows how closely the describing-function stability analysis of nonlinear systems parallels the Nyquist or Bode-plot analysis of linear systems. In particular, oscillations are predicted for linear systems at frequencies where the loop transmission is $+1$, while describing-function analysis indicates possible oscillations for amplitude-frequency combinations that produce the nonlinear-system equivalent of unity loop transmission.

The basic approximation of describing-function analysis is now evident. It is assumed that under conditions of steady-state oscillation, the input to the nonlinear element consists of a single-frequency sinusoid. While this assumption is certainly not exactly satisfied because the nonlinear element generates harmonics that propagate around the loop, it is often a useful approximation for two reasons. First, many nonlinearities generate harmonics with amplitudes that are small compared to the fundamental. Second, since many linear elements in feedback systems are low-pass in nature, the harmonics in the signal returned to the nonlinear element are often attenuated to a greater degree than the fundamental by the linear elements. The second reason indicates a better approximation for higher-order low-pass systems.

The existence of the relationship indicated in Eqn. 6.35 is often determined graphically. The transfer function of the linear element is plotted in gain-phase form. The function $-1/G_D(E, \omega)$ is also plotted on the same graph. If G_D is frequency independent, $-1/G_D(E)$ is a single curve with E a parameter along the curve. The necessary condition for oscillation is satisfied if an intersection of the two curves exists. The frequency can be determined from the $a(j\omega)$ curve, while amplitude of the fundamental component of the signal into the nonlinearity is determined from the $-1/G_D(E)$ curve. If the nonlinearity is frequency dependent, a family of curves $-1/G_D(E, \omega_1)$, $-1/G_D(E, \omega_2)$, ... , is plotted. The oscillation condition is satisfied if the $-1/G_D(E, \omega_i)$ curve intersects the $a(j\omega)$ curve at the point $a(j\omega_i)$.

The satisfaction of Eqn. 6.35 is a necessary though not sufficient condition for a limit cycle to exist. It is also necessary to insure that the oscillation predicted by the intersection is stable in amplitude. In order to test for amplitude stability, it is assumed that the amplitude E increases slightly, and the point corresponding to the perturbed value of E is found on the $-1/G_D(E, \omega)$ curve. If this point lies to the left of the $a(j\omega)$ curve, the geometry implies that the system poles[3] lie in the left-half plane for an increased value of E, tending to restore the amplitude to its original value. Alternatively, if the perturbed point lies to the right of the $a(j\omega)$ curve, a growing-amplitude oscillation results from the perturbation and a limit cycle with parameters predicted by the intersection is not possible. These relationships can be verified by applying the Nyquist stability test to the loop transmission, which includes the linear transfer function and the describing function of interest.

It should be noted that the stability of arbitrarily complex nonlinear systems that combine a multiplicity of nonlinear elements in a loop with linear elements can, at least in theory, be determined using describing functions. For example, numerous Nyquist plots corresponding to the nonlinear loop transmissions for a variety of signal amplitudes might be constructed to determine if the possibility for instability exists. Unfortunately, the effort required to complete this type of analysis is generally prohibitive.

6.3.3 Examples

Since describing-function analysis predicts the existence of stable-amplitude limit cycles, it is particularly useful for the investigation of oscillators, and for this reason the two examples in this section involve oscillator circuits.

The discussion of Section 4.2.2 showed that it is possible to produce sinusoidal oscillations by applying negative feedback around a phase-shift network with three identically located real-axis poles. If the magnitude of the low-frequency loop transmission is exactly 8, the system closed-loop

---

[3] The concept of a pole is strictly valid only for a linear system. Once we apply the describing-function approximation (which is a particular kind of linearization about an operating point defined by a signal amplitude), we take the same liberty with the definition of a pole as we do with systems that have been linearized by other methods.

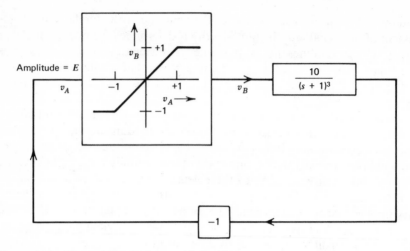

Figure 6.12   Phase-shift oscillator with limiting.

poles are on the imaginary axis and, thus, resultant oscillations are stable in amplitude. It is possible to control the magnitude of the loop transmission precisely by means of an auxiliary feedback loop that measures the amplitude of the oscillation and adjusts loop transmission to regulate this amplitude. This approach to amplitude control is discussed in Section 12.1.4.

An alternative and simpler approach that is often used is illustrated in Fig. 6.12. The loop transmission of the system for small signal levels is made large enough (in this case 10) to insure growing-amplitude oscillations if signal levels are such that the limiter remains linear. As the peak amplitude of the signal v_A increases beyond one, the limiter reduces the magnitude of the loop transmission (in a describing-function sense) so as to stabilize the amplitude of the oscillations.

The describing function for the limiter in Fig. 6.12 is (see Table 6.1)

$$G_D(E) = 1 \not\!\angle\, 0° \qquad E \le 1 \tag{6.36a}$$

$$G_D(E) = \frac{2}{\pi}\left(\sin^{-1}\frac{1}{E} + \frac{1}{E}\sqrt{1 - \frac{1}{E^2}}\right) \not\!\angle\, 0° \qquad E > 1 \tag{6.36b}$$

This function decreases monotonically as E increases beyond one. Thus the quantity $-1/G_D(E)$ increases monotonically for E greater than one and has an angle of $-180°$. The general behavior of $-1/G_D(E)$ and the transfer function of the linear portion of the oscillator circuit are sketched on the gain-phase plane of Fig. 6.13.

The intersection shown is seen to represent a stable-amplitude oscillation when the test proposed in the last section is used. An increase in E

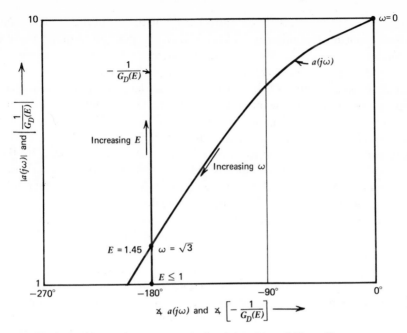

Figure 6.13 Describing-function analysis of the phase-shift oscillator.

from the value at the intersection moves the $-1/G_D(E)$ point to the left of the $a(j\omega)$ curve. The physical significance of the rule is as follows. Assume the system is oscillating with the value of E necessary to make $G_D(E)\,a(j\sqrt{3}) = -1$. An incremental increase in the value of E decreases the magnitude of $G_D(E)$ and thus decreases the loop transmission below the value necessary to maintain a constant-amplitude oscillation. The amplitude decreases until E is restored to its original value. Similarly, an incremental decrease in E leads to a growing-amplitude oscillation until E reaches its equilibrium value.

The magnitude of E under steady-state conditions can be determined directly from Eqn. 6.36. The magnitude of $a(j\omega)$ at the frequency where its phase shift if $-180°$, $(\omega = \sqrt{3})$, is 1.25. Thus oscillations occur with $G_D(E) = 0.8$. Solving Eqn. 6.36 for the required value of E by trial and error results in $E \simeq 1.45$, and this value corresponds to the amplitude of the fundamental component of v_A.

The validity of the describing-function assumption concerning the purity of the signal at the input of the nonlinear element is easily demonstrated for this example. If a sinusoid is applied to the limiter, only odd harmonics are present in its output signal, and the amplitudes of higher harmonics decrease monotonically. The usual Fourier-series calculations show that the ratio of the magnitude of the third harmonic to that of the fundamental at the output of the limiter is 0.14 for a 1.45-volt peak-amplitude sinusoid as the limiter input. The linear elements attenuate the third harmonic of a $\sqrt{3}$ radian-per-second sinusoid by a factor of 18 greater than the fundamental. Thus the ratio of third harmonic to fundamental is approximately 0.008 at the input to the nonlinear element. The amplitudes of higher harmonics are insignificant since their magnitudes at the limiter output are smaller and since they are attenuated to a greater extent by the linear element. As a matter of practical interest, the attenuation provided by the phase-shift network to harmonics is the reason that good design practice dictates the use of the signal out of the phase-shift network rather than that from the limiter as the oscillator output signal.

Figure 6.14a shows another oscillator configuration that is used as a second example of describing-function analysis. This circuit, which combines a Schmitt trigger and an integrator, is a simplified representation of that used in several commercially available function generators. It can be shown by direct evaluation that the signal at the input to the nonlinear element is a two-volt peak-to-peak triangle wave with a four-second period and that the signal at the output of the nonlinear element is a two-volt peak-to-peak square wave at the same frequency. Zero crossings of these two signals are displaced by one second as shown in Fig. 6.14b. The ratio of the third harmonic to the fundamental at the input to the nonlinear element is 1/9, a considerably higher value than in the previous example.

Table 6.1 shows that the describing function for this nonlinearity is

$$G_D(E) = \frac{4}{\pi E} \angle -\sin^{-1} \frac{1}{E} \qquad E \geq 1 \qquad (6.37)$$

The quantity $-1/G_D(E)$ and the transfer function for the linear element are plotted in gain-phase form in Fig. 6.15. The intersection occurs for a value of E that results in the maximum phase lag of 90° from the nonlinear element. The parameters predicted for the stable-amplitude limit cycle implied by this intersection are a peak-to-peak amplitude for v_A of two volts and a period of oscillation of approximately five seconds. The correspondence between these parameters and those of the exact solution is excellent considering the actual nature of the signals involved.

6.3.4 Conditional Stability

The system shown in block-diagram form in Fig. 6.16 combines a saturating nonlinearity with linear elements. The negative of the loop trans-

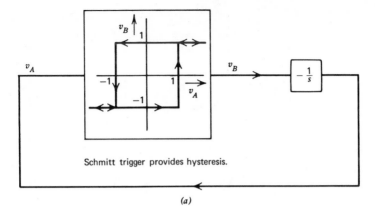

(a)

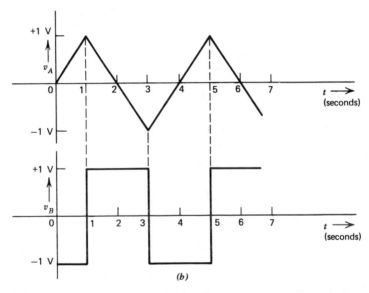

(b)

Figure 6.14 Function generator. (*a*) Configuration. (*b*) Waveforms.

mission for this system, assuming that the amplitude of the signal at v_A is less than 10^{-5} volts so that the nonlinearity provides a gain of 10^5, is determined by breaking the loop at the inverting block, yielding

$$-L(s) = 10^5 a(s) = \frac{5 \times 10^5 (0.02s + 1)^2}{(s + 1)^3 (10^{-3}s + 1)^2} \tag{6.38}$$

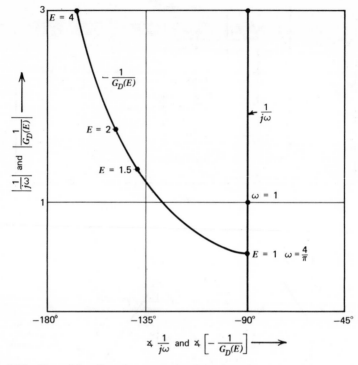

Figure 6.15   Describing-function analysis of the function generator.

A Nyquist diagram for this function is shown in Fig. 6.17. The plot reveals a phase margin of 40° combined with a gain margin of 10, implying moderately well-damped performance. The plot also shows that if the magnitude of the low-frequency loop transmission is *lowered* by a factor of between 8 and 6 × 10⁴, the system becomes unstable. Systems having the property that a decrease in the magnitude of the low-frequency loop transmission from its design-center value converts them from stable to unstable performance are called *conditionally stable* systems.

The nonlinearity can produce the decrease in gain that results in instability. The system shown in Fig. 6.16 is stable for sufficiently small values of the signal v_A. If the amplitude of v_A becomes large enough, possibly because of an externally applied input (not shown in the diagram) or because of the transient that may accompany the turn-on, the system may start to oscillate because the describing-function gain decreases.

The common characteristic of conditionally stable systems is a phase curve that drops below −180° over some range of frequencies and then recovers so that positive phase margin exists at crossover. These phase

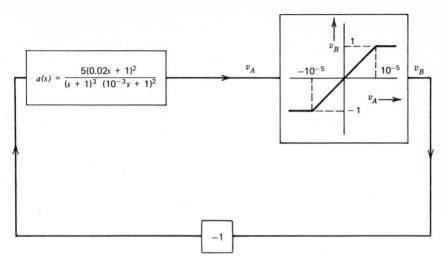

Figure 6.16 Conditionally stable system.

characteristics can result when the amplitude falls off more rapidly than $1/\omega^2$ over a range of frequencies below crossover. The high-order rolloff is used in some systems since it combines large loop transmissions at moderate frequencies with a limited crossover frequency. For example, the transfer function

$$-L'(s) = \frac{5 \times 10^5}{(2.5 \times 10^3s + 1)(10^{-3}s + 1)^2} \tag{6.39}$$

has the same low-frequency gain and unity-gain frequency as does Eqn. 6.38. However, the desensitivity associated with Eqn. 6.38 exceeds that of 6.39 at frequencies between 4×10^4 radians per second and 50 radians per second because of the high-order rolloff associated with Eqn. 6.38. The gain advantage reaches a maximum of approximately 10^3 at one radian per second. This higher gain results in significantly greater desensitivity for the loop transmission of Eqn. 6.38 over a wide range of frequencies.

Quantitative information about the performance of the system shown in Fig. 6.16 can be obtained using describing-function analysis. The describing-function for the nonlinearity for $E > 10^{-5}$ is

$$G_D(E) = \frac{2 \times 10^5}{\pi} \left(\sin^{-1} \frac{10^{-5}}{E} + \frac{10^{-5}}{E} \sqrt{1 - \frac{10^{-10}}{E^2}} \right) \measuredangle\, 0° \tag{6.40}$$

where E is the amplitude of the (assumed sinusoidal) signal v_A. The quantities $-1/G_D(E)$ and $a(j\omega)$ are plotted in gain-phase form in Fig. 6.18, and two intersections are evident. The intersection at $\omega \simeq 50$ radians per sec-

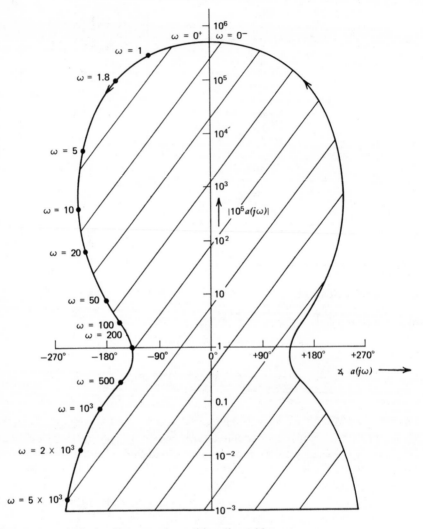

Figure 6.17 Nyquist diagram of conditionally stable system.

ond, $E \simeq 10^{-4}$ volt does not represent a stable limit cycle. If the system is assumed to be oscillating with these parameters, an incremental decrease in the amplitude of the signal v_A leads to a further decrease in amplitude and the system returns to stable operation. This result follows from the rule mentioned in Section 6.3.2. In this case, a decrease in E causes the $-1/G_D(E)$ curve to lie to the left of the $a(j\omega)$ curve, and thus the system poles move from the imaginary axis to the left-half plane as a consequence

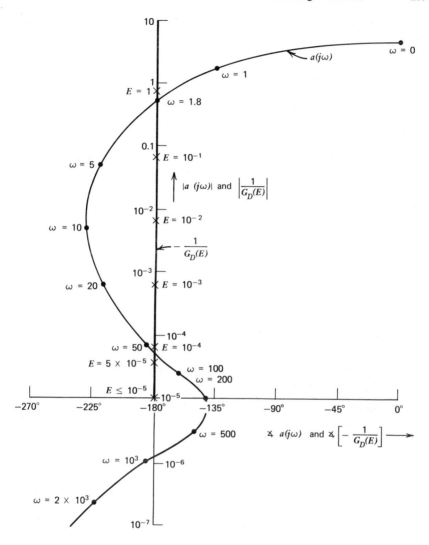

Figure 6.18 Describing function analysis of conditionally stable system.

of the perturbation. The same conclusion is reached if we consider the Nyquist plot for the system when the amplitude of v_A is 10^{-4} volt. The gain attenuation of the limiter then shifts the curve of Fig. 6.17 downward so that the point corresponding to $\omega = 50$ radians per second intersects the -1 point. An incremental decrease in E moves the curve upward slightly, and the resulting Nyquist diagram is that of a stable system.

Similar reasoning shows that a small increase in amplitude at the lower intersection leads to further increases in amplitude. Following this type of perturbation, the system eventually achieves the stable-amplitude limit cycle implied by the upper intersection with $\omega \simeq 1.8$ radians per second and $E \simeq 0.73$ volt. (The reader should convince himself that the upper intersection satisfies the conditions for a stable-amplitude limit cycle.)

It should be noted that the concept of conditionally stable behavior aids in understanding the large-signal performance of systems for which the phase shift approaches but does not exceed $-180°$ well below crossover, and then recovers to a more reasonable value at crossover. While these systems can exhibit excellent performance for signal levels that constrain operation to the linear region, performance generally deteriorates dramatically when some element in the loop saturates. For example, the recovery of this type of system following a large-amplitude step may include a number of large-signal overshoots, even if the small-signal step response of the system is approximately first order.

Although a detailed analysis of such behavior is beyond the scope of this book, examples of the large-signal performance of systems that approach conditional stability are included in Chapter 13.

6.3.5 Nonlinear Compensation

As we might suspect, the techniques for compensating nonlinear systems using either linear or nonlinear compensating networks are not particularly well understood. The method of choice is frequently critically dependent on exact details of the linear and nonlinear elements included in the loop. In some cases, describing-function analysis is useful for indicating compensation approaches, since systems with greater separation between the $a(j\omega)$ and $-1/G_D(E)$ curves are generally relatively more stable. This section outlines one specific method for the compensation of nonlinear systems.

As mentioned earlier, fast-rolloff loop transmissions are used because of the large magnitudes they can yield at intermediate frequencies. Unfortunately, if the phase shift of this type of loop transmission falls below $-180°$ at a frequency where its magnitude exceeds one, conditional stability can result. Nonlinear compensation can be used to eliminate the possibility of oscillations in certain systems with this type of loop transmission.

As one example, consider a system with a linear-region loop transmission

$$-L(s) = \frac{200}{(s + 1)(10^{-3}s + 1)^2} \qquad (6.41)$$

This loop transmission has a monotonically decreasing phase shift as a function of increasing frequency, and exhibits a phase margin of approximately 65°. Consequently, unconditional stability is assured even when some element in the loop saturates.

In an attempt to improve the desensitivity of the system, series compensation consisting of gain and two lag transfer functions might be added to the loop transmission of Eqn. 6.41, leading to the modified loop transmission

$$-L'(s) = \left[\frac{200}{(s+1)(10^{-3}s+1)^2}\right]\left[\frac{2.5 \times 10^3(0.02s+1)^2}{(s+1)^2}\right] \quad (6.42)$$

This loop transmission is of course the one used to illustrate the possibility of conditional stability (Eqn. 6.38).

Consider the effect of implementing one or both of the lag transfer functions with a network of the type shown in Fig. 6.19. If the magnitude of voltage v_C is less than V_B, the diodes do not conduct and the transfer function of the network is

$$\frac{V_o(s)}{V_i(s)} = \frac{R_2Cs + 1}{(R_1 + R_2)Cs + 1} \quad (6.43)$$

Element values can be selected to yield the lag parameters included in Eqn. 6.42.

The bias voltage V_B is chosen so that when the signal applied to the network is that which exists when the loop oscillates, the diodes clip the capacitor voltage during most of the cycle. Under these conditions, the gain of the nonlinear network (in a describing-function sense) is

$$\frac{v_O}{v_I} \simeq \frac{R_2}{R_1 + R_2} \quad (6.44)$$

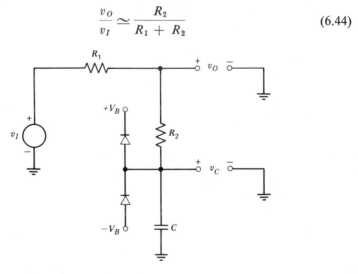

Figure 6.19   Nonlinear compensating network.

Note that if both lag transfer functions are realized this way, the loop transmission can be made to automatically convert from that given by Eqn. 6.42 to that of Eqn. 6.41 under conditions of impending instability. This type of compensation can eliminate the possibility of conditionally stable performance in certain systems. The signal levels that cause saturation also remove the lag functions, and thus the possibility of instability can be eliminated.

PROBLEMS

P6.1

One of the difficulties involved in analyzing nonlinear systems is that the order of nonlinear elements in a block diagram is important. Demonstrate this relationship by comparing the transfer characteristics that result when the two nonlinear elements shown in Fig. 6.20 are used in the order

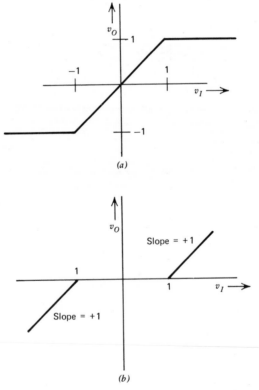

Figure 6.20   Nonlinear elements. (*a*) Limiter. (*b*) Deadzone.

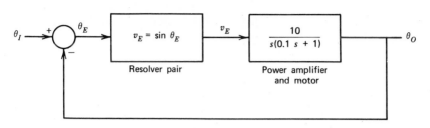

Figure 6.21 Positional servomechanism.

ab with the transfer characteristics that result when the order is changed to *ba*.

P6.2

Resolvers are essentially variable transformers that can be used as mechanical-angle transducers. When two of these devices are used in a servomechanism, the voltage obtained from the pair is a sinusoidal function of the difference between the input and output angles of the system. A model for a servomechanism using resolvers is shown in Fig. 6.21.

(a) The voltage applied to the amplifier-motor combination is zero for $\theta_O - \theta_I = n\pi$, where n is any integer. Use linearized analysis to determine which of these equilibrium points are stable.

(b) The system is driven at a constant input velocity of 7 radians per second. What is the steady-state error between the output and input for this excitation?

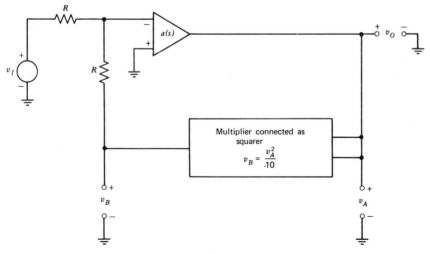

Figure 6.22 Square-rooting circuit.

(c) The input rate is charged from 7 to 7.1 radians per second in zero time. Find the corresponding output-angle transient.

P6.3

An analog divider was described in Section 6.2.2. Assume that the transfer function of the operational amplifier shown in Fig. 6.2 is

$$a(s) = \frac{3 \times 10^5}{(s + 1)(10^{-5}s + 1)^2}$$

Is the divider stable over the range of inputs $-10 < v_A < +10, 0 < v_B < +10$?

A square-rooting circuit using a technique similar to that of the divider is shown in Fig. 6.22. What is the ideal input-output relationship for this circuit? Determine the range of input voltages for which the square-rooter is stable, assuming $a(s)$ is as given above.

P6.4

Figure 6.23 defines variables that can be used to describe the motion of an inverted pendulum. Determine a transfer function that relates the angle θ to the position x_B, which is valid for small values of θ. *Hint.* You may find that a relatively easy way to obtain the required transfer function is to use the two simultaneous equations (or the corresponding block diagram) which relate x_T to θ and θ to x_B and x_T.

Assume that you are able to drive x_T as a function of θ. Find a transfer function, $X_t(s)/\theta(s)$, such that the inverted pendulum is stabilized.

P6.5

A diode-capacitor network is shown in Fig. 6.24. Plot the output voltage that results for a sinewave input signal with a peak value of E. You may assume that the diodes have an ideal threshold of 0.5 volt (i.e., no conduction until a forward-bias voltage of 0.5 volt is reached, any forward current possible without increasing the diode voltage above 0.5 volt). Evaluate the magnitude and angle of $G_D(1)$ for this network. (You may, of course,

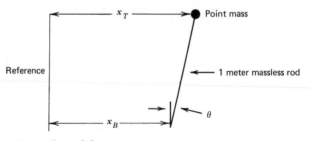

Figure 6.23   Inverted pendulum.

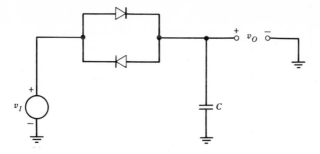

Figure 6.24 Diode-capacitor network.

work out $G_D(E)$ in general if you wish, but it is a relatively involved expression.)

P6.6

Determine the describing function for an element with the transfer characteristics shown in Fig. 6.25.

P6.7

Analyze the loop shown in Fig. 6.26. In particular, find the frequency of oscillation and estimate the levels of the signals v_A and v_B. Also calculate the ratio of third harmonic to first harmonic at the input to the nonlinear element.

P6.8

Can the system shown in Fig. 6.27 produce a stable amplitude limit cycle? Explain.

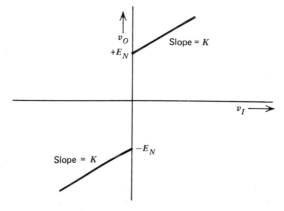

Figure 6.25 Nonlinear transfer relationship.

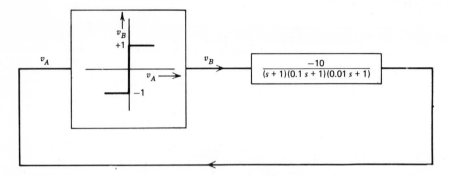

Figure 6.26 Nonlinear oscillator.

P6.9
Find a transfer function that, when combined with a limiter, can produce stable-amplitude limit cycles at two different frequencies. Design an
operational-amplifier network that realizes your transfer function.

P6.10
The transfer characteristics for a three-state, relay-type controller are
illustrated in Fig. 6.28.

(a) Show that the describing function for this element is

$$G_D(E) = \frac{2}{\pi E} \sqrt{2 + 2 \sqrt{1 - \frac{1}{E^2}}} \; \measuredangle \; -\tan^{-1} \frac{1}{E\left(1 + \sqrt{1 - \frac{1}{E^2}}\right)}$$

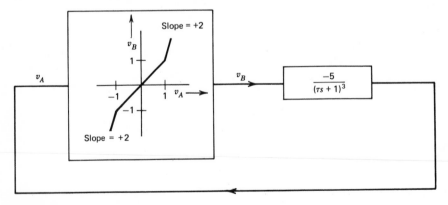

Figure 6.27 Nonlinear system.

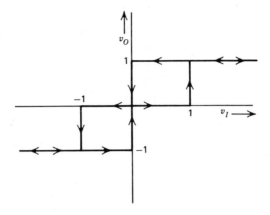

Figure 6.28 Controller transfer characteristics.

(b) The controller is combined in a negative-feedback loop with linear
elements with a transfer function

$$a(s) = \frac{a_0}{(s + 1)(0.1s + 1)}$$

What is the range of values of a_0 for stable operation?

(c) For a_0 that is twice the critical value, find the amplitude of the fundamental component of the signal applied to the controller.

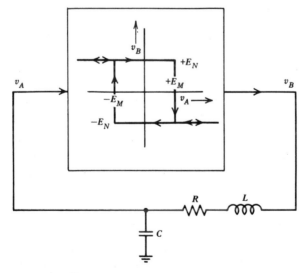

Figure 6.29 R-L-C oscillator.

P6.11

One possible configuration for a sinusoidal oscillator combines a Schmitt trigger with an *R-L-C* circuit as shown in Fig. 6.29. Find the relationship between E_M, E_N, and the damping ratio of the network that insures that oscillations can be maintained. (You may assume negligible loading at the input and output of the Schmitt trigger.)

P6.12

Three loop-transmission values, given by Eqns. 6.38, 6.39, and 6.41 were considered as part of the discussion of conditionally stable systems. Assume that three negative-feedback systems are constructed with $f(s) = 1$ and loop transmissions given by the expressions referred to above. Compare performance by calculating the first three error coefficients for each of the three systems.

CHAPTER VII

DIRECT-COUPLED AMPLIFIERS

7.1 INTRODUCTION

Operational amplifiers incorporate circuit configurations that may be relatively unfamiliar to the circuit designer with a background in other areas. An understanding of these special techniques is necessary for the most effective use of operational amplifiers.

One of the more challenging problems arises in the design of the input stage of an operational amplifier. One important consideration is that this stage provides gain to zero frequency. Thus the usual biasing techniques which incorporate capacitors that reduce low-frequency gain cannot be used. Circuits that provide useful gain at zero frequency are called *direct-coupled* or *direct-current* (d-c) *amplifiers*. The design of the direct-coupled input stage[1] of an operational amplifier is further complicated by the fact that it should have low input current.

Direct-coupled amplifiers are also useful other than as the input stage of an operational amplifier. Applications include processing certain signals of biological or geological origin that may contain significant components at a fraction of a hertz. While bandpass amplifiers can theoretically be used for such signals, the various capacitors required may become prohibitively large or expensive. Furthermore, the recovery time associated with large capacitors following overload or turn on is intolerable in some applications. In other cases, signals of interest contain frequencies of cycles per week, and response to zero frequency is mandatory in these situations. Alternatively, the designer may be interested in realizing a high-frequency amplifier, where minimization of capacitance to ground at certain critical nodes is of primary concern. If a large coupling capacitor is used, its stray capacitance to ground can deteriorate high-frequency performance.

The design of d-c amplifiers poses new problems because of the *drift* associated with such amplifiers. Drift is a phenomena whereby the output

---

[1] It is obviously necessary that all stages of an operational amplifier be direct coupled if the complete circuit is to provide useful gain at zero frequency. Emphasis here is given to the input stage because it represents the most challenging design problem.

of an amplifier changes not because of a change in the input voltage applied to the amplifier but rather in response to changes in circuit elements. In direct-coupled circuits, it is not possible to distinguish between an output that is a result of an applied input signal and one that occurs in response to drift. For this reason, drift limits the minimum input signal that can be detected.

A new circuit technique is required for the design of an amplifier that provides sufficiently low drift to be useful in d-c applications. In this chapter we shall concentrate on one circuit, the differential amplifier, which is used almost exclusively for d-c amplification. This circuit is particularly valuable when realized with bipolar transistors, since their highly predictable characteristics are readily exploited to yield low-drift performance.[2]

The discussion in this chapter focuses on the techniques used to reduce the drift and input current of a d-c amplifier, and thus the techniques described are useful in a range of applications. Toward the end of expanding the applicability of the techniques described in this chapter, certain aspects are covered in greater detail than is necessary for a basic understanding of operational amplifiers. Thus, as is the case with the material on feedback systems, operational amplifiers are used as a vehicle for illustrating technology valuable in a variety of electronic circuit and system design problems. The specific ways that these design techniques are incorporated into operational amplifiers are reserved for discussion in subsequent sections.

7.2   DRIFT REFERRED TO THE INPUT

The most useful measure of the drift of an amplifier is a quantity called *drift referred to the input*, and unless specifically stated otherwise, this quantity is the one implied when the term drift is used. Drift referred to the input is defined with reference to Fig. 7.1. This figure shows an amplifier with an assumed desired output voltage of zero for zero input voltage. The amplifier is initially *balanced* by making $v_I = 0$, and adjusting some amplifier parameter (shown diagrammatically in Fig. 7.1 as a variable resistor) until $v_O = 0$. An external quantity, such as temperature, supply voltage, or time, is then changed and, if the amplifier is sensitive to this quantity, its output voltage changes. An input voltage is then applied to the amplifier, and v_I is adjusted until v_O again equals zero. The drift referred to the

---

[2] A humorous comment on the difficulty of achieving acceptable d-c amplifier performance before modern bipolar transistors were developed is provided in L. B. Argumbau and R. B. Adler, *Vacuum-Tube Circuits and Transistors*, Wiley, New York, 1956. Chapter III, section 15 of this book is titled "Direct-Voltage Amplifiers—Why to Avoid Building Them."

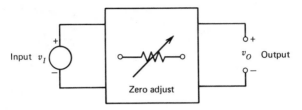

Figure 7.1 System used to define drift referred to the input.

input of the amplifier is equal to the value of v_I necessary to zero the output. The resultant magnitude is often normalized and specified, for example, as volts per degree Centigrade, volts per volt (of supply voltage), or volts per week. The minimum-detectable-signal aspect of this definition is self-evident.

In many situations we are concerned not only with the variability of the circuit as some external influencing factor is changed, but also with uncertainties that arise from the manufacturing process. In these cases, rather than initially balancing the circuit, the voltage that must be applied to its input to make its output zero may be specified as the *offset referred to the input*. The specifications related to drift and offset are at times combined by listing the maximum input offset that will result from manufacturing variations and over a range of operating conditions.

There is a tendency to use an alternative (incorrect) definition of drift, which involves dividing the drift measured at the output of the amplifier by the amplifier gain. The difficulty in this approach arises since the gain is frequently dependent on the drift-stimulating variable.

While alternative measurements of drift or offset may be equivalent in special cases, and are often used in the laboratory to simplify a measurement procedure, it is necessary to insure equivalence of other methods for each circuit. We shall normally use the original definitions for our calculations.

Figure 7.2 shows a very simple amplifier, which will be used to illustrate drift calculations and to determine how the base-to-emitter voltage of a bipolar transistor changes with temperature. It is assumed that the drift of the circuit with respect to temperature is required, and that the initial temperature is 300° K. It is further assumed that for the transistor used, $i_C = 1$ mA at $v_{BE} = 0.6$ V and $T = 300°$ K. With $v_I = 0$, these parameters show that it is necessary to adjust the potentiometer to its midposition to make $v_O = 0$. The temperature is then changed to 301° K, and it is observed the v_O is negative. (The amount is unimportant for our purposes.) In order to return v_O to zero (required by our definition of drift), it is necessary to return the transistor collector current to its original value. The change in

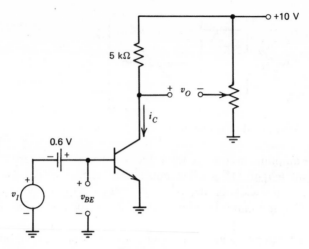

Figure 7.2   Circuit illustrating drift calculation.

v_{BE} required to restore collector current is identically equal to the required change in v_I and is therefore, by definition, the drift referred to the input of the amplifier. This discussion shows that drift for this circuit can be evaluated by determining how v_{BE} must vary with temperature to maintain constant collector current.

Drift for the circuit shown in Fig. 7.2 can be determined from the relationship between transistor terminal variables and temperature. If ohmic drops are negligible and the collector current is large compared to the saturation current I_S[3]

$$i_C = I_S e^{q v_{BE}/kT} = AT^3 e^{qV_{go}/kT} e^{qV_{BE}/kT} = AT^3 e^{q(v_{BE}-V_{go})/kT} \qquad (7.1)$$

where A is a constant dependent on transistor type and geometry, q is the charge on an electron, k is Boltzmann's constant, T is the temperature, and V_{go} is the width of the energy gap extrapolated to absolute zero divided by the electron charge ($V_{go} = 1.205$ volts for silicon).[4] It is possible to verify the exponential dependence of collector current on base-to-emitter voltage experimentally over approximately nine decades of operating current for many modern transistors.

---

[3] P. E. Gray et al., *Physical Electronics and Models of Transistors*, Wiley, New York, 1964.

[4] There is disagreement among authors concerning the exponent of T in Eqn. 7.1, with somewhat lower values used in some developments. As we shall see, the quantity has relatively little effect on the final result. (The exponent appears only as a multiplying factor in the final term of Eqn. 7.5 and as a coefficient in Eqn. 7.8). Furthermore, two similar transistors should have closely matched values for this exponent, and the degree of match between a pair is the most important quantity in anticipated applications.

Solving Eqn. 7.1 for v_{BE} yields

$$v_{BE} = \frac{kT}{q} \ln \frac{i_C}{AT^3} + V_{go} \tag{7.2}$$

The partial derivative of v_{BE} with respect to temperature at constant i_C is the desired relationship, and

$$\left. \frac{\partial v_{BE}}{\partial T} \right|_{i_C = \text{const}} = \frac{k}{q} \ln \frac{i_C}{AT^3} - \frac{3k}{q} \tag{7.3}$$

However, from Eqn. 7.2

$$\frac{k}{q} \ln \frac{i_C}{AT^3} = \frac{v_{BE} - V_{go}}{T} \tag{7.4}$$

Substituting Eqn. 7.4 into Eqn. 7.3 yields

$$\left. \frac{\partial v_{BE}}{\partial T} \right|_{i_C = \text{const}} = \frac{v_{BE} - V_{go}}{T} - \frac{3k}{q} \tag{7.5}$$

The quantity $v_{BE} - V_{go}/T$ is -2 mV/°C at $T = 300°$ K for the typical v_{BE} value of 0.6 volt. The term $3k/q = 0.26$ mV/°C; therefore to a good degree of approximation

$$\left. \frac{\partial v_{BE}}{\partial T} \right|_{i_C = \text{const}} \simeq \frac{v_{BE} - V_{go}}{T} \tag{7.6}$$

The approximation of Eqn. 7.6 links the two rule-of-thumb values of 0.6 V and -2 mV/°C for the magnitude and temperature dependence, respectively, of the forward voltage of a silicon junction.

It is valuable to note two relationships that are exploited in the design of transistor d-c amplifiers. First, with no approximations beyond those implied by Eqn. 7.1, it is possible to determine the required transistor base-to-emitter voltage variation for constant collector current knowing only the voltage, the temperature, and the material used to fabricate the transistor. Furthermore, if two silicon (or two germanium) transistors have identical base-to-emitter voltages at one temperature and at certain (not necessarily identical) operating currents, the temperature coefficients of the base-to-emitter voltages must be equal. Second, the base-to-emitter temperature coefficient at any one operating current is very nearly independent of temperature as shown by the following development. The variation of temperature coefficient with temperature is found by differentiating Eqn. 7.5 with respect to temperature, yielding

$$\frac{\partial}{\partial T} \left[\frac{\partial v_{BE}}{\partial T} \right]_{i_C = \text{const}} = \frac{-(v_{BE} - V_{go}) + T(\partial v_{BE}/\partial T)}{T^2} \tag{7.7}$$

Substituting from Eqn. 7.5 for the $\partial v_{BE}/\partial T$ term in Eqn. 7.7, we obtain

$$\frac{\partial}{\partial T}\left(\frac{\partial v_{BE}}{\partial T}\right) = -3k/qT \qquad (7.8)$$

Evaluating Eqn. 7.8 at 300° K shows that the magnitude of the change in base-to-emitter voltage temperature coefficient with temperature is less than 1 μV/°C/°C.[5]

It is now possible to determine the drift referred to the input of our original amplifier. In order to return v_O in Fig. 7.2 to zero at the elevated temperature, it is necessary to decrease i_C to its original value of 1 mA, and this decrease requires a -2.26 mV change in v_I (Eqn. 7.5). The drift referred to the input of our amplifier is by definition -2.26 mV/°C, and Eqn. 7.8 insures that this drift is essentially constant over a wide range of temperatures.

7.3 THE DIFFERENTIAL AMPLIFIER

The highly predictable temperature coefficient of the base-to-emitter voltage of a bipolar transistor offers the possibility that some type of compensation can be used to produce low-drift amplifiers. It is evident that the use of one transistor junction to compensate for voltage variations of a second similar junction should provide excellent results since both devices vary in a similar way. This section describes a connection that exploits the characteristics of a pair of bipolar transistors to provide low drift combined with several other useful features.

7.3.1 Topology

Consider the connection shown in Fig. 7.3. Here transistor Q_2 is connected as a common-base amplifier, while transistor Q_1 is connected as an emitter follower. Assume that initially $v_{I1} = 0$, that the two transistors are at the same temperature and that they are matched in the sense that they have identical saturation currents. In this case the voltages at the emitters of the two transistors will be equal, or $v_{O1} = v_{I2}$. The connection shown as a dotted line can then be completed with no change in any voltage level. If the magnitude of the voltage V_2 is much larger than anticipated variations in base-to-emitter voltage, the current through parallel resistor combination is virtually temperature independent. The matched transistor characteristics insure that this constant current divides equally between the

---

[5] An interesting alternative development of this relationship is given in "An Exact Expression for the Thermal Variation of the Emitter Base Voltage of Bi-Polar Transistors," R. J. Widlar, National Semiconductor Corp., Technical Paper TP-1, March, 1967.

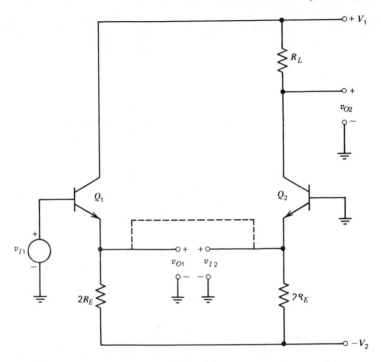

Figure 7.3 Circuit illustrating development of the differential amplifier.

two transistors. If we also assume that the common-base current gain of transistor Q_2 is one, changes in temperature result in negligible changes in the collector current of this device. Thus the drift referred to the input of this connection can be close to zero. In addition to providing temperature compensation, the current gain and input resistance of transistor Q_1 increases the input-resistance of the circuit by a factor of 2β above that seen at the emitter of Q_2.

The circuit that results when the dotted connection in Fig. 7.3 is completed is shown in Fig. 7.4. The inherent symmetry of the differential amplifier has been emphasized by including a collector-load resistor for Q_1 and permitting input signals to be applied to either base. A second output signal is indicated between the collectors of the two transistors in Fig. 7.4, so that both *differential* (between collector) or *single-ended* (either collector to ground) outputs are available.

7.3.2 Gain

The output of the circuit of Fig. 7.4 for any particular input voltage can be calculated by the usual methods. However, an alternative and useful

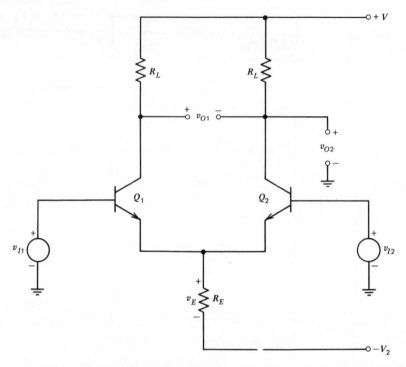

Figure 7.4 The differential amplifier.

analytic technique is available[6] that simplifies the calculations and gives greater insight into the operation of the circuit. The gain of the circuit is calculated for two particular types of inputs, a *differential* input with $v_{I1} = -v_{I2}$, and a *common-mode* input with $v_{I1} = v_{I2}$.

Figure 7.5 shows a schematic where the transistors have been replaced by appropriate, identical circuit models. Consider initially a pure differential input, of sufficiently small size so that the linear-region model remains valid. It is easily shown that in this case the voltage v_e does not change and that the common emitter connection may therefore be considered an incrementally grounded point. The incremental model for either half circuit reduces to that shown in Fig. 7.6. The incremental gain to the single-ended output, v_{o2}, is simply that of a common-emitter amplifier:

$$\frac{v_{o2}}{v_{i2}}\bigg|_{v_{i1} = -v_{i2}} = \frac{-g_m R_L r_\pi}{r_x + r_\pi} \tag{7.9}$$

---

[6] An essentially identical analysis is given for vacuum-tube differential amplifiers in T. S. Gray, *Applied Electronics*, 2nd Ed., Wiley, New York, 1954, pages 504–509.

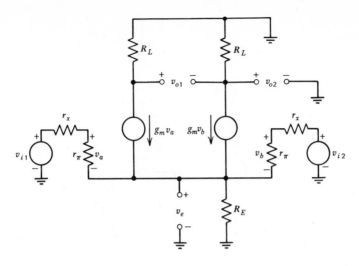

Figure 7.5 Incremental model for a differential amplifier.

The differential output component of v_{o1} for the left-hand half circuit is identical in magnitude but opposite in sign to that of the right-hand half circuit; therefore, $v_{o2} = -\frac{1}{2} v_{o1}$. The incremental gain to a differential output is then

$$\frac{v_{o1}}{v_{i2}}\bigg|_{v_{i1} = -v_{i2}} = \frac{2g_m R_L r_\pi}{r_x + r_\pi} \qquad (7.10)$$

It is conventional to consider gains calculated for a differential input signal applied between two bases of the amplifier, rather than by assuming a signal applied to one base and its negative applied to the other. If the signal between the bases is $e_d = 2\,v_{i1} = -2\,v_{i2}$ the gains become

$$\frac{v_{o2}}{e_d} = \frac{g_m R_L r_\pi}{2(r_x + r_\pi)} \qquad (7.11)$$

and

$$\frac{v_{o1}}{e_d} = \frac{-g_m R_L r_\pi}{r_x + r_\pi} \qquad (7.12)$$

For a pure common-mode input the voltage ($v_{i1} = v_{i2}$), symmetry insures that voltage v_{o1} (Fig. 7.5) remains zero and that $v_a = v_b$. Therefore, it is possible to "fold" the circuit about its vertical midline and parallel corresponding components. The resulting incremental model is shown in

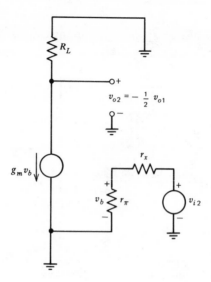

Figure 7.6 Right-hand half circuit for a differential input.

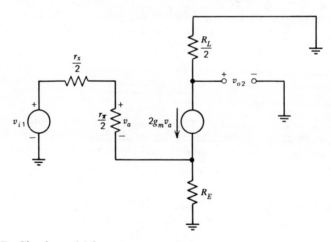

Figure 7.7 Circuit model for common-mode inputs.

Fig. 7.7. The gain to a single-ended output is identical to that of a common-emitter amplifier with emitter degeneration:

$$\frac{v_{o2}}{v_{i1}}\bigg|_{v_{i1}\,=\,v_{i2}} = \frac{-\,g_m R_L r_\pi}{2[r_\pi/2 + r_x/2 + (\beta + 1)R_E]} \qquad (7.13)$$

The common-mode input to differential-output gain is zero since v_{o1} does not change in response to a common-mode input signal.

While the gain of the differential amplifier has been calculated only for two specific types of input signals, any input can be decomposed into a sum of differential and common-mode signals. The output to each individual component can be calculated and, because of linearity, the output is the sum of the responses to the two individual inputs. For example, assume inputs e_a and e_b are applied to the left- and right-hand inputs of the circuit, respectively. The decomposition yields a common-mode component $e_{cm} = (e_a + e_b)/2$, and a differential component (applied between inputs) $e_d = e_a - e_b$. The physical implication is clear. It is assumed that any combination of input voltage levels is actually the sum of two signals: a common-mode signal (the two bases are incremented by equal amounts) equal to the average level, and a differential signal (the two bases are incremented by equal-magnitude, opposite-polarity signals) equal to the voltage applied between inputs.

7.3.3   Common-Mode Rejection Ratio

The evolution of the name differential amplifier is evident when we realize that circuit element values are typically such that the gain to a differential signal is significantly higher than that to a common-mode signal. The ratio of differential gain to common-mode gain is called the *common-mode rejection ratio* (CMRR), and many applications require high CMRR. For example, an electrocardiogram is a recording of the signal that results as the heart contracts, and is useful for the diagnosis of certain types of heart disease. The desired signal, detected by means of two electrodes attached to the body, has an amplitude of approximately 1 mV. In addition to the desired signal, a noise component at the power-line frequency with an amplitude of as much as 0.1 volt may be present as a common-mode signal on both electrodes. An amplifier with sufficiently high CMRR can be used to separate the desired signal from the interfering noise.

The analysis of Section 7.3.2 indicates that the common-mode rejection ratio of a differential amplifier with the output taken between collectors should be infinite. (As we shall see, this result is a consequence of the idealized model used.) The CMRR for a single-ended-output differential amplifier is obtained by dividing Eqn. 7.11 by Eqn. 7.13 yielding the magnitude

$$\text{CMRR} = \frac{r_\pi/2 + r_x/2 + (\beta + 1)R_E}{r_\pi + r_x} \tag{7.14}$$

Typically, $(\beta + 1)R_E \gg r_\pi \gg r_x$, so that

$$\text{CMRR} \simeq \frac{(\beta + 1)R_E}{r_\pi} \simeq g_m R_E \tag{7.15}$$

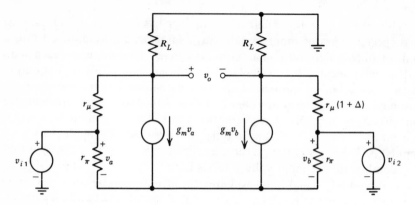

Figure 7.8 Circuit illustrating effect of unequal r_μ's.

Since the quiescent current through R_E (Fig. 7.4) is equal to twice the emitter current of either transistor, the CMRR can be related to V_E, the quiescent voltage across R_E, by

$$\text{CMRR} = \frac{q}{kT}\frac{V_E}{2R_E}R_E \simeq 20V_E \qquad (7.16)$$

Equation 7.16 shows that one way to achieve high common-mode rejection ratios for single-ended-output differential amplifiers is to use a large bias voltage. An attractive alternative (which allows more moderate supply voltage) is the use of a current source (realized with a transistor with emitter degeneration) in place of R_E. This approach has the further advantage that the quiescent current level is independent of the common-mode input signal, and for these reasons most high-performance d-c amplifiers include an emitter-circuit current source.

If the simplified transistor model used up to now were strictly valid, the CMRR for an amplifier with an emitter-circuit current source would be infinite regardless of whether a single-ended or a differential output is used, since the incremental resistance of the current source (which replaces R_E in Eqn. 7.15) is infinite. Analysis based on a more complete model shows that it is not possible to achieve infinite CMRR with a single-ended output, but that CMRR can be made arbitrarily high for a differential-output amplifier by matching all transistor parameters sufficiently closely. It is useful to illustrate the degradation that results from imperfect matching by example. Figure 7.8 shows a linear-region equivalent circuit for a differential amplifier. A collector-to-base resistance has been included in the transistor

model.[7] The physical reason for the presence of this element in the model
is described in Section 8.3.1. The magnitude of this resistance is r_μ for one
transistor, while that of the second device differs by a fraction Δ. All other
circuit parameters are identically matched. It is assumed that r_x is negligibly
small compared to r_π.[8] It is further assumed that the circuit has been constructed with an ideal emitter-circuit current source. Since $r_\mu \gg R_L$, the
gain for a differential input is

$$\frac{v_o}{(v_{i2} - v_{i1})}\bigg|_{v_{i1} = -v_{i2}} = g_m R_L \tag{7.17}$$

The gain for a common-mode input is

$$\frac{v_o}{v_{i1}}\bigg|_{v_{i1} = v_{i2}} = \frac{\Delta R_\mu R_L}{(R_L + r_\mu)[(1 + \Delta)r_\mu + R_L]} \tag{7.18}$$

Again invoking the inequality $r_\mu \gg R_L$ leads to

$$\frac{v_o}{v_{i1}}\bigg|_{v_{i1} = v_{i2}} \simeq \frac{\Delta R_L}{(1 + \Delta)r_\mu} \tag{7.19}$$

The resultant CMRR is obtained by dividing Eqn. 7.17 by Eqn. 7.19, yielding

$$\text{CMRR} = \frac{g_m(1 + \Delta)r_\mu}{\Delta} \tag{7.20}$$

A similar approach can be used to calculate common-mode errors that
arise from other sources such as unequal transistor collector-to-emitter
resistance or unequal values of r_x. It can be shown that since each of these
effects is small, there is little interaction among them, and it is valid to
compute each error separately.

As a matter of practical interest, it is possible to obtain well enough
matched transistors to obtain low-frequency values for CMRR on the order
of 10^4 to 10^6 with a simple differential-amplifier connection.

[7] We shall also see that an additional resistor between collector and emitter is necessary
to complete the model. This second resistor is omitted from the present discussion since
the simplified model illustrates the point adequately.

[8] This assumption is frequently valid in the analysis of d-c amplifiers because the transistors are usually operated at low currents to decrease input current and to minimize offsets
from differential self-heating. The resistance r_π grows approximately inversely with collector
current, while the value of r_x is bounded, with a usual maximum value of 100 to 200 Ω.
A typical value for r_π for transistors such as the 2N5963 is 2.5 MΩ at an operating current
of 10 μA.

7.3.4 Drift Attributable to Bipolar Transistors

The reason for the almost exclusive use of the differential amplifier for
d-c amplifier circuits is because of the inherent drift cancellation afforded
by symmetrical components. The purpose of this section is to indicate how
the circuit should be balanced for minimum drift.

If a differential amplifier such as that shown in Fig. 7.4 is constructed
with symmetrical components, the differential output voltage v_{O1} is zero
for $v_{I1} = v_{I2}$. While resistors are available with virtually perfectly matched
characteristics, selection of well-matched transistors is a significant problem.

It has been assumed up to this point that the transistors used in a differential amplifier are matched in the sense that they have equal saturation
currents. One measure of the degree of match is to specify the ratio of the
saturation currents for a pair of transistors. This ratio is exactly the same
as the ratio of the collector currents of the two transistors when operated
at equal base-to-emitter voltages, since at a base-to-emitter voltage V_{BE}
(assuming operation at currents large compared to I_S), the collector current
of one transistor is

$$I_{C1} = I_{S1} e^{q V_{BE}/kT} \qquad (7.21)$$

while that of the second transistor is

$$I_{C2} = I_{S2} e^{q V_{BE}/kT} \qquad (7.22)$$

Alternatively, the degree of match can be indicated by specifying the difference ΔV between the base-to-emitter voltages of the two transistors
when both are operated at some collector current I_C. This specification
implies that at some base-to-emitter voltage V_{BE}

$$I_{C1} = I_{S1} e^{q V_{BE}/kT} = I_C = I_{C2} = I_{S2} e^{q(V_{BE}+\Delta V)/kT} \qquad (7.23)$$

This measure of match is easily related to the degree of match between
saturation currents, since Eqn. 7.23 shows that

$$\frac{I_{S1}}{I_{S2}} = e^{q \Delta V/kT} \qquad (7.24)$$

Equation 7.24 also shows that the base-to-emitter voltage mismatch, ΔV,
is independent of the operating current level selected for the test.

If the circuit of Fig. 7.4 is used as a d-c amplifier, the quantity ΔV for
the transistor pair is exactly the offset referred to the input of the amplifier,
since this differential voltage must be applied to the input to equalize collector currents and thus make v_{O1} zero. For this reason, semiconductor
manufacturers normally specify the degree of match between two transistors

in terms of their base-to-emitter voltage differential at equal currents rather than as the ratio of saturation currents.

Several options are available to the designer to obtain well-matched pairs for use in differential amplifiers. Matched transistors are available from many manufacturers at a cost of from 2 to 10 times that of the two individual devices. These transistors are frequently mounted in a single can so that the differential temperature of the two chips is minimized. The best specified match available in a particular series of devices is typically a 3-mV base-to-emitter voltage differential when the devices operate at equal collector currents.

An alternative involves user matching of the transistors. This possibility is attractive for several reasons. There are economic advantages, particularly if large numbers of matched pairs are required, since relatively modest equipment suffices and since the effort required is not prohibitive. Better matches for a greater number of parameters are possible than with purchased matched pairs. However, lack of money, patience, and environmental control (remember the typical temperature coefficient of $-2\,\mathrm{mV/°C}$) generally limits achievable base-to-emitter voltage matches to the order of 0.5 mV. It is also necessary to provide some sort of thermal coupling to keep the matched devices at equal temperatures during operation.

A third possibility is the use of a monolithic integrated-circuit differential pair. Through proper control of processing, all transistor parameters are simultaneously matched, and differential base-to-emitter voltages on the order of 1 mV are possible with present technology. Excellent thermal equality is obtained because of the proximity of the two devices. This approach is used as an integral part of all monolithic operational amplifiers. There are also a number of single and multiple monolithic matched pairs available for use in discrete designs. Several more sophisticated monolithic designs are available[9] that include temperature sensing and heating elements on the chip to keep its temperature relatively constant. The effects of ambient temperature variations are largely eliminated by this technique.

Regardless of the matching procedure used, some type of trimming is required to reduce the offset of the amplifier to zero at one temperature. One popular technique is to include a potentiometer in the emitter circuit as shown in Fig. 7.9. The two bases are shorted together and the pot is adjusted until the two collector currents are equal so that $v_O = 0$. This adjustment is possible for $R > 2\Delta V/I$, where ΔV is the base-to-emitter voltage differential of the pair at equal collector currents. (The use of too

[9] Examples include the Fairchild Semiconductor μA726 and μA727.

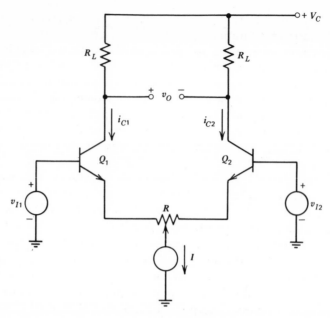

Figure 7.9   Balancing with emitter-circuit potentiometer.

large a potentiometer is undesirable since it lowers the transconductance[10] of the pair, and we shall see that this quantity becomes important when the effect of other circuit components on drift is considered.) While this balance method is frequently used, it is fundamentally in error if minimization of drift with temperature is the design objective. The approach equalizes collector currents and thus insures that one transistor operates at a quiescent base-to-emitter voltage of v_{BE1}, while the other operates at a voltage of $v_{BE1} + \Delta V$. The required difference in base-to-emitter voltages is obtained by adjusting the pot so that the voltages across its two segments differ by ΔV. Since the voltages across the pot segments are the same whenever the input voltage is adjusted to make v_O zero (assuming the common-base current gain of the transistors is one, the current through each pot segmen must be $I/2$ when $v_O = 0$), the drift referred to the input with respect to

---

[10] The transconductance of a differential pair is defined as the ratio of the incremental change in either collector current to the incremental differential input voltage. Assuming that both transistors have large values for β and negligible base resistance, the transconductance for the configuration shown in Fig. 7.9 is

$$\left| \frac{i_{c1}}{v_{i1} - v_{i2}} \right| = \left| \frac{i_{c2}}{v_{i1} - v_{i2}} \right| \simeq \frac{1}{1/g_{m1} + 1/g_{m2} + R}$$

temperature for this design is identically equal to the differential change in the transistor base-to-emitter voltages with temperature. From Eqn. 7.5,

$$\frac{\partial}{\partial T} (v_{BE1} - v_{BE2}) \bigg|_{i_{C1} = i_{C2} = \text{const}} = \left(\frac{v_{BE1} - V_{go}}{T} - \frac{3k}{q} \right)$$

$$- \left(\frac{v_{BE2} - V_{go}}{T} - \frac{3k}{q} \right) = \frac{v_{BE1} - v_{BE2}}{T} \qquad (7.25)$$

Since the difference $v_{BE1} - v_{BE2}$ is ΔV,

$$\frac{\partial}{\partial T} (v_{BE1} - v_{BE2}) = \frac{\Delta V}{T} \qquad (7.26)$$

For example, a 3-mV mismatch at room temperature leads to a drift of 10 μV/$°$C.

An alternative is to operate the transistors with equal base-to-emitter voltages. This condition requires that the quiescent collector-current ratio be equal to the ratio of the transistor saturation currents, or

$$\frac{I_{C1}}{I_{C2}} = \frac{I_{S1}}{I_{S2}} = e^{q \Delta V / kT} \qquad (7.27)$$

where, as defined above, ΔV is the difference between the base-to-emitter voltages of the two devices *when they are operated at equal collector currents*. In this case, a 3-mV value for ΔV requires a 12% difference in collector currents to equalize base-to-emitter voltages. A possible circuit configuration is shown in Fig. 7.10. The two bases are shorted together, which forces equal base-to-emitter voltages and zero differential input voltage. The potentiometer is then adjusted to make $v_O = 0$. The results of earlier analysis indicate that the temperature drift attributable to the transistors should be zero following this adjustment. While very low values are attainable by this method, there are other detailed effects, neglected in our simplified analysis, which lead to nonzero drift. It is possible to adjust the relative base-to-emitter voltages to compensate for these effects.[11] In practice, even the simplified balancing technique can result in drifts of a fraction of a microvolt per degree Centigrade.

It is stressed that this balancing technique should not be considered a substitute for careful matching of the devices, but rather as a final trim following matching. If a large base-to-emitter voltage mismatch is compensated for by this method, there is a large differential power dissipation with associated differential heating, base currents will differ by a large amount,

[11] A. H. Hoffait and R. D. Thornton, "Limitations of Transistor DC Amplifiers," *Proceedings Institute of Electrical and Electronic Engineers*, February, 1964.

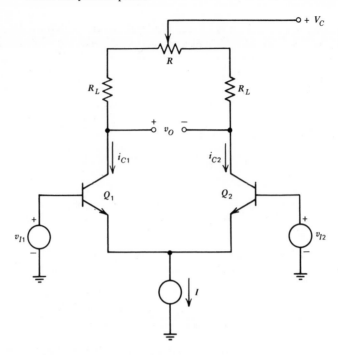

Figure 7.10 Method for balancing with equal base-to-emitter voltages.

and the transconductance of the pair will be significantly lower than if well-matched devices are used. For example, compensation for a 60-mV mismatch requires collector currents with a 10 to 1 ratio and lowers transconductance by a factor of five compared with a well-matched pair operated at the same total emitter current. Operation with severely unbalanced collector currents also mismatches all current-dependent transistor parameters.

7.3.5 Other Drift Considerations

It is interesting to note that the excellent compensation afforded by even the simplified balancing technique described above emphasizes the drift contribution of other components in circuit. Consider the circuit shown in Fig. 7.11. (For simplicity it is assumed that inputs are applied to only one side of the circuit.) Assume that the transistors are perfectly matched so that when the collector resistors are equal $v_O = 0$ for $v_I = 0$. A drift results if the relative collector-resistor values change as a result of differential changes with temperature or aging. The drift attributable to a collector-resistor fractional unbalance Δ can be calculated as follows. With $v_I = 0$, $i_{C1} = i_{C2} \simeq I/2$. As v_i is increased, $i_{C1} = I/2 + (g_m/2)v_i$ and $i_{C2} = I/2 - (g_m/2)v_i$, where g_m is the transconductance of either transistor. (It is as-

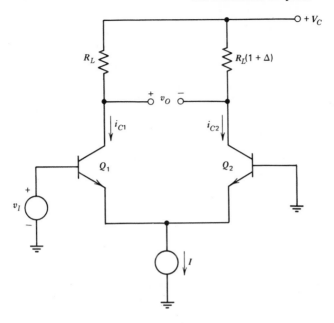

Figure 7.11   Circuit with unequal load resistors.

sumed that $r_\pi \gg r_x$ for the transistors.) In order to return v_O to zero, it is necessary to have

$$\left(\frac{I}{2} + \frac{g_m}{2}\, v_i\right)R_L = \left(\frac{I}{2} - \frac{g_m}{2}\, v_i\right)(1 + \Delta)R_L \qquad (7.28)$$

or

$$g_m v_i = \frac{\Delta I}{2} \qquad (7.29)$$

(A term containing the small cross product $g_m v_i\, \Delta R_L$ has been dropped.) Since each device is operating at a quiescent current level $I/2$, $g_m = qI/2kT \simeq 20I$ at room temperature. Thus the input voltage required to return the output voltage to zero (by definition the drift referred to the input) is $\Delta/40$. The significance of this sensitivity is appreciated when one considers that two ordinary equal-value carbon-composition resistors can have temperature coefficients that differ by as much as one part per thousand per degree Centigrade. Use of such resistors would result in an amplifier drift of 25 $\mu V/°C$! It is clear that the quality of the resistors used is an important factor when a 1 $\mu V/°C$ amplifier is designed.

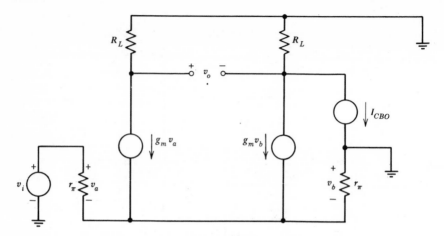

Figure 7.12 Equivalent circuit for finding drift as a function of I_{CBO}.

A similar conclusion is reached when the effects of collector-to-base leakage current I_{CBO}[12] are considered. An equivalent circuit that can be used to predict the drift from I_{CBO} is shown in Fig. 7.12. Since the magnitude of I_{CBO} is likely to be significantly different for two otherwise well-matched transistors, only one leakage current generator is shown in Fig. 7.12. Its value can be made the difference between the leakages if one component is not negligible. Proceeding as before, the value of v_i required to reduce the output to zero is given by solving

$$\frac{g_m v_i}{2} = \frac{-g_m v_i}{2} + I_{CBO} \tag{7.30}$$

for v_i, yielding

$$v_i = \frac{I_{CBO}}{g_m} \tag{7.31}$$

The transconductance of either input transistor g_m can be related to the bias level for the differential pair (each member operates at $I/2$) as $g_m = 20I$. Therefore, the offset expressed in volts is $I_{CBO}/20I$. Typical values are again evoked to illustrate the problem. The FT107A (an attractive choice for the input stage of a d-c amplifier since its specifications include a typical β of

---

[12] The assumptions often used to simplify device physics to the contrary, this quantity is not related to the saturation current in the transistor equation. The magnitude of I_S is dominated by effects within the body of the semiconductor, while the dominant component of I_{CBO}, at least at room temperature, results from surface effects. Temperature coefficients are significantly different. While I_S doubles every $6°$ C, I_{CBO} near room temperature typically doubles every $10°$ C.

1100 at $10 \mu A$ of collector current!) has a specified maximum leakage current that increases from essentially zero at 25° C to 1 μA at 125° C. The resultant average drift over the 100° C temperature range for the device operating at a collector current level of 10 μA ($I = 20 \mu A$) is therefore bounded by 25 $\mu V/°C$. Fortunately the typical value for I_{CBO} is 2% of the maximum specified value, but additional screening procedures are required to insure this lower level is met by any particular device.

It is worth emphasizing the importance of proper thermal design for low-drift d-c amplifiers. A temperature differential of 0.001° C results in an offset of 2 μV for a differential pair that is perfectly matched when the temperatures of the transistors are identical. Several factors influence the temperature differential of a pair. Good thermal contact between the members of the pair is mandatory. This required contact can be achieved by locating the two chips close together on a thermally conductive plate, or via monolithic integrated-circuit construction.

It is also necessary to minimize heating effects that disturb the pair. Self-heating as a consequence of the power dissipated in the pair is particularly important. Differential self-heating is reduced by operating the two members of the pair at matched, low collector currents and at low collector voltage. The location of other heat sources that can establish thermal gradients across the pair must also be considered. These sources are easily isolated in discrete-component designs, but impose severe constraints on component placement in integrated circuits.

Another aspect of the thermal problem involves the way in which the differential-amplifier transistors are connected to the input signal or to other circuit components. A thermocouple with an approximately 20 $\mu V/°C$ coefficient is formed when kovar, an alloy frequently used for transistor leads, is connected to copper. Thus thermal gradients across the circuit, which result in different temperatures for series-connected thermocouple junctions in the signal path, can contribute significant offset voltage.

7.4 INPUT CURRENT

The discussion of input-circuit errors up to this point has focused on voltage drift referred to the input. Additional input offset signals arise from input current if the signal source resistance is high. In many d-c amplifiers constructed using bipolar transistors, offsets from input current dominate. One alternative is the use of junction-gate or metal-oxide-semiconductor (MOS) field-effect transistors that exhibit substantially lower input currents. Unfortunately, the voltage drift of junction-gate field-effect transistors is about one order of magnitude worse than that of bipolar devices. Mos devices, with threshold voltages dependent on trapped surface charge, are even more unstable. The techniques used to stabilize the operation of these devices are significantly different than those used with bipolar transistors and are not discussed here.[13]

In contrast to the base-to-emitter voltage, which varies in a highly predictable fashion with temperature, the temperature dependence of base current is a complex function of transistor structure. Furthermore, matching most parameters of two transistors, including β at one temperature, does not insure equal current gain at some different temperature. As a matter of practical interest, the fractional change in current gain with temperature, $(1/\beta)(\partial\beta/\partial T)$, is typically 0.5 to 1% per degree Centigrade, with somewhat higher values measured at low collector currents and low temperatures.

While these unpredictable variations in β make input-current compensation schemes less precise than voltage-drift compensation, several useful methods are available for lowering input current.

7.4.1 Operation at Low Current

In spite of manufacturers' reluctance to admit it, there are many types of transistors that exhibit useful current gains at low collector currents. It is not unusual to find units with a value for β in excess of 10 at $I_C = 10^{-11}$ A, and devices with current gains of 100 at $I_C = 10^{-9}$ A are easily selected from several families. Clearly, operation at reduced collector current is one approach to low input current. A disadvantage of this technique is that collector-to-base leakage current may dominate input current, particularly at high temperatures, or may contribute to excessive voltage drift (see Section 7.3.5). However, I_{CBO} can be eliminated by operating a transistor at zero collector-to-base voltage, and there are several circuit techniques that keep this voltage low yet permit operation over a wide range of input voltages.

A more fundamental problem is the low f_T (current gain-bandwidth product) of devices operating at low collector currents. Below some current level the base-to-emitter capacitance C_π is dominated by a space-charge-layer capacitance, and this quantity is independent of current. Since collector-to-base capacitance C_μ is independent of operating current and g_m is directly proportional to current,

$$f_T = \frac{g_m}{2\pi(C_\pi + C_\mu)} \tag{7.32}$$

is directly proportional to current at low operating currents. A typical value for f_T at a collector current of 1 nA is 1 kHz.

---

[13] L. Orchard and T. Hallen, "Fet Amplifier Design Precautions," *EDN*, August, 1968.

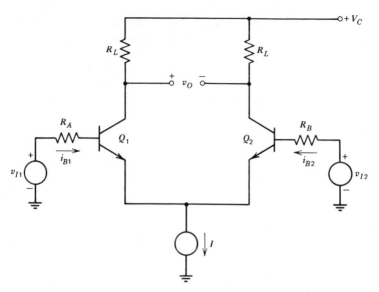

Figure 7.13   Method to eliminate effects of input current.

7.4.2   Cancellation Techniques

While the variation of input current with temperature is not as predictable as that of the base-to-emitter voltage, several compensation techniques take advantage of matching this quantity. Figure 7.13 shows one possibility. Here it is assumed that the source impedances associated with the two input signals are resistive and fixed. If

$$i_{B1}R_A = i_{B2}R_B \tag{7.33}$$

the drop across each source resistor is equal and the net effect is simply to apply a common-mode input signal to the amplifier.[14] Similarly, if

$$R_A \frac{\partial i_{B1}}{\partial T} = R_B \frac{\partial i_{B2}}{\partial T} \tag{7.34}$$

the effects of temperature-dependent input currents are eliminated. Both Eqns. 7.33 and 7.34 are satisfied if the resistors are selected to equalize voltage drops at one temperature and if the fractional change in β with temperature is equal for both devices. The technique of equalizing the re-

---

[14] It is assumed in this discussion that the input currents are independent of differential input voltage. This is not true for large signals, but in many applications the signals applied to a differential amplifier are sufficiently small to make base-current variations with signal level negligible. A technique to compensate for varying input current with signal levels is indicated in Section 7.4.3.

sistances connected to the two inputs (effectively assuming equal input currents) is frequently used in operational-amplifier connections.

In some applications, it is important to reduce the magnitude of one or both base currents of an amplifier, not simply insure that the two input currents to a differential amplifier are equal. Clearly one very simple approach is to provide the amplifier bias currents via resistors connected to an appropriate-polarity supply voltage. Unfortunately, the bias current supplied by this method is temperature independent, and thus the variation in amplifier input current with temperature is not decreased. Figure 7.14 shows one way to provide a degree of cancellation. If the β's of corresponding NPN and PNP transistors are equal, the current seen at either input is zero when the collector currents of the two NPN's are equal. The use of current sources in the emitters of the PNP's provides a compensating current that is independent of common-mode level.

Another technique is to use the temperature-dependent forward-voltage characteristics of a diode to generate a temperature-dependent compensating current, as shown in Fig. 7.15. The amplifier itself is shown diagrammatically in this figure, and only one input, close to ground potential, is indicated. Resistor R_1 establishes a bias current through the diode. It is assumed that this current is constant since it is selected to be much larger than i_A and that V_C is much greater than V_A and v_F. The temperature dependence of v_F, $\partial v F/\partial T$, is identical to that of a transistor (Eqn. 7.5) and is approxi-

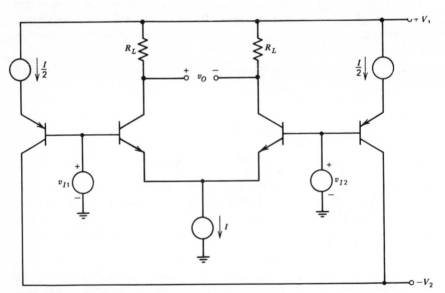

Figure 7.14   Input-current cancellation with transistors.

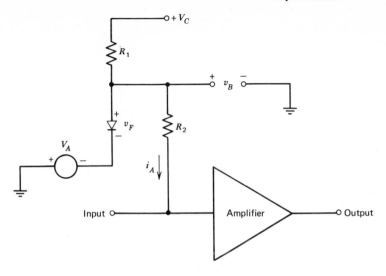

Figure 7.15   Use of a diode for input-current compensation.

mately constant with temperature.[15] The compensating current i_A is equal to v_B/R_2, and has a fractional change with temperature equal to

$$\frac{1}{i_A}\frac{\partial i_A}{\partial T} = \frac{1}{v_B}\frac{\partial v_B}{\partial T} = \frac{1}{(v_F + V_A)}\frac{\partial v_F}{\partial T} \tag{7.35}$$

The two degrees of freedom represented by the selection of V_A and R_2 can be used to cancel at one temperature both the input current and its first derivative with respect to temperature.

There are several variations on this basic topology that effectively bootstrap the reference voltage for the compensating diode from a node referenced to the common-mode input level such as the emitter connection of differential pair. The compensating current provided can be made relatively independent of common-mode level in this way, thus allowing the technique to be used with input voltages at arbitrary levels with respect to ground.

7.4.3   Compensation for Infinite Input Resistance

The compensation methods introduced up to this point have been intended to compensate for temperature variations of the input-transistor bias current. It has been assumed that the input signals are small enough

---

[15] Carrier recombination in a diode can multiply the $3k/q$ term in Eqn. 7.5 by a factor between one and two. This modification does not significantly alter the basic dependence.

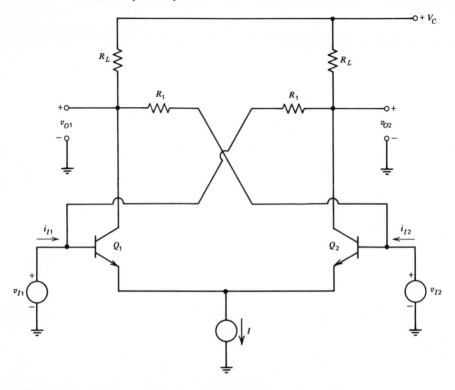

Figure 7.16   Circuit that can yield infinite differential input resistance.

so that the input-current component attributable to the input resistance of the amplifier is negligible. While this inequality is generally satisfied in applications (such as operational amplifiers) where the input circuit is followed by additional stages of voltage amplification, many differential-amplifier stages operate with appreciable differential signals applied to their input.

Figure 7.16 shows a connection that can be adjusted to provide infinite input resistance to differential signals. Consider a differential input signal, $v_{I1} = -v_{I2}$. A positive v_{I1} increases the current flowing into the base of Q_1 and causes a positive change in v_{O2}. By proper choice of parameters it is possible to supply the required base current through the right-hand R_1 so that the change in i_{I1} is zero[16]. The necessary value for R_1 is computed with the aid of the incremental model of Fig. 7.17. (The usual approximations

[16] This technique, which involves positive feedback, is not without its hazards. The topology of the circuit is essentially identical to that of a flip-flop, and if the circuit is overcompensated and driven from high impedance sources, bistable operation is possible.

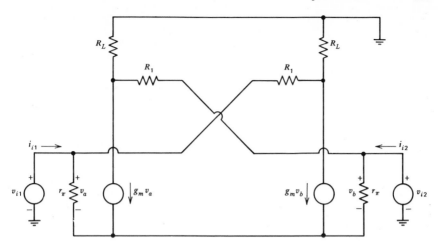

Figure 7.17 Increment model for circuit of Fig. 7.16.

have been included in developing the model.) Normally $R_1 \gg R_L$ so that
the loading by R_1 can be neglected. With this assumption, the incremental
input current i_{i1} that results for a pure differential input is

$$i_{i1} = v_{i1} \left[\frac{1}{r_\pi} - \left(\frac{g_m R_L - 1}{R_1} \right) \right] \qquad (7.36)$$

If the voltage gain of the circuit is large so that $g_m R_L \gg 1$, the differential
input resistance is infinite for

$$\frac{g_m r_\pi R_L}{R_1} = 1 \qquad \text{or } R_1 = \beta R_L \qquad (7.37)$$

The common-mode input resistance is lowered by the compensating resistors, since Fig. 7.17 shows that

$$\left. \frac{v_{i1}}{i_{i1}} \right|_{v_{i1} = v_{i2}} = R_1 \qquad (7.38)$$

High common-mode input resistance can be restored by including PNP
transistors in this compensating circuit as shown in Fig. 7.18. In addition
to supplying the compensating current from a high-resistance source, selection of the bias voltage gives an additional degree of freedom in controlling the quiescent level of the compensating current.

7.4.4 Use of a Darlington Input

One obvious way to lower input current is to use transistors with higher
current gains. As mentioned earlier, transistors with current gains in ex-

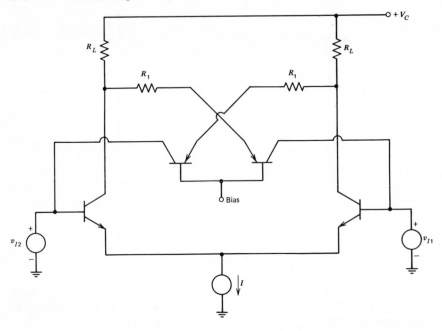

Figure 7.18 Use of common-base transistors to increase common-mode input resistance.

cess of 1000 are available, and this value should increase as processing techniques improve. It is also possible to use two transistors in the *Darlington connection* shown in Fig. 7.19. It is easy to show that at low frequencies this connection approximates a single transistor between terminals *B, C,* and *E* with current gain given by

$$\beta = \beta_2(\beta_1 + 1) + \beta_1 \simeq \beta_1\beta_2 \qquad (7.39)$$

and a transconductance

$$g_m = \frac{q}{2kT} I_C \qquad (7.40)$$

Current gains in excess of 10^5 are possible with available devices.

Figure 7.20 shows a differential amplifier with Darlington-connected input transistors. While a connection of this type yields low values for input current, the voltage drift for this configuration usually exceeds that of the conventional differential amplifier. The problem stems from differential changes in the base currents of transistors Q_1 and Q_2. (Remember that current gain varies in a relatively unpredictable way with temperature.) Since

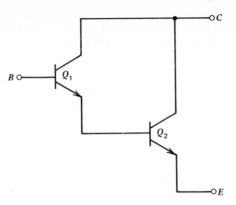

Figure 7.19 Darlington-connected transistors.

the resistance seen at the emitters of transistors Q_3 and Q_4 is relatively high, current changes produce significant changes in voltages v_A and v_B. A differential change in v_A and v_B results in drift equal in value to this change.

In order to compute drift referred to the input from this effect, it is necessary to determine how v_I must vary with i_A and i_B to keep $v_O = 0$.

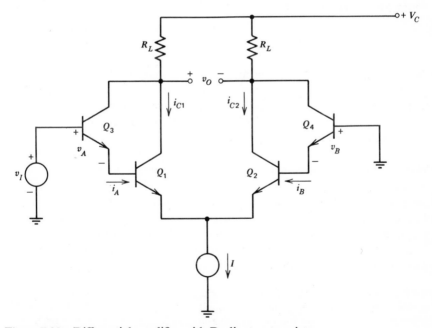

Figure 7.20 Differential amplifier with Darlington transistors.

Assume the operating point values for the two emitter currents are I_A and I_B. The incremental changes in these two currents that arise from changes in the current gains of transistors Q_1 and Q_2 are related to I_A and I_B by

$$i_a = -I_A \frac{\Delta\beta_1}{\beta_1} \tag{7.41a}$$

$$i_b = -I_B \frac{\Delta\beta_2}{\beta_2} \tag{7.41b}$$

where $\Delta\beta/\beta$ is recognized as the fractional change in current gain for a transistor.

The incremental output resistance of an emitter follower is approximately equal to the reciprocal of its transconductance. Thus the incremental differential change between v_A and v_B caused by changes in i_A and i_B, which is identically equal to the change in v_I required to keep v_O equal to zero is

$$v_a - v_b = \frac{i_a}{g_{m3}} - \frac{i_b}{g_{m4}} = \frac{I_B \Delta\beta_2}{g_{m4}\beta_2} - \frac{I_A \Delta\beta_1}{g_{m3}\beta_1} \tag{7.42}$$

Since the transconductances are proportional to operating-point currents, Eqn. 7.42 reduces to

$$v_a - v_b = \frac{I_B \Delta\beta_2}{(qI_B/kT)\beta_2} - \frac{I_A \Delta\beta_1}{(qI_A/kT)\beta_1} = \frac{kT}{q}\left(\frac{\Delta\beta_2}{\beta_2} - \frac{\Delta\beta_1}{\beta_1}\right) \tag{7.43}$$

Note that the drift component attributable to this effect is dependent only on the differential changes in the fractional current gains of the inner transistors. A typical value for the fractional change in current gain with temperature is 0.6% per degree Centigrade. If transistors Q_1 and Q_2 have this value matched to within 10%,[17] the resultant drift is 15 μV/°C.

Another potential difficulty with the use of the Darlington input connections is that its fractional change in input current with temperature is approximately a factor of two greater than that of an individual transistor because two devices are cascaded in the Darlington connection. Thus the low bias current of the Darlington configuration does not result in correspondingly low changes in bias current with temperature.

It is possible to trade input current for drift by increasing the emitter currents of Q_3 and Q_4 above the base currents of Q_1 and Q_2, for example

[17] This degree of match is realistic for discrete transistors selected for matched base-to-emitter voltages and current gains. Better results are normally achieved with monolithic matched transistors where the manufacuring process for the two devices is highly uniform.

by placing resistors from base to emitter of Q_1 and Q_2. Changes in base current have less effect since the output resistances of Q_3 and Q_4 are lower as a consequence of increased bias current. This technique is frequently used in the design of amplifiers with Darlington input transistors.

7.5 DRIFT CONTRIBUTIONS FROM THE SECOND STAGE

Thus far the discussion has focused on single-stage direct-coupled amplifiers. No consideration has been given to situations that require a second stage either to provide greater voltage gain or to isolate a low-resistance load. The use of a second stage is mandatory in the design of operational amplifiers and thus must be investigated.

There is a popular misconception that the dominant source of voltage drift for a d-c amplifier is always associated with its input stage. The argument supporting this view is that drift arising in the second stage is divided by the gain of the first stage when referred to the input of the amplifier, and is negligible if the first-stage gain is high. This assumption is not always justified because of the extraordinarily low values of drift that can be achieved with a properly balanced first stage. Balancing techniques similar to those used for the input stage are not effective for the second stage, since its drift contribution is often attributable to variations in input current rather than in base-to-emitter voltage.

7.5.1 Single-Ended Second Stage

Figure 7.21 shows a differential first stage (with two matched transistors collectively labeled Q_1) driving a common-emitter PNP second stage. Two perturbation sources are shown, which will be used later to calculate drift. In addition to providing gain, the second stage shifts level so that the output voltage can swing both positive and negative with respect to ground. If the base resistance of all transistors is negligibly small, the voltage gain of this amplifier is

$$\frac{v_o}{v_i} = \frac{-g_{m1}R_{L1}\beta_2 R_{L2}}{2(r_{\pi2} + R_{L1})} \tag{7.44}$$

Drift referred to the input for this two-stage amplifier is calculated as before by determining how v_I must vary to keep v_O equal to zero. Note that in order to maintain a fixed output voltage, it is necessary for i_{C2} to remain constant. There are a number of sources of drift for this amplifier. In this development only changes in i_{B2} and v_{EB2} that arise as the parameters of Q_2 vary are considered. These changes can be modeled by the perturbation generators shown in Fig. 7.21. If the changes are small compared

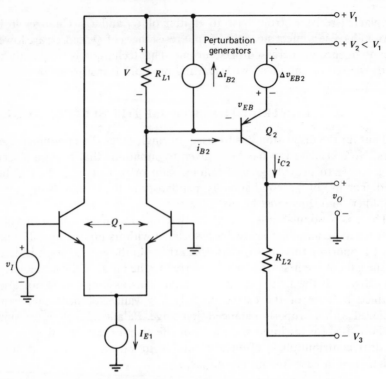

Figure 7.21 Two-stage d-c amplifier.

to operating-point values, linear analysis methods can be used to determine the drift referred to the input of the amplifier.

The results of this analysis show

$$v_I \Big|_{i_{C2}\,=\,\text{const}} = \frac{-2\,\Delta v_{EB2}}{g_{m1}R_{L1}} - \frac{2\,\Delta i_{B2}}{g_{m1}} \tag{7.45}$$

The gain portion of the first term on the right of Eqn. 7.45 can be expressed in terms of V, the quiescent voltage across R_{L1}. Similarly, the second term can be expressed in terms of I_{E1}, I_{C2}, the current gain of Q_2, and its fractional change. These substitutions yield

$$v_I \Big|_{i_{C2}\,=\,\text{const}} = \frac{-2kT\,\Delta v_{EB2}}{qV} + \frac{4kTI_{C2}\,\Delta\beta_2}{qI_{E1}\beta_2^2}$$

$$\simeq \frac{-\Delta v_{EB2}}{20V} + \frac{I_{C2}\,\Delta\beta_2}{10I_{E1}\beta_2^2} \tag{7.46}$$

at room temperature.

Typical values are used to illustrate magnitudes of these drift components with temperature. The voltage V is constrained by available supply voltages, and a value of 5 volts is assumed. The typical Δv_{EB} value of -2 mV/°C is used. A current gain of 300 coupled with a temperature coefficient of 0.6% per degree Centigrade is assumed for Q_2. Because the quiescent current level normally increases from the first stage to the second, a ratio of 5 is used for I_{C2}/I_{E1}. Substituting these values into Eqn. 7.46 shows that the drift attributable to changes in v_{EB2} is approximately 20 μV/°C, while the component arising from i_{B2} changes is 10 μV/°C. These values contrast dramatically with the drift that can be obtained from a properly designed first stage, and indicate the dominant effect that the second stage can have on drift performance.

The drift calculations of this section apply even if current gain only is required from the second stage. It is easy to show that the calculated values of drift are the same if an emitter follower is used in place of the grounded-emitter stage.

The final term in Eqn. 7.46 indicates the importance of changes in second-stage input current on drift performance. This term indicates that the drift performance deteriorates as the ratio of the quiescent operating current of the second stage to that of the first stage is increased. This result is one example of how certain design considerations (in particular, the desire to increase quiescent currents from the first to subsequent stages) must be compromised to achieve low drift performance.

7.5.2 Differential Second Stage

It is evident from the typical values calculated in the last section that unless care is taken in the design of the second stage of a d-c amplifier, this stage can easily dominate the drift performance of the circuit. One approach to the design of low-drift multistage d-c amplifiers is to use a differential second stage so that reflected drift is determined by differential rather than absolute changes in second-stage parameters.

Figure 7.22 shows a two-stage differential amplifier. Individual members of the first- and second-stage pairs are assumed matched. It is -further assumed that a single-ended output is desired, so one collector of the second-stage pair is grounded.

Normally a resistor is used in place of the current source I_{E2}. Since only differential input signals can be applied to the second stage, and therefore the common-emitter point of the second stage is incrementally grounded, the impedance connected to this point is irrelevant. However, the calculations are somewhat more convenient if a current source is included.

It is interesting to note that the voltage gain of this amplifier is identical to that of Fig. 7.21. Since the common-emitter connection of the second

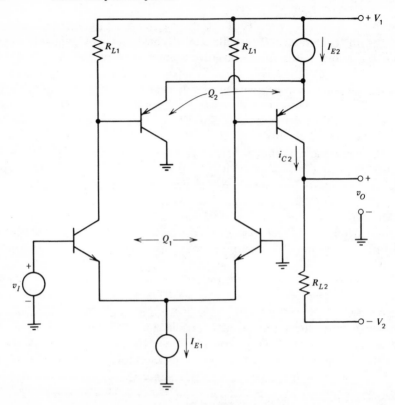

Figure 7.22 Amplifier with two differential stages.

stage is incrementally grounded for any possible input signal, no gain increase results from the left-hand member of the PNP pair.

Input drift attributable to second-stage differential base-to-emitter voltage changes is generally negligible if any degree of match exists. The drift referred to the input of the second stage is equal to the ratio $\Delta v_{BE2}/T$ per degree Centigrade (see Section 7.3.4). This value (typically on the order of 10 to 100 μV/°C) is divided by the unloaded differential voltage gain of the first stage (twice the single-ended value calculated in the preceding section) when reflected to the input.

The drift attributable to differential fractional changes in second-stage current gain is (assuming initially matched values for second-stage current gains)

$$v_I \bigg|_{i_{C_2}\ =\ \text{const}} = \frac{kTI_{E2}}{\beta_2 q I_{E1}} \left(\frac{\Delta \beta_{2A}\ -\ \Delta \beta_{2B}}{\beta_2} \right) \tag{7.47}$$

where the A and B subscripts indicate the two members of the second-stage pair. (The factor of four compared with the calculation of the last section occurs since each second-stage transistor is operating at $I_{E2}/2$ and since the differential connection requires that only half the differential current change be offset at either side.) The quantity $(\Delta\beta_{2A} - \Delta\beta_{2B})/\beta_2$ is typically 0.1% per degree Centigrade for well-matched discrete components, and is often lower for integrated-circuit pairs. It is interesting to note that this component of drift dominates many amplifier designs, particularly if the current gains and the temperature coefficients of the second stage are not well matched, or if the operating current level of the second stage is high relative to that of the first stage.

The use of a Darlington second stage with its lower input current offers some improvement, since the higher voltage drift of the Darlington is tolerable in this stage. Another possibility is to adjust the relative collector currents of the second stage so that the differential change in second-stage base current with temperature is zero. Unfortunately, this adjustment is difficult to make.

7.6 CONCLUSIONS

The successful design of low-drift direct-coupled amplifiers depends on exploiting the unique tracking properties of the differential amplifier, and the application of a number of drift reducing tricks that have evolved. In view of the many possible pitfalls, it is reassuring to realize that the drift of several commercially available integrated-circuit operational amplifiers is on the order of 3 μV per degree Centigrade or lower, and that at least one discrete-component design achieves a drift of 0.5 μV per degree Centigrade.

The purpose of the simple but somewhat tedious derivations and examples of this section has not been to permit exact evaluation of the drift of a circuit, but rather to emphasize that "little things mean a lot," and to indicate the dominant drift sources of a particular design so that they may be reduced.

PROBLEMS

P7.1

Figure 7.23 shows several amplifying connections that consist of ideal amplifiers and passive components. Offset sources are shown as batteries. Calculate the offset referred to the input (the input voltage required to make $v_O = 0$), the output offset (the output voltage with $v_I = 0$), and the gain (v_o/v_i) for each connection.

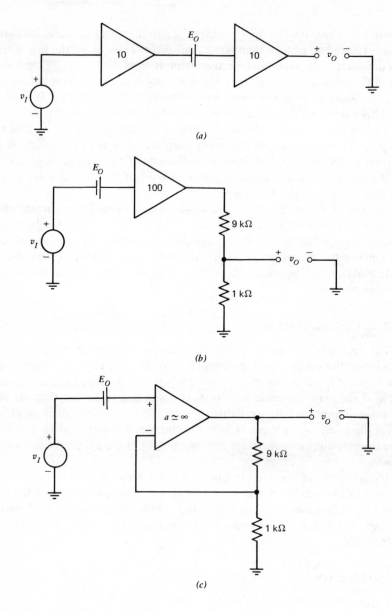

Figure 7.23 Amplifier Connections.

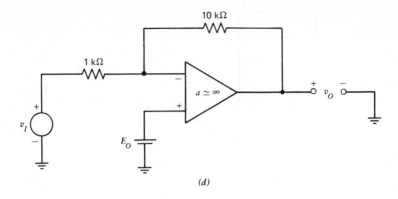

(d)

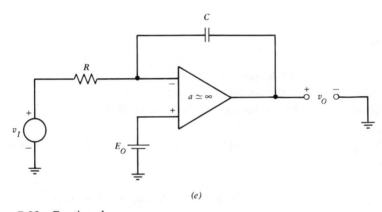

(e)

Figure 7.23—Continued

P7.2

Consider an operational amplifier with a particular value of offset E_O referred to its input. Compare the offset referred to the input of amplifier connections that combine this amplifier with passive components to provide inverting or noninverting gains with a magnitude of A.

P7.3

Figure 7.24 shows a circuit that can provide a temperature-independent output voltage. Assume that the transistor has very high β and that $i_O = 0$. The diode variables are related as

$$i_D = A_d T^3 \, e^{\, q(\, v_D - 0.782)/kT}$$

while the transistor relationship is

$$i_C = A_t T^3 \, e^{\, q(\, v_{BE} - 1.205)/kT}$$

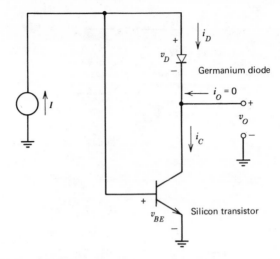

Figure 7.24 Voltage reference.

(a) For what ratio of A_d to A_t does $\partial v_O/\partial T = 0$?
(b) What is v_O with the condition of part a satisfied?
(c) What is the output resistance of this connection?

P7.4

The current-voltage relationship for a family of diodes can be approximated as

$$i_D = K\, e^{q(v_D - 1.2)/kT}$$

where K is a (temperature-independent) constant that may vary from diode to diode.

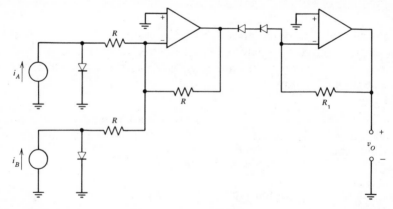

Figure 7.25 Nonlinear circuit.

(a) Four of these diodes with identical values for K are connected as shown in Fig. 7.25. Find v_O as a function if i_A and i_B. You may assume that the currents through all resistors R are much smaller than i_A or i_B and that both operational amplifiers are ideal.

(b) Determine an expression for

$$\frac{\partial v_D}{\partial T}\bigg|_{i_D \,=\, \text{const}} \quad \text{for these diodes.}$$

(c) Assume that, because of incredibly poor control of the process used to make these diodes, it is possible to find two diodes which, at $T = 300°$ K and 1 mA of forward current, have forward voltages of 0.3 V and 0.9 V, respectively. These diodes are connected as shown in Fig. 7.26, and the pot is adjusted so that $\partial v_O/\partial T = 0$. What is v_O with this pot setting?

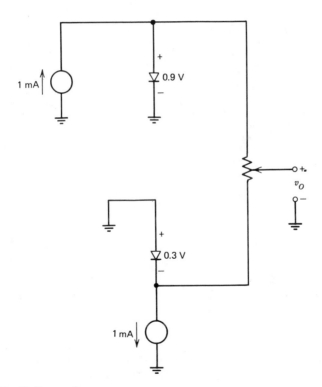

Figure 7.26 Voltage reference.

P7.5

The current-voltage relationship for a particular diode is

$$i_D = AT^{2.5}e^{q(v_D-1.205)/kT}$$

The value of the constant A is such that at 300° K and $v_D = 0.6$ V, $i_D = 1$ mA.

(a) Determine $\dfrac{\partial v_D}{\partial T}\bigg|_{i_D\,=\,\text{const}}$

(b) Seven identical diodes are connected as shown in Fig. 7.27. By appropriate choice of i_B, it is possible to make v_O temperature independent over a limited range of temperature. Determine the required value of v_O so that

$$\frac{\partial v_O}{\partial T}\bigg|_{i_B\,=\,\text{const}} = 0 \qquad \text{at } T = 300° \text{ K}$$

Approximate the value of I_B necessary to obtain the required value of v_O.

(c) Calculate the second derivative of v_O with respect to temperature. Use this value to estimate the temperature range over which v_O remains within one part in 10^5 of its 300° K value.

(d) Repeat part b assuming that the magnitude of the right-hand current source is increased to 10 mA.

The type of voltage reference that results from this topology is called a *band-gap reference*. The underlying principle is used as a voltage reference in several available integrated circuits.

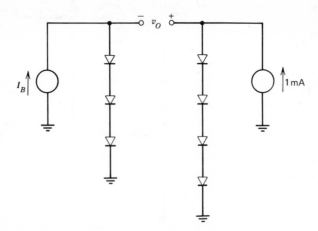

Figure 7.27 Band-gap standard.

P7.6

A differential amplifier is built with the topology shown in Fig. 7.11, with the exception that signals may also be applied to the base of the right-hand transistor. The value of the current source is 20 μA, and the incremental output resistance of this element is 10 MΩ. (The reasons for finite output resistance from current sources are discussed in Section 8.3.5.) Calculate the common-mode rejection ratio of this amplifier as a function the fractional unbalance in collector load resistors, Δ, assuming all transistor parameters are perfectly matched.

P7.7

An operational amplifier is built using a bipolar-transistor differential input stage. It is found that when the inverting input of the amplifier is grounded, the output voltage of the amplifier is zero at 25° C when a positive voltage of magnitude ΔV is applied to the noninverting input of the amplifier. You may assume that this offset and any temperature-dependent drift of the operational amplifier are caused only by a mismatch between the quantities I_S of the input-transistor pair, and that transistor variables are related by Eqn. 7.1.

The operational amplifier is intended for use in an inverting-amplifier connection, and therefore it is possible to reduce the effective offset at the inverting input to zero at 25° C by applying a voltage ΔV to the noninverting input. Three techniques for obtaining this bias voltage are indicated in Fig. 7.28. Comment on the effectiveness of these three balancing methods in reducing the temperature drift of the amplifier. Assume that the diode forward-voltage variation with temperature is given by

$$\left. \frac{\partial v_D}{\partial T} \right|_{i_D \,=\, \text{const}} = \frac{(v_D - V_{go})}{T} - \frac{3k}{q}$$

in parts b and c.

P7.8

A differential amplifier is constructed and balanced as shown in Fig. 7.10. Following balancing, it is found that transistor Q_1 is operating at a quiescent collector current of 1.1 mA, while Q_2 operates at a collector current of 0.9 mA. The transistors used are discrete devices mounted in reasonably close thermal proximity, and have a differential thermal resistance of 20° C per watt (i.e., if one member of the pair operates at a power level ΔP watts above that of the other, its temperature is 20 $\times$ ΔP degrees Centigrade higher). Estimate the offset referred to the input that results for a one-volt change in power-supply voltage.

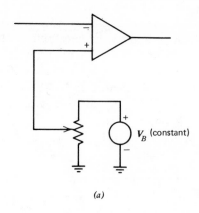

(a)

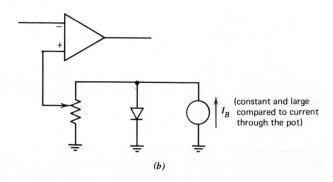

(constant and large
I_B compared to current
through the pot)

(b)

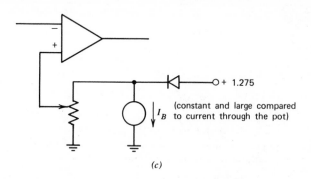

+ 1.275

(constant and large compared
I_B to current through the pot)

(c)

Figure 7.28 Methods to reduce offset at inverting terminal to zero. (Potentiometer set to make voltage at noninverting input ΔV at 300° K in all cases.)

P7.9

A differential amplifier that can provide low input capacitance, and, by proper control of bias voltage V_B, high common-mode rejection ratio, is shown in Fig. 7.29. Assume that Q_1 and Q_2 are perfectly matched. Further assume that $\beta_3 = \beta_4 = 100$ at 25° C. The output voltage is then zero for $v_I = 0$. Assume that the fractional change in β_3 is 0.5% per degree Centigrade, while that of β_4 is 1% per degree Centigrade. Calculate the offset referred to the input for a 1° C temperature change.

P7.10

An operational amplifier is found to have a bias-current requirement at its noninverting input that is 10% higher than that at its inverting input at all temperatures of interest. The amplifier is connected as shown in Fig. 7.30. Select the value of R that minimizes the effect of input current on circuit performance.

P7.11

The current at the inverting input of a certain operational amplifier is found to be equal to $10^{-3} A/T^2$ where T is the temperature in degrees Kelvin. The amplifier is to be used in an inverting connection; conse-

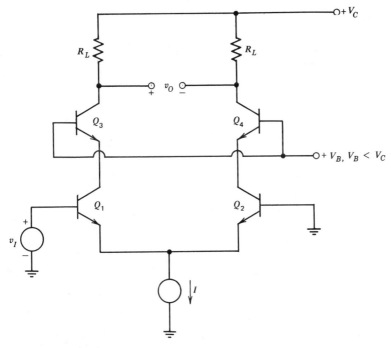

Figure 7.29 Cascoded differential amplifier.

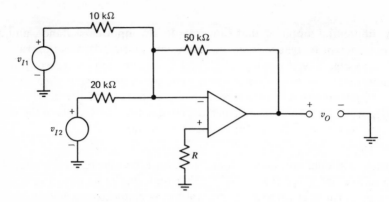

Figure 7.30 Summing amplifier.

quently the technique illustrated in Fig. 7.15 can be employed for input-current compensation. Parameters are selected so that the diode operates at a very nearly constant 1 mA, and its forward voltage at 300° K is 600 mV at this current. The diode current-voltage characteristics are of the general form

$$i_D = AT^3 e^{q(V_D - V_{go})/kT}$$

Select resistor R_2 and bias source V_A in Fig. 7.15 so that the input current and its derivative with respect to temperature are cancelled at 300° K. What is the maximum compensated input current over the temperature range of 250 to 350° K using this form of compensation? Contrast this range with the corresponding quantity obtained with no compensation and by cancelling the input current at 300° K with a fixed bias current.

P7.12

The use of Darlington-connected input-stage transistors is discussed in Section 7.4.4. An alternative high-gain connection is the complementary Darlington connection shown in Fig. 7.31a. A differential amplifier employing this connection is shown in Fig. 7.31b. Determine the voltage drift of this connection as a function of relative current-gain changes of the Q_1-Q_2 pair by an argument similar to that used for Fig. 7.20.

P7.13

A regulated power supply is constructed as shown in Fig. 7.32. This supply uses feedback around a very simple d-c amplifier in an attempt to make $v_O = V_R$.

(a) Determine the output voltage for circuit values as shown.

(b) How much does the output voltage change for a small fractional change in the current gain of Q_2?

(c) Suggest a circuit modification that will reduce the dependence of v_O on the fractional change in β_2.

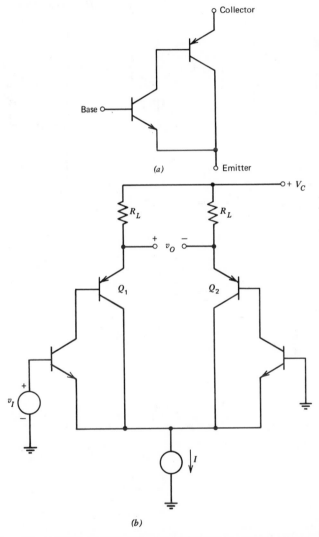

(a)

(b)

Figure 7.31 Differential amplifier using complementary Darlington-connected input transistors. (a) Base, collector, and emitter refer to terminals of the compound transistor. (b) Connection.

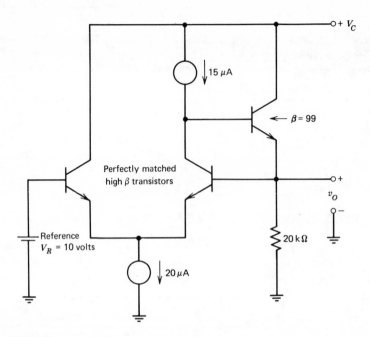

Figure 7.32 Power supply.

CHAPTER VIII

OPERATIONAL-AMPLIFIER DESIGN TECHNIQUES

8.1 INTRODUCTION

This chapter introduces some of the circuit configurations that are used for the design of high-performance operational amplifiers. This brief exposure cannot make operational-amplifier designers of us all, since considerable experience coupled with a sprinkling of witchcraft seems essential to the design process. Fortunately, there is little need to become highly proficient in this area, since a continuously updated assortment of excellent designs is available commercially. However, the optimum performance can only be obtained from these circuits when their capabilities and limitations are appreciated. Furthermore, this is an area where good design practice has evolved to a remarkable degree, and the techniques used for operational-amplifier design are often valuable in other applications.

The input stage of an operational amplifier usually consists of a bipolar-transistor differential amplifier that provides the differential input connection and the low drift essential in many applications. The design of this type of amplifier was investigated in detail in Chapter 7. The input stage is normally followed by one or more intermediate stages that combine with it to provide the voltage gain of the amplifier. Some type of buffer amplifier that isolates the final voltage-gain stage from loads and provides low output impedance completes the design. Configurations that are used for the intermediate and output stages are described in this chapter.

The interplay between a number of conflicting design considerations leads to a complete circuit that reflects a number of engineering compromises. For example, one simple way to provide the high voltage gain characteristic of operational amplifiers is to use several voltage-gain stages. However, we shall see that the use of multiple gain stages complicates the problem of insuring stability in a variety of feedback connections. Similarly, the dynamics of an amplifier are normally improved by operation at higher quiescent current levels, since the frequency response of transistors increases with increasing bias current until quite high levels are reached. However,

295

operation at higher current levels deteriorates d-c performance characteristics. Some of the guidelines used to resolve these and other design conflicts
are outlined in this chapter and illustrated by the example circuit described
in Chapter 9.

8.2 AMPLIFIER TOPOLOGIES

Requirements usually constrain the input and output stages of an operational amplifier to be a differential amplifier and some type of buffer
(normally an emitter-follower connection), respectively.

It is in the intermediate stage or stages that design flexibility is evident,
and the difference in performance between a good and a poor circuit often
reflects the differences in intermediate-stage design. The primary performance objective is that this portion of the circuit provide high voltage gain
coupled with a transfer function that permits stable, wide-band behavior
in a variety of feedback connections. Furthermore, the flexibility of easily
and predictably modifying the amplifier open-loop transfer function in
order to optimize it for a particular feedback connection is desirable for a
general-purpose design.

8.2.1 A Design with Three Voltage-Gain Stages

One much-too-frequently used design is shown in simplified form in Fig.
8.1. The path labeled feedforward is one technique used to stabilize the
amplifier, and is not essential to the initial description of operation. The
basic circuit uses a differential input since this connection is mandatory for
low drift and high common-mode rejection ratio. Two common-emitter
stages (transistors Q_3 and Q_4) are used to provide the high voltage gain
characteristic of operational amplifiers. Some sort of buffer amplifier
(shown diagrammatically as the unity-gain amplifier in the output portion)
is used to provide the required output characteristics.

Casual inspection indicates some merit for the design of Fig. 8.1. Low
drift is possible and d-c gains in excess of 10^5 can be achieved. The difficulty
is evident only when the dynamics of the amplifier are examined. The transfer function $V_o(s)/[V_{i2}(s) - V_{i1}(s)]$ determines stability in feedback connections. With typical element values, this transfer function has three or four
poles located within a two-to-three decade range of frequency. It is not
possible to achieve large loop-transmission magnitude and simultaneously
to maintain stability with this type of transfer function. The designer of
this type of amplifier should be discouraged when he compares his circuit
with that of a phase-shift oscillator, where negative feedback is applied
around three or more closely spaced poles.

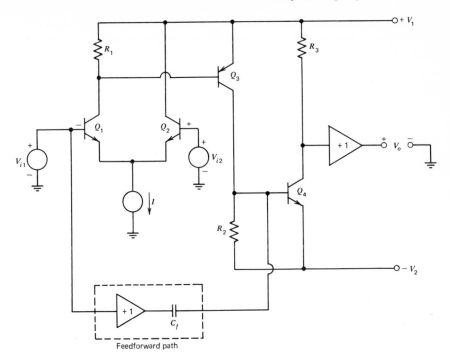

Figure 8.1   One approach to operational-amplifier design.

The problem can be illustrated by computing the transfer function for the amplifier shown in Fig. 8.1 with component values listed in Table 8.1. The reasons for selecting these component values are as follows. Fifteen-volt supplies are used since this value has become the standard for many solid-state operational amplifiers. The quiescent operating current of the first stage is low to reduce input bias current.

Relatively modest increases in quiescent currents from stage-to-stage are used to minimize loading effects. At these levels, circuit impedances are such that little change in the transfer function results if r_x is assumed equal to zero. However, r_x has been retained for completeness. Junction capacitances are dominated by space-charge layer effects at low operating currents, so equal values for all transistor capacitances have been assumed. Clearly any equal change in all capacitances simply frequency scales the transfer function. The resistors in the base circuits of Q_3 and Q_4 are assumed large to maximize d-c gain. In practice, current sources can be used to maintain high incremental resistance and to establish bias currents. Resistor R_3 is chosen to yield a quiescent output voltage equal to zero.

Table 8.1.   Parameter Values for Example Using Amplifier of Fig. 8.1

Supply voltages:

   ± 15 V

Bias currents:

   $I_{C1} = I_{C2} = 10 \ \mu A$
   $I_{C3} = 50 \ \mu A$
   $I_{C4} = 250 \ \mu A$

Transconductances[a] implied by bias currents:

   $g_{m1} = g_{m2} = 4 \times 10^{-4}$ mho
   $g_{m3} = 2 \times 10^{-3}$ mho
   $g_{m4} = 10^{-2}$ mho

Other transistor parameters:

   β = 100 (all transistors)
   $r_{\pi 1} = r_{\pi 2} = 250$ kΩ
   $r_{\pi 3} = 50$ kΩ
   $r_{\pi 4} = 10$ kΩ
   r_x = 100 Ω (all transistors)
   $C_\mu = C_\pi = 10$ pF (all transistors)

Reisistors:

   R_1 and R_2 large compared to $r_{\pi 3}$ and $r_{\pi 4}$, respectively.
   $R_3 = 60$ kΩ

   (Satisfying the inequalities normally requires that current sources be used rather than resistors in practical designs.)

Buffer amplifier assumed to have infinite input impedance.

---

   [a] Recall that for any bipolar transistor operating at current levels where ohmic drops are unimportant, the transconductance is related to quiescent collector current by $g_m = q|I_C|/kT \simeq 40 \ V^{-1}|I_C|$ at room temperature.

A computer-generated transfer function $V_o(j\omega)/[V_{i2}(j\omega) - V_{i1}(j\omega)]$ for this amplifier is shown in Bode-plot form in Fig. 8.2.[1] Two important features of this transfer function are easily related to circuit parameters. The low-frequency gain can be determined by inspection. Invoking the usual assump-

---

   [1] The gains of the amplifier for signals applied to its two inputs are not identical at high frequencies because a fraction of the signal applied to the base of Q_1 is coupled directly to the base of Q_3 via the collector-to-base capacitance of Q_1. This effect, which is insignificant until frequencies approaching the f_T's of the transistors used in the circuit, has been ignored in calculating the amplifier transfer function so that a true differential gain expression results.

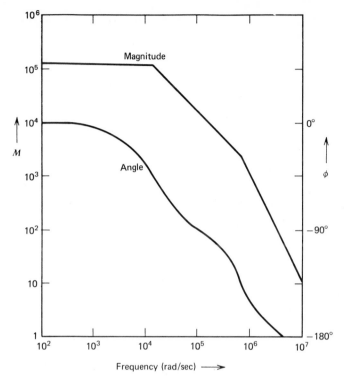

Figure 8.2   Transfer function for amplifier of Fig. 8.1.

tions, the incremental changes in first-stage collector current is related to an incremental change in differential input voltage as

$$i_{c1} = -\left(\frac{v_{i2} - v_{i1}}{1/g_{m1} + 1/g_{m2}}\right) \tag{8.1}$$

Since R_1 is large compared to the input resistance of Q_3, all of this incremental current flows into the base of Q_3. This base current is amplified by a factor of β_3, and resulting incremental current flows into the base of Q_4. The incremental output voltage becomes

$$v_o = -i_{c1}\beta_3\beta_4 R_3 \tag{8.2}$$

combining Eqns. 8.1 and 8.2 shows that the low-frequency voltage gain is

$$\frac{v_o}{v_{i2} - v_{i1}} = \frac{\beta_3\beta_4 R_3}{(1/g_{m1} + 1/g_{m2})} \tag{8.3}$$

Substituting parameter values from Table 8.1 into this equation shows that the incremental d-c gain is 1.2×10^5.

The lowest frequency pole plotted in Fig. 8.1 has a break frequency of 1.36×10^4 radians per second. This pole results from feedback through the collector-to-base capacitance of Q_4 (sometimes called Miller effect), as shown by the following development. An incremental model that can be used to evaluate the transimpedance of the final common-emitter stage is shown in Fig. 8.3. This transimpedance is a multiplicative term in the complete amplifier transfer function.

Node equations for this circuit are

$$-I_{c3} = [g_{\pi4} + (C_{\mu4} + C_{\pi4})s]V_a - C_{\mu4}sV_o$$

$$0 = (g_{m4} - C_{\mu4})sV_a + (G_3 + C_{\mu4}s)V_o \qquad (8.4)$$

Solving for the transimpedance shows that

$$\frac{V_o(s)}{I_{c3}(s)} = \frac{\beta R_3[-(C_{\mu4}/g_{m4})s + 1]}{r_{\pi4}R_3C_{\mu4}C_{\pi4}s^2 + r_{\pi4}\{[(g_{m4} + g_{\pi4})R_3 + 1]C_{\mu4} + C_{\pi4}\}s + 1} \qquad (8.5)$$

The denominator of Eqn. 8.5 is normally dominated by the term that includes the factor $g_{m4}R_3C_{\mu4}$, reflecting the importance of feedback through $C_{\mu4}$. Substituting values from Table 8.1 into Eqn. 8.5 and factoring the denominominal polynominal results in

$$\frac{V_o(s)}{I_{c3}(s)} = \frac{6 \times 10^6(-10^{-9}s + 1)}{(10^{-9}s + 1)(6.08 \times 10^{-5}s + 1)} \qquad (8.6)$$

This development shows that the output stage would have a dominant pole with a 1.64×10^4 radians-per-second break frequency in its transfer function if the other components in the circuit did not alter the location of this pole. This value agrees with the location of the dominant pole for the complete amplifier within approximately 20%.

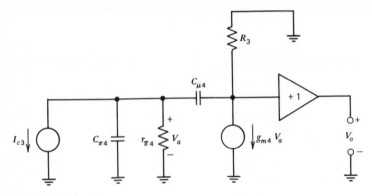

Figure 8.3 Model used to determine dynamics of final common-emitter stage of three-stage amplifier.

The algebra involved in getting this result can be circumvented by recognizing that a one-pole[2] (or Miller-effect) approximation to the input capacitance of transistor Q_4 predicts a value

$$C_T = C_{\pi4} + C_{\mu4}(1 + g_{m4}R_3) \tag{8.7}$$

The break frequency estimated at this node is

$$\omega_h = \frac{1}{r_{\pi4}C_T} = 1.66 \times 10^4 \text{ rad/sec} \tag{8.8}$$

While the d-c gain and the dominant pole location for this configuration are easily estimated, the location of other transfer-function singularities are related to amplifier parameters in a more complex way.

The essential feature to be gained from the Bode plot of Fig. 8.2 is that this transfer function is far from ideal for use in many feedback connections. The amplifier is hopelessly unstable if it is operated with its noninverting input connected to an incremental ground and a wire connecting its output to its inverting input, creating a loop with a as shown in the Bode plot and $f = 1$. In fact, if frequency-independent feedback is applied around the amplifier, it is necessary to reduce the magnitude of the loop transmission by a factor of 50 below the gain of the amplifier itself to make it stable in an absolute sense, and by a factor of 2000 to obtain 45° of phase margin. The required attenuation could be obtained by means of resistively shunting the input of the amplifier or through the use of a lag network (see Section 5.2.4). Either of these approaches severely compromises desensitivity and noise performance in many applications because of the large attenuation necessary for stability. Better results can normally be obtained by modifying the dynamics of the amplifier itself.

8.2.2 Compensating Three-Stage Amplifiers

At least two methods are often used to improve the dynamics of an amplifier similar to that described in the previous section. One of these approaches recognizes that the poles in the amplifier can be modeled as occurring because of R-C circuits located at various amplifier nodes. This type of association was made in the previous section for the dominant amplifier pole. The transfer function for a gain stage includes a multiplicative term of the general form $R_e/(R_eC_es + 1)$, where R_e and C_e are the effective resistance and capacitance at a particular node (see Fig. 8.4). If a com-

---

[2] P. E. Gray and C. L. Searle, *Electronic Principles: Physics, Models, and Circuits*, Wiley, New York, 1969, pp. 497–503.

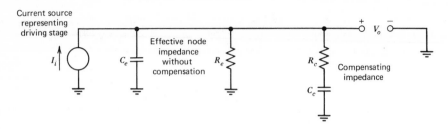

Figure 8.4 Compensation by adding a shunt impedance.

pensating series R-C network to ground consisting of a resistor $R_c \ll R_e$ and a capacitor $C_c \gg C_e$ is added, the transfer function becomes

$$\frac{V_o(s)}{I_i(s)} \simeq \frac{R_e(R_cC_cs + 1)}{(R_eC_cs + 1)(R_cC_es + 1)} \tag{8.9}$$

The single pole has been replaced by two poles and a zero. (Note that asymptotic behavior at high and low frequencies, which is controlled by R_e and C_c, has not been changed.) Component values are chosen so that one pole occurs at a much lower frequency than the original pole and the other at a frequency above the unity-gain frequency of the complete amplifier, as illustrated in Fig. 8.5. The positive phase shift of the zero often can improve the phase margin of the amplifier. This type of compensation can be viewed as one of combining the uncompensated transfer function with appropriately located lag and lead transfer functions. While the singularities must be related so that the compensated and uncompensated transfer functions are identical at very low and very high frequencies, the second pole can always be moved to arbitrarily high frequencies by locating the first pole at a sufficiently low frequency.

An alternative way to view this type of compensation is shown in the s-plane diagrams of Fig. 8.6. It is assumed that the three-stage amplifier has three poles at frequencies of interest. The lowest-frequency pole of the triad is replaced by two poles and a zero by means of a shunt R-C network. One possible way to choose singularity locations is to use the zero to cancel the second pole in the original transfer function and to locate the high-frequency pole that results from compensation above the highest-frequency original pole. The net effect of this type of compensation is to increase the separation of the poles so that greater desensitivity can be achieved for a given relative stability.

Several variations of the basic compensation scheme exist. It is possible to realize similar kinds of transfer functions by connecting a series R-C network from collector to base of a transistor rather than from its base to

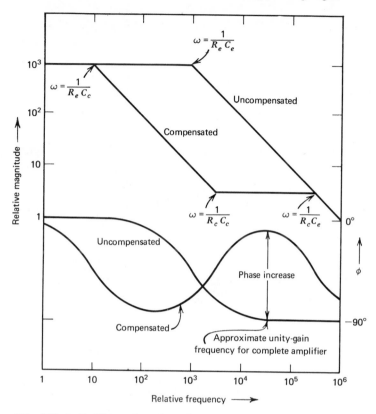

Figure 8.5   Effect of adding a shunt impedance on the transfer function of one stage.

an incrementally-grounded point. The same kind of compensation can be used at more than one node, and this multiple compensation is frequently required in more complex amplifiers.

While this general type of compensation is effective and has been successfully applied to a number of amplifier designs, it is less than ideal for several reasons. One of the more important considerations is that the determination of element values that result in a given transfer function requires rather involved calculations. This difficulty tends to discourage the user from finding the optimum compensating-element values for use in other than standard applications. This type of compensation also requires large capacitors (typically 1000 pF to 0.1 μF) when the network is shunted from base to an incremental ground. The energy storage of a large capacitor can delay recovery following an amplifier overload that charges the capacitor to the wrong voltage level.

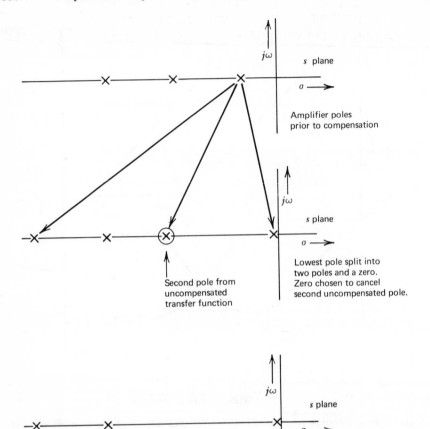

Figure 8.6 *s*-plane plots illustrating effect of shunt impedance on three-stage amplifier transfer function.

An alternative type of compensation that may be used alone or in conjunction with a shunt impedance is to "feed forward" around one or more amplifier stages as shown in Fig. 8.1. Here a unity-voltage-gain buffer amplifier (not essential but included in some designs to prevent loading at the inverting input terminal) couples the input signal to the base of Q_4 through capacitor C_f. Since the first stages are bypassed at high frequency, the high-frequency dynamics of the operational amplifier should be essentially those of the output stage. The hope is that the output stage has only one pole at frequencies of interest, and therefore will be stable with any amount of frequency-independent feedback.

Feedforward is not without its disadvantages. The frequency response of a feedforward amplifier is significantly lower for signals applied to the non-inverting input than for signals applied to its inverting input. Thus the amplifier has severely reduced bandwidth when used in noninverting connections. There are also problems that stem from the type of transfer functions that result from feedforward compensation. There is usually a second- or third-order rolloff at low frequencies, with the transfer function recovering to first order in the vicinity of the unity-gain frequency. Since this transfer function resembles those obtained with lag compensation, the settling time may be relatively long because of the small amplitude "tails" that can result with lag compensation (see Section 5.2.6). It is also possible to have these amplifiers become conditionally stable in certain connections (Section 6.3.4). This topic is investigated in Problem P8.3.

Before leaving the subject of three-stage amplifiers, the liberty that has been taken in the definition of a stage is worth noting. The stages are never as simple as those shown in Fig. 8.1. The essential feature that characterizes a voltage-gain stage is that it generally introduces one pole at moderate frequencies. The 709 (Fig. 8.7) is an example of an early integrated-circuit amplifier that is a three-stage design. While we do not intend to investigate the operation of this circuit in detail (several modern and more useful amplifiers are described in Chapter 10), the basic signal-flow path illustrates the three-stage nature of this design. Transistors Q_1 and Q_2 form a differential amplifier. The main second-stage amplification occurs through the Q_4-Q_6 Darlington-connected pair. Transistors Q_3 and Q_5 complete a differential second stage with the Q_4-Q_6 pair and are included primarily to reduce amplifier drift. Transistors Q_8 and Q_9 are used for level shifting, with common-emitter stage Q_{12} the final stage of voltage gain. Emitter followers Q_{13} and Q_{14} function as a buffer amplifier. There is some minor-loop feedback applied around the output stage to linearize its performance and to modify its dynamics via R_{15}.

Compensation is implemented by connecting a series R-C network from the output to the input of the second stage. It is also necessary to use capacitive feedback from the amplifier output to the base of Q_{12} (essentially around the output stage) to obtain acceptable stability in most applications.

8.2.3 A Two-Stage Design

While a number of operational-amplifier designs with three (or even more) voltage amplifying stages exist, it is hard to escape the conclusion that one is fighting nature when he tries to stabilize an amplifier with three

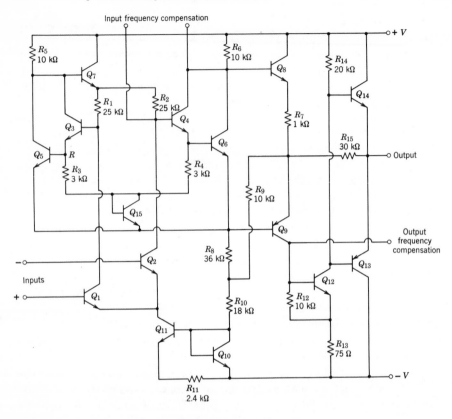

Figure 8.7   The 709 integrated-circuit operational amplifier.

or more closely spaced poles. The key to successful operational-amplifier design is to realize that the only really effective way to eliminate poles in an amplifier transfer function is to reduce the number of voltage-gain producing stages. Stages that provide current gain only, such as emitter followers, generally have poles located at high enough frequencies to be ignored.

An amplifier with two voltage-gain stages results if one of the common-emitter stages of Fig. 8.1 is eliminated, as shown in Fig. 8.8.[3] Again, transistors Q_1 and Q_2 function as a differential amplifier. However, in contrast to the previous amplifier, note that the base of transistor Q_1 is the inverting input of the complete amplifier, while the first-stage output is the collector

---

[3] The great value and versatility of this basic amplifier and its many variations were first pointed out to me by Dr. F. W. Sarles, Jr.

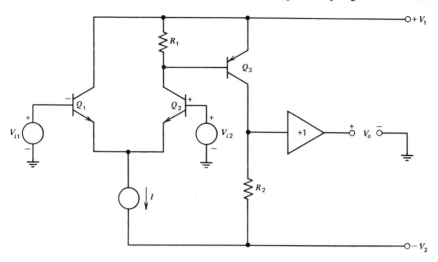

Figure 8.8 Basic two-stage amplifier.

of transistor Q_2. This emitter-coupled connection assures low input capacitance (approximately $C_{\mu 1} + C_{\pi 1}/2$) at the base of Q_1 since this device
is operating as an emitter follower. Low input capacitance is an advantage
in many applications since feedback is normally applied from the output
of the amplifier to its inverting input terminal. The input capacitance at the
inverting input can introduce an additional moderate-frequency pole in the
loop transmission of the amplifier-feedback network combination with attendant stability problems. Thus low input capacitance increases the range
of feedback impedances that can be used without deteriorating the loop
transmission.

The transfer function for this amplifier calculated using the parameter
values in Table 8.2 is[4]

$$\frac{V_o(s)}{V_{i2}(s) - V_{i1}(s)} = \frac{6 \times 10^3}{(3 \times 10^{-4}s + 1)(1.1 \times 10^{-8}s + 1)} \qquad (8.10)$$

with all other singularities above 5×10^8 sec^{-1}. The corresponding Bode
plot (Fig. 8.9) shows that a phase margin of 75° results even when the out-

---

[4] As in the case of the three-stage amplifier, the slight input-stage unbalance that occurs
at high frequencies because of signals fed directly to the base of Q_3 via the collector-to-base
capacitance of Q_2 has been ignored in the analysis that leads to this transfer function. The
error introduced by this simplification is insignificant at frequencies below the unity-gain
frequency of the amplifier. Furthermore, the transfer function of interest in most feedback
applications where the feedback signal is applied to the base of Q_1 does not include the
feed-forward term associated with $C_{\mu 2}$.

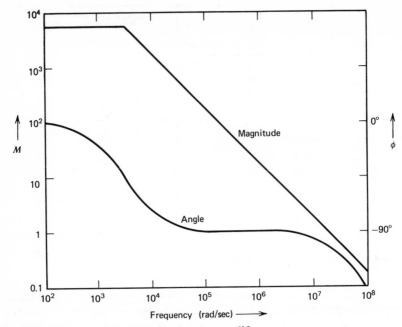

Figure 8.9 Transfer function of two-stage amplifier.

put of the amplifier is fed directly back to its inverting input. This type of transfer function, obtained without including any additional compensation components, contrasts sharply with the uncompensated three-stage-amplifier transfer function of the previous section.

It is informative to see why the transfer function of this amplifier is dominated by a single pole and why the second pole is separated from the dominant pole by a factor of approximately 30,000. This separation, which permits excellent desensitivity in feedback applications while maintaining good relative stability, is a major advantage attributable to the two-stage design. The dominant pole is primarily a result of energy storage in the collector-to-base capacitance of transistor Q_3. A C_T approximation to the input capacitance of this transistor is (see the discussion associated with Eqn. 8.7)

$$C_T = C_{\pi 3} + C_{\mu 3}(1 + g_{m3}R_2) = 6.02 \times 10^{-9} \text{ F} \qquad (8.11)$$

The corresponding time constant

$$\tau_{B3} = C_T r_{\pi 3} = 3.01 \times 10^{-4} \text{ sec} \qquad (8.12)$$

agrees with the dominant time constant in Eqn. 8.10. The essential point is that the feedback through $C_{\mu 3}$, which is actually a form of minor loop compensation (see Section 5.3), controls the transfer function of the com-

Table 8.2 Parameter Values for Example Using Amplifier of Fig. 8.8

Supply voltages:

   ± 15 V

Bias currents:

   $I_{C1} = I_{C2} = 10 \ \mu$A
   $I_{C3} = 50 \ \mu$A

Transconductances implied by bias currents:

   $g_{m1} = g_{m2} = 4 \times 10^{-4}$ mho
   $g_{m3} = 2 \times 10^{-3}$ mho

Other transistor parameters:

   $\beta \ = 100$ (all transistors)
   $r_{\pi 1} = r_{\pi 2} = 250$ kΩ
   $r_{\pi 3} = 50$ kΩ
   $r_x \ = 100 \ \Omega$ (all transistors)
   $C_\mu = C_\pi = 10$ pF (all transistors)

Resistors:

   $R_1 \gg r_{\pi 3}$
   $R_2 = 300$ kΩ

Buffer amplifier assumed to have infinite input impedance.

---

plete amplifier at frequencies between approximately 3.3×10^3 and 10^8 radians per second. As we shall see, the minor-loop feedback mechanism that dominates amplifier performance in this case can be used to advantage for compensation of more complex amplifiers that share the topology of this circuit.

Most modern high-performance operational amplifiers represent relatively straightforward extensions of the circuit shown in Fig. 8.8, and this popularity is a direct consequence of the excellent dynamics associated with the topology. An important modification included in most designs is the use of a more complex second stage than the simple common-emitter amplifier shown in Fig. 8.8 in order to achieve higher d-c open-loop gain. Other options exist in the way the output buffer circuit is realized and the drift-reducing modifications that may be incorporated into the first and second stages.

8.3 HIGH-GAIN STAGES

As mentioned in the previous section, a high-gain second stage is usually used to provide the basic amplifier with the voltage gain normally required from an operational amplifier. As we shall see, high current gain or high power gain alone is insufficient. It is necessary to have stages with high voltage gain, high transresistance (ratio of incremental output voltage to incremental input current), or both included in an operational-amplifier circuit. Note that there is no restriction on the number of transistors used in the stage. The implication in our definition of stage is that its dynamics are similar to that of a single common-emitter amplifier, that is, it introduces only one pole at frequencies that are low compared to the f_T of the devices used.

Use of the usual hybrid-pi model for the analysis of the simple common-emitter amplifier of Fig. 8.10 shows that the low-frequency incremental voltage is $v_o/v_i = -g_m R_L$ and the incremental transistance is $v_o/i_i = -\beta R_L$. The magnitude of either of these quantities can be increased (seemingly without limit) by increasing R_L. In order to obtain high gains without high supply voltages [the voltage gain of the circuit of Fig. 8.10 is $(q/kT)(V_C - V_O) \simeq 40(V_C - V_O)$], a current source can be used as the collector load. We realize that this technique will not result in infinite voltage gain and transresistance in an actual circuit because the simplified hybrid-pi model does not accurately predict the behavior of circuits with voltage gains in excess of several hundred. In order to proceed it is necessary to develop a more complete hybrid-pi model.

8.3.1   A Detailed Low-Frequency Hybrid-Pi Model[5]

The simplified hybrid-pi model predicts that both the base current and the collector current of a transistor are independent of changes in collector-to-base voltage. Actually, both currents are voltage-level dependent because of an effect called *base-width modulation*, as illustrated by the following argument. Consider an NPN transistor operating at moderate current levels with fixed base-to-emitter voltage V_{BE} and collector-to-base voltage V_{CB}. The approximate charge distribution in the base region for this transistor is shown by the solid line in Fig. 8.11. In this figure, n_p is the minority-carrier concentration in the base region; N_{po} is the equilibrium concentration of electrons in the base region; and x is the distance into the base region with $x = 0$ at the base edge of the emitter-base space-charge layer. The charge distribution drops linearly from its value $n_p(0)$ at $x = 0$ to essentially zero (if the collector-to-base junction is reverse biased by at least several hundred millivolts) at the edge of the collector space-charge layer. However, the width of the collector space-charge layer is

[5] This material is covered in greater detail in P. E. Gray et al., *Physical Electronics and Circuit Models for Transistors*, Wiley, New York, 1964, Chapter 8, and C. L. Searle et al., *Elementary Circuit Properties of Transistors*, Wiley, New York, 1964, Chapter 4.

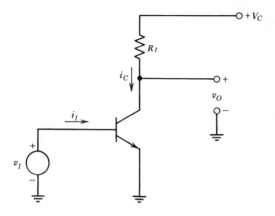

Figure 8.10 Common-emitter amplifier.

monotonically increasing function of collector-to-base voltage. Thus, if the collector-to-base voltage is reduced, the collector space-charge layer becomes narrower. This narrowing increases the effective width of the base region from its original value of W to a new value $W + \Delta W$. The resultant new charge distribution is shown by the dotted line in Fig. 8.11.

Two changes in terminal variables result from this change in base width. First, the collector current (proportional to the slope of the distribution) becomes smaller. Second, the base current increases, since the total rate at which charge recombines in the base region is directly proportional to the total charge in this region. The magnitudes of these changes are calculated as follows.

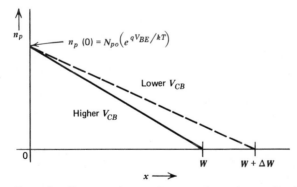

Figure 8.11 Effect of collector-to-base voltage on base-charge distribution (NPN transistor).

The collector current of an NPN transistor is related to transistor and physical constants by

$$I_C = \frac{qN_{po}AD_e}{W} e^{qV_{BE}/kT} \tag{8.13}$$

where

N_{po} is the equilibrium concentration of electrons in the base region.
A is the cross-sectional area of the base.
D_e is the diffusion constant for electrons in the base region.

The assumptions necessary to derive this relationship include operation under conditions of low-level injection but at current levels large compared to leakage currents, and that the ohmic drops in the base region are negligible. The assumption of negligible ohmic voltage drop in the base region results in no loss of generality, since a base resistance can be added to the model which evolves from Eqn. 8.13.

Under conditions of constant base-to-emitter voltage and temperature, Eqn. 8.13 reduces to

$$I_C = \frac{K}{W} \tag{8.14}$$

where the constant K includes all other terms from Eqn. 8.14. Differentiating yields

$$\frac{dI_C}{dW} = -\frac{K}{W^2} \tag{8.15}$$

Differential changes in W are related to incremental changes in collector-to-base voltage as

$$\Delta W = \frac{dW}{dV_{CB}} v_{cb} \tag{8.16}$$

Incremental changes in collector current can thus be expressed in terms of incremental changes in collector-to-base voltage as

$$i_c = -\frac{K}{W^2} \frac{dW}{dV_{CB}} v_{cb} \tag{8.17}$$

Solving Eqn. 8.14 for K and substituting into Eqn. 8.17 yields

$$i_c = -\frac{I_C}{W} \frac{dW}{dV_{CB}} v_{cb} \tag{8.18}$$

The transconductance of a transistor is related to quiescent collector current as

$$g_m = \frac{qI_C}{kT} \tag{8.19}$$

Solving Eqn. 8.19 for I_C and substituting this result into Eqn. 8.18 shows that

$$i_c = \left[-\frac{kT}{qW} \frac{dW}{dV_{CB}} \right] g_m v_{cb} \tag{8.20}$$

The bracketed quantity in Eqn. 8.20 is called the *base-width modulation factor* and is denoted by the symbol η. Introducing this notation and adding the familiar relationship between incremental components of collector current and base-to-emitter voltage to Eqn. 8.20 yields

$$i_c = g_m v_{be} + \eta g_m v_{cb} \tag{8.21}$$

The quantity η is typically 10^{-3} to 10^{-4}, indicating that the collector current is much more strongly dependent on base-to-emitter voltage than on collector-to-base voltage. This is, of course, the reason we are able to ignore the effect of collector-to-base voltage variations except in high-gain situations.

The change in base current as a function of collector-to-base voltage can be calculated with the aid of Fig. 8.11. If reverse injection from the base into the emitter region is assumed small, the base current is directly proportional to the area of the triangle, since the total number of minority carriers that recombine per unit time and thus contribute to base current is proportional to the total number of these carriers in the base region. The geometry of Fig. 8.11 shows that the magnitude of the fractional change in the area of the triangle is equal to the magnitude of the fractional change in slope of the distribution for small changes in W. Furthermore, an increase in W decreases collector current and increases base current. Equating fractional changes yields

$$\frac{i_b}{I_B} = -\frac{i_c}{I_C} = -\frac{\eta g_m v_{cb}}{I_C} \tag{8.22}$$

Rearranging Eqn. 8.22 and recognizing that $I_C/I_B = \beta$ yields for the incremental dependence of base current on collector-to-base voltage at constant base-to-emitter voltage

$$i_b = -\frac{\eta g_m v_{cb}}{\beta} \tag{8.23}$$

Adding the incremental relationship between base current and base-to-emitter voltage to Eqn. 8.23 results in

$$i_b = \frac{g_m}{\beta} v_{be} - \frac{\eta g_m}{\beta} v_{cb} \tag{8.24}$$

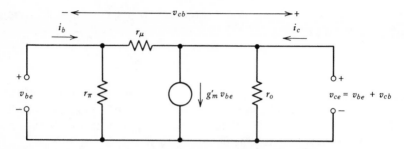

Figure 8.12   Intrinsic hybrid-pi model that includes base-width modulation effects.

It is necessary to augment the familiar hybrid-pi transistor model to include the effects of base-width modulation when the model is used for the analysis of high-gain circuits. While there are several model modifications that would accurately represent base-width-modulation phenomena, convention dictates that the model be augmented by the addition of a collector-to-emitter resistor r_o and a collector-to-base resistor r_μ as shown in Fig. 8.12. The objective is to choose the four elements of the model so that the terminal relationships dictated by Eqns. 8.21 and 8.24 are obtained. Note that, since four degrees of freedom are required to match arbitrary two-port relationships, it may be necessary to have the dependent current-generator scale factor in Fig. 8.12 differ from g_m, and this possibility is indicated by calling this scale factor g_m'.

The terminal relationships developed from the analysis of the effects of base-width modulation are repeated here for convenience:

$$i_c = g_m v_{be} + \eta g_m v_{cb} \tag{8.21}$$

$$i_b = \frac{g_m}{\beta} v_{be} - \frac{\eta g_m}{\beta} v_{cb} \tag{8.24}$$

The equations relating the same variables for the model of Fig. 8.12 are[6]

$$i_c = g_m' v_{be} + g_\mu v_{cb} + g_o(v_{be} + v_{cb})$$

$$= (g_m' + g_o)v_{be} + (g_o + g_\mu)v_{cb} \tag{8.25}$$

$$i_b = g_\pi v_{be} - g_\mu v_{cb} \tag{8.26}$$

Equationing coefficients in these two sets of equations yields

$$g_m' + g_o = g_m \tag{8.27}$$

---

[6] Recall that corresponding r's and g's are reciprocally related. Thus, for example, $g_o = 1/r_o$.

$$g_o + g_\mu = \eta g_m \tag{8.28}$$

$$g_\pi = \frac{g_m}{\beta} \tag{8.29}$$

$$g_\mu = \frac{\eta g_m}{\beta} \tag{8.30}$$

These equations are readily solved to determine model element values:

$$g'_m = g_m \left[1 - \eta \left(1 - \frac{1}{\beta} \right) \right] \tag{8.31}$$

$$r_\pi = \frac{1}{g_\pi} = \frac{\beta}{g_m} \tag{8.32}$$

$$r_o = \frac{1}{g_o} = \frac{1}{\eta g_m [1 - (1/\beta)]} \tag{8.33}$$

$$r_\mu = \frac{1}{g_\mu} = \frac{\beta}{\eta g_m} \tag{8.34}$$

Since for any well-designed transistor $|\eta| \ll 1$ (typical values are 10^{-3} to 10^{-4}) and $\beta \gg 1$, the approximations

$$g'_m \simeq g_m = \frac{q|I_C|}{kT} \tag{8.35}$$

and

$$r_o \simeq \frac{1}{\eta g_m} \tag{8.36}$$

usually replace Eqns. 8.31 and 8.33, respectively.

It is instructive to examine the relative magnitudes of the model parameters for a transistor under typical conditions of operation. Assume that a transistor with $\beta = 200$ and $\eta = 4 \times 10^{-4}$ is operated at $I_C = 1$ mA at room temperature. Then $g_m = 40$ mmho, $g_\pi = 200$ μmho or $r_\pi = 5$ kΩ, $g_o = 16$ μmho or $r_o = 62.5$ kΩ and $g_\mu = 0.08$ μmho or $r_\mu = 12.5$ MΩ. Note that all conductances in the intrinsic model are proportional to g_m and therefore to quiescent collector current.

8.3.2 Common-Emitter Stage with Current-Source Load

In spite of the internal loading of r_o and r_μ, high voltage gain is possible with a current-source load for a common-emitter stage, and this connection is used in many operational-amplifier designs. Figure 8.13a shows a sche-

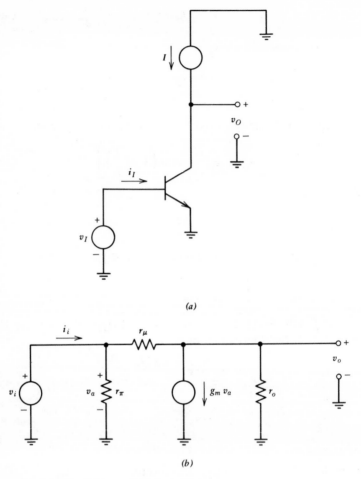

(a)

(b)

Figure 8.13 Current-source-loaded common-emitter stage. (a) Schematic. (b) Incremental equivalent circuit (r_x negligibly small).

matic for such a stage and Fig. 8.13b is the corresponding low-frequency equivalent circuit. It is assumed that the incremental resistance of the current source is infinite. (The problems associated with realizing a high-resistance current source will be described in Section 8.3.5.) It is also assumed that the base resistance of the transistor can be neglected. This assumption is best justified by considering a complete amplifier where the resistances at various nodes are known. In most anticipated applications r_x will either be small enough so that it can be neglected even for voltage-source drives at the base of the transistor in question, or the value of r_x will be masked by a large driving resistance connected in series with it.

The equivalent circuit of Fig. 8.13b is easily analyzed by solving the output-node equation:

$$g_m v_i + g_o v_o + g_\mu (v_o - v_i) = 0 \qquad (8.37)$$

Since $g_\mu \ll g_o$ (see Eqns. 8.34 and 8.36) and $g_\mu \ll g_m$,

$$\frac{v_o}{v_i} \simeq -g_m r_o \qquad (8.38)$$

With the equivalence of Eqn. 8.36, $r_o = 1/\eta g_m$, the voltage-gain of the circuit becomes simply $-1/\eta$. As mentioned earlier typical values for η are 10^{-3} to 10^{-4}, and therefore a voltage-gain magnitude of 10^3 to 10^4 is possible.

The incremental input current can be calculated as follows.

$$i_i = (g_\pi + g_\mu)v_i - g_\mu v_o \qquad (8.39)$$

Substituting from Eqn. 8.38 yields

$$i_i = (g_\pi + g_\mu + g_m r_o g_\mu)v_i \qquad (8.40)$$

Recognizing that

$$g_m r_o g_\mu = g_\pi \qquad (8.41)$$

simplifies Eqn. 8.40 to

$$i_i = (2g_\pi + g_\mu)v_i \simeq 2g_\pi v_i \qquad (8.42)$$

This relationship indicates that the use of a current-source load halves the input resistance of a common-emitter amplifier compared to the value when loaded with a moderate-value resistor, since the currents flowing through r_π and r_μ are equal in this high-gain connection.

Combining Eqns. 8.42 and 8.38 shows that the transresistance is

$$\frac{v_o}{i_i} = -\frac{r_\pi g_m r_o}{2} = -\frac{\beta r_o}{2} = -\frac{r_\mu}{2} \qquad (8.43)$$

The dominant pole for this amplifier, at least for realistic values of driving-source resistance, occurs at the input. Because of the high voltage gain, the input capacitance includes a component several thousand times larger than C_μ, and this effective input capacitance is the primary energy-storage element.

8.3.3 Emitter-Follower Common-Emitter Cascade

The current-source-loaded common-emitter stage analyzed in the preceding section can be driven with an emitter follower to increase transresistance. Figure 8.14 illustrates this connection. Analysis is simplified by applying the results of the last section. Since the input resistance of the

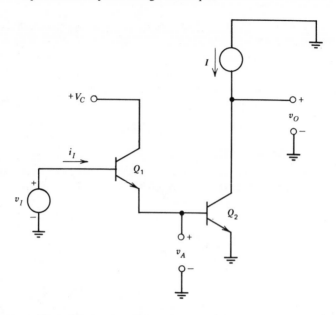

Figure 8.14   Emitter-follower common-emitter cascade.

common-emitter amplifier is $r_\pi/2$ (Eqn. 8.42), the transfer ratios v_a/v_i and v_a/i_i can be calculated by replacing the input circuit of Q_2 with a resistor equal to $r_{\pi2}/2$. These results are combined with Eqns. 8.38 and 8.42 to determine gain and transresistance. Furthermore, it is not necessary to consider elements r_o and r_μ in the model for transistor Q_1 since the voltage gain of this device is low. An incremental equivalent circuit that relates v_a to v_i is shown in Fig. 8.15.

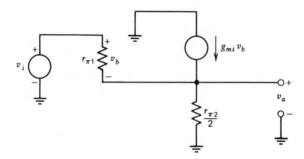

Figure 8.15   Equivalent circuit used to determine v_a/v_i for circuit of Fig. 8.14.

The voltage-transfer ratio is

$$\frac{v_a}{v_i} = 1 - \frac{1}{1 + r_{\pi 2}/2r_{\pi 1} + g_{m1}r_{\pi 2}/2} \tag{8.44}$$

For the circuit of Fig. 8.14 the quiescent collector current of Q_2 is I, while that of Q_1 is approximately I/β_2. Therefore,

$$r_{\pi 2} = \frac{\beta_2}{g_{m2}} = \frac{\beta_2 kT}{qI} \tag{8.45}$$

and

$$r_{\pi 1} = \frac{\beta_1}{g_{m1}} = \frac{\beta_1\beta_2 kT}{qI} = \beta_1 r_{\pi 2} \tag{8.46}$$

Equation 8.46 shows that for reasonable values of β_1, the term $r_{\pi 2}/2r_{\pi 1}$ in Eqn. 8.44 can be dropped.

Introducing this simplification and noting that $g_{m2} = \beta_2 g_{m1}$, so that $r_{\pi 2} = 1/g_{m1}$ reduces Eqn. 8.44 to

$$\frac{v_a}{v_i} = \frac{1}{3} \tag{8.47}$$

Therefore

$$\frac{v_o}{v_i} = -\frac{1}{3\eta_2} \tag{8.48}$$

Since $v_a = \frac{1}{3}\,v_i$, the input resistance is

$$\frac{v_i}{i_i} = \frac{3}{2}\,r_{\pi 1} \tag{8.49}$$

Combining Eqns. 8.48 and 8.49 shows that the transresistance is

$$\frac{v_o}{i_i} = -\frac{r_{\pi 1}}{2\eta_2} \tag{8.50}$$

This equation can be compared with Eqn. 8.43 by noting that $r_{\pi 1} = \beta_1\beta_2/g_{m2}$. Thus

$$\frac{v_o}{i_i} = -\frac{\beta_1\beta_2}{2g_{m2}\eta_2} = -\frac{\beta_1 r_{\mu 2}}{2} \tag{8.51}$$

Transistor Q_1 simply improves the transresistance of the circuit by a factor of β_1.

The dominant pole for this circuit is associated with the input of Q_2, since the incremental resistance to ground at this point remains high even with the emitter follower included.

8.3.4 Current-Source-Loaded Cascode

The gain limitations of the common-emitter amplifier stem from an internal negative-feedback mechanism related to transistor operation. As the collector-to-base voltage changes, the effective width of the base region also changes and resulting variations in collector- and base-terminal current oppose the original change. This effect is similar to that of the collector-to-base capacitance C_μ that supplies charge to both the collector and base terminals in such a direction as to oppose rapid variations in collector voltage. The cascode connection, which is useful because it minimizes feedback through C_μ at high frequencies, can also be used to minimize the effects of base-width modulation on circuit performance.

A connection that combines a cascode amplifier with a current-source load is shown in Fig. 8.16. This circuit can be analyzed by brute-force techniques, or a little thought can be traded for a page of calculations. We have already shown that the voltage gain of a current-source-loaded common-emitter amplifier is $-1/\eta$.

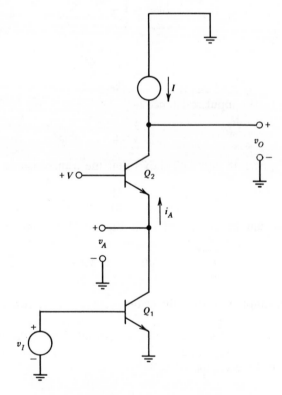

Figure 8.16 Cascode amplifier with current-source load.

Therefore the transfer ratio v_o/v_a in Fig. 8.16 is

$$\frac{v_o}{v_a} = \frac{1}{\eta_2} + 1 \simeq \frac{1}{\eta_2} \qquad (8.52)$$

We have also shown that the input resistance for the common-emitter amplifier is $r_\pi/2$. Observe that since the incremental collector current of Q_2 cannot change in the connection of Fig. 8.16, the incremental ratio v_a/i_a must be the same as the input resistance of the common-emitter amplifier, or

$$\frac{v_a}{i_a} = \frac{r_{\pi 2}}{2} \qquad (8.53)$$

The voltage gain of Q_1 can be calculated by simply assuming it is loaded with a resistor equal $r_{\pi 2}/2$. Accordingly,

$$\frac{v_a}{v_i} = -g_{m1} \frac{r_{\pi 2}}{2} \qquad (8.54)$$

providing this gain is small enough so that $r_{\mu 1}$ and r_{o1} are negligible. Equation 8.54 can be simplified by noting that $r_{\pi 2} = \beta_2/g_{m2}$, and that $g_{m1} = g_{m2}$ since both devices are operating at virtually identical quiescent currents. With this relationship the voltage gain of the current-source-loaded cascode becomes

$$\frac{v_o}{v_i} = -\frac{\beta_2}{2\eta_2} \qquad (8.55)$$

Since the input resistance of Q_1 is $r_{\pi 1}$, the transresistance for the circuit is

$$\frac{v_o}{i_i} = -\frac{\beta_2 r_{\pi 1}}{2\eta_2} = -\frac{\beta_2 \beta_1}{2\eta_2 g_{m1}} = -\frac{\beta_2 \beta_1}{2\eta_2 g_{m2}} = -\frac{\beta_1 r_{\mu 2}}{2} \qquad (8.56)$$

Comparing the cascode with the two previous circuits, we see that it provides the same transresistance as the circuit including the emitter follower and has significantly higher voltage gain than either of the other circuits. It is of practical interest to note that transistors are available that can provide voltage gains in excess of 10^5 in this connection.

The dominant pole occurs at the collector of Q_2 because the incremental resistance at this node is extremely high. The use of the cascode reduces the capacitance seen at the base of Q_1 so that even with a high source resistance, the time constant at this node is typically between 100 and 10,000 times shorter than the collector-circuit time constant.

8.3.5   Related Considerations

The circuits described in the last three sections offer at least one further advantage that is useful for the design of operational amplifiers. The current source included in all of these circuits insures that the transistors operate at quiescent current levels that are essentially independent of output voltage. Large output-voltage swings are therefore possible without altering any current-dependent transistor parameters.

Care may be required in the design of a current source with sufficiently high output resistance to prevent significant loading of the high-gain stages. Figure 8.17a shows a transistor connected as a current source. The output resistance for this connection determined from the incremental circuit model is

$$\frac{v_o}{i_o} = r_\mu \left\| \left[\frac{1 + (g_m + g_o)(r_\pi \| R_E)}{g_o} \right] \simeq r_\mu \right\| \left[\frac{1 + g_m(r_\pi \| R_E)}{g_o} \right] \quad (8.57)$$

The output resistance varies from

$$\frac{v_o}{i_o} \simeq r_o \qquad \text{for} \qquad R_E = 0 \tag{8.58}$$

to

$$\frac{v_o}{i_o} \simeq r_\mu \left\| \frac{g_m r_\pi}{g_o} = \frac{r_\mu}{2} \right. \qquad \text{for} \qquad R_E \gg r_\pi \tag{8.59}$$

This analysis indicates that it is not possible to build a current source of this type with an output resistance in excess of $r_\mu/2$.

Since r_μ is current dependent and since the current source operates at a current level equal to that of its driving transistor in the high-gain circuits, r_μ and r_o for a current-source transistor will be comparable to those of the driving transistor. The analysis of Section 8.3.2 can be extended to show that the output resistance of the common-emitter stage is r_o when driven from a voltage source and is $r_o/2$ when driven from a high impedance source. Thus use of a common-emitter current source ($R_E = 0$ in Fig. 8.17) can reduce the gain of this stage by as much as a factor of two. Since the output resistance of the emitter-follower common-emitter cascode is $2r_o/3$ when driven from a voltage source, the susceptibility of this stage to loading is comparable to that of the common-emitter stage.

The output resistance of the cascode is $r_\mu/2$, so even the highest output resistance that can be achieved with a bipolar-transistor current source will halve the unloaded gain of this stage. A further practical difficulty is that approaching a current-source resistance of $r_\mu/2$ requires $R_E \gg r_\pi$ (Eqn. 8.57). If we assume the base-to-emitter voltage of the transistor is small compared to V in Fig. 8.17a,

$$R_E \simeq \frac{V}{I_E} = \frac{qV}{kTg_m} \simeq \frac{40Vr_\pi}{\beta} \tag{8.60}$$

In order to satisfy the inequality $R_E \gg r_\pi$, it is necessary to have $V \gg \beta/40$.

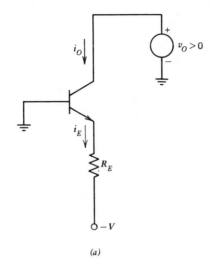

(a)

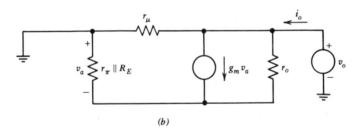

(b)

Figure 8.17 Current source. (a) Schematic. (b) Equivalent circuit.

The use of low β transistors is not the answer, since such transistors also have low r_μ. One way to avoid the requirement for high supply voltage is to use the connection of Fig. 8.18. Cascoding serves the same function as it does in the amplifier, and provides an output resistance of approximately $r_\mu/2$ with a total supply voltage of several volts.

The analysis presented above shows that the output resistance of a bipolar-transistor current source is bounded by $r_\mu/2$, and that this maximum value occurs only when the base of the transistor is connected to a low resistance level relative to the emitter-circuit resistance. Field-effect transistors (FET's) can be used in the interesting connection shown in Fig. 8.19a to increase the output resistance of a current source. A model that can be used for the linear-region analysis of the FET is shown in Fig. 8.19b. An incremental equivalent circuit of the cascoded source, assuming that the finite output resistance of the current source $R_S = v_a/i_a$ completely de-

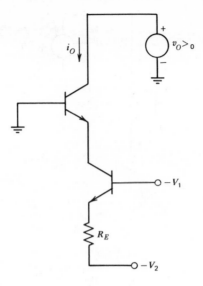

Figure 8.18 Cascoded current source.

scribes this element, is shown in Fig. 8.19c. This equivalent circuit shows that the relationship between v_o and i_o is

$$v_o = i_o R_S + \frac{i_o}{y_{os}} + \frac{i_o R_S y_{fs}}{y_{os}} \tag{8.61}$$

or that

$$\frac{v_o}{i_o} = \frac{1}{y_{os}} + R_S \left(1 + \frac{y_{fs}}{y_{os}}\right) \tag{8.62}$$

Since the quantity y_{fs}/y_{os} can be several hundred or more for certain FET's, this connection greatly increases the incremental resistance of the current source itself. For example, by using a bipolar-transistor current source cascoded with a FET, incremental resistances in excess of 10^{12} Ω can be obtained at a quiescent current of 10 μA. It is theoretically possible to further increase current-source output resistance by using multiple cascoding with FET's, although stray conductance limits the ultimate value in actual circuits.

Another problem that occurs in the design of high-gain stages is that the output of the stage must be isolated with a very high-input-resistance buffer to prevent loading that can cause a severe reduction in the voltage gain of the stage. One approach is to use a FET as a source follower, since the input resistance of this connection is essentially infinite. The use of a FET as a buffer or to cascode a current source is frequently the best technique

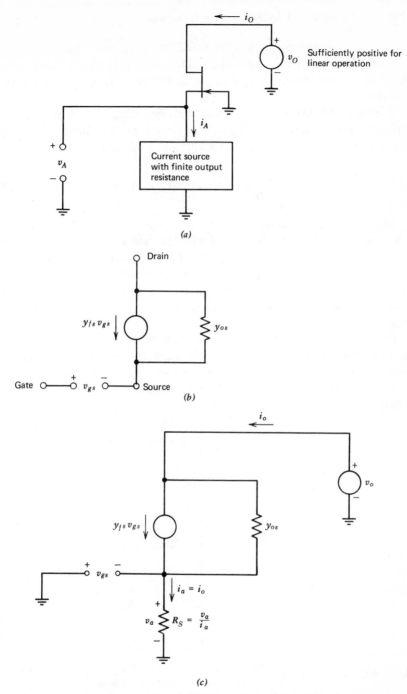

Figure 8.19 Current source cascoded with a field-effect transistor. (*a*) Circuit. (*b*) Linear model for field-effect transistor. (*c*) Incremental equivalent circuit.

in discrete-component designs. However, it is presently difficult to fabricate high-quality bipolar and field-effect transistors simultaneously in monolithic integrated-circuit designs; thus alternatives are necessary for these circuits.

If a bipolar-transistor emitter follower (Fig. 8.20) is used, care must be taken to insure sufficiently high input resistance. The incremental input resistance for this circuit with no additional loading is

$$\frac{v_i}{i_i} \simeq r_\mu \| [r_\pi + \beta(r_o \| R_E)] \tag{8.63}$$

In order to approach the maximum input resistance of $r_\mu/2$ (particularly important if the buffer is to be used with the cascode amplifier), it is necessary to have $R_E \gg r_o$. This inequality normally cannot be satisfied with reasonable supply voltages, so a current source is frequently used in place of R_E. A further advantage of the current source is that the drive current that can be supplied to any following stage becomes independent of voltage level.

One design constraint for an emitter follower intended for use with the current-source-loaded cascode amplifier is that the quiescent operating current of this stage should not be large compared with that of the cascode or else the gain of the stage will be determined primarily by r_μ of the emitter follower.

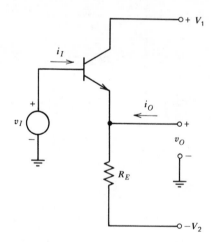

Figure 8.20 Emitter follower.

8.4 OUTPUT AMPLIFIERS

Factors that influence the design of the differential amplifier normally used as the input stage of an operational amplifier were investigated in Chapter 7, and the design of stages that provide high voltage gain was covered in earlier sections of this chapter. Modern operational amplifiers that combine a differential-amplifier input stage (often current-source loaded) with a current-source-loaded second stage require a final amplifier to supply output current and to provide additional isolation for the preceding high-gain stage. The dividing line between the devices used primarily to supply output current and those used to isolate the high-resistance node of the high-gain stage is often hazy. The emphasis in this section is on the power-handling aspect of the output amplifier. The guidelines of the previous section are used when isolation is the major objective.

Some type of emitter-follower circuit is almost always used as the output stage of an operational amplifier, since this configuration combines the necessary current gain with dynamics that can usually be ignored until frequencies above the unity-gain frequency of the complete amplifier are reached.

The simplest emitter-follower connection is shown in Fig. 8.21, and this circuit is powered from the ±15-volt supplies that have become relatively standard for operational amplifiers. While this circuit can provide the necessary output current and isolation, it requires high quiescent power relative to the maximum power it can supply to the load. If the circuit is designed so that the output voltage can swing to at least − 10 volts (a typical value

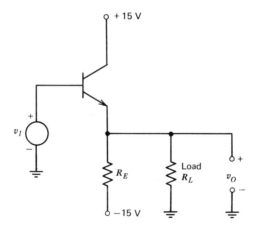

Figure 8.21 Emitter follower with resistive biasing.

for operation from 15-volt supplies), it is necessary to make R_E equal to half the minimum expected load resistance, since at the most negative output voltage the transistor will be cut off and the load current must be supplied via R_E. If, for example, $R_L = 500\ \Omega$, R_E must be less than or equal to $250\ \Omega$ to insure that a -10-volt output level can be obtained. The power delivered to the load is 200 mW at $v_O = \pm\ 10$ volts, while the total power required from the supplies under quiescent conditions ($v_O = 0$) is 1.8 watts, or power nine times as large as the maximum output power for negative output voltage. This low ratio of peak output power to quiescent power is intolerable in many applications. A second and related problem is that the input resistance to the stage will be only $\beta R_L/3$ when R_E is selected to guarantee a -10-volt output.

The situation improves significantly if the biasing resistor is replaced by a current source as shown in Fig. 8.22. A -10-volt output is obtained with $I = 10$ volts$/R_L$. If we use the earlier value of $500\ \Omega$ for R_L, a 200-mW peak output for negative output voltage results with 600 mW of quiescent power consumption. The input resistance to the circuit is similarly increased by a factor of three.

Further improvement results if a complementary emitter follower (Fig. 8.23) is used. Neither transistor in this connection is forward biased with $v_I = v_O = 0$, and thus the quiescent power consumption of the circuit is zero. The NPN supplies output current for positive output voltages, while the PNP supplies the current for negative output voltages. In either case only one transistor conducts, so that the load current only is required from the loaded power supply.

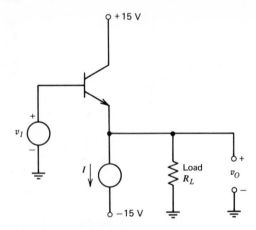

Figure 8.22   Emitter follower with current-source biasing.

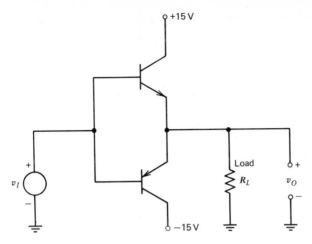

Figure 8.23 Complementary emitter follower.

As might be expected, the complementary emitter follower has its own design problems; the most difficult of these involve establishing appropriate quiescent levels. If the circuit is constructed as shown in Fig. 8.23, it exhibits *crossover distortion* since it is necessary to forward bias either transistor base-to-emitter junction by approximately 0.6 volt to initiate conduction. Consequently, there is a 1.2-volt range of input voltage for which the output remains essentially zero. The idealized transfer characteristics as well as representative input and output waveforms for this circuit are shown in Fig. 8.24. We might initially feel that, since this circuit is intended for use as the output stage of an operational amplifier, the effect of this nonlinearity would be reduced to insignificant levels by the gain that precedes it in most feedback applications. In fact, the example presented in Section 2.3.2 showed that feedback virtually eliminated the distortion from this type of dead zone in one system. Unfortunately, the moderation of the nonlinearity depends on the gain of the linear elements in the loop, and is often insufficient at higher frequencies where this gain is reduced. As a result, while an output stage as simple as the one shown in Fig. 8.23 is at times successfully used in high-power low-frequency applications, it must normally be linearized to yield acceptable performance in moderate- to high-frequency situations.

The required linearization is accomplished by forward biasing the base-to-emitter junctions of the transistors so that both are conducting at low levels with zero input signal. One conceptually possible biasing scheme is shown in Fig. 8.25. If each of the two batteries is selected to just turn on its respective transistor, the input and output voltages of circuit will be identi-

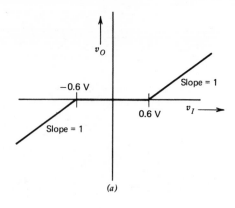

(a)

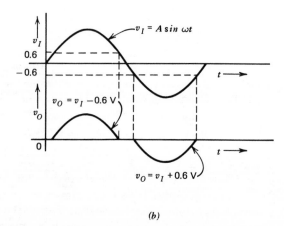

(b)

Figure 8.24 Input-output relationships for the complementary emitter follower. (a) Transfer characteristics. (b) Waveforms.

cal. Ignoring the practical difficulties involved in realizing the floating voltage sources (which can be resolved), two types of difficulties are probable: the biasing voltages will either be too small or too large. These problems occur because of the exponential and highly temperature-dependent relationship between collector current and base-to-emitter voltage. If too small bias voltages are used, a fraction of the crossover distortion remains, while if the bias voltages are too large, the circuit can conduct substantial quiescent current through the two transistors, and there is the probability of *thermal runaway*.

Thermal runaway is a potentially destructive process that is most easily understood by considering a transistor biased with a fixed base-to-emitter

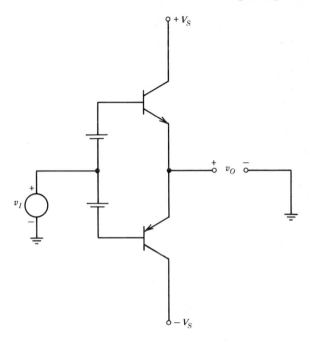

Figure 8.25 One approach to biasing the complementary emitter follower.

voltage so that it conducts some collector current. The power dissipation that results heats the transistor, and since the device is operating at fixed base-to-emitter voltage, the resultant temperature increase leads to a larger collector current, which results in higher power dissipation, etc. If the gain around this thermal positive-feedback loop exceeds one, the collector current increases until the transistor dies. (See Problem P8.13.)

In order to avoid these difficulties, forward-biased junctions are normally used to provide the bias voltages. If these biasing junctions are matched to the output-transistor base-to-emitter junctions and located in close thermal proximity to them, excellent control of bias current results. This approach is particularly attractive for monolithic integrated-circuit designs because of the ease of obtaining matched, isothermal devices with this construction technique. Further insurance against thermal runaway is often obtained by including resistors in series with the emitters of the output transistors. Voltage drops across these resistors reduce base-to-emitter voltage and thus tend to stabilize bias currents as these currents increase. The value of these resistors represents a compromise between the increased operating-point stability that results from higher-value resistors and the lower output re-

sistance associated with smaller resistors. A compromise value of approximately 25 Ω is frequently used for designs with peak output current in the
20-mA range.

One interesting bias-circuit variation for a complementary emitter-follower connection is used in the 741 integrated-circuit operational amplifier.
This circuit is shown in simplified form along with quiescent current levels
in Fig. 8.26. The circled components function as a diode and a half (or more
precisely a diode and three-fifths) to establish a conservative bias-voltage
value. Because the base current of the transistor is small compared to the
currents through the two resistors, this negative-feedback connection forces
the voltages across the resistors to be proportional to their relative values.

While forward-biasing techniques make the use of complementary connections practical, minor nonlinearities usually remain. For this reason,
operational amplifiers intended for use at very high frequencies occasionally
use a current-source-biased emitter follower (Fig. 8.22) in order to achieve
improved linearity.

It is often necessary to incorporate current limiting in the design of an
output stage intended for general-purpose applications. While it would be
ideal if the current limit protected the amplifier for shorts from the output
to ground or either supply voltage, this requirement often severely compromises maximum output current. Consequently, the current limit is at times
designed for protection from output-to-ground shorts only.

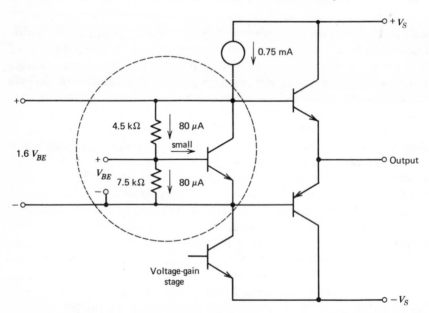

Figure 8.26   Bias circuit used in 741 amplifier.

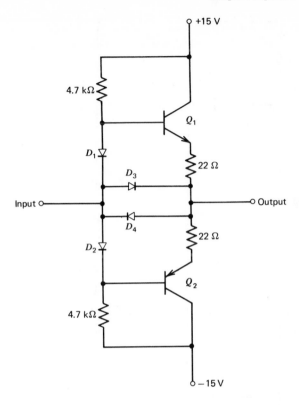

Figure 8.27   Resistively biased complementary emitter follower.

Figure 8.27 shows a discrete-component output stage that illustrates some of the concepts introduced above. Assume that the input and output voltage levels are both zero, and that no current is drawn from the output. Under these conditions, approximately 3 mA flows through diodes D_1 and D_2 and the two 4.7-kΩ resistors. If diodes D_1 and D_2 are matched to the base-to-emitter junctions of Q_1 and Q_2, respectively, the quiescent bias current of the transistor pair is slightly more than 1 mA. (The details of this type of calculation are given in Section 10.3.1.) The 22-Ω resistors effectively protect against thermal runaway. Assume, for example, that the temperatures of the transistor junctions each rise 50° C above their respective diodes. As a result of this temperature differential, the voltage across each 22-Ω resistor increases by at most 100 mV, and thus the quiescent-current increase is limited to less than 5 mA.

Base drive for the transistors is supplied from the 4.7-kΩ resistors rather than directly from the input-signal source. The current limit occurs when this required drive current is eliminated in the following way. Assume that

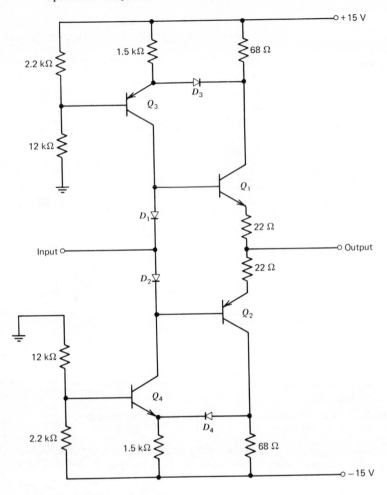

Figure 8.28 Current-source biased complementary emitter follower.

the input voltage is positive and that transistor Q_1 is supplying an output current of approximately 25 mA. Under these conditions diode D_3 is on the verge of conduction, since with approximately the same voltages across D_1 and the base-to-emitter junction of Q_1, the voltages across the top 22-Ω resistor (22 Ω × 25 mA = 550 mV) and D_3 are nearly equal. If the input-signal source is limited to low current output, diode D_3 clamps the input voltage level, preventing further increases in base drive. Because the limiting current level is proportional to the forward voltage of a diode, the limiting level decreases with increasing ambient temperature. This dependence is advantageous, since the power-handling capacity of the output transistors also decreases with increasing temperature.

This relatively simple circuit is often an adequate output stage. One deficiency is that the input resistance of the circuit is dominated by the parallel combination of the biasing resistors. Since the output current is limited to approximately 25 mA, minimum load resistors on the order of 400 Ω are anticipated. The current gain of the output pair insures that the input loading attributable to this value of load resistor is insignificant compared to that of the biasing resistors. Increasing the value of the biasing resistors can result in insufficient base drive at maximum output voltages.

The circuit shown in Fig. 8.28 can be used when maximum input resistance to the buffer amplifier is required. Diodes D_1 and D_2 function as they did in the previous circuit. However they are biased with 1-mA current sources formed by transistors Q_3 and Q_4 rather than by resistors. The high incremental resistance of these current sources minimizes loading at the amplifier input. Since the current sources supply base drive for the output transistors, turning these current sources off limits output current. The limiting occurs as follows for a positive input voltage. When the output current is approximately 30 mA, the voltage at the cathode end of diode D_3 equals the voltage at the base of Q_3. Further increases in output current lower the upper current-source magnitude, thereby reducing drive.

PROBLEMS

P8.1

Consider an operational amplifier built with n identical stages, and an open-loop transfer function

$$a(s) = \frac{a_o}{(\tau s + 1)^n}$$

This amplifier is used in a noninverting unity-gain connection. Determine the maximum stable value of a_o for $n = 3$ and $n = 4$. What is the limiting stable value for a_o as $n \to \infty$?

P8.2

Figure 8.29 illustrates a model for a multiple-stage operational amplifier. The output impedance of the input section of the amplifier is very high, and the transfer admittance is

$$y(s) = \frac{I_a(s)}{V_i(s)} = \frac{0.67 \times 10^{-2}}{(10^{-6}s + 1)(10^{-7}s + 1)}$$

The quiescent collector current of the transistor is 100 μA. Transistor parameters include $\beta = 100$, $C_\mu = 5$ pF, and $C_\pi = 10$ pF. You may as-

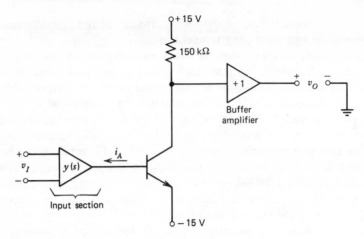

Figure 8.29 Multiple-stage operational amplifier.

sume that a one-pole approximation adequately characterizes the common-emitter stage, and that the input impedance of the buffer amplifier is very high. Ignore base-width-modulation effects.

(a) Find the transfer function $V_o(s)/V_i(s)$ for this amplifier. What is the magnitude of this transfer function at the frequency where it has a phase shift of $-180°$?
(b) Determine a compensating impedance that can be placed between base and emitter of the transistor so that the second pole of the compensated transfer function occurs near its unity-gain frequency. What is the open-loop transfer function with your compensation?
(c) Find a compensating impedance that can be placed between collector and base of the transistor to yield a transfer function similar to that obtained in part *b*.

P8.3

A model for an operational amplifier incorporating feedforward compensation is shown in Fig. 8.30. Approximate the open-loop transfer func-

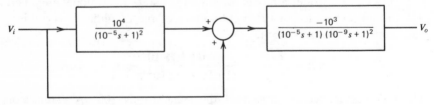

Figure 8.30 Block diagram for feedforward amplifier.

tion $V_o(s)/V_i(s)$ for this amplifier. (Note that you should be able to estimate the transfer function of interest fairly accurately without having to factor any polynomials.) What is the amplifier phase shift at its unity-gain frequency? Draw a Bode plot of the transfer function. Comment on possible difficulties with this amplifier.

P8.4

Do you expect the base-width modulation factor η of a bipolar transistor to be more strongly dependent on quiescent collector current or quiescent collector-to-emitter voltage? Explain.

P8.5

Figure 8.31 shows the characteristics of a certain NPN transistor as displayed on a curve tracer when the base current is 10 μA. Find values for g_m, r_π, r_o, and r_μ for this device valid at $I_C = 1$ mA, $V_{CE} = 10$ volts. Estimate η for this transistor.

P8.6

Assume that the transistor connection shown in Fig. 8.14 is modified to include a bias current source that increases the value of the emitter current of Q_1. Express the voltage gain and transresistance of the resulting circuit in terms of the value of the bias source and other circuit parameters.

P8.7

A current-source-loaded Darlington connection is shown in Fig. 8.32. Find the low-frequency voltage gain and transresistance of this circuit, assuming that both transistors have identical values for β and η.

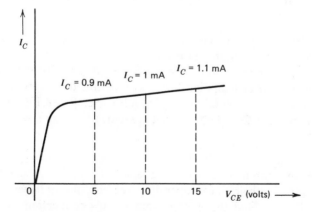

Figure 8.31 Transistor *I-V* characteristics.

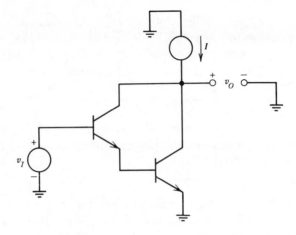

Figure 8.32   Current-source-loaded Darlington amplifier.

P8.8

Determine the low-frequency gain v_o/v_i and transresistance v_o/i_i for the current-source-loaded differential amplifier shown in Fig. 8.33. Assume both transistors are identical and characterized by β and η.

P8.9

A bipolar transistor is used in a current-source connection with its emitter connected to ground. Compare the output resistances that result when the base of the transistor is biased with a high or a low resistance source. Show that the same values result for the output resistance of a common-emitter amplifier loaded with an ideal current source as a function of the driving-source resistance.

P8.10

A transistor is available with $\beta = 200$ and $\eta = 5 \times 10^{-4}$. This device is used as the common-emitter portion of a current-source-loaded cascode connection operating at a quiescent current of 10 μA. The second cascode transistor can either be a bipolar device with parameters as given above or a FET with $y_{fs} = 10^{-4}$ mho and $y_{os} = 10^{-6}$ mho. (See Fig. 8.19b for an incremental FET model.) Compare the voltage gain that results with these two options.

P8.11

Consider the amplifier shown in Fig. 8.34. The biasing is such that when all devices are in their linear operating regions, the quiescent operating current is 10 μA. Find the voltage gain of this connection assuming all four bipolar transistors have identical parameter values as do both FET's.

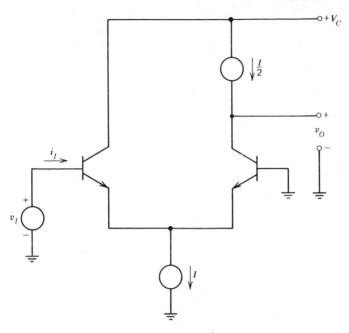

Figure 8.33   Current-source-loaded differential amplifier.

Use the values given in Problem P8.10. Estimate the break frequency of the dominant pole in the amplifier transfer function assuming that both FET's have drain-to-gate capacitances of 2 pF and that these capacitances dominate the frequency response.

P8.12

Determine the input resistance of the emitter-follower connection shown in Fig. 8.35 as a function of transistor parameters and quiescent operating levels. You may assume both transistors are identical.

P8.13

Thermal runaway is a potentially destructive process that can result when a transistor operates at fixed base-to-emitter and collector-to-emitter voltage because of the following sequence of events. The device heats up as a consequence of power dissipated in it.   This heating leads to a higher collector current, a correspondingly higher power dissipation, and consequently a further increase in temperature. The objective of this problem is to determine the conditions under which unbounded thermal runaway results.

The transistor in question is biased with a fixed collector-to-emitter voltage of 10 volts, and fixed base-to-emitter voltage that yields a quiescent collector current I_C.   You may assume the transistor has a large value for β,

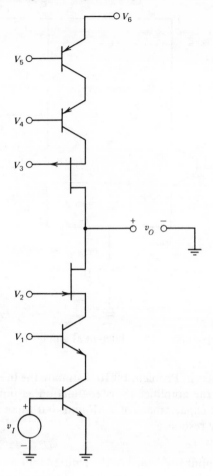

Figure 8.34 High-gain amplifier.

and that transistor base-to-emitter voltage, collector current, and temperature are related by Eqn. 7.1. The constant A in this equation is such that the transistor collector current is 10 mA at 0° C chip temperature with a base-to-emitter voltage of 650 mV.

The device is operating at an ambient temperature of 0° C. Measurements indicate that chip temperature is linearly related to power dissipation. The transfer function relating these two quantities is

$$\frac{T_j(s)}{P_d(s)} = 100 \left(\frac{1}{10^{-3}s + 1} + \frac{1}{100s + 1} \right)$$

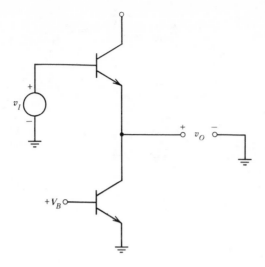

Figure 8.35   Emitter follower.

where T_j is the junction temperature in degrees Centigrade and P_d is the device power dissipated in watts.

Form a linearized block diagram that allows you to investigate the possibility of thermal runaway. Determine the quiescent value of I_C that results in transistor destruction. Now modify your block diagram to show how the inclusion of a transistor emitter resistor increases the safe region of operation of the connection.

P8.14

A certain operational amplifier can supply an output current of ± 5 mA over an output voltage range of ± 12 volts. Design a unity-voltage-gain stage that can be added to the output of the operational amplifier to increase the output capability of the combination to at least ± 100 mA over a ± 10-volt range. Available power-supply voltages are ± 15 volts. Assume that complementary transistors with a minimum β of 50 and a power dissipation capability of 2.5 watts are available. A reasonable selection of low power devices is also available. Your design should include current limiting to protect it for shorts from the output of the stage to ground.

CHAPTER IX

AN ILLUSTRATIVE DESIGN

9.1 CIRCUIT DESCRIPTION

The purpose of this section is to illustrate by example one way that the basic two-stage amplifier can be expanded into a complete, useful operational amplifier. Later sections of this chapter analyze the circuit to determine its performance, show how it can be compensated in order to tailor its open-loop transfer function for use in specific applications, and indicate how design alternatives might affect performance.

No attempt is made to justify this particular implementation of the two-stage amplifier other than to point out that the circuit was designed at least in part for its educational value. An appreciation of the salient features of this particular circuit leads directly to improved understanding of other operational amplifiers, including a number of integrated-circuit designs, which have evolved from the basic topology. The modifications incorporated into the basic design are certainly not the only possible ones, nor are they all likely to be required in any given application. The circuit does illustrate how a designer might resolve some of the tradeoffs available to him, and also provides a background for much of the material in later sections.

9.1.1 Overview

The complete circuit and important quiescent levels are shown in Fig. 9.1. The circuit represents a modification of the basic amplifier that combines a differential amplifier incorporating several of the drift minimizing techniques described in Chapter 7 with a high-gain stage consisting of a current-source-loaded cascode amplifier. A unity-voltage-gain buffer amplifier isolates the high-resistance node at the output of the cascode amplifier and provides high current output drive capability. The amplifier is designed to provide a ±10-volt maximum output signal and operate from standard ±15-volt supplies. The supply voltages are both bypassed with a parallel combination of an electrolytic and a ceramic capacitor, since this combination is effective over a wide frequency range.

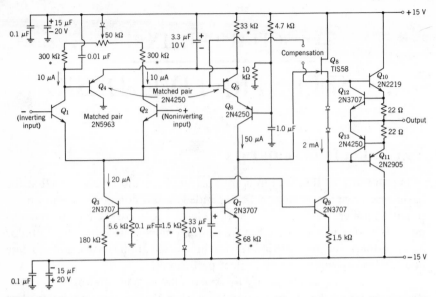

Figure 9.1 Discrete-component operational amplifier. *Note.* *Indicates 1% metal-film resistor.

This circuit shares a characteristic with a number of other moderately involved designs, which is often disturbing to novice circuit designers since there is some difficulty in determining which transistors are actually in the signal path. It is important to resolve this uncertainty prior to any detailed discussion of the circuit. Referring to Fig. 9.1, we see that transistors Q_1 and Q_2 are the differential-amplifier input stage. As we shall see, the second-stage topology constrains the emitter connection of the Q_4-Q_5 pair to be incrementally grounded. Thus Q_5 and Q_6 form a cascode amplifier. This current-source-loaded cascode provides the largest fraction of the amplifier gain, with analysis to be presented indicating a voltage gain of 180,000 in this portion of the circuit.

The high-resistance node at the output of the cascode amplifier is isolated with source-follower-connected FET Q_8. The source follower drives transistors Q_{10} and Q_{11}, which are connected as a complementary emitter follower.

The amplifier can be compensated by connecting an appropriate network between the indicated terminals, thereby forming a minor loop that includes the high-gain stage. Details of this process are given in Section 9.2.3.

The above discussion shows that the signal path includes only transistors Q_1, Q_2, Q_5, Q_6, Q_8, Q_{10}, and Q_{11}. The remaining transistors are used either as current sources (Q_3, Q_7, and Q_9), or to reduce voltage drift referred to the input by forming a differential second stage at d-c (Q_4), or to limit output current (Q_{12} and Q_{13}).

9.1.2   Detailed Considerations

Once the topology of the circuit is selected, a decision concerning approximate bias-current levels is a necessary first step in the detailed design process. Low current levels give improved d-c performance since input currents and input-stage self-heating are reduced. However, the frequency response of the amplifier is reduced by operation at low currents. (See Section 9.3.3 for a description of power-speed tradeoffs.)

A compromise collector current level of 10 μA, which can provide excellent d-c performance combined with closed-loop frequency response of several MHz, was selected for the first-stage transistors. Transistor Q_3 is a current source that provides the total 20-μA quiescent current of the first stage and insures high common-mode rejection ratio. This current source shares a common bias network with two other current sources. The bias network includes a diode that provides approximate temperature compensation for the current sources, and also includes capacitive bypassing to the negative supply. Bypassing to the negative supply rather than to ground is preferable in this case since it insures that the current-source output is independent of high-speed transients on the negative supply line.

The differential input stage is a matched pair of 2N5963 transistors. The devices are selected to have base-to-emitter voltages matched to within 3 mV at equal collector currents and, furthermore, to have current gains matched to within 10% at the operating current level. They are mounted in close thermal proximity to reduce temperature differentials. Wrapping wire around the pair or mounting them in an aluminum block drilled to accept the transistors improves the thermal bond. The 2N5963 is selected because it is inexpensive and provides a typical current gain of 1100 at a collector current of 10 μA. The resultant bias current required at either input is approximately 10 nA without any form of current compensation. Compensating techniques such as these described in Section 7.4.2 can be used to lower this bias current to less than 1 nA over a 50° C temperature range.

Transistors Q_5 and Q_6 are the cascode-amplifier transistors. An additional PNP transistor, Q_4, is used to improve d-c performance by forming a differential amplifier with transistor Q_5. While this transistor lowers drift, it does not affect the operation of the Q_5-Q_6 pair in any way as shown by the following discussion. It is evident that at low frequencies the common-emitter point of pair Q_4-Q_5 is incrementally grounded since only differential signals can be applied to this pair by the input stage. The capacitor[1] included across the 33-kΩ emitter-circuit resistor guarantees that the emitter of Q_5 also remains incrementally grounded at high frequencies. Since transistor Q_4 is included only to improve d-c performance and is not required for gain at any frequency, its base circuit can be bypassed at moderate and high frequencies. Bypassing insures that Q_1 operates as a common-collector stage at these frequencies. It was mentioned in the last chapter that operation in this mode is advantageous since it minimizes the input capacitance seen at the base of Q_1 (the inverting input of the complete amplifier), and thus allows a wider range of feedback networks to be used without significant high-frequency loading.

The amplifier is balanced by changing relative collector load resistor values in the first stage. Since the input-stage transistors are matched for a maximum base-to-emitter voltage differential of 3 mV at equal collector currents, the ratio of the collector currents will be at most $e^{3mV(q/kT)} \simeq$ 1.12 at equal base-to-emitter voltages. The 50-kΩ potentiometer that allows a maximum collector-resistor ratio of 1.17:1 is therefore adequate for balancing even if some mismatch of second-stage base currents exists. The diode included in the Q_1-Q_2 collector circuit provides a degree of compensation for the base-to-emitter voltage changes of transistors Q_4-Q_5 with temperature in order to stabilize their quiescent current.

The 2N4250 transistors used in the second stage are one of the highest-gain PNP types available, with a typical current gain in excess of 300 at 50 μA of collector current. This gain permits a five-to-one increase in quiescent operating level between the first and second stages (valuable since this increase improves the bandwidth of the second-stage devices) without seriously compromising drift performance. It also contributes to high overall amplifier gain. While it is not necessary to use the same transistor type for both members of a cascode amplifier pair, the 2N4250 is also used in the common-base section of the cascode (Q_6) since it has high r_μ, a necessary condition for high voltage gain. The 2N3707 used as the current-source load for the cascode is also selected in part because of high r_μ.

All critical resistors associated with the first two stages are precision metal film types. These are preferred since their low temperature coefficients reduce voltage drift and because of their low noise characteristics.

A field-effect transistor is used to isolate the high-impedance node at the cascode output. The virtually infinite input resistance of the FET improves

---

[1] As a matter of practical interest, eliminating this capacitor has only a minor effect on the overall performance of the amplifier, but complicates the analysis. This is an example of a component included primarily for educational purposes.

voltage gain. Component economy is also achieved, since an additional stage of current gain would probably be required for isolation if bipolar transistors were used. A current source is used for FET bias so that the bias current is independent of output-voltage level. The quiescent level of this stage is chosen to meet maximum drive requirements for the following stage.

A complementary emitter-follower pair (Q_{10}-Q_{11}) is used to provide large positive or negative output currents with minimum quiescent power dissipation. Metal-can rather than epoxy-cased transistors are used in this stage for increased power-handling capability. The two diodes included in the base circuit of the emitter-follower pair reduce crossover distortion, while the 22-Ω resistors eliminate the possibility of thermal runaway that accompanies this connection.

Transistors Q_{12} and Q_{13} combine with the 22-Ω resistors to limit the output current of the amplifier to approximately 30 mA. This limiter circuit, which is similar in operation to the diode limiter described in connection with Fig. 8.27, is used since it is identical in form to one frequently used in integrated-circuit designs. Consider the limiting process when the amplifier output voltage is negative. If the sink current exceeds 25 to 30 mA, transistor Q_{13} conducts, since its base-to-emitter voltage approximates 600 mV. This conduction reduces base drive for Q_{11}. The current that must be conducted by Q_{13} in order to eliminate base drive to Q_{11} is at most 2 mA, the output level of current source Q_9.

When the amplifier output voltage is positive, transistor Q_{12} conducts to limit output current. This situation is potentially hazardous, since it is conceivable that the driving transistor (Q_8) could be destroyed if no mechanism limited its drain current. However, the geometry of the TIS58 is such that its drain current is the order of 5 mA when the gate-to-source voltage of this device reaches the forward-conduction value. Thus, while transistor Q_{12} may conduct approximately 3 mA in positive output current limit, destruction of Q_8 is not possible. Note also that since the maximum collector current of Q_6 is limited to modest values by the 33-kΩ emitter-circuit resistor associated with Q_4-Q_5, the maximum current from Q_6 cannot injure any devices.

No attempt is made to control internal amplifier voltages, such as the emitter potential of Q_5, during current overload. The charge stored on the 3.3-μF capacitor delays recovery from overload, but since current limit is not anticipated during normal operation (overload protection is included primarily to protect us from our own errors during system breadboarding), this delay is unimportant.

9.2 ANALYSIS

In order to demonstrate the performance features of the amplifier introduced in the previous section, it is necessary to approximate analytically some of its more important characteristics. While the exact details of the analysis are specific to this amplifier, several significant features, particularly those concerning dynamics and compensation, are common to all two-stage operational amplifiers. Thus the conclusions we shall reach extend beyond this particular circuit.

We should realize that certain aspects of the following analysis are likely to be in error by a factor of two or more, since the uncertainty of some of the parameter values associated with the transistors limits accuracy. Another type of difficulty is encountered in the analysis of the dynamics of the amplifier, since a number of poles are predicted in the vicinity of the f_T of the transistors used in the amplifier. Such results are always suspect because transistor-model deficiencies prevent accurate analysis in this frequency range. Fortunately, these inaccuracies are of little concern since our objective is not so much precise prediction of the performance of this particular amplifier as it is an understanding of the important features of this general type of amplifier.

9.2.1 Low-Frequency Gain

One important characteristic of an operational amplifier is its d-c open-loop gain. Calculation of the gain of this amplifier is necessary because accurate measurement of the signal levels that would permit experimental gain determination is precluded by noise and drift.

By far the largest fraction of the low-frequency gain of the amplifier occurs in the cascode stage for this particular implementation of the basic topology. The analysis of the complete amplifier is facilitated by initially developing a low-frequency equivalent circuit for the cascode amplifier. The analysis of Section 8.3.4 showed that the voltage gain of an unloaded cascode amplifier is

$$ -\frac{\beta_6}{2\eta_6} = -\frac{g_{m6}r_{\mu6}}{2} $$

while its input resistance is $r_{\pi5}$. (Subscripts differentiating between the two transistors in the cascode connection refer to Fig. 9.1.) While the output resistance of the cascode connection was not specifically calculated, a result from Section 8.3.5 can be used to determine this quantity. Equation 8.59 gives $r_{\mu}/2$ as the output resistance of a common-base current source with a large incremental emitter-circuit resistance. The output resistance of the cascode must be identical since its output consists of a common-base connection with a large emitter-circuit resistance. These results show that the low-frequency performance of the cascode portion of the amplifier can be modeled by the equivalent circuit of Fig. 9.2.

The d-c gain of the circuit shown in Fig. 9.1 is determined using the parameter values shown in Table 9.1 for the transistors. The calculation is performed assuming that the noninverting input of the amplifier is incrementally grounded. This assumption yields the same value for d-c gain that would be obtained considering a true differential input voltage. Incrementally grounding the noninverting input does eliminate an insignificant high-frequency term in the transfer function that results from signals fed through the collector-to-base capacitance of Q_2 (see Section 8.2.3).

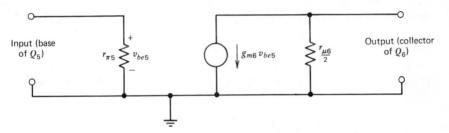

Figure 9.2 Equivalent circuit for cascode amplifier at low frequencies.

Table 9.1 Transistor Parameters for Circuit of Fig. 9.1

Transistor Number	Type	I_C or I_D (μA)	g_m (mmho)	β	r_π (kΩ)	r_μ (MΩ)	r_o (MΩ)	C_μ or C_{gd} (pF)	C_π or C_{gs} (pF)
Q_1, Q_2	2N5963	10	0.4	1100	2750	*	*	6	10
Q_3	2N3707	20	*	*	*	*	*	8	10
Q_4, Q_5, Q_6	2N4250	50	2	350	175	500	1.4	10	15
Q_7	2N3707	50	2	200	100	500	2.5	8	10
Q_8	TIS58	2 mA	*	—	—	—	—	2	*
Q_9	2N3707	2 mA	*	*	*	*	*	*	*
Q_{10}	2N2219	*	*	200	*	*	*	*	*
Q_{11}	2N2905	*	*	200	*	*	*	*	*
Q_{12}	2N3707	0	*	*	*	*	*	*	*
Q_{13}	2N4250	0	*	*	*	*	*	*	*

— Not relevant.
* Value unimportant in included analysis.

Overall gain is found by first calculating the transfer relationships for various portions of the circuit. An incremental input voltage applied to the base of Q_1, v_i, causes a change in the collector current of Q_2 given by

$$i_{c2} = -\frac{v_i g_{m1}}{2} \tag{9.1}$$

(It has been assumed that both input transistors are operating at equal currents so that $g_{m1} = g_{m2}$.)

The previously developed cascode equivalent circuit shows that the change in base voltage of Q_5 is related to the Q_2 collector-current change by

$$v_{be5} = -i_{c2}(325 \text{ k}\Omega \| r_{\pi5}) \tag{9.2}$$

(The collector-circuit potentiometer has been assumed set to center position so that the load resistor of transistor Q_2 is equal to 325 kΩ.) In order to determine the voltage gain of the cascode amplifier, it is necessary to calculate the load applied to it. The input resistance of field-effect transistor Q_8 is essentially infinite, while the output resistance for the current source Q_7 is

$$r_{\mu7} \left\| \left[\frac{1 + g_{m7}(r_{\pi7} \| 68 \text{ k}\Omega)}{g_{o7}} \right] \right. \tag{9.3}$$

(See Eqn. 8.57.) It is computationally convenient to reduce this equation now and to introduce the experimentally verifiable assumption that $r_{\mu7} \simeq r_{\mu6}$. This value is reasonable, since both devices are operating at identical currents, and are fabricated using similar (though complementary) processing. The 2N3707 has a typical β of 200 at 50 μA, so that $r_{\pi7}$ is typically 100 kΩ at this current. Therefore, $r_{\pi7} \| 68 \text{ k}\Omega \simeq 0.4 r_{\pi7}$. Accordingly, the output resistance of Q_7 becomes

$$r_{\mu7} \left\| \left[\frac{1 + g_{m7}(0.4 r_{\pi7})}{g_{o7}} \right] \simeq r_{\mu7} \left\| \left[\frac{0.4\beta_7}{g_{o7}} \right] \right. = r_{\mu7} \| 0.4 r_{\mu7} \simeq 0.28 r_{\mu7} \right. \tag{9.4}$$

Using this relationship, the assumed equivalence of $r_{\mu7}$ and $r_{\mu6}$, and the model of Fig. 9.2 shows that the loaded cascode voltage gain is

$$\frac{v_{cb6}}{v_{be5}} \simeq -g_{m6} \left(\frac{r_{\mu6}}{2} \| 0.28 r_{\mu6} \right) \simeq -g_{m6}(0.18 r_{\mu6}) \tag{9.5}$$

Recognizing that the unloaded voltage gain from the collector of Q_6 to the amplifier output is unity and combining Eqns. 9.1, 9.2, and 9.5 yields

$$\frac{v_o}{v_i} = -\frac{g_{m1}}{2} (325 \text{ k}\Omega \| r_{\pi5}) g_{m6}(0.18 r_{\mu6}) \tag{9.6}$$

Substituting parameter values from Table 9.1 into Eqn. 9.6 predicts a d-c open-loop gain magnitude of 4×10^6. The gain is dominated by the contribution of 1.8×10^5 from the cascode amplifier (see Eqn. 9.5).

9.2.2 Transfer Function

The locations of all poles and zeros of the amplifier could be predicted for the complete circuit by substituting appropriate incremental models for the active devices, although this would be a formidable task even with the aid of a computer. The approach used here is to make relatively crude approximations to gain insight into the controlling dynamics of the amplifier and then to verify the approximate results with a more detailed (though still incomplete) computer analysis.

The unloaded low-frequency voltage gain of the buffer amplifier (transistors Q_8 through Q_{11}) is unity. Amplifier loads as low as several hundred ohms do not appreciably alter its performance. If the load applied to the amplifier is not capacitive, the frequency response of the buffer approaches the f_T of the devices used in it. Furthermore, the input impedance of Q_8, which loads the cascode amplifier, is independent of any load applied to the amplifier output since the FET is unilateral. Thus the influence of the buffer can be modeled by simply using the input capacitance of Q_8, C_{gd8}, as a load for the cascode. Similarly, the loading of transistor Q_7 can be represented as a parallel impedance consisting of its output capacitance $C_{\mu7}$ and output resistance $0.28r_{\mu7}$ (Eqn. 9.4).

An incremental model that reflects these simplifications is shown in Fig. 9.3. The base resistances (r_x's) of all transistors, as well as r_μ and r_o of transistors other than Q_6 and Q_7 (the transistors in the high-gain portion of the circuit) have also been ignored. An argument based on the concept of open-circuit time constants[2] is used to further simplify this model. The open-circuit resistances[3] facing capacitors $C_{\pi1}$, $C_{\mu1}$, $C_{\pi2}$, $C_{\mu3}$, and $C_{\pi6}$ are all on the order of $1/g_m$ for the related transistor or lower. Thus these capacitors do not affect the dynamics of the amplifier at frequencies low compared to the f_T's of the various transistors and are eliminated for the initial approximation. As a result of this approximation the only contribution of the input stage to amplifier dynamics is a consequence of the loading $C_{\mu2}$ applies to the base of Q_5, and the stage itself can be modeled as a single dependent current source.

[2] See P. E. Gray and C. L. Searle, *Electronic Principles: Physics, Models, and Circuits*, Wiley, New York, 1969, Chapters 15 and 16.
[3] The open-circuit resistance facing a capacitor is the incremental resistance at the terminal pair in question calculated with all other capacitors in the circuit removed or open-circuited.

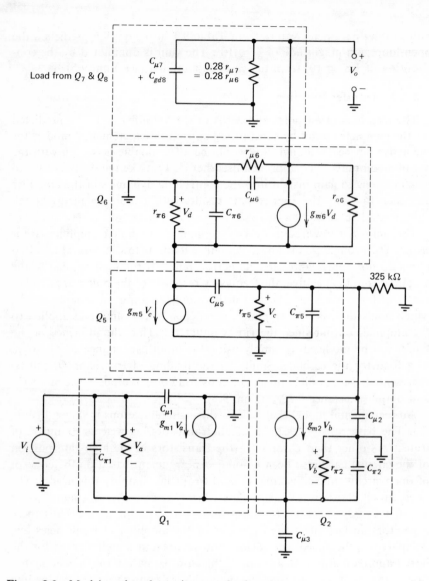

Figure 9.3 Model used to determine transfer function.

The further-simplified incremental model incorporating the approximations introduced above and shown in Fig. 9.4 is used to approximate the
location of the two low-frequency amplifier poles. The node equations for
this circuit are

$$\frac{g_{m1}V_i}{2} = [(C_1 + C_{\mu5})s + G_1]V_a - C_{\mu5}sV_b$$

$$0 = (-C_{\mu5}s + g_{m5})V_a + (C_{\mu5}s + g_{m6} + g_{\pi6} + g_{o6})V_b - g_{o6}V_o$$

$$0 = (-g_{m6} - g_{o6})V_b + (C_2s + g_{o6} + G_2)V_o \qquad (9.7)$$

(See Fig. 9.4 for the definition of parameters in this equation.)

The poles are found by equating the determinant of the matrix of coefficients of Eqn. 9.7 to zero, yielding

$$\frac{C_1 C_2 C_{\mu5}}{g_{m6}G_1(G_2 + g_{\mu6})} s^3 + \frac{C_2(C_1 + 2C_{\mu5})}{G_1(G_2 + g_{\mu6})} s^2 + \frac{C_2}{G_2 + g_{\mu6}} s + 1 = 0 \qquad (9.8)$$

In reducing Eqn. 9.7 to 9.8, small terms have been dropped. However,
only terms that are small because of transistor and topological inequalities
such as $g_m \gg g_\pi \gg g_o \gg g_\mu$, and $C_2 > C_{\mu6}$ since one component of C_2 is
$C_{\mu6}$ have been eliminated. Thus the conclusions that will be drawn from
Eqn. 9.8 are applicable to a variety of circuits that share this topology

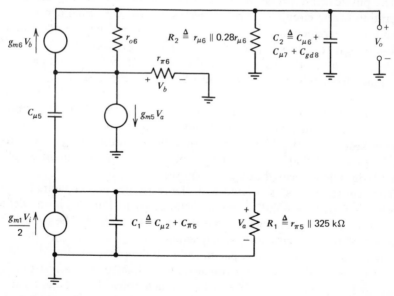

Figure 9.4 Simplification of Fig. 9.3.

rather than being limited to the specific choice of element values shown in Fig. 9.1. Fundamental relationships among parameter values also insure that the three poles represented by Eqn. 9.8 will be real and widely spaced. Consequently, this cubic equation can be easily factored, since

$$(\tau_a s + 1)(\tau_b s + 1)(\tau_c s + 1) \simeq \tau_a \tau_b \tau_c s^3 + \tau_a \tau_b s^2 + \tau_a s + 1$$

$$\text{for} \qquad \tau_a \gg \tau_b \gg \tau_c \quad (9.9)$$

Equation 9.9 allows us to write Eqn. 9.8 as

$$\left(\frac{C_2}{G_2 + g_{\mu 6}} s + 1 \right)\left(\frac{C_1 + 2C_{\mu 5}}{G_1} s + 1 \right)\left(\frac{C_1 C_{\mu 5}}{g_{m6}(C_1 + 2C_{\mu 5})} s + 1 \right) = 0 \quad (9.10)$$

indicating that

$$\tau_a = \frac{C_2}{G_2 + g_{\mu 6}}$$

$$\tau_b = \frac{C_1 + 2C_{\mu 5}}{G_1}$$

$$\tau_c = \frac{C_1 C_{\mu 5}}{g_{m6}(C_1 + 2C_{\mu 5})} \qquad (9.11)$$

The physical interpretation of the time constants lends insight into the operation of the circuit. The resistance associated with time constant τ_a is simply the incremental resistance from the high resistance node (the collector of Q_6) to ground. [Recall that $1/(G_2 + g_{\mu 6}) = 0.28 r_{\mu 6} \| r_{\mu 6} \| r_{\mu 6} = 0.18 r_{\mu 6}$, the value obtained earlier and used in Eqn. 9.5 for the incremental resistance from this node to ground.] Similarly, capacitance $C_2 = C_{\mu 6} + C_{\mu 7} + C_{gd8}$ is the capacitance from the high resistance node to ground. Since the capacitance of all amplifier nodes is the same order of magnitude, it is not surprising that the dominant amplifier pole is associated with energy storage at the highest resistance node. Substituting values from Table 9.1 shows that $\tau_a = 1.8$ ms, implying that the dominant amplifier open-loop pole is located at $s = -550 \text{ sec}^{-1}$.

Time constant τ_b is associated with the resistance and capacitance from the base of Q_5 to ground. The conductance G_1 in Eqn. 9.11 was defined previously as the conductance from this node to ground. The capacitance consists of the collector-to-base capacitance of Q_2 that shunts this node and the total effective input capacitance (including that attributed to Miller effect) Q_5 would display if this transistor were loaded with a resistive load equal to $1/g_{m5}$. Note that at frequencies much above $1/\tau_a$ radians per second, the capacitive loading at the collector of Q_6 has reduced the voltage

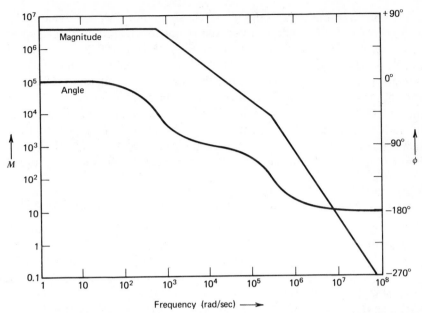

Figure 9.5 Amplifier open-loop transfer function based on two lowest-frequency poles (no compensation).

gain of this transistor; as a result, there is no significant feedback to the emitter of Q_6 through r_{o6} at these frequencies. Thus transistor Q_6 provides the $1/g_{m6} = 1/g_{m5}$ load for Q_5. The time constant τ_b is equal to 4.5 μs, implying that the second amplifier pole is located at $s = -2.2 \times 10^5$ sec^{-1}. Time constant τ_c corresponds to a frequency that approximates f_T for the transistors in the circuit, and thus to one of many high-frequency poles that are ignored in the simplified analysis.

Combining the d-c gain (Eqn. 9.6) with the dynamics predicted above yields

$$\frac{V_o(s)}{V_i(s)} = \frac{-4 \times 10^6}{(1.8 \times 10^{-3}s + 1)(4.5 \times 10^{-6}s + 1)} \tag{9.12}$$

Equation 9.12 is shown as a Bode plot[4] in Fig. 9.5.

---

[4] The transfer function plotted in Fig. 9.5 is actually the negative of Eqn. 9.12. This modification is made because we anticipate using the amplifier in negative-feedback connections. Since the loop transmission has the same sign as the gain calculated for the amplifier in these applications, plotting the negative of the amplifier gain follows the convention of plotting the negative of the loop transmission of a feedback system. Viewed alternatively, the transfer function plotted in Fig. 9.5 would result if the input signal were applied to the noninverting input terminal of the amplifier.

The pole locations for this design were also predicted by computer analysis, in order to verify some of the assumptions introduced in the preceding development. The equivalent circuit of Fig. 9.3 with 100-Ω base resistors added to the circuit model for each transistor was analyzed. Thus only the buffer amplifier was eliminated from the computer calculations. The locations of the two dominant poles predicted by the computer were -520 sec^{-1} and -2.15×10^5 sec^{-1}. All other poles had break frequencies in excess of 10^7 radians per second. In spite of the seemingly drastic approximations included in the analysis of this circuit, the predicted locations of the two dominant poles are confirmed by the computer calculation to within round-off errors.

9.2.3 A Method for Compensation

The transfer function of this amplifier (Eqn. 9.12) has the poles separated by a factor of 400, and in many feedback amplifiers this amount of separation would seem ideal from a stability point of view. Unfortunately, with the massive low-frequency open-loop gain characteristic of operational amplifiers (4×10^6 in this design), greater separation is required to insure adequate stability in many applications. For example, if the amplifier is used as a unity-gain follower by connecting its output to its inverting input, a loop is formed with $a(j\omega)$ as shown in Fig. 9.5 and $f = 1$. The Bode plot shows that the phase margin of the system is approximately $0.5°$ in this case, clearly an unsatisfactory value. In practice, this configuration would be unstable, since the negative phase shift associated with neglected open-loop singularities is far greater than $0.5°$ at the amplifier unity-gain frequency. It is clear that some method must be used to modify the open-loop transfer function of the amplifier in order to achieve acceptable performance in this and many other connections.

One of the significant advantages of the amplifier configuration described in this section and of all amplifiers that share its topology is that it is possible to use internal feedback to provide easily predicted and well-controlled compensation. The compensation is implemented by connecting a network between the terminals marked compensation in Fig. 9.1. This network completes a minor loop that includes the high-gain stage. Since both dominant amplifier poles are included inside the local feedback loop, it is possible to alter the location of the most important poles in the amplifier transfer function by this type of internal feedback. The degree of control that minor-loop feedback can exercise on the transfer function of a two-stage amplifier was hinted at in Section 5.3 and in the discussion of the effects of C_μ of the high-gain stage in Section 8.2.3.

There are at least two important limitations to this type of compensation. First, since this compensation is a form of negative feedback, the magnitude of the compensated open-loop amplifier transfer function will be less than or equal to the magnitude of the uncompensated transfer function at most frequencies. While resonances introduced by the minor feedback loop may give a gain increase at one or two particular frequencies, the bandwidth over which such increases exist is necessarily limited. Second, there is some maximum frequency for which this is an effective method of compensation, since beyond this frequency the influence of other singularities, some of which are outside the compensating loop and therefore cannot be controlled, become important. While these singularities are all at frequencies comparable to the f_T's of the transistors, they do set the ultimate bandwidth limitation of the amplifier because of the phase shift that they contribute to its open-loop transfer function at frequencies of interest. For example, at $1/10$ of its break frequency, a 10th-order pole contributes $57°$ of negative phase shift to a transfer function but only changes the magnitude by 5%. In practice, the unity-gain frequency of the amplifier-feedback network combination is normally chosen to limit the phase contribution of the high-frequency singularities to less than $30°$ at this frequency so that stability is not compromised. It is often necessary to determine the frequency at which the phase shift of higher-order singularities becomes important experimentally because of the difficulties associated with accurate analytic prediction of their locations.

An incremental model for the amplifier of Fig. 9.1 that can be used to analyze the effects of the internal feedback used for compensation is shown in Fig. 9.6. The development of this model relies heavily on the analysis of Section 9.2.2. The input impedance of the amplifier, which is unimportant for purposes of this calculation, is Z_i. An input voltage forces a proportional current at the node including the base of Q_5.[5]

The impedance at the base of Q_5 is modeled as a parallel R-C network with a time constant equal to τ_b in Eqn. 9.11. The remainder of the cascode is modeled as an impedance equal to the impedance from the collector of Q_6 to ground driven by a dependent-current source supplying a current $g_{m6}V_{be5}$. The impedance transformation of the field-effect transistor is represented as a unity-voltage-gain buffer amplifier. The complementary emit-

---

[5] This representation assumed an input voltage applied to the inverting input of the amplifier. If voltages are applied to both inputs, the differential voltage is used for V_i. An advantage of this type of amplifier is that the dynamics of the first stage do not significantly influence the transfer function at frequencies of interest; thus it functions as a true differential-input amplifier.

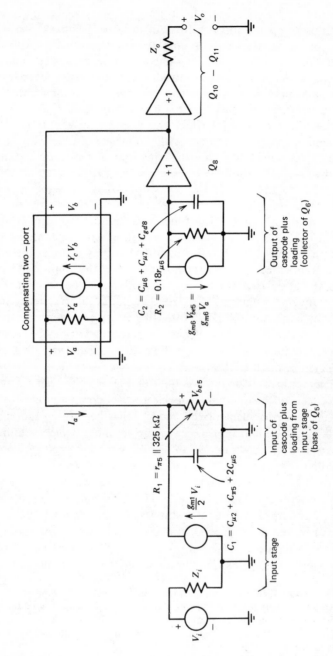

Figure 9.6 Model used to illustrate method of compensation.

ter-follower pair is modeled as a second buffer amplifier with an output impedance Z_o.

The compensating minor loop is formed by connecting a two-port network between the output of the source follower and the base of Q_5. Since the right-hand port of the network is driven by the low-impedance source follower, the voltage V_b is independent of V_a; thus the two-port can be completely represented in this application by the two admittances[6]

$$Y_a = -\frac{I_a}{V_a} \qquad V_b = 0 \qquad\qquad (9.13a)$$

$$Y_c = \frac{I_a}{V_b} \qquad V_a = 0 \qquad\qquad (9.13b)$$

Node equations for the model of Fig. 9.6 are

$$\frac{g_{m1}}{2} V_i = (Y_1 + Y_a)V_a - Y_c V_b \qquad\qquad (9.14)$$

$$0 = g_{m6}V_a + Y_2 V_b$$

where

$$Y_1 = \frac{1}{R_1} + C_1 s$$

$$Y_2 = \frac{1}{R_2} + C_2 s$$

Recognizing that output voltage V_o is identical to V_b in the absence of load allows us to determine the gain of the amplifier from Eqn. 9.14 as

$$\frac{V_o}{V_i} = \frac{V_b}{V_i} = -\frac{(g_{m1}/2)g_{m6}/[(Y_1 + Y_a)Y_2]}{1 + g_{m6}Y_c/[(Y_1 + Y_a)Y_2]} \qquad\qquad (9.15)$$

The quantity $g_{m6}Y_c/[(Y_1 + Y_a)Y_2]$ is identified as the negative of the loop transmission of the inner loop formed when the amplifier is compensated. In many cases of practical interest, the phase angle of this expression is close to plus or minus 90° when its magnitude is unity. The 90° phase margin of the compensating loop then insures that there is no peaking in its response. In these cases a very simple approximation serves to determine the magnitude of the open-loop transfer function of the amplifier, and the

---

[6] These definitions differ from those conventionally used to describe two-port networks in that the reference direction for I_a is out of the network. This choice reduces the number of minus signs in the following equations.

approximation yields a result that is correct within a factor of 0.707 at all frequencies. The implication from 9.15 is that

$$\frac{V_o(j\omega)}{V_i(j\omega)} \simeq -\frac{g_{m1}}{2Y_c(j\omega)} \tag{9.16}$$

at frequencies where

$$\left| \frac{g_{m6} Y_c(j\omega)}{[Y_1(j\omega) + Y_a(j\omega)]Y_2(j\omega)} \right| > 1$$

and

$$\frac{V_o(j\omega)}{V_i(j\omega)} \simeq -\frac{g_{m1}}{2} \frac{g_{m6}}{[Y_1(j\omega) + Y_a(j\omega)]Y_2(j\omega)} \tag{9.17}$$

at all other frequencies. Thus, when the minor-loop transmission magnitude is large, the open-loop transfer function of the amplifier is controlled by the minor-loop feedback element.

This approximation is particularly easy to apply graphically. The open-loop transfer function of the amplifier without compensation, but with the compensating network loading the base of Q_5, is plotted on log-magnitude vs. log-frequency coordinates. The proper loading is realized by connecting one side of the network to the base of Q_5 in the usual manner, and by disconnecting the other side of the network from the source of Q_8 and connecting it instead to an incremental ground. This first plot is particularly easy to obtain if a single capacitor is used as the compensating element (the most frequent case because this compensation leads to an approximately single pole open-loop transfer function) since only the location of the higher-frequency pole in Eqn. 9.12 is changed. The magnitude of the expression $g_{m1}/2Y_c(j\omega)$ is also plotted on the same coordinates. The magnitude of the amplifier open-loop transfer function at any frequency is then approximately equal to the lower magnitude of the two plotted curves. This relationship is easily developed from Eqns. 9.16 and 9.17, by noticing that the gain of the amplifier with the shorted compensating network connected to the base of Q_5 is

$$\frac{g_{m1}}{2} \frac{g_{m6}}{(Y_1 + Y_a)Y_2}$$

and that if

$$\left| \frac{g_{m1}}{2Y_c} \right| < \left| \frac{g_{m1}}{2} \frac{g_{m6}}{(Y_1 + Y_a)Y_2} \right|$$

then

$$\left| \frac{g_{m6} Y_c}{(Y_1 + Y_a)Y_2} \right| > 1$$

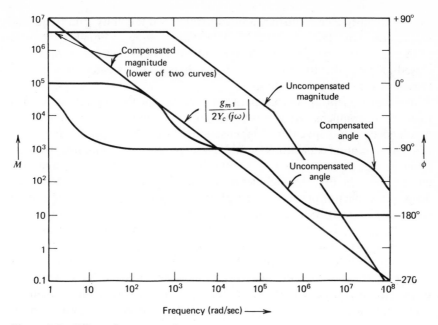

Figure 9.7 Effect of compensation.

Figure 9.7 illustrates the effects of compensating the amplifier shown in Fig. 9.1 with a 20-pF capacitor. The quantities Y_c and Y_a for this compensating network are both equal to $2 \times 10^{-11}s$. One of the two curves is obtained directly from the uncompensated transfer function of Fig. 9.5 by moving the second pole from 2.2×10^5 radians per second to 1.5×10^5 radians per second, since loading by the compensating capacitor increases the total capacitance at the base of Q_5 by 50%. The second plot is

$$\left| \frac{g_{m1}}{2 Y_c(j\omega)} \right| = \frac{10^7}{\omega}$$

The curve for the compensated amplifier is the lower of the two plots at all frequencies.

The advantages of this compensation for certain applications are obvious. It was shown earlier that operation with $f = 1$ would cause the uncompensated amplifier to oscillate. If a 20-pF compensating capacitor is used, the phase margin of the amplifier with direct feedback is greater than 45°.

Note that this compensation lowers the first amplifier open-loop pole to 2.5 radians per second. The location of the low-frequency pole cannot be independently chosen if we insist on a single-pole rolloff at frequencies below the unity-gain frequency and constrain both the unity-gain frequency and the d-c gain. The pole must be located at a frequency equal to the ratio of the unity-gain frequency to the d-c gain. This pole does not compromise closed-loop bandwidth, since closed-loop bandwidth is determined by the crossover frequency of the loop.

It is worth mentioning that parameter values for this amplifier are such that the uncompensated open-loop transfer function will be noticeably modified by any capacitive compensation in excess of approximately 0.1 pF! The minimum capacitor value necessary to modify the amplifier transfer function can be determined by noting that the uncompensated magnitude curve shown in Fig. 9.5 includes a region where its value is $2 \times 10^9/\omega$. Thus, if a capacitor in excess of 0.1 pF is used for compensation, the magnitude $|g_{m1}/2Y_c(j\omega)|$ will be smaller than the uncompensated magnitude over some frequency range. Furthermore, it is evident that feedback from any high level part of the circuit (from the collector of Q_6 on) back to the base circuit of Q_5 has approximately the same effect as feedback via the compensation terminals. Inevitable stray capacitance between these two parts of the circuit is usually on the order of 1 pF, and it is therefore concluded that the "uncompensated" curve of Fig. 9.7 can probably never be measured for an actual amplifier.

As indicated above, feedback from any portion of the circuit from the collector of Q_6 on modifies performance in much the same way as feedback from the source of Q_8, and in certain applications it may be advantageous to compensate by feeding back from an alternate point. For example, feedback from the output terminal includes more of the amplifier inside the compensating loop and thus with the control of this loop. Unfortunately, compensating-loop stability is less certain for this type of minor-loop feedback. Similarly, if large capacitors are used for compensation, greater inner-loop stability may be achieved by compensating from the collector Q_6.

Some of the reasons for selecting an amplifier topology with the possibility for this type of compensation should now be clear. The compensation is normally chosen so that it, rather than uncompensated amplifier dynamics, dominates amplifier performance at all frequencies of interest. Thus the open-loop transfer function of the amplifier with compensation becomes quite reliable. A wide variety of open-loop transfer functions can be obtained (several examples will be given in Chapter 13) with the main limitation being the requirement of maintaining the stability of the compensating loop. Furthermore, it is easy to determine what compensating network should be used to produce a given open-loop transfer function.

9.3   OTHER CONSIDERATIONS

A myriad of performance characteristics combine to determine the overall utility of an operational amplifier. The possibilities for modifications that compromise one characteristic in order to enhance another are numerous in this type of complex circuit. While the major advantage of the two-stage design centers on its easily controlled dynamics, the topology can be readily tailored to specific applications by other types of modifications. This section indicates a few of the "hidden" features of the two-stage design and points out the possibility of certain types of design compromises.

9.3.1   Temperature Stability

The last section shows that the use of internal feedback to compensate the amplifier under discussion yields an open-loop transfer function inversely proportional to the transfer admittance of the compensating network over a wide range of frequencies. The constant of proportionality for this and other variations of the two-stage design includes the transconductance of either input transistor, and is thus inversely related to temperature if the collector current of these transistors is temperature independent. This relatively mild variation with temperature is tolerable in many applications.

If greater transfer-function stability is required, the input-stage bias current can be made directly proportional to the absolute temperature. As a result, input-stage transconductance, and therefore the open-loop transfer function, will be temperature independent. A further advantage of this type of bias-current variation is that it partially compensates for input-transistor current-gain variations with temperature and thus reduces input-current changes.

The required bias-current temperature dependence can be implemented by appropriate selection of the total voltage applied to the base-to-emitter junction and the emitter resistor of the input-stage current source (Q_3 in Fig. 9.1). It can be shown that the output current from the source will be directly proportional to temperature if this voltage is constant and is approximately equal to the energy-band-gap voltage V_{go} (see Problem P9.11).

9.3.2   Large-Signal Performance

The analysis of the effects of compensation on amplifier performance has been limited up to now to linear-region operation. It is clear that compensation also effects large-signal behavior. For example, an open-loop transfer function similar to that obtained using a 20-pF compensating capacitor could be obtained by connecting a series-connected 3.6-μF capacitor and 500-Ω resistor from the base of Q_5 to ground. However, recovery from overload might be greatly delayed with this type of compensation because of the time required to change the voltage on a 3.6-μF capacitor with the limited current available at this node.

The compensation also limits the *slew rate*, or maximum time rate of change of output voltage of the amplifier. Consider an output voltage time rate of change $\dot{v}_O$. If a compensating capacitor C_c is used, the capacitor current required at the node including the base of Q_5 is $C_c\dot{v}_O$. The maximum magnitude of the current that can be supplied to this node by the first stage and that is available to charge the capacitor is approximately equal to the quiescent bias current of either input transistor I_{C1}. Thus the slew rate is $\dot{v}_O(\max) = I_{C1}/C_c$. However, the ratio I_{C1}/C_c also controls the unity-gain frequency of the amplifier, since this frequency is $g_{m1}/2C_c = qI_{C1}/2kTC_c$. The important point is that if some consideration, such as the phase shift from high-frequency singularities, limits the unity-gain frequency, it also limits the slew rate if a single capacitor is used to compensate the amplifier.

One way to circumvent this relationship is to add equal-value emitter resistors to both input transistors so that the transconductance of the input stage is lower than $g_{m1}/2$. Unfortunately, emitter degeneration also degrades the drift of the amplifier. Another more attractive possibility is the use of more involved compensation than that provided by a single capacitor. This alternative will be discussed in Chapter 13.

9.3.3   Design Compromises

There are many variations of the basic amplifier topology that result in useful designs, and some of these variations will be illustrated in Chapter 10. Other degrees of freedom are possible by varying quiescent operating current and by changing transistor types. The purpose of this section is to indicate how these variations influence amplifier performance.

Consider the changes that result from increasing all quiescent operating currents by a factor K. This change can be effected by decreasing all circuit resistors by the same factor. In response to the current change, all internal transistor resistances will decrease by the same factor, since all are multiples of $1/g_m$. Current gains of the various transistors do not change significantly if K is not grossly different from one. Thus the d-c voltage gain, which is a ratio of transistor and circuit conductances of the amplifier, will not change in response to changes in quiescent current. Input current will increase directly with quiescent current, and drift may increase somewhat because of increased self-heating in the first stage.

The dynamics for the design in question (at least without compensation) are determined primarily by the resistance and capacitance values at the base of Q_5 and at the collector of Q_6. The resistance values at these nodes decrease by an amount K, since they consist of combinations of transistor and circuit resistances. The capacitance values remain constant, at least for moderate changes from the levels used in the last sections, for the following reason. The capacitances involved are transistor-junction capacitances C_{gd}, C_μ, and C_π. Capacitances C_{gd} and C_μ are current-level independent, while C_π is the sum of a constant term plus a component linearly proportional to current. For transistor types likely to be used in this circuit, the current-proportional term is not important at levels below 1 mA. Thus an increase in current levels by as much as a factor of 10 from the values indicated in Fig. 9.1 does not significantly change critical node capacitances.

The argument above shows that moderate increases in operating current cause proportional increases in the locations of uncompensated open-loop poles. The form of the amplifier uncompensated open-loop transfer function remains unchanged and is simply shifted toward higher frequency. The possibility for increased bandwidth after compensation as a result of this modification is evident.

A second alternative is to change the relative ratios of first- and second-stage currents. An increase in second-stage current relative to that of the first stage has three major effects:

1. Drift increases because second-stage loading becomes more significant.
2. Gain decreases because the input resistance of the second stage decreases.
3. Bandwidth increases because the second-stage resistances decrease.

Significant flexibility is afforded by the choice of the active devices. The transistor types shown in Fig. 9.1 were selected primarily for high values of β and $1/\eta$. These types result in an amplifier design with high d-c voltage gain, low input current, and low drift. Unfortunately, because of compromises necessary in transistor fabrication, these types may have relatively high junction capacitances.

Clearly higher-frequency transistors can be used in the design. In fact, amplifiers with this topology have been operated with closed-loop band-widths in excess of 100 MHz by appropriately selecting transistor types and operating currents. However, the d-c voltage gain for a design using high-frequency transistors is usually one to two orders of magnitude lower than that of the design shown in Fig. 9.1. Input current and voltage drift are also severely degraded. Furthermore, many high-frequency transistors have breakdown voltages on the order of 10 to 15 volts, resulting in limited dynamic range for an amplifier using such transistors.

At times high-frequency types are used for transistors Q_4 and Q_5, with high-gain types used in other locations. This change improves the bandwidth of the amplifier, but compromises voltage gain and drift because of the lower current gain typical of high-frequency transistors. Since transistors Q_4 and Q_5 operate at low voltage levels, dynamic range is not altered.

9.4   EXPERIMENTAL RESULTS

While the amplifier described in this chapter was designed primarily as an educational vehicle, it has been built and tested, and can be used to demonstrate certain performance features of the two-stage design. Although a detailed description of the experimentally measured performance of this amplifier is of questionable value since it is not a commercially available

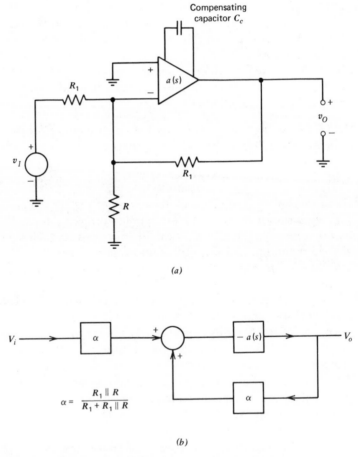

(a)

$$\alpha = \frac{R_1 \parallel R}{R_1 + R_1 \parallel R}$$

(b)

Figure 9.8   Inverting amplifier. (*a*) Circuit. (*b*) Block diagram.

design, the presentation of several transient responses seems a worthwhile prelude to the more detailed experimental evaluation of compensation included in Chapter 13.

The amplifier was connected as shown in Fig. 9.8a. This connection, which results in the block diagram shown in Fig. 9.8b, is useful for demonstrations since it permits control of the loop transmission both by selection of the value of C_c [which influences $a(s)$] and by choice of R. The ideal closed-loop gain of the connection is minus one independent of R.

The magnitude of the loop transmission for this system, with only the lowest-frequency pole included, is shown in Bode-plot form in Fig. 9.9. As anticipated, the crossover frequency is dependent on the ratio α/C_c.

The output of the amplifier in response to -20-mV step input signals with $R = \infty$ ($\alpha = 1/2$) for four different values of compensating capacitor is shown in Fig. 9.10. Note that for the larger values of C_c, the response is very nearly first order, and that the 10 to 90% rise time agrees closely with the value predicted for single-pole systems, $t_r = 2.2/\omega_c$. Smaller compensating-capacitor values change the character of the response as the system becomes relatively less stable and faster. The highly oscillatory response that results for $C_c = 5$ pF indicates that the phase shift added at the crossover frequency by the second- and higher-frequency poles is very nearly $90°$ in this case.

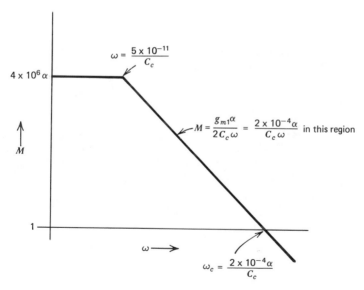

Figure 9.9 Loop-transmission magnitude for inverting amplifier.

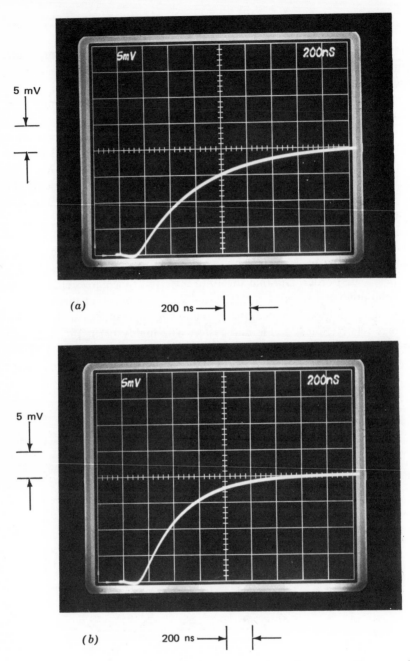

5 mV

200 ns →

(a)

5 mV

200 ns →

(b)

Figure 9.10 Closed-loop step response as a function of compensating capacitor (input-step amplitude is −20 mV). (*a*) $C_c = 47$ pF. (*b*) $C_c = 33$ pF. (*c*) $C_c = 10$ pF. (*d*) $C_c = 5$ pF.

368

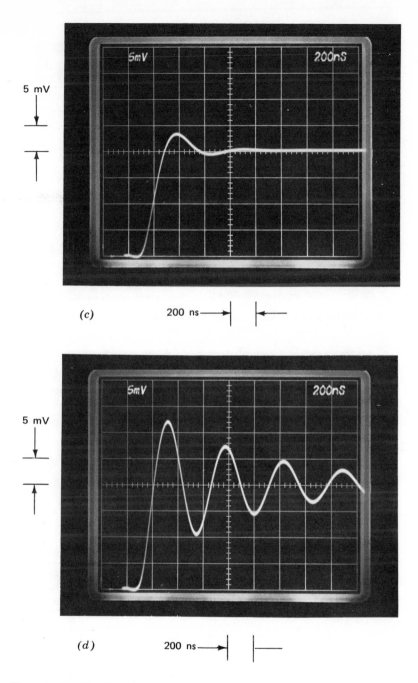

5 mV

(c)   200 ns

5 mV

(d)   200 ns

Figure 9.10—Continued

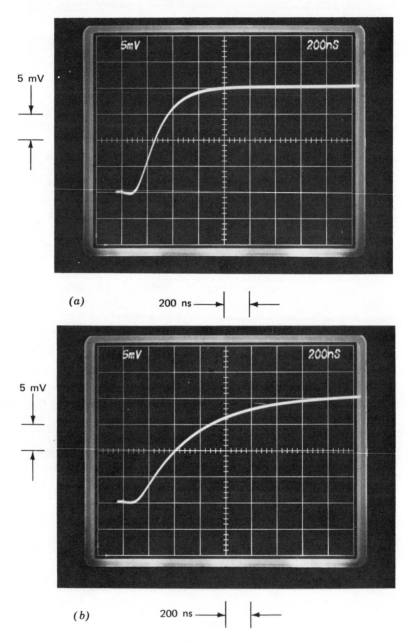

5 mV

200 ns →| |←

(a)

5 mV

200 ns →| |←

(b)

Figure 9.11 Step response as a function of compensating capacitor and α (input-step amplitude is -20 mV). (a) $C_c = 20$ pF, $\alpha = 1/2$. (b) $C_c = 20$ pF, $\alpha = 1/4$. (c) $C_c = 10$ pF, $\alpha = 1/4$.

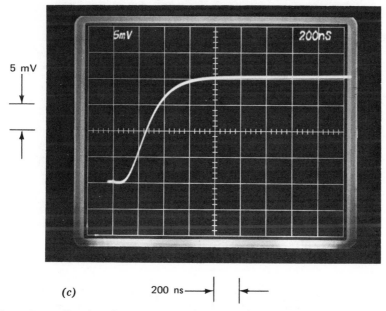

(c)   200 ns

Figure 9.11—Continued

The step response shown in Fig. 9.11 shows how this design allows the effects of changing attenuation inside the loop to be offset by altering compensation. While the attenuation is changed by changing the value of R in this demonstration, it depends on the ideal closed-loop gain in many practical connections. Figure 9.11a shows the step response for $\alpha = 1/2$ ($R = \infty$) and $C_c = 20$ pF. The response for $\alpha = 1/4$ ($R = \frac{1}{2}R_1$) and $C_c = 20$ pF is shown in Fig. 9.11b. The rise time is approximately twice as long in Fig. 9.11b, anticipated since the crossover frequency is a factor of two lower in this connection (see Fig. 9.9). The crossover frequency can be restored to its original value by lowering C_c to 10 pF. The transient response for this value of compensating capacitor (Fig. 9.11c) is virtually identical to that shown in part a of this figure.

Figure 9.12 demonstrates the slew rate of the amplifier by showing its slew-rate limited response to 20-volt peak-to-peak square wave signals. The parameter values for Fig. 9.12a are $\alpha = 1/2$ and $C_c = 20$ pF, while those of Fig. 9.12b are $\alpha = 1/4$ and $C_c = 10$ pF. These are the values that gave the virtually identical small-signal responses shown in Figs. 9.11a and 9.11c, respectively. The large-signal responses show that the slew rate is inversely proportional to compensating-capacitor value, as predicted in Section 9.3.2.

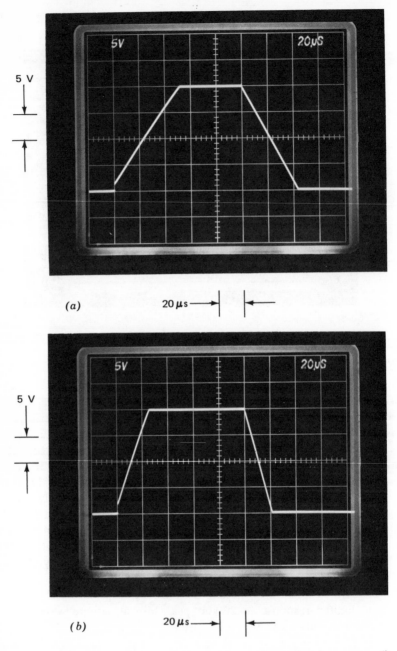

Figure 9.12 Effect of compensating capacitor on large-signal response (input square-wave amplitude is 20 volts peak-peak). (*a*) $C_c = 20$ pF, $\alpha = 1/2$. (*b*) $C_c = 10$ pF, $\alpha = 1/4$.

PROBLEMS

P9.1

Figure 9.13 shows schematics for several available integrated circuits. Determine the transistors that actually contribute to signal amplification for each of these circuits.

P9.2

Assume that measurements made on an operational amplifier of the type described in this chapter indicate a bias current required at either input terminal equal to $9 \times 10^{-4} A/T^2$, where T is the temperature in degrees Kelvin. We intend to use the amplifier connected for a noninverting gain of two. Design a temperature-dependent network that can partially compensate the input current seen at the noninverting input of the amplifier. Note that since an input voltage range of ± 5 volts is anticipated, the incremental resistance of the compensating source must be the order of $10^{10} \Omega$ to achieve good compensation.

P9.3

The input transistors of the amplifier described in this chapter are matched such that the difference between the base-to-emitter voltages of these two devices is less than 3 mV when they operate at equal collector currents. Assume that this matching is not performed, and consequently that the base-to-emitter voltage of Q_2 (see Fig. 9.1) is 50 mV lower than that of Q_1 when the two devices operate at equal currents. The amplifier can still be balanced by replacing the collector-circuit resistor network of the pair with a 650-kΩ potentiometer, and possibly changing the 33-kΩ resistor in the emitter circuit of the Q_4-Q_5 pair so that the quiescent operating level of these devices remains 50 μA following balancing. Calculate the effect that balancing an amplifier with this degree of mismatch between input devices has on the open-loop gain of the amplifier.

P9.4

Figure 9.14 shows a simplified representation for an operational amplifier. You may assume that the current sources have infinite output impedance and that the buffer amplifier has infinite input resistance. All transistors are characterized by $\beta = 200$ and $\eta = 5 \times 10^{-4}$.
(a) Estimate the low-frequency open-loop gain of this configuration.
(b) What is the input offset voltage of the amplifier, assuming that the two input transistors have identical values for I_S?
(c) What is the common-mode rejection ratio of this amplifier?
(d) Estimate the time constant associated with the dominant amplifier pole, assuming all transistors have $C_\pi = 10$ pF, $C_\mu = 5$ pF.
(e) Suggest at least three circuit changes (aside from simply using better transistors) that can increase the value of the d-c open-loop gain.

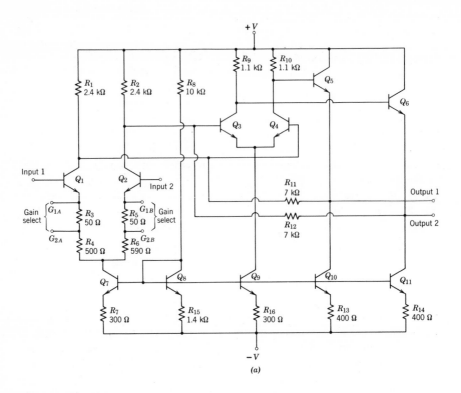

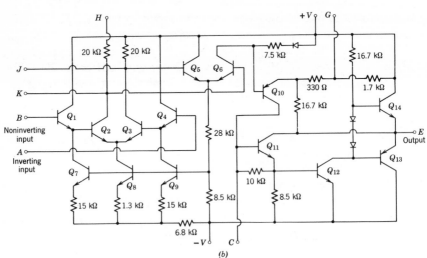

Figure 9.13 Integrated-circuit amplifiers. (*a*) μA733. (*b*) MC1533. (*c*) μA741. (*d*) MC1539.

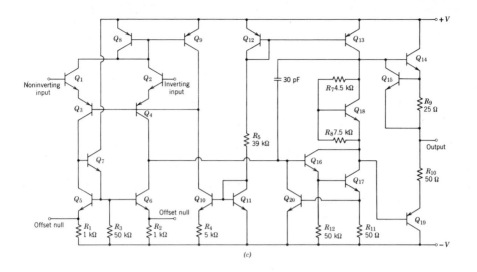

(c)

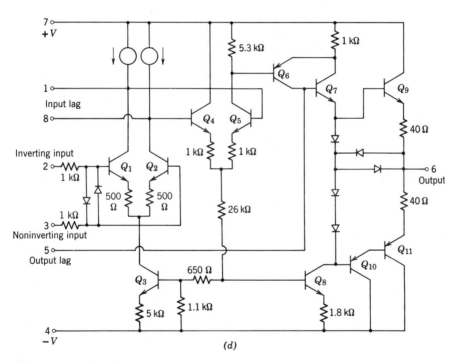

(d)

Figure 9.13—Continued

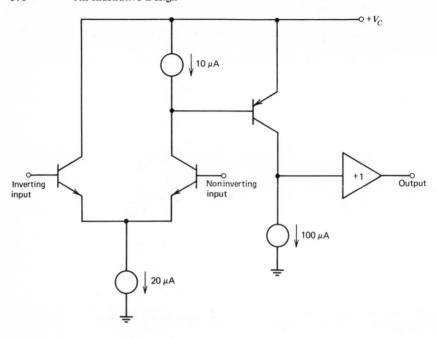

Figure 9.14   Operational amplifier.

P9.5

An interesting amplifier topology that can be used for operational amplifiers intended to be connected as unity-gain voltage followers is shown in Fig. 9.15. (Note that the amplifier is shown connected as a voltage follower.) You may assume that the current sources have infinite output impedance and that all transistors are characterized by $\beta = 100$ and $\eta = 2 \times 10^{-4}$.

(a) How many voltage-gain stages does this amplifier have?

(b) Estimate the unloaded, low-frequency open-loop gain of the amplifier.

(c) Estimate the low-frequency closed-loop output impedance of the circuit.

P9.6

Assume that the field-effect transistor (Q_8 in Fig. 9.1) in the amplifier described in this chapter is replaced with a 2N3707. Use values given in Table 9.1, with appropriate modifications reflecting operation at 2 mA, to determine values for g_m, r_π, r_o, and r_μ. You may assume that the value of C_π at 2 mA is 50 pF. Determine the changes in amplifier d-c open-loop gain and the changes in uncompensated dynamics that result from this design change.

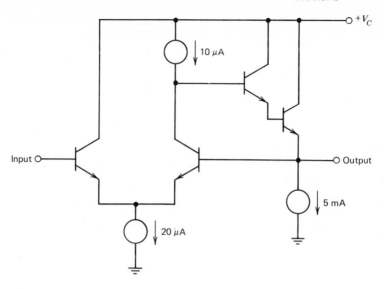

Figure 9.15 Follower-connected amplifier.

P9.7

A detailed analysis of a certain operational amplifier shows that its open-loop transfer function contains a single low-frequency pole, and that the location of this pole is easily controlled by appropriate compensation. In addition to this dominant pole, the open-loop transfer function includes 7 poles at $s = -10^8$ sec^{-1} and two right-half-plane zeros at $s = 2 \times 10^8$ sec^{-1}. Show that, at least at frequencies up to several megahertz, the net effect of these higher-frequency singularities can be modeled as a single time delay. Determine the delay time of an approximating transfer func-

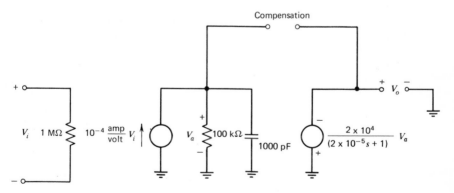

Figure 9.16 Operational-amplifier model.

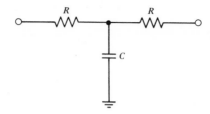

Figure 9.17   Low-pass T network.

tion. Use the time-delay approximation to describe the effect of the higher-order singularities on the maximum crossover frequency of feedback connections that include this amplifier inside the loop. If the d-c open-loop gain of the amplifier is 10^5, how should the dominant pole be located in order to achieve 45° of phase margin when the amplifier is connected as a unity-gain inverter?

P9.8

A model for an operational amplifier is shown in Fig. 9.16. This amplifier is connected as a unity-gain voltage follower.

(a) What is the phase margin with no compensation?

(b) If a capacitor is used between the compensating terminals, how large a value is required to double the uncompensated phase margin?

(c) How large a capacitor should be used to obtain 45° of phase margin in the follower connection?

(d) An alternative compensating technique involves shunting a series R-C network across the 100-kΩ resistor and 1000-pF capacitor combination shown in Fig. 9.16. Find parameter values for this type of compensation that yields results similar to those obtained in part c.

P9.9

The amplifier described in Problem P9.8 is used in a loop where an approximate open-loop transfer function of $10 \, (10^{-2}s + 1)$ is required. It

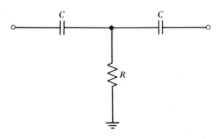

Figure 9.18   High-pass T network.

is suggested that the required transfer function be obtained by compensating the amplifier with the T network shown in Fig. 9.17. Determine network-parameter values that might reasonably be expected to approximate the required transfer function.

When the amplifier is tested with this type of compensation, we find that our first guess was incorrect. Explain.

P9.10

Another class of application involves the use of the T network shown in Fig. 9.18 to compensate the amplifier described in Problem P9.8. This network can be used without encountering the type of difficulties that occur using the network described in Problem P9.9. Determine the type of transfer function that results using the high-pass T, and comment on the value of this type of compensation.

P9.11

It was mentioned in Section 9.3.1 that the temperature stability of the amplifier described in this chapter could be improved by making the bias current source of the first stage have an output current directly proportional to temperature. This proportionality can be accomplished by means of the circuit shown in Fig. 9.19. Assume that the transistor current-voltage characteristic is

$$i_C = AT^3 \, e^{q(V_{BE}-V_{go})/kT}$$

Determine the value of V_B that results in an output current directly proportional to temperature at $300°$ K.

P9.12

A two-stage operational amplifier can be modeled as shown in Fig. 9.20. In this representation, the high-gain second stage itself is modeled

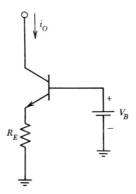

Figure 9.19   Temperature-dependent current source.

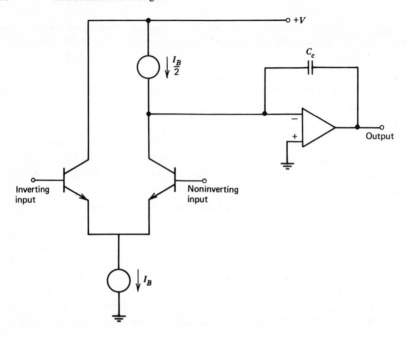

Figure 9.20   Model for two-stage operational amplifier.

as an operational amplifier with a minor-loop feedback element connected around it. You may assume that the second stage has ideal characteristics (i.e., infinite gain and input impedance, zero output impedance,
etc.).

(a) Determine the unity-gain frequency of this amplifier as a function of
I_B and C_c.

(b) Express the slew rate of the amplifier in terms of the same parameters.

(c) Find a design modification that allows an increase in slew rate without
increasing unity-gain frequency.

CHAPTER X

INTEGRATED-CIRCUIT OPERATIONAL AMPLIFIERS

10.1 INTRODUCTION

The trend toward the use of operational amplifiers as general-purpose analog building blocks began when modular, solid-state discrete-component designs became available to replace the older, more expensive vacuum-tube circuits that had been used primarily in analog computers. As cost decreased and performance improved, it became advantageous to replace specialized circuits with these modular operational amplifiers.

This trend was greatly accelerated in the mid 1960s as low-cost monolithic integrated-circuit operational amplifiers became available. While the very early monolithic designs had sadly deficient specifications compared with discrete-component circuits of the era, present circuits approach the performance of the best discrete designs in many areas and surpass it in a few. Performance improvements are announced with amazing regularity, and there seem to be few limitations that cannot be overcome by appropriately improving the circuit designs and processing techniques that are used. No new fundamental breakthrough is necessary to provide performance comparable to that of the best discrete designs. It seems clear that the days of the discrete-component operational amplifier, except for special-purpose units where economics cannot justify an integrated-circuit design, are numbered.

In spite of the clear size, reliability, and in some respects performance advantages of the integrated circuit, its ultimate impact is and always will be economic. If a function can be realized with a mass-produced integrated circuit, such a realization will be the cheapest one available. The relative cost advantage of monolithic integrated circuits can be illustrated with the aid of the discrete-component operational amplifier used as a design example in the previous chapter. The overall specifications for the circuit are probably slightly superior to those of presently available general-purpose integrated-circuit amplifiers, since it has better bandwidth, d-c gain, and open-loop output resistance than many integrated designs. Unfortunately, economic reality dictates that a company producing the circuit would probably have to sell it for more than $20 in order to survive. General-purpose integrated-circuit operational amplifiers are presently available for approximately $0.50 in quantity, and will probably become cheaper in the future. Most system designers would find a way to circumvent any performance deficiencies of the integrated circuits in order to take advantage of their dramatically lower cost.

The tendency toward replacing even relatively simple discrete-component analog circuits with integrated operational amplifiers will certainly increase as we design the ever more complex electronic systems of the future that are made economically feasible by integrated circuits. The challenge to the designer becomes that of getting maximum performance from these amplifiers by devising clever configurations and ways to tailor behavior from the available terminals. The basic philosophy is in fundamental agreement with many areas of design engineering where the objective is to get the maximum performance from available components.

Prior to a discussion of integrated-circuit fabrication and designs, it is worth emphasizing that when compromises in the fabrication of integrated circuits are exercised, they are frequently slanted toward improving the economic advantages of the resultant circuits. The technology exists to design monolithic operational amplifiers with performance comparable to or better than that of the best discrete designs. These superior designs will become available as manufacturers find the ways to produce them economically. Thus the answer to many of the "why don't they" questions that may be raised while reading the following material is "at present it is cheaper not to."

10.2 FABRICATION

The process used to make monolithic integrated circuits dictates the type and performance of components that can be realized. Since the probabilities of success of each step of the fabrication process multiply to yield the probability of successfully completing a circuit, manufacturers are understandably reluctant to introduce additional operations that must reduce yields and thereby increase the cost of the final circuit. Some manufacturers do use processes that are more involved than the one described here and thus increase the variety and quality of the components they can form, but unfortunately the circuits made by these more complex processes can usually be easily recognized by their higher costs.

The most common process used to manufacture both linear and digital integrated circuits is the six-mask planar-epitaxial process. This technology evolved from that used to make planar transistors. Each masking operation itself involves a number of steps, the more important of which are as

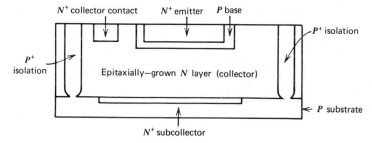

Figure 10.1 NPN transistor made by the six-mask epitaxial process.

follows. A silicon-dioxide layer is first formed by exposing the silicon integrated-circuit material to steam or oxygen at elevated temperatures. This layer is photosensitized, and regions are defined by photographically exposing the wafer using a specific pattern, developing the resultant image, and removing unhardened photosensitive material to expose the oxide layer. This layer is then etched away in the unprotected regions. The oxide layer itself thus forms a mask which permits N- or P-type dopants to be diffused into the silicon wafer. Following diffusion, the oxide is reformed and the masking process repeated to define new areas.

While the operation described above seems complex, particularly when we consider that it is repeated six times, a large number of complete circuits can be fabricated simultaneously. The circuits can be tested individually so localized defects can be eliminated. The net result is that a large number of functioning circuits are obtained from each successfully processed silicon wafer at a low average cost per circuit.

10.2.1 NPN Transistors

The six-mask process is tailored for making NPN transistors, and transistors with characteristics similar to those of virtually all discrete types can be formed by the process. The other components necessary to complete the circuit must be made during the same operations that form the NPN transistors.

A cross-sectional view of an NPN transistor made by the six-mask planar-epitaxial process is shown in Fig. 10.1.[1] Fabrication starts with a P-type

---

[1] It is cautioned that in this and following figures, relative dimensions have been grossly distorted in order to present clearly essential features. In particular, vertical dimensions in the epitaxial layer have been expanded relative to other dimensions. The minimum horizontal dimension is constrained to the order of 0.001 inch by uncertainties associated with the photographic definition of adjacent regions. Conversely, vertical dimensions in the epitaxial layer are defined by diffusion depths and are typically a factor of 10 to 100 times smaller.

substrate (relatively much thicker than that shown in the figure) that provides mechanical rigidity to the entire structure. The first masking operation is used to define heavily doped N-type (designated as N^+) regions in the substrate. The reason for these subcollector or buried-layer regions will be described subsequently. A relatively lightly doped N layer that will be the collector of the complete transistor is then formed on top of the substrate by a process of epitaxial growth.

The next masking operation performed on the epitaxial layer creates heavily doped P-type (or P^+) regions that extend completely through the epitaxial layer to the substrate. These isolation regions in conjunction with the substrate separate the epitaxial layer into a number of N regions each surrounded by P material. The substrate (and thus the isolation regions) will be connected to the most negative voltage applied to the circuit. Since the N regions adjacent to the isolation and substrate cannot be negatively biased with respect to these regions, the various N regions are electrically isolated from each other by reverse-biased P-N junctions. Subsequent steps in the process will convert each isolated area into a separate component.

The P-type base region is formed during the next masking operation. The transistor is completed by diffusing an N^+ emitter into the base. A collector contact, the need for which is described below, is formed in the collector region during the emitter diffusion. The oxide layer is regrown for the last time, and windows that will allow contact to the various regions are etched into this oxide. The entire wafer is then exposed to vaporized aluminum, which forms a thin aluminum layer over the surface. The final masking operation separates this aluminum layer into the conductor pattern that interconnects the various components.

The six masking operations described above can be summarized as follows:

1. Subcollector or buried layer
2. Isolation
3. Base
4. Emitter
5. Contact window
6. Conductor pattern

The buried layer and the heavily doped collector-contact regions are included for the following reasons. Recall that in order to reduce reverse injection from the base of a transistor into its emitter which lowers current gain, it is necessary to have the relative doping level of the emitter significantly greater than that of the base. It is also necessary to dope the collector lightly with respect to the base so that the collector space-charge layer extends dominantly into the collector region in order to prevent low collector-to-base breakdown voltage. As a result of these cascaded inequalities, the collector region is quite lightly doped and thus has high resistivity. If collector current had to flow laterally through this high-resistivity material, a transistor would have a large resistor in series with its collector. The low-resistivity subcollector acts as a shorting bar that connects the active collector region immediately under the base to the collector contact. The length of the collector current path through the high resistivity region is shortened significantly by the subcollector. (Remember that the vertical dimensions in the epitaxial region are actually much shorter than horizontal dimensions.)

The heavily doped N^+ collector contact is necessary to prevent the collector material from being converted to P type by the aluminum that is a P-type dopant. It is interesting to note that the Schottky-diode junction that can form when aluminum is deposited on lightly doped N material is used as a clamp diode in certain digital integrated circuits.

As mentioned earlier, excellent NPN transistors can be made by this process, and the performance of certain designs can be better than that of their discrete-component counterparts. For example, the collector-to-base capacitance of modern high-speed transistors can be dominated by lead rather than space-charge-layer capacitance. The small geometries possible with integrated circuits reduce interconnection capacitance. Furthermore, NPN transistors are extremely economical to fabricate by this method, with the incremental increase in selling price attributable to adding one transistor to a circuit being a fraction of a cent.

Since all transistors on a particular wafer are formed simultaneously, all must have similar characteristics (to within the uniformity of the processing) on a per-unit-area basis. This uniformity is in fact often exploited for the fabrication of matched transistors. A degree of design freedom is retained through adjustment of the relative active areas of various transistors in a circuit, since the collector current of a transistor at fixed base-to-emitter voltage is proportional to its area. This relationship is frequently used to control the collector-current ratios of several transistors (see Section 10.3). Alternatively, the area of a transistor may be selected to optimize current gain at its anticipated quiescent current level. Thus transistors used in the output stage of an operational amplifier are frequently larger than those used in its input stage.

A recent innovation[2] used in some high-performance designs incorporates two emitter diffusions to significantly increase the current gain of certain transistors in the circuit. The oxide layer is first etched away in the emitter

---

[2] R. J. Widlar, "Super Gain Transistors for IC's," National Semiconductor Corporation, Technical Paper TP-11, March, 1969.

region of selected transistors, and the first emitter diffusion is completed. Then, without any oxide regrowth, the emitter regions of the remaining transistors are exposed and the second emitter diffusion is completed. The transistors that have received both emitter diffusions are sometimes called "super-β" transistors since the narrow base width that results from the two diffusions can yield current gains between 10^3 and 10^4. The narrow base region also lowers collector-to-base breakdown voltage to several volts, and precautions must be taken in circuits that use these devices to insure that the breakdown voltage is not exceeded. A second problem is that an overzealous diffusion schedule can easily reduce the base width to zero, and the price of amplifiers using super-β transistors usually reflects this possibility.

10.2.2   PNP Transistors

The six-mask epitaxial process normally used for monolithic integrated circuits is optimized for the fabrication of NPN transistors, and any other circuit components are compromised in that they must be made compatible with the NPN fabrication. One of the limitations of the process is that high-quality PNP transistors cannot be made by it. This limitation is particularly severe in view of the topological advantages associated with the use of complementary transistors. For example, the voltage level shifting required to make input and output voltage ranges overlap in an operational amplifier is most easily accomplished by using one polarity device for the input stage combined with the complementary type in the second stage. Similarly, designs for output stages that do not require high quiescent current are cumbersome unless complementary devices are used.

One type PNP transistor that can be made by the six-mask process is called a lateral PNP. This device is made using the NPN base diffusion for both the emitter and collector regions. The N-type epitaxial layer is used as the base region. Figure 10.2 shows a cross-sectional view of one possible geometry.[3] Current flows laterally from emitter to collector in this structure, in contrast to the vertical flow that results in a conventional design.

There are a number of problems associated with the lateral PNP transistor. The relative doping levels of its emitter, base, and collector regions are far from optimum. More important, however, is the fact that the base width for the structure is controlled by a masking operation rather than a diffusion depth, and is one to two orders of magnitude greater than that of a conventional transistor. There is also parasitic current gain to the substrate that acts as a second collector for the transistor. These effects originally

---

[3] Practical geometries usually surround the emitter stripe with a collector region. This refinement does not alter the basic operation of the device.

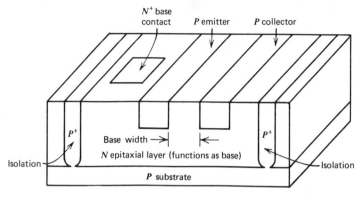

Figure 10.2   Lateral-PNP transistor.

combined to produce very low current gain, with values for β of less than unity common in early lateral PNP's. More recently, process refinements primarily involving the use of the buried layer to reduce parasitic current gain have resulted in current gains in excess of 100.

A more fundamental limitation is that the extremely wide base leads to excessive charge storage in this region and consequently very low values for f_T. The phase shift associated with this configuration normally limits to 1 to 2 MHz the closed-loop bandwidth of an operational amplifier that includes a lateral PNP in the gain path.

One interesting variation of the lateral-PNP transistor is shown in Fig. 10.3. The base-to-emitter voltage applied to this device establishes the per-unit-length current density that flows in a direction perpendicular to the emitter. The relative currents intercepted by the two collectors are thus equal to the relative collector lengths. The concept can be extended, and lateral-PNP transistors with three or more collectors are used in some designs.

One advantage of the lateral-PNP structure is that the base-to-emitter breakdown voltage of this device is equal to the collector-to-base breakdown voltage of the NPN transistors that are formed by the same process. This feature permits nonlinear operation with large different input voltages for operational amplifiers that include lateral PNP's in their input stage. (Two examples are given in Section 10.4.)

A second possible PNP structure is the vertical or substrate PNP illustrated in Fig. 10.4. This type of transistor consists of an emitter formed by the NPN base diffusion and a base of NPN collector material, with the substrate forming the P-type collector. The base width is the difference between the depth of the P-type diffusion and the thickness of the epitaxial layer and can be controlled moderately well. Current gain can be reasonably high and

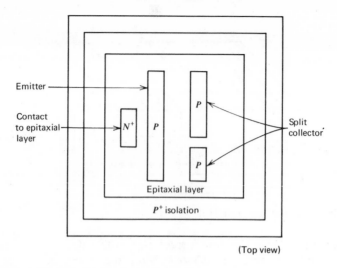

(Top view)

Figure 10.3   Split-collector lateral-PNP transistor.

bandwidth is considerably better than that of a lateral design. One undesirable consequence of the necessary compromises is that large-area transistors must be used to maintain gain at moderate current levels. Another
more serious difficulty is that the collectors of all substrate PNP's are common and are connected to the negative supply voltage. Thus substrate
PNP's can only be used as emitter followers.

10.2.3   Other Components

The P-type base material is normally used for resistors, and the resistivity
of this material dictated by the base-region doping level is typically 100 to
200 ohms per square. Problems associated with achieving high length-to-
width ratios in a reasonable area and with tolerable distributed capacitance
usually limit maximum resistance values to the order of 10 kilo-ohms.
Similarly, other geometric considerations limit the lower value of resistors

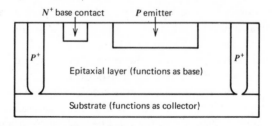

Figure 10.4   Vertical or substrate PNP transistor.

made using the base diffusion to the order of 25 ohms. Higher-value resistors (up to approximately 100 kilo-ohms) can be made using the higher-resistivity collector material, while lower-value resistors are formed from the heavily doped emitter material.

Practical considerations make control of absolute resistance values to better than 10 to 20% uneconomical, and the temperature coefficient of all integrated-circuit resistors is high by discrete-component standards. However, it is possible to match two resistors to 5% or better, and all resistors made from one diffusion have identical temperature coefficients.

It is possible to make large-value, small-geometry resistors by diffusing emitter material across a base-material resistor (see Fig. 10.5). The cross-sectional area of the current path is decreased by this diffusion, and resistance values on the order of 10 kΩ per square are possible. The resultant device, called a pinched resistor, has the highly nonlinear characteristics illustrated in Fig. 10.6. The lower-current portion of this curve results from field-effect transistor action, with the P-type resistor material forming a channel surrounded by an N-type gate. The potential of the gate region is maintained close to that of the most positive end of the channel by conduction through the P-N junction. Thus, if the positively biased end of the pinched resistor is considered the source of a P-channel FET, the characteristics of the resistor are the drain characteristics of a FET with approximately zero gate-to-source voltage. When the voltage applied across the structure exceeds the reverse breakdown voltage of the N^+ and P junction, the heavily doped N^+ region forms a low-resistance path across the resistor. The high-conductance region of the characteristics results from this effect.

In addition to the nonlinearity described above, the absolute value of a pinched resistor is considerably harder to control than that of a standard base-region resistor. In spite of these limitations, pinched resistors are used

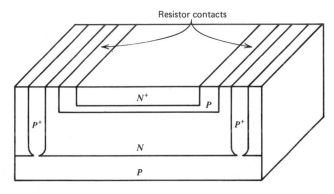

Figure 10.5 Pinched resistor.

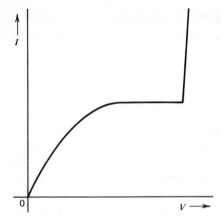

Figure 10.6   Pinched-resistor current-voltage characteristic.

in integrated circuits, often as shunt paths across base-to-emitter junctions of bipolar transistors. The absolute value of such a shunt path is relatively unimportant in many designs, and the voltage applied to the resistor is limited to a fraction of a volt by the transistor junction.

An alternative high-resistance structure that has been used as a bias current source in some integrated-circuit designs is the collector FET shown in Fig. 10.7. This device, which acts as an N-channel FET with its gate biased at the negative supply voltage of circuit, does not have the breakdown-voltage problems associated with the pinched resistor.

Integrated-circuit diodes are readily fabricated. The collector-to-base junction of NPN transistors can be used when moderately high reverse breakdown voltage is necessary. The diode-connected transistor (Fig. 10.8) is used when diode characteristics matched to transistor characteristics are required. If it is assumed that the transistor terminal relationships are

$$I_C = I_S \, e^{q V_{BE}/kT}$$

we can write for the diode-connected transistor

$$I_D = I_B + I_C = \left(1 + \frac{1}{\beta}\right) I_C$$

$$= \left(1 + \frac{1}{\beta}\right) I_S \, e^{q V_D/kT} \simeq I_S \, e^{q V_D/kT} \qquad (10.1)$$

The base-to-emitter junction is used as a Zener diode in some circuits. The reverse breakdown voltage of this junction is determined by transistor processing, with a typical value of six volts.

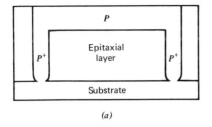

(a)

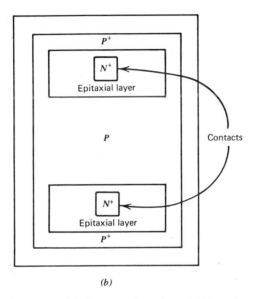

(b)

Figure 10.7   Collector FET. (a) Cross-section view. (b) Top view.

Reverse-biased diode junctions can be used as capacitors when the non-linear characteristics of the space-charge-layer capacitance are acceptable. An alternative linear capacitor structure uses the oxide as a dielectric, with the aluminum metalization layer one plate and the semiconductor material the second plate. This type of metal-oxide-semiconductor capacitor has the further advantage of bipolar operation compared with a diode. The capacitance per unit area of either of these structures makes capacitors larger than 100 pF impractical.

10.3   INTEGRATED-CIRCUIT DESIGN TECHNIQUES

Most high-volume manufacturers of integrated circuits have chosen to live with the limitations of the six-mask process in order to enjoy the

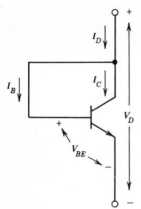

Figure 10.8   Diode-connected transistor.

associated economy. This process dictates circuit considerations beyond those implied by the limited spectrum of component types. For example, large-value base-material resistors or capacitors require a disproportionate share of the total chip area of a circuit. Since defects occur with a per-unit-area probability, the use of larger areas that decrease the yield of the process and thus increase production cost are to be avoided.

The designers of integrated operational amplifiers try to make maximum use of the advantages of integrated processing such as the large number of transistors that can be economically included in each circuit and the excellent match and thermal equality that can be achieved among various components in order to circumvent its limitations. The remarkable performance of presently available designs is a tribute to their success in achieving this objective. This section describes some of the circuit configurations that have evolved from this type of design effort.

10.3.1   Current Repeaters

Many linear integrated circuits use a connection similar to that shown in Fig. 10.9, either for biasing or as a controlled current source. Assume that both transistors have identical values for saturation current I_S and that β is high so that base currents of both transistors can be neglected. In this case, the collector current of Q_1 is equal to i_I. Since the base-to-emitter voltages of Q_1 and Q_2 are identical, currents i_I and i_O must be equal.[4] An

---

[4] In the discussion of this and other current-repeater connections it is assumed that the output terminal voltage is such that the output transistor is in its forward operating region. Note that it is not necessary to have the driving current i_I supplied from a current source. In many actual designs, this current is supplied from a voltage source via a resistor or from another active device.

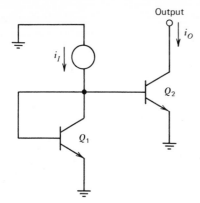

Figure 10.9 Current repeater.

alternative is to change the relative areas of Q_1 and Q_2. This geometric change results in a directly proportional change in saturation currents, so that currents i_I and i_O become a controlled multiple of each other. If i_I is made constant, transistor Q_2 functions as a current source for voltages to within approximately 100 mV of ground. This performance permits the dynamic voltage range of many designs to be nearly equal to the supply voltage.

The split-collector lateral PNP transistor described earlier functions as a current repeater when connected as shown in Fig. 10.10. The constant K that relates the two collector currents in this connection depends on the relative sizes of the collector segments. Since the base current for the

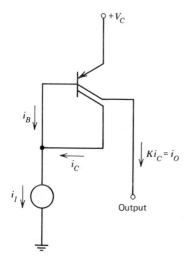

Figure 10.10 Split-collector PNP transistor connected for controlled gain.

lateral PNP is equal to the sum of the two collector currents divided by its current gain β_P, we can write

$$i_I = i_B + i_C = i_C \frac{(1 + K)}{\beta_P} + i_C \qquad (10.2)$$

and

$$i_O = Ki_C \qquad (10.3)$$

Combining Eqns. 10.2 and 10.3 shows that the current gain for this connection is

$$\frac{i_O}{i_I} = \frac{K}{1 + [(1 + K)/\beta_P]} \qquad (10.4)$$

If values are selected so that $1 + K \ll \beta_P$, the feedback inherent to this connection makes its input-output transfer ratio relatively insensitive to changes in β_P. This desensitivity is advantageous since the quantity K, determined by mask geometry, is significantly better controlled than is β_P. The feedback also increases the current-gain half-power frequency of the controlled-gain PNP above the β cutoff frequency of the lateral-PNP transistor itself.

The simple current repeater shown in Fig. 10.9 is frequently augmented to make its current transfer ratio less sensitive to changes in transistor parameters. Equal-value emitter resistors can be included to stabilize the transfer ratio of the connection for changes in the base-to-emitter voltages of the two transistors. While this technique is sometimes used for discrete-component current repeaters, it is of questionable value in many integrated designs because matched resistors are as difficult to fabricate as matched transistors.

Other modifications are intended to reduce the dependency of the current transfer ratio on the transistor current gain. It is easily shown that the current transfer ratio for Fig. 10.9, assuming perfectly matched transistors, is

$$\frac{i_O}{i_I} = \frac{1}{1 + 2/\beta} \qquad (10.5)$$

Figure 10.11 shows two somewhat more complex current-repeater connections assumed constructed with perfectly matched transistors. Intermediate currents that facilitate calculation of current transfer ratios are included in these diagrams. The circuit of Fig. 10.11a uses an emitter follower to buffer the base currents of a conventional current repeater. The resultant current transfer ratio is

$$\frac{i_O}{i_I} = \frac{1}{1 + 2/[\beta(\beta + 1)]} = \frac{\beta^2 + \beta}{\beta^2 + \beta + 2} \qquad (10.6)$$

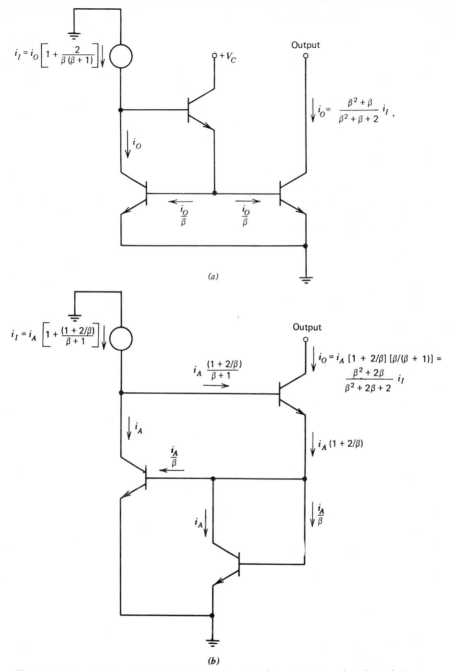

$$i_I = i_O \left[1 + \frac{2}{\beta(\beta+1)} \right]$$

$+V_C$

Output

$$i_O = \frac{\beta^2 + \beta}{\beta^2 + \beta + 2} i_I .$$

i_O

$\frac{i_O}{\beta}$ $\frac{i_O}{\beta}$

(a)

$$i_I = i_A \left[1 + \frac{(1 + 2/\beta)}{\beta+1} \right]$$

Output

$$i_O = i_A [1 + 2/\beta][\beta/(\beta+1)] =$$
$$\frac{\beta^2 + 2\beta}{\beta^2 + 2\beta + 2} i_I$$

$i_A \dfrac{(1 + 2/\beta)}{\beta+1}$

i_A

$i_A (1 + 2/\beta)$

$\dfrac{i_A}{\beta}$

i_A

$\dfrac{i_A}{\beta}$

(b)

Figure 10.11 Improved current-repeater connections. (a) Use of emitter follower. (b) Use of current compensation.

The connection of Fig. 10.11b uses an interesting current cancellation technique to obtain a transfer ratio

$$\frac{i_O}{i_I} = \frac{[1 + 2/\beta] \, [\beta/(\beta + 1)]}{1 + (1 + 2/\beta)/(\beta + 1)} = \frac{\beta^2 + 2\beta}{\beta^2 + 2\beta + 2} \tag{10.7}$$

Either of these currents repeaters has a transfer ratio that differs from unity by a factor of approximately $(1 + 2/\beta^2)$ compared with a factor of $(1 + 2/\beta)$ for the circuit of Fig. 10.9, and are thus considerably less sensitive to variations in β. It can also be shown (see Problem P10.5) that the output resistance of the circuit illustrated in Fig. 10.11b is the order of $r\mu$ while that of either of the other circuits is the order of r_o. This difference is significant in some high-gain connections.

A clever modification of the current repeater, first used in the 709 design, yields a low-value constant-current source using only moderate-value resistors. Assuming high β and a large value of V relative to V_{BE1} in Fig. 10.12,

$$I_{C1} \simeq \frac{V}{R_1} \tag{10.8}$$

so that

$$V_{BE1} \simeq \frac{kT}{q} \ln \frac{V}{R_1 I_{S1}} \tag{10.9}$$

However,

$$I_{C2}R_2 + \frac{kT}{q} \ln \frac{I_{C2}}{I_{S2}} = V_{BE1} \tag{10.10}$$

If it is assumed that saturation currents are equal, combining Eqns. 10.9 and 10.10 yields

$$I_{C2}R_2 = \frac{kT}{q} \ln \frac{V}{R_1 I_{C2}} \tag{10.11}$$

The resultant transcendental equation can be solved for any particular choice of constants. For example, if Eqn. 10.11 is evaluated at room temperature ($kT/q \simeq 26$ mV) for $V/R_1 = 1$ mA and $R_2 = 12$ kΩ, $I_{C2} \simeq 10$ μA.

10.3.2   Other Connections

Most operational-amplifier designs require both NPN and PNP transistors in order to provide voltage level shifting. Several connections effectively augment the low gain of many lateral PNP designs by combining the PNP transistor with an NPN transistor as shown in Fig. 10.13. (This connection is also used in discrete-component circuits and is called the *complementary*

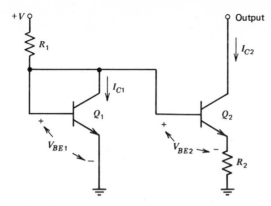

Figure 10.12 Low-level current source.

Darlington connection.) At low frequencies this combination appears as a single PNP transistor with the base, emitter, and collector terminals as indicated. The current gain of this compound transistor is approximately equal to the product of the gains of the two individual devices, while transconductance is related to collector current of the combination as in a conventional transistor.

An ingenious connection using lateral PNP transistors, shown in Fig. 10.14, was introduced in the LM101 amplifier design. Assume that the two NPN transistors have identical saturation currents, as do the PNP's. Further assume that the current gains of both PNP transistors are β_P. The total output collector current, $i_{C3} + i_{C4}$, must be equal to $\beta_P I$. If the input voltages

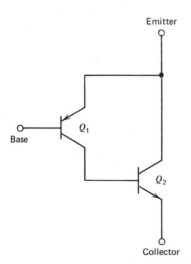

Figure 10.13 Complementary Darlington connection.

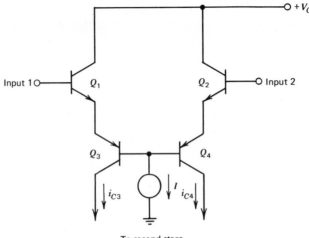

Figure 10.14   Differential input stage.

are equal, i_{C3} and i_{C4} must be equal because of the matched saturation currents. As a differential input signal is applied, the relative collector currents change differentially; therefore this stage can be used to perform the circuit function of a differential pair of PNP transistors. However, the ratio of input current to collector current depends on the current gain of the high-gain NPN's. Another advantage is that the input capacitance is low since the input transistors are operating as emitter followers. Furthermore, the low-bandwidth PNP devices are operating in an incrementally grounded-base connection for differential input signals, and this connection maximizes their bandwidth in the circuit. One disadvantage is that the series connection of four base-to-emitter junctions lowers transconductance by a factor of two compared to a standard differential amplifier operating at the same quiescent current level.

It is interesting to note that the successful operation of this circuit is actually dependent on the low gain characteristic of the lateral-PNP transistors used. If high-gain transistors were used, capacitive loading at the bases of the two PNP transistors would cause large collector currents as a function of the time rate of change of common-mode level. The controlled gain PNP shown in Fig. 10.10 is used in this connection in some modern amplifier designs.

Several connections are used to double the effective transconductance of an input differential pair and thus increase the gain provided by this portion of an operational amplifier. One such circuit is shown in Fig. 10.15. Assume equal operating currents for Q_1 and Q_2. If Q_4 were a constant

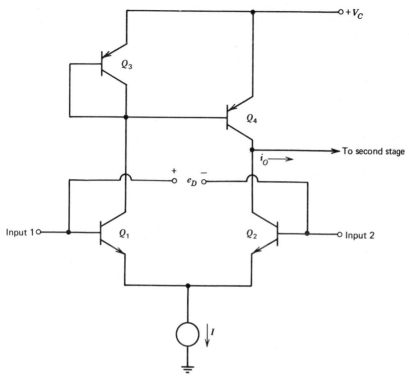

Figure 10.15   Use of current repeater to increase stage transconductance.

current source, the incremental output current would be related to a differential input voltage e_d as $i_o/e_d = g_m/2$. The differential connection of Q_1 and Q_2 insures that incremental changes in collector currents of these devices are equal in magnitude but opposite in polarity, and the current repeater connection of Q_3 and Q_4 effectively subtracts the change in collector current of Q_1 from that of Q_2. (The more sophisticated current repeaters described in the last section are often substituted.) The gain is increased by a factor of two so that $i_o/e_d = g_m$. Another advantage is that the impedance level at the circuit output is high so that this stage can provide high voltage gain if required. We will see that some integrated-circuit operational amplifiers exploit this possibility to distribute the total gain more equally between the two stages than was done with the discrete-component design discussed in the last chapter.

Another approach is illustrated in Fig. 10.16. A differential input causes equal-magnitude changes in the collector currents of Q_1 and Q_2. However, the high gain of the Q_3-Q_5 loop changes the voltage at the emitter of Q_5 in

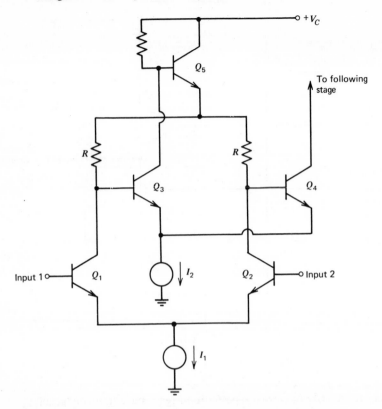

Figure 10.16   Use of local feedback to increase stage transconductance.

such a way as to minimize current changes at the base of Q_3. Thus the current through the load resistor for Q_1 is changed by an amount approximately equal to the change in i_{C1}. A corresponding change occurs in the current through the load resistor for Q_2, doubling the current into the base of Q_4.

10.4   REPRESENTATIVE INTEGRATED-CIRCUIT OPERATIONAL AMPLIFIERS

A number of semiconductor manufacturers presently offer a variety of integrated-circuit operational amplifiers. While an exhaustive study of available amplifiers is beyond the scope of this book, an examination of several representative designs demonstrates some of the possible variations of the basic topology described in Chapters 8 and 9 and serves as a useful prelude to the material on applications.

It should be mentioned that most of the circuits described are popular enough to be built, often with minor modifications, by a number of manufacturers. These "second-source" designs usually retain a designation that maintains an association with the original. Another factor that contributes to the proliferation of part numbers is that most manufacturers divide their production runs into two or three categories on the basis of measured parameters such as input bias current and offset voltage as well as the temperature range over which specifications are guaranteed. For example, National Semiconductor uses the 100, 200, and 300 series to designate whether military, intermediate, or commercial temperature range specifications are met, while Fairchild presently suffixes a C to designate commercial temperature range devices.

We should observe that no guarantee of inferior performance is implied when the less splendidly specified devices are used. Since all devices in one family are made by an identical process and since yields are constantly improving, a logical conclusion is that many commercially specified devices must in fact be meeting military specifications. These considerations coupled with a dramatic cost advantage (the order of a factor of three) suggest the use of the commercial devices in all but the most exacting applications.

10.4.1   The LM101 and LM101A Operational Amplifiers

The LM101 operational amplifier[5] occupies an important place in the history of integrated-circuit amplifiers since it was the first design to use the two-stage topology combined with minor-loop feedback for compensation. Its superiority was such that it stimulated a variety of competing designs as well as serving as the ancestor of several more advanced National Semiconductor amplifiers.

The schematic diagram for the amplifier is shown in Fig. 10.17, and specifications are included in Table 10.1. (The definitions of some of the specified quantities are given in Chapter 11.) As was the case with the discrete-component amplifier described in the last chapter, it is first necessary to identify the functions of the various transistors, with emphasis placed on the transistors in the gain path. Transistors Q_1 through Q_4 form a differential input connection as described in the last section. The Q_5 through Q_7 triad is a current-repeater load for the differential stage. Transistors Q_8 and Q_9 are connected as an emitter follower driving a high voltage gain common-emitter stage. The voltage gains of the first and second stages

[5] R. J. Widlar, "A New Monolithic Operational Amplifier Design," National Semiconductor Corporation, Technical Paper TP-2, June, 1967.

Table 10.1 LM101 Specifications: Electrical Characteristics

Parameter	Conditions	Min	Typ	Max	Units
Input offset voltage	$T_A = 25°$ C, $R_S \leq 10$ kΩ		1.0	5.0	mV
Input offset current	$T_A = 25°$ C		40	200	nA
Input bias current	$T_A = 25°$ C		120	500	nA
Input resistance	$T_A = 25°$ C	300	800		kΩ
Supply current	$T_A = 25°$ C, $V_S = \pm20$ V		1.8	3.0	mA
Large-signal voltage gain	$T_A = 25°$ C, $V_S = \pm15$ V				
	$V_{out} = \pm10$ V, $R_L \geq 2$ kΩ	50	160		V/mV
Input offset voltage	$R_S \leq 10$ kΩ			6.0	mV
Average temperature	$R_S \leq 50$ Ω		3.0		μV/$°$C
coefficient of input					
offset voltage	$R_S \leq 10$ kΩ		6.0		μV/$°$C
Input offset current	$T_A = +125°$ C		10	200	nA
	$T_A = -55°$ C		100	500	nA
Input bias current	$T_A = -55°$ C		0.28	1.5	μA
Supply current	$T_A = +125°$ C, $V_S = \pm20$ V		1.2	2.5	mA
Large-signal voltage gain	$V_S = \pm15$ V, $V_{out} = \pm10$ V				
	$R_L \geq 2$ kΩ	25			V/mV
Output voltage swing	$V_S = \pm15$ V, $R_L = 10$ kΩ	±12	±14		V
	$R_L = 2$ kΩ	±10	±13		V
Input voltage range	$V_S = \pm15$ V	±12			V
Common-mode rejection					
ratio	$R_S \leq 10$ kΩ	70	90		dB
Supply-voltage rejection					
ratio	$R_S \leq 10$ kΩ	70	90		dB

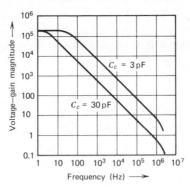

Open — loop freqnency response
±15 — volt supplies, 25°C

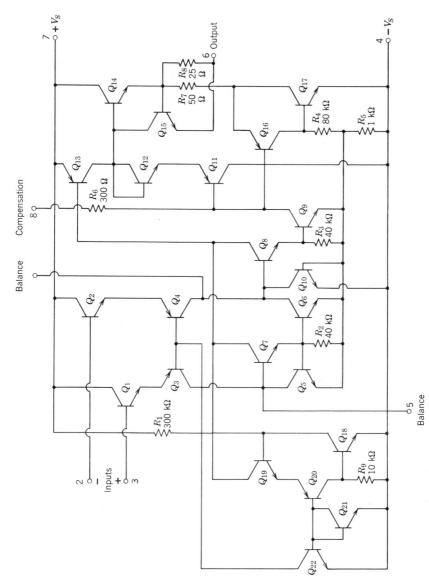

Figure 10.17 LM101 schematic diagram.

403

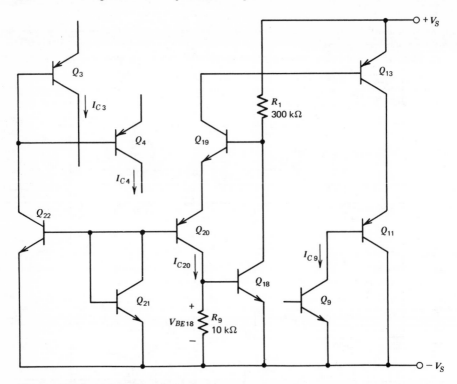

Figure 10.18 LM101 bias circuitry.

of this amplifier are both proportional to the reciprocal of the base-width modulation factor and thus are comparable in magnitude.

The complementary Darlington connection Q_{16} and Q_{17} supplies negative output current. The use of this connection augments the low gain of the lateral PNP. (Recall that this amplifier was manufactured when current gains of 5 to 10 were anticipated from lateral-PNP transistors.) While a vertical-PNP transistor could have been used in the output stage, the designer of the 101 elected the complementary Darlington since it reduced total chip area[6] and since processing was simplified.

Positive output current is supplied by Q_{14}. The gain path from the collector of Q_9 to the emitter of Q_{14} includes transistor Q_{11}, another lateral PNP. This device matches the current gain from the collector of Q_9 to the output for positive output swings with the gain for negative output swings. By locating current source Q_{13} in the emitter circuit of Q_{11}, this current source provides bias for Q_{11} as well as a high-resistance load for Q_9. Diode-

---

[6] It is interesting to note that the size of the LM101 chip is 0.045 inch square, smaller than many single transistors.

connected transistor Q_{12} is included in the output circuit to reduce cross-over distortion.

The operation of the biasing circuit for the LM101 depends on achieving equal current gains from certain lateral-PNP transistors. This approach was used since while low, unpredictable gains characterized the lateral PNP's of the era, the performance was highly uniform from device to device on one chip. The transistors used for biasing are shown in Fig. 10.18. The loop containing transistors Q_{18}, Q_{19}, and Q_{20} controls I_{C20} so that $I_{C20} \simeq V_{BE18}/R_9 \simeq 60 \ \mu A$.

The high-value resistor, R_1, included in this circuit is a collector FET. The characteristics of this resistor make the current supplied by it relatively independent of supply voltage. The base current of Q_{20} is repeated by transistors Q_{21} and Q_{22} and applied to the common-base connection of Q_3 and Q_4. If the areas of Q_{21} and Q_{22} and the current gains of Q_3, Q_4, and Q_{20} were equal, the total first-stage collector current, $I_{C3} + I_{C4}$, would be equal to I_{C20}. The area of Q_{21} is actually made larger than that of Q_{22} so that each input transistor operates at a quiescent collector current of 10 μA.

Biasing for transistor Q_9 includes transistors Q_{11}, Q_{13}, Q_{19}, and Q_{20}. Assuming high gain from Q_{19},

$$I_{C9} = \frac{I_{C20}(\beta_{20} + 1)}{\beta_{20}} \frac{\beta_{13}}{(1 + \beta_{11})} \qquad (10.12)$$

Thus $I_{C9} = I_{C20}$ for equal PNP gains.

The actual circuit (Fig. 10.17) shows that the collectors of Q_7 and Q_8 are connected in parallel with that of Q_{19}. This doesn't significantly alter operation since $I_{C19} \gg I_{C7} \simeq I_{C8}$, and allows a smaller geometry chip since Q_7, Q_8, and Q_{19} can all be located in the same isolation diffusion.

Positive output current is limited by transistor Q_{15} (Fig. 10.17) when the voltage across R_8 becomes approximately 0.6 volt. The negative current limit is more involved. When the voltage across R_7 reaches approximately 1.2 volts, the collector-to-base junction of Q_{15} becomes forward biased, and further increases in output current are supplied by Q_{11}. Since this lateral PNP has low gain, the emitter current of Q_9 increases significantly when the limiting value of output current is reached. The emitter current of Q_9 flows through R_5, and when the drop across this resistor reaches 0.6 volt, transistor Q_{10} limits base drive for Q_8, preventing further increases in output current.

There are two reasons for this unusual limiting circuit. First, the peculiarities of lateral PNP Q_{16} make it advantageous to have relatively high resistance between the emitter of this transistor and the output of the circuit to insure stability with capacitive loads. Second, this limit also protects Q_9 if its collector is clamped to some voltage level. Such clamping applied to point 8 can be used to limit the output voltage of the amplifier.

The amplifier can be balanced to reduce input offset voltage by connecting a high-value resistor (typically 20 MΩ to 100 MΩ) from either point 5 or point 1 to ground. This type of balancing results in minimum voltage drift from the input transistors.

Compensating minor-loop feedback around the high-gain portion of the circuit is applied between points 1 to 8. The 300-Ω resistor in this circuit provides a zero at a frequency approximately one decade above the amplifier unity-gain frequency when a capacitor is used for compensation. The positive phase shift associated with this zero improves amplifier stability.

Measurements made on the amplifier show that the transconductance from the input terminals to the base of Q_8 is approximately 2×10^{-4} mho so that the open-loop transfer function of the amplifier at frequencies of interest is approximately $2 \times 10^{-4}/Y_c$, where Y_c is the short-circuit transfer admittance of the compensating network as defined in Section 9.2.3. This value of transconductance is consistent with the four series-connected input transistors operating at 10 μA of quiescent current. The transconductance to either output of the differential pair is $qI_C/4kT \simeq 10^{-4}$ mho, and this value is doubled by the current-repeater load used for the input stage. While the compensating network does load the high-impedance node at the collector of Q_9, such loading is usually insignificant.

The open-loop transfer function included as part of the specifications shows that the amplifier has a single-pole response with a unity-gain frequency of approximately 1 MHz when compensated with a 30-pF capacitor. This result can also be obtained from the analytic expression given above. The amplifier dynamics other than those which result from the inner loop limit the crossover frequency of loops using this amplifier to between 1 and 2 MHz. The phase shift that leads to instability for higher crossover frequencies results primarily from the lateral PNP transistors in the input stage.

Evolutionary modifications changed the LM101 amplifier to the LM101A shown in Fig. 10.19, and this amplifier is (as of this writing) still the standard to which all other general-purpose, externally compensated integrated operational amplifiers are compared. The differences reflect primarily the increased performance of components available at the time the LM101A was designed. Better matching tolerances reduced the maximum input offset voltage to 2 mV at 25° C and improved common-mode rejection ratio and power-supply rejection modestly. Improved input-transistor current gain and a modified bias circuit reduced the maximum input bias current over the full −55° C to +125° C temperature range to 100 nA and reduced the typical room-temperature offset current to 1.5 nA.

A detailed discussion of the bias circuit of the LM101A (transistors Q_{18} through Q_{22} in Fig. 10.19) is beyond the scope of the book.[7] Its most im-

[7] R. J. Widlar, "I. C. Op Amp with Improved Input-Current Characteristics," *EEE*, pp. 38–41, December, 1968.

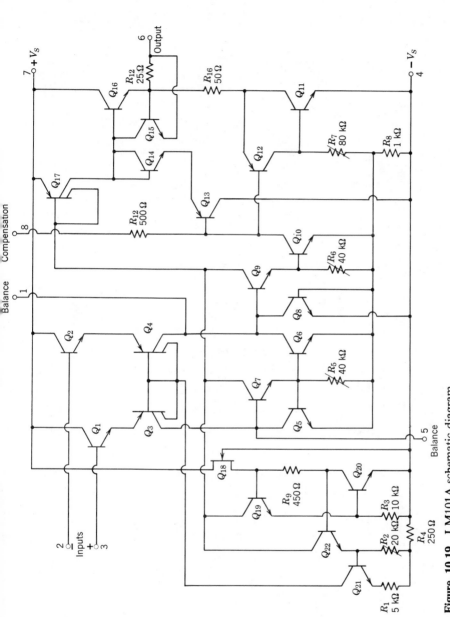

Figure. 10.19 LM101A schematic diagram.

407

Table 10.2 μA776 Specifications: ± 15 Volt Operation for 776; Electrical Characteristics (T_A is 25° C, unless otherwise specified)

Parameter	Conditions	$I_{SET} = 1.5\ \mu A$			$I_{SET} = 15\ \mu A$			Units
		Min	Typ	Max	Min	Typ	Max	
Input offset voltage	$R_S \leq 10$ kΩ		2.0	5.0		2.0	5.0	mV
Input offset current			0.7	3.0		2.0	15	nA
Input bias current			2.0	7.5		15	50	nA
Input resistance			50			5.0		MΩ
Input capacitance			2.0			2.0		pF
Offset voltage adjustment range			9.0			18		mV
Large-signal voltage gain	$R_L \geq 75$ kΩ, $V_{\text{out}} = \pm 10$ V	200	400					V/mV
	$R_L \geq 5$ kΩ, $V_{\text{out}} = \pm 10$ V				100	400		V/mV
Output resistance			5.0			1.0		kΩ
Output short-circuit current			3.0			12		mA
Supply current			20	25		160	180	μA
Power consumption				0.75			5.4	mW
Transient response Rise time (unity gain)	$V_{\text{in}} = 20$ mV, $R_L \geq 5$ kΩ, $C_L = 100$ pF		1.6			0.35		μs
Overshoot			0			10		%
Slew rate	$R_L \geq 5$ kΩ		0.1			0.8		V/μs
Output voltage swing	$R_L \geq 75$ kΩ	± 12	± 14		± 10	± 13		V
	$R_L \geq 5$ kΩ							V

The following specifications apply: $-55^\circ\ \mathrm{C} \leq T_A \leq +125^\circ\ \mathrm{C}$

Parameter	Conditions	Min	Typ	Max	Units
Input offset voltage	$R_S \leq 10\ \mathrm{k\Omega}$			6.0	mV
Input offset current	$T_A = +125^\circ\ \mathrm{C}$		5.0	15	nA
	$T_A = -55^\circ\ \mathrm{C}$		10	40	nA
Input bias current	$T_A = +125^\circ\ \mathrm{C}$		7.5	50	nA
	$T_A = -55^\circ\ \mathrm{C}$		20	120	nA
Input-voltage range		± 10	± 10		V
Common-mode rejection ratio	$R_S \leq 10\ \mathrm{k\Omega}$	70	90		dB
Supply-voltage rejection ratio	$R_S \leq 10\ \mathrm{k\Omega}$	25	150	150	μV/V
Large-signal voltage gain	$R_L \geq 75\ \mathrm{k\Omega},\ V_{out} = \pm 10\ \mathrm{V}$	100	150	75	V/mV
Output voltage swing	$R_L \geq 75\ \mathrm{k\Omega}$	± 10		± 10	V
Supply current			30	200	μA
Power consumption			0.9	6.0	mW

portant functional characteristic is that the quiescent collector current of the input stage is made proportional to absolute temperature. As a result, the transconductance of the input stage (which has a direct effect on the compensated open-loop transfer function of the amplifier) is made virtually temperature independent. A subsidiary benefit is that the change in quiescent current with temperature partially offsets the current-gain change of the input transistors so that the temperature dependence of the input bias current is reduced. The modified bias circuit became practical because the improved gain stability of the controlled-gain lateral PNP's used in the LM101A eliminated the requirement for the bias circuit to compensate for gross variations in lateral-PNP gain.

We shall get a greater appreciation for the versatility of the LM101A, particularly with respect to the control of its dynamics afforded by various types of compensation, in Chapter 13.

10.4.2   The μA776 Operational Amplifier

The LM101A circuit described in the previous section can be tailored for use in a variety of applications by choice of compensation. An interesting alternative way of modifying amplifier performance by changing its quiescent operating currents is used in the μA776 operational amplifier. Some of the tradeoffs that result from quiescent current changes were discussed in Section 9.3.3, and we recall that lower operating currents compromise bandwidth in exchange for reduced input bias current and power consumption.

The schematic diagram for this amplifier is shown in Fig. 10.20, with performance specifications listed in Table 10.2. Several topological similarities between this amplifier and the LM101 are evident. Transistors Q_1 through Q_6 form a current-repeater-loaded differential input stage. Transistors Q_7 and Q_9 are an emitter-follower common-emitter combination loaded by current source Q_{12}. Diode-connected transistors Q_{21} and Q_{22} forward bias the Q_{10}-Q_{11} complementary output pair. Capacitor C_1 compensates the amplifier.

The unique feature of the μA776 is that all quiescent operating currents are referenced to the current labeled I_{SET} in the schematic diagram by means of a series of current repeaters. Thus changing this set current causes proportional changes in all quiescent currents and scales the current-dependent amplifier parameters.

The collector current of Q_{19} is proportional to the set current because of the Q_{16}-Q_{18}-Q_{19} connection. The difference between this current and the collector current of Q_{15} is applied to the common-base connection of the Q_3-Q_4 pair. The collector current of Q_{15} is proportional to the total quiescent operating current of the differential input stage, since Q_{14} and Q_{15} form a current repeater for the sum of the collector currents of Q_1 and Q_2.

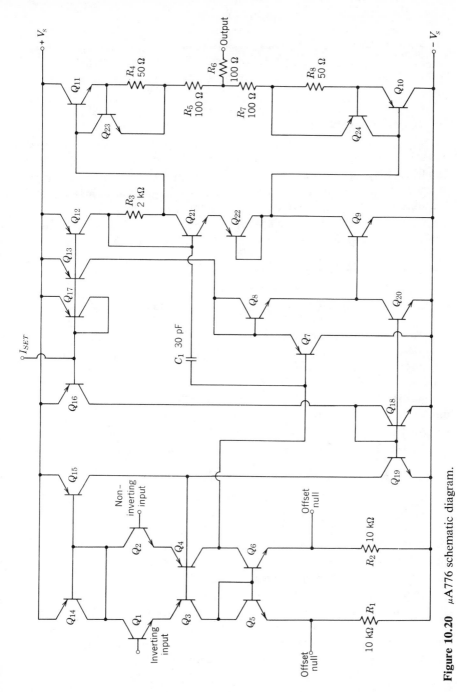

Figure 10.20 μA776 schematic diagram.

411

The resultant negative feedback loop stabilizes quiescent differential-stage current. The geometries of the various transistors are such that the quiescent collector currents of Q_1, Q_2, Q_3, and Q_4 are each approximately equal to I_{SET}.

The amplifier can be balanced by changing the relative values of the emitter resistors of the Q_5-Q_6 current-repeater pair via an external potentiometer. While this balance method does not equalize the base-to-emitter voltages of the Q_5-Q_6 pair, any drift increase is minimal because of the excellent match of first-stage components. An advantage is that the external balance terminals connect to low-impedance circuit points making the amplifier less susceptible to externally-generated noise.

One of the design objectives for the μA776 was to make input- and output-voltage dynamic ranges close to the supply voltages so that low-voltage operation became practical. For this purpose, the vertical PNP Q_7 is used as the emitter-follower portion of the high-gain stage. The quiescent voltage at the base of Q_7 is approximately the same as the voltage at the base of Q_9 (one diode potential above the negative supply voltage) since the base-to-emitter voltage of Q_7 and the forward voltage of diode-connected transistor Q_8 are comparable. (Current sources Q_{13} and Q_{20} bias Q_7 and Q_8.) Because the operating potential of Q_7 is close to the negative supply, the input stage remains linear for common-mode voltages within about 1.5 volts of the negative supply.

Transistor Q_{21} is a modified diode-connected transistor which, in conjunction with Q_{22}, reduces output stage crossover distortion. At low set-current levels (resulting in correspondingly low collector currents for Q_9 and Q_{12}) the drop across R_3 is negligible, and the potential applied between the bases of Q_{10} and Q_{11} is equal to the sum of the base-to-emitter voltages of Q_{21} and Q_{22}. At higher set currents, the voltage drop across R_3 lowers the ratio of output-stage quiescent current to that of Q_9 as an aid toward maintaining low power consumption.

A vertical-PNP transistor is used in the complementary output stage, and this stage, combined with its driver (Q_9 and Q_{12}), permits an output voltage dynamic range within approximately one volt of the supplies at low output currents. Current limiting is identical to that used in the discrete-component amplifier described in Chapter 9.

The ability to change operating currents lends itself to rather interesting applications. For example, operation with input bias currents in the pico-ampere region and power consumption at the nanowatt level is possible with appropriately low set current if low bandwidth is tolerable. The amplifier can also effectively be turned into an open circuit at its input and output terminals by making the set current zero, and thus can be used as an analog switch. Since the unity-gain frequency for this amplifier is $g_m/(2 \times 30$ pF) where g_m is the (assumed equal) transconductance of transistors Q_1 through Q_4, changes in operating current result in directly proportional changes in unity-gain frequency.

This amplifier is inherently a low-power device, even at modest set-current levels. For example, many performance specifications for a μA776 operating at a set current of 10 μA are comparable to those of an LM101A when compensated with a 30-pF capacitor. However, the power consumption of the μA776 is approximately 3 mW at this set current (assuming operation from 15-volt supplies) while that of the LM101A is 50 mW. The difference reflects the fact that the operating currents of the second and output stage are comparable to that of the first stage in the μA776, while higher relative currents are used in the LM101A. One reason that this difference is possible is that the slew rate of the μA776 is limited by its fixed, 30-pF compensating capacitor. Higher second-stage current is necessary in the LM101A to allow higher slew rates when alternate compensating networks are used.

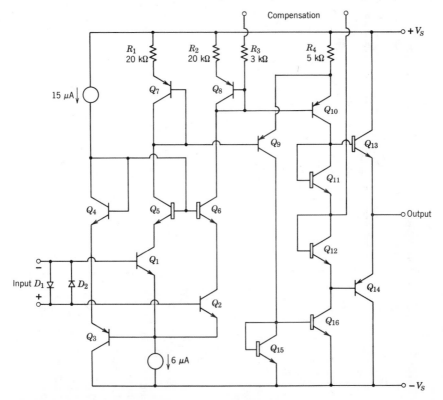

Figure 10.21   LM108 simplified schematic diagram.

Table 10.3 LM108 Specifications: Electrical Characteristics

Parameter	Conditions	Min	Typ	Max	Units
Input offset voltage	$T_A = 25°C$		0.7	2.0	mV
Input offset current	$T_A = 25°C$		0.05	0.2	nA
Input bias current	$T_A = 25°C$		0.8	2.0	nA
Input resistance	$T_A = 25°C$	30	70		MΩ
Supply current	$T_A = 25°C$		0.3	0.6	mA
Large-signal voltage gain	$T_A = 25°C, V_S = ±15$ V, $V_{out} = ±10$ V, $R_L \geq 10$ kΩ	50	300		V/mV
Input offset voltage				3.0	mV
Average temperature coefficient of input-offset voltage			3.0	15	μV/°C
Input offset current				0.4	nA
Average temperature coefficient of input offset current			0.5	2.5	pA/°C
Input bias current				3.0	nA
Supply current	$T_A = +125°C$		0.15	0.4	mA
Large-signal voltage gain	$V_S = ±15$ V, $V_{out} = ±10$ V, $R_L \geq 10$ kΩ	25			V/mV
Output voltage swing	$V_S = ±15$ V, $R_L = 10$ kΩ	±13	±14		V
Input voltage range	$V_S = ±15$ V	±14			V
Common-mode rejection ratio		85	100		dB
Supply-voltage rejection ratio		80	96		dB

10.4.3   The LM108 Operational Amplifier[8]

The LM108 operational amplifier was the first general-purpose design to use super β transistors in order to achieve ultra-low input currents. While a detailed discussion of the operation of this circuit is beyond the scope of this book, the LM108 does illustrate another of the many useful ways that the basic two-stage topology can be realized.

A simplified schematic diagram that illustrates some of the more important features of the design is shown in Fig. 10.21, with specifications given in Table 10.3. (The complete circuit, which is considerably more complex, is described in the reference given in the footnote.) The schematic diagram indicates two types of NPN transistors. Those with a narrow base (Q_1, Q_2, and Q_4) are super β transistors with current gains of several thousand and low breakdown voltage. The wide-base NPN transistors are conventional devices.

The input differential pair operates at a quiescent current level of 3 μA per device. This quiescent level combined with the high gain of Q_1 and Q_2 results in an input bias current of less than one nanoampere, and thus the LM108 is ideally suited to use in high-impedance circuits.

In order to prevent voltage breakdown of the input transistors, their collectors are bootstrapped via cascode transistors Q_5 and Q_6. Operating currents and geometries of transistors Q_3, Q_4, Q_5, and Q_6 are chosen so that the input transistors operate at nearly zero collector-to-base voltage. Thus collector-to-base leakage current (which can dominate input current at elevated temperatures) is largely eliminated. It is also necessary to diode clamp the input terminals to prevent breaking down input transistors under large-signal conditions. This clamping, which deteriorates performance in some nonlinear applications, is one of the prices paid for low input current.

Transistors Q_9 and Q_{10} form a second-stage differential amplifier. Diode-connected transistors Q_7 and Q_8 compensate for the base-to-emitter voltages of Q_9-Q_{10}, so that the quiescent voltage across R_4 is equal to that across R_1 or R_2. Resistor values are such that second-stage quiescent current is twice that of the first stage. Transistors Q_{15} and Q_{16} connected as a current repeater reflect the collector current of Q_9 as a load for Q_{10}. This connection doubles the voltage gain of the second stage compared with using a fixed-magnitude current source as the load for Q_{10}. The high-resistance node is buffered with a conventional output stage.

Compensation can be effected by forming an inner loop via collector-to-base feedback around Q_{10}. Circuit parameters are such that single-pole compensation with dynamics comparable to the feedback-compensated case results when a dominant pole is created by shunting a capacitor from

[8] R. J. Widlar, "I. C. Op Amp Beats FET's on Input Current," National Semiconductor Corporation, Application note AN-29, December, 1969.

the high-resistance node to ground. This alternate compensation results in superior supply-voltage noise rejection. (One disadvantage of capacitive coupling from collector to base of a second-stage transistor is that this feedback forces the transistor to couple high-frequency supply-voltage transients applied to its emitter directly to the amplifier output.)

The dynamics of the LM108 are not as good as those of the LM101A. While comparable bandwidths are possible in low-gain, resistively loaded applications, the bandwidth of the LM101A is substantially better when high closed-loop voltage gain or capacitive loading is required. The slower dynamics of the LM108 result in part from the use of the lateral PNP's in the second stage where their peculiarities more directly affect bandwidth and partially from the low quiescent currents used to reduce the power consumption of the circuit by a factor of five compared with that of the LM101A.

10.4.4   The LM110 Voltage Follower

The three amplifiers described earlier in this section have been general-purpose operational amplifiers where one design objective was to insure that the circuit could be used in a wide variety of applications. If this requirement is relaxed, the resultant topological freedom can at times be

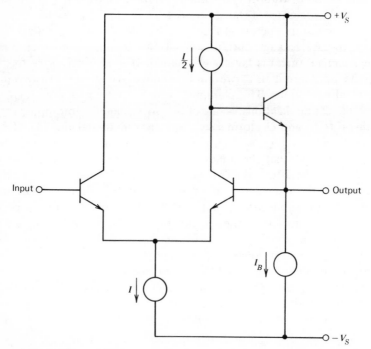

Figure 10.22   Voltage follower.

exploited. Consider the simplified amplifier shown in Fig. 10.22. Here a current-source-loaded differential amplifier is used as a single high-gain stage and is buffered by an emitter follower. The emitter follower is biased with a current source. This very simple operational amplifier is connected in a unity-gain noninverting or voltage-follower configuration. Since it is known that the input and output voltage levels are equal under normal operating conditions, there is no need to allow for arbitrary input-output voltage relationships. One very significant advantage is that only NPN transistors are included in the gain path, and the bandwidth limitations that result from lateral PNP transistors are eliminated.

This topology is actually a one-stage amplifier, and the dynamics associated with such designs are even more impressive than those of two-stage amplifiers. While the low-frequency open-loop voltage gain of this design may be less than that of two-stage amplifiers, open-loop voltage gains of several thousand result in adequate desensitivity when direct output-to-input feedback is used.

The LM110 voltage follower (Fig.10-23) is an integrated-circuit operational amplifier that elaborates on the one-stage topology described above. Perfomance specifications are listed in Table 10.4. Note that this circuit, like the LM108, uses both super β (narrow base) and conventional (wide base) NPN transistors. The input stage consists of transistors Q_8 through Q_{11} connected as a differential amplifier using two modified Darlington pairs. Pinch resistors R_8 and R_9 increase the emitter current of Q_8 and Q_{11} to reduce voltage drift. (See Section 7.4.4 for a discussion of the drift that can result from a conventional Darlington connection.) Transistor Q_{15} supplies the operating current for the input stage. Transistor Q_{16} supplies one-half of this current (the nominal operating current of either side of the differential pair) to the current repeater Q_1 through Q_3 that functions as the first-stage load.

Transistors Q_5 and Q_6 form a Darlington emitter follower that isolates the high-resistance node from loads applied to the amplifier. The emitter of Q_6, which is at approximately the output voltage, is used to bootstrap the collector voltage of the Q_{10}-Q_{11} pair. The resultant operation at nominally zero collector-to-base voltage results in negligible leakage current from Q_{11}. The Q_8-Q_9 pair is cascoded with transistor Q_4. Besides protecting Q_8 and Q_9 from excessive voltages, the cascode results in higher open-loop voltage gain from the circuit.

Diode D_1 and diode-connected transistor Q_{13} limit the input-to-output voltage difference for a large-signal operation to protect the super β transistors and to speed overload recovery. Transistor Q_7 is a current limiter, while Q_{14} functions as a current-source load for the output stage. The single-ended emitter follower is used in preference to a complementary

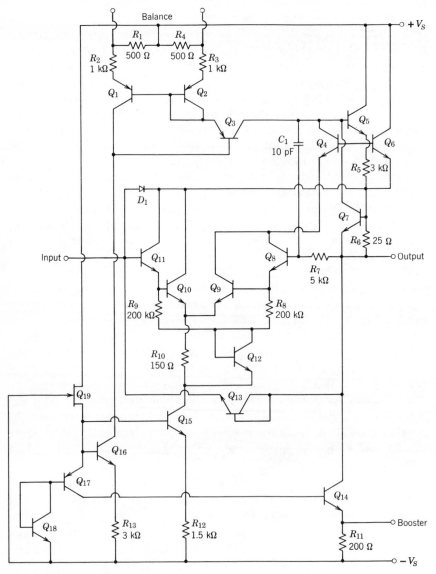

Figure 10.23 LM110 schematic diagram.

Table 10.4 LM110 Specifications: Electrical Characteristics

Parameter	Conditions	Min	Typ.	Max	Units
Input offset voltage	$T_A = 25°$ C		1.5	4.0	mV
Input bias current	$T_A = 25°$ C		1.0	3.0	nA
Input resistance	$T_A = 25°$ C	10^{10}	10^{12}		Ω
Input capacitance			1.5		pF
Large-signal voltage gain	$T_A = 25°$ C, $V_S = \pm15$ V, $V_{out} = \pm10$ V, $R_L = 8$ kΩ	0.999	0.9999		V/V
Output resistance	$T_A = 25°$ C		0.75	2.5	Ω
Supply current	$T_A = 25°$ C		3.9	5.5	mA
Input offset voltage	$-55°$ C $\leq T_A \leq 85°$ C			6.0	mV
Offset voltage temperature drift	$T_A = 125°$ C		6		μV/°C
			12		μV/°C
Input bias current				10	nA
Large-signal voltage gain	$V_S = \pm15$ V, $V_{out} = \pm10$ V, $R_L = 10$ kΩ	0.999			V/V
Output voltage swing	$V_S = \pm15$ V, $R_L = 10$ kΩ	±10			V
Supply current	$T_A = 125°$ C		2.0	4.0	mA
Supply-voltage rejection ratio	5 V $\leq V_S \leq 18$ V	70	80		dB

connection since it is more linear and thus better suited to high-frequency applications. An interesting feature of the design is that the magnitude of the current-source load for the emitter follower can be increased by shunting resistor R_{11} via external terminals. This current can be increased when it is necessary for the amplifier to supply substantial negative output current. The use of boosted output current also increases the power consumption of the circuit, raises its temperature, and can reduce input current because of the increased current gain of transistor Q_{11} at elevated temperatures.

The capacitive feedback from the collector of Q_4 to the base of Q_8 stabilizes the amplifier. Since the relative potentials are constrained under normal operating conditions, a diode can be used for the capacitor.

The small-signal bandwidth of the LM110 is approximately 20 MHz. This bandwidth is possible from an amplifier produced by the six-mask process because, while lateral PNP's are used for biasing or as static current sources, none are used in the signal path.

It is clear that special designs to improve performance can often be employed if the intended applications of an amplifier are constrained. Unfortunately, most special-purpose designs have such limited utility that fabrication in integrated-circuit form is not economically feasible. The LM110 is an example of a circuit for which such a special design is practical, and it provides significant performance advantages compared to general-purpose amplifiers connected as followers.

10.4.5   Recent Developments

The creativity of the designers of integrated circuits in general and monolithic operational amplifiers in particular seems far from depleted. Innovations in processing and circuit design that permit improved performance occur with satisfying regularity. In this section some of the more promising recent developments that may presage exciting future trends are described.

The maximum closed-loop bandwidth of most general-purpose monolithic operational amplifiers made by the six-mark process is limited to approximately 1 MHz by the phase shift associated with the lateral-PNP transistors used for level shifting. While this bandwidth is more than adequate for many applications, and in fact is advantageous in some because amplifiers of modest bandwidth are significantly more tolerant of poor decoupling, sloppy layout, capacitive loading, and other indiscretions than are faster designs, wider bandwidth always extends the application spectrum. Since it is questionable if dramatic improvements will be made in the frequency response of process-compatible PNP transistors in the near future, present efforts to extend amplifier bandwidth focus on eliminating the lateral PNP's from the gain path, at least at high frequencies.

One possibility is to capacitively bypass the lateral PNP's at high frequencies. This modification can be made to an LM101 or LM101A by connecting a capacitor from the inverting input to terminal 1 (see Figs. 10.17 and 10.19). The capacitor provides a feedforward path (see Section 8.2.2) that bypasses the input-stage PNP transistors. Closed-loop bandwidths on the order of 5 MHz are possible, and this method of compensation is discussed in greater detail in a later section. Unfortunately, feedforward does not improve the amplifier speed for signals applied to the noninverting input, and as a result wideband differential operation is not possible.

The LM118 pioneered a useful variation on this theme. This operational amplifier is a three-stage design including an NPN differential input stage, an intermediate stage of lateral PNP's that provides level shifting, and a final NPN voltage-gain stage. The intermediate stage is capacitively bypassed, so that feedforward around the lateral-PNP stage converts the circuit to a two-stage NPN design at high frequencies, while the PNP stage provides the gain and level shifting required at low frequencies. Since the feedforward is used following the input stage, full bandwidth differential operation is retained. Internal compensation insures stability with direct feedback from the output to the inverting input and results in a unity-gain frequency of approximately 15 MHz and a slew rate of at least 50 volts per microsecond. External compensation can be used for greater relative stability.

A second possibility is to use the voltage drop that a current source produces across a resistor for level shifting. It is interesting to note that the μA702, the first monolithic operational amplifier that was designed before the advent of lateral PNP's, uses this technique and is capable of closed-loop bandwidths in excess of 20 MHz. However, the other performance specifications of this amplifier preclude its use in demanding applications. The μA715 is a more modern amplifier that uses this method of level shifting. It is an externally compensated amplifier capable of a closed-loop bandwidth of approximately 20 MHz and a slew rate of 100 V/μs in some connections.

It is evident that improved high-speed amplifiers will evolve in the future. The low-cost availability of these designs will encourage the use of circuits such as the high-speed digital-to-analog converters that incorporate them.

A host of possible monolithic operational-amplifier refinements may stem from improved thermal design. One problem is that many presently available amplifiers have a d-c gain that is limited by thermal feedback on the chip. Consider, for example, an amplifier with a d-c open-loop gain of 10^5, so that the input differential voltage required for a 10-volt output is 100 μV. If the thermal gradient that results from the 10-volt change in output level changes the input-transistor pair temperature differentially by 0.05° C (a real possibility, particularly if the output is loaded), differential input voltage is dominated by thermal feedback rather than by limited d-c gain. Several modern instrumentation amplifiers use sophisticated thermal-design techniques such as multiple, parallel-connected input transistors located to average thermal gradients and thus allow usable gains in the range of 10^6. These techniques should be incorporated into general-purpose operational amplifiers in the future.

An interesting method of output-transistor protection was originally developed for several monolithic voltage regulators, and has been included in the design of at least one high-power monolithic operational amplifier. The level at which output current should be limited in order to protect a circuit is a complex function of output voltage, supply voltage, the heat sink used, ambient temperature, and the time history of these quantities because of the thermal dynamics of the circuit. Any limit based only on output current level (as is true with most presently available operational amplifiers) must be necessarily conservative to insure protection. An attractive alternative is to monitor the temperature of the chip and to cut off the output before this temperature reaches destructive levels. As this technique is incorporated in more operational-amplifier designs, both output current capability and safety (certain present amplifiers fail when the output is shorted to a supply voltage) will improve. The high pulsed-current capability made possible by thermal protection would be particularly valuable in applications where high-transient capacitive changing currents are encountered, such as sample-and-hold circuits.

Another thermal-design possibility is to include temperature sensors and heaters on the chip so that its temperature can be stabilized at a level above the highest anticipated ambient value. This technique has been used in the μA726 differential pair and μA727 differential amplifier. Its inclusion in a general-purpose operational-amplifier design would make parameters such as input current and offset independent of ambient temperature fluctuations.

10.5   ADDITIONS TO IMPROVE PERFORMANCE

Operational amplifiers are usually designed for general-purpose applicability. For this reason and because of limitations inherent to integrated-circuit fabrication, the combination of an integrated-circuit operational amplifier with a few discrete components often tailors performance advantageously for certain applications. The use of customized compensation networks gives the designer a powerful technique for modifying the dynamics of externally compensated operational amplifiers. This topic is discussed in Chapter 13. Other frequently used modifications are intended to improve either the input-stage or the output-stage characteristics of

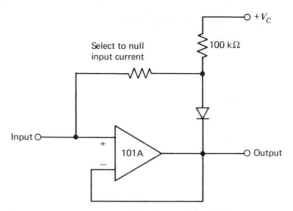

Figure 10.24   Input current compensation for voltage follower.

monolithic amplifiers, and some of these additions are mentioned in this section.

One advantage that many discrete-component operational amplifiers have compared with some integrated-circuit designs is lower input current. This improvement usually results because the input current of the discrete-component design is compensated by one of the techniques described in Section 7.4.2. These techniques can reduce the input current of discrete-component designs, particularly at one temperature, to very low levels. The same techniques can be used to lower the input current of integrated-circuit amplifiers. Many amplifiers can be well compensated using transistors as shown in Fig. 7.14. Transistor types, such as the 2N4250 or 2N3799, which have current-gain versus temperature characteristics similar to those of the input transistors of many amplifiers, should be used.

The connection shown in Fig. 10.24 can be used to reduce the input current of a follower-connected LM101A. As a consequence of the temperature-dependent input-stage operating current of this amplifier (see Section 10.4.1), the temperature coefficient of its input current is approximately 0.3% per degree Centigrade, comparable to that of a forward-biased silicon diode at room temperature.

Another possibility involves the use of the low input current LM110 as a preamplifier for an operational amplifier. Since the bandwidth of the LM110 is much greater than that of most general-purpose operational amplifiers, feedback-loop dynamics are unaffected by the addition of the preamplifier. While this connection increases voltage drift, the use of self heating (see Section 10.4.4) and simple input current compensation can result in an input current under 0.5 nA over a wide temperature range.[9]

---

[9] While the LM108 has comparably low input current, its dynamics and load-driving capability are inferior to those of many other general-purpose amplifiers. As a result the connection described here is advantageous in some applications.

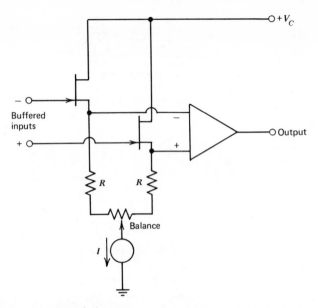

Figure 10.25   Use of FET followers.

Field-effect transistors[10] can be connected as source followers in front of an operational amplifier as shown in Fig. 10.25. Input current of a fraction of a nA at moderate temperatures is obtained at the expense of increased drift and poorer common-mode rejection ratio. The use of relatively inexpensive dual field-effect transistors yields typical drift figures of 10 to 100 $\mu V/°C$. The product of source-follower output resistance and amplifier input capacitance is normally small enough so that dynamics remain unchanged. If this capacitive loading is a problem, the gate and source terminals of the FET's can be shunted with small capacitors.

It is possible to reduce the drift of an operational amplifier by preceding it with a differential-amplifier stage, since the drift of a properly designed discrete-component differential amplifier can be made a fraction of a microvolt per degree Centigrade (see Chapter 7). This method is most effective when relatively high voltage gain is obtained from the differential stage and when its operating current is high compared with the input

[10] While the best-matched field-effect transistors are made by a monolithic process, the process cannot simultaneously fabricate high-quality bipolar transistors. Some manufacturers offer hybrid integrated circuits that combine two chips in one package to provide a FET-input operational amplifier. The $\mu A740$ is a monolithic FET-input amplifier, but its performance is not as good as that of the hybrids.

current of the operational amplifier. A recently developed integrated circuit (the LM121) is also intended to function as a preamplifier for operational amplifiers. The bias current of this preamplifier can be adjusted, and combined drift of less than one microvolt per degree Centigrade is possible. The use of a preamplifier that provides voltage gain often complicates compensation because the increased loop transmission that results may compromise stability in some applications.

The output current obtainable from an integrated-circuit operational amplifier is limited by the relatively small geometry of the output transistors and by the low power dissipation of a small chip. These limitations can be overcome by following the amplifier with a separate output stage.

There is a further significant performance advantage associated with the use of an external output stage. If output current is supplied from a transistor included on the chip, the dominant chip power dissipation is that associated with load current when currents in excess of several milliamperes are supplied. As a consequence, chip temperature can be strongly dependent on output voltage level. As mentioned earlier, thermal feedback to the input transistors deteriorates performance because of associated drift and input current changes. A properly designed output stage can isolate the amplifier from changes in load current so that chip temperature becomes virtually independent of output voltage and current.

Output-stage designs of the type described in Section 8.4 can be used. The wide bandwidth of emitter-follower circuits normally does not compromise frequency response. The output stage can be a discrete-component design, or any of several monolithic or hydrid integrated circuits may be used. The MC1538R is an example of a monolithic circuit that can be used as a unity-gain output buffer for an operational amplifier. This circuit is housed in a relatively large package that permits substantial power dissipation, and can provide output currents as high as 300 mA. Its bandwidth exceeds 8 MHz, considerably greater than that of most general-purpose operational amplifiers.

Another possibility in low output power situations is to use a second operational amplifier connected as a noninverting amplifier (gains between 10 and 100 are commonly used) as an output stage for a preceding operational amplifier. Advantages include the open-loop gain increase provided by the noninverting amplifier, and virtual elimination of thermal-feedback problems since the maximum output voltage required from the first amplifier is the maximum output voltage of the combination divided by the closed-loop gain of the noninverting amplifier. It is frequently necessary to compensate the first amplifier very conservatively to maintain stability in feedback loops that use this combination.

PROBLEMS

P10.1

You are the president of Single-Stone Semiconductor, Inc. Your best-selling product is a general-purpose operational amplifier that has chip dimensions of 0.05 inch square. Experience shows that you make a satisfactory profit if you sell your circuits at a price equal to 10 times the cost at the wafer level. You presently fabricate your circuits on wafers with a usable area of 3 square inches. The cost of processing a single wafer is $40, and yields are such that you currently sell your amplifier for one dollar. Your chief engineer describes a new amplifier that he has designed. It has characteristics far superior to your present model and can be made by the same process, but it requires a chip size of 0.05 inch by 0.1 inch. Explain to your engineer the effect this change would have on selling price. You may assume that wafer defects are randomly distributed.

P10.2

A current repeater of the type shown in Fig. 10.9 is investigated with a transistor curve tracer. The ground connection in this figure is connected to the curve-tracer emitter terminal, the input is connected to the base terminal of the tracer, and the repeater output is connected to the collector terminal of the curve tracer. Assume that both transistors have identical values for I_S and very high current gain. Draw the type of a display you expect on the curve tracer.

P10.3

Assume that it is possible to fabricate lateral-PNP transistors with a current gain of 100 and a value of C_π equal to 400 pF at 100 μA of collector current. A unity-gain current repeater is constructed by bisecting the collector of one of these transistors and connecting the device as shown in Fig. 10.10. Calculate the 0.707 frequency of the current transfer function for this structure operating at a total collector current of 100 μA. You may neglect C_μ and base resistance for the transistor. Contrast this value with the frequency at which the current gain of a single-collector lateral-PNP transistor with similar parameters drops to 0.707 of its low-frequency value.

P10.4

Figure 10.26 shows a connection that can be used as a very low level current source. Assume that all transistors have identical values of I_S, high β, and are at a temperature of 300° K. Find values for R_1, R_2, and R_3 such that the output current will be 1 μA subject to the constraint that the sum of the resistor values is less than 100 kΩ.

P10.5

Consider the three current repeater structures shown in Figs. 10.9, 10.11a, and 10.11b. Assume perfectly matched transistors, and calculate the

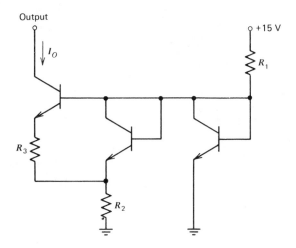

Figure 10.26   Very low level current source.

incremental output resistance of each connection in terms of β and η of the
transistors and the operating current level.

P10.6

The final voltage-gain stage and output buffer of an integrated-circuit
operational amplifier is shown in Fig. 10.27. Calculate the quiescent collector-current level of the complementary emitter follower. You may assume
that the current gain of the NPN transistors is 100, while that of PNP's is 10.
You may further assume that all NPN's have identical characteristics, as do
all PNP's.

P10.7

An operational amplifier input stage is shown in Fig. 10.28. Calculate
the drift referred to the input of this amplifier attributable to changes in
current gain of the transistors used in the current repeater. You may assume
that these transistors are perfectly matched, have a common-emitter current
gain of 100, and a fractional change in current gain of 0.5% per degree
Centigrade. Suggest a circuit modification that reduces this drift.

P10.8

A portion of the biasing circuitry of the LM101A operational amplifier
is shown in Fig. 10.29. This circuit has the interesting property that for a
properly choosen operating current level, the bias voltage is relatively insensitive to changes in the current level. Determine the value of I_B that
makes $\partial V_O / \partial I_B$ zero. You may assume high current gain for both transistors.

P10.9

Determine how unity-gain frequency and slew rate are related to the
first-stage bias current for the μA776 operational amplifier. Use the specifi-

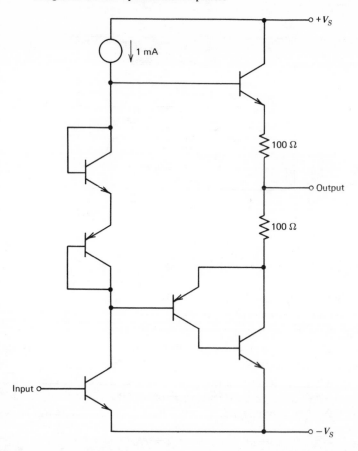

Figure 10.27 Amplifier output stage.

cations for this amplifier to estimate input-stage quiescent current at a
set-current value of 1.5 μA. Assuming that the ratio of bias current to set
current remains constant for lower values of set current, estimate the 10 to
90% rise time in response to a step for the μA776 connected as a noninverting gain-of-ten amplifier at a set current of 1 nA. Also estimate slew
rate and power consumption for the μA776 at this set current.

P10.10
A μA776 is connected in a loop with a LM101A as shown in Fig. 10.30.

(a) Show that this is one way to implement a function generator similar to
    that described in Section 6.3.3.
(b) Plot the transfer characteristics of the LM101A with feedback (i.e., the
    voltage v_A as a function of v_B).

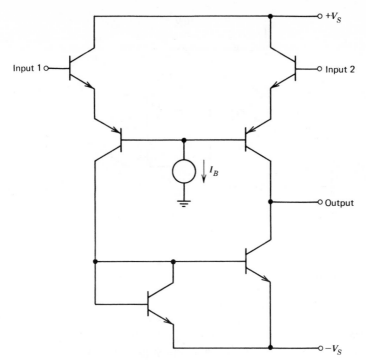

Figure 10.28 Amplifier input stage.

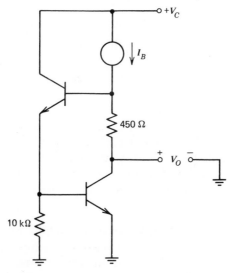

Figure 10.29 Bias circuit.

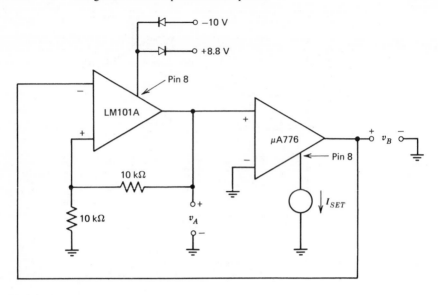

Figure 10.30   Nonlinear oscillator.

(c) Draw the waveforms $v_A(t)$ and $v_B(t)$ for this circuit. You may assume that the slew rate of the 101A is much greater than that of the μA776.

(d) How do these waveforms change as a function of I_{SET}?

P10.11

A simplified schematic of an operational amplifier is shown in Fig. 10.31.

(a) How many stages has this amplifier?

(b) Make the (probably unwarranted) assumption that all transistors have identical (high) values for β and identical values for η. Further assume that appropriate pairs have matched values of I_S. Calculate the low-frequency gain of the amplifier.

(c) Calculate the amplifier unity-gain frequency and slew rate as a function of the current I, the capacitor C_c, and any other quantities you need. You may assume that this capacitor dominates amplifier dynamics.

(d) Suggest a circuit modification that retains essential features of the amplifier performance, yet increases its low-frequency voltage gain.

P10.12

Specifications for the LM101A operational amplifier indicate a maximum input bias current of 100 nA and a maximum temperature coefficient of input offset current (input offset current is the difference between the bias current required at the two amplifier inputs) of 0.2 nA per degree Centigrade. These specifications apply over a temperature range of $-55°$ C to $+125°$ C. Our objective is to precede this amplifier with a matched pair of 2N5963

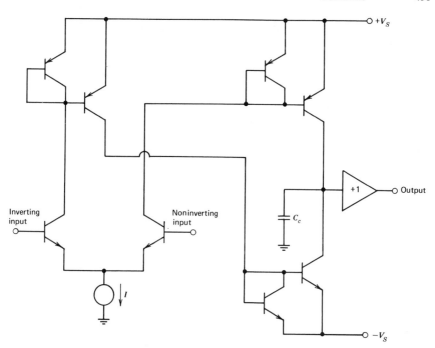

Figure 10.31 Operational amplifier.

transistors connected as emitter followers so that it can be used in applications that require very low input currents. Assume that you are able to
match pairs of 2N5963 transistors so that the difference in base-to-emitter
voltages of the pair is less than 2 mV at equal collector currents. Design
an emitter-follower circuit using one of these pairs and any required bias-
circuit components with the following characteristics:

1. The bias current required at the input of the emitter followers is
relatively independent of common-mode level over the range of ± 10 volts.

2. The drift referred to the input added to the complete circuit by the
emitter followers is less than ± 2 μV per degree Centigrade. Indicate how
you plan to balance the emitter-follower pair in conjunction with the
amplifier to achieve this result.

Estimate the input current for the modified amplifier with your circuit,
assuming that the common-emitter current gain of the 2N5963 is 1000.
Estimate the differential input resistance of the modified amplifier. You
will probably need to know the differential input resistance of the LM101A
to complete this calculation. In order to determine this quantity, show that
for the LM101A input-stage topology, differential input resistance can be
determined from input bias current *alone*.

CHAPTER XI

BASIC APPLICATIONS

11.1 INTRODUCTION

The operational amplifier is an extremely versatile general-purpose linear circuit. Clearly a primary reason for studying this device is to determine how it can best be used to solve design problems. Since this book is intended as a text book and not as a handbook, we hope to accomplish our objective by giving the reader a thorough understanding of the behavior of the operational amplifier so that he can innovate his own applications, rather than by giving him a long list of connections that others have found useful. Furthermore, there are a number of excellent references[1] available that provide extensive collections of operational-amplifier circuits, and there is little to be gained by competing with these references for completeness.

We have already seen several operational-amplifier connections in the examples used in preceding sections. In this and the following chapter we shall extend our list of applications in order to illustrate useful basic techniques. We hope that the reader finds these topologies interesting, and that they help provide the concepts necessary for imaginative, original design efforts. Some of the common hazards associated with the use of operational amplifier are discussed, as is the measurement and specification of performance characteristics. The vitally important issue of amplifier compensation for specific applications is reserved for Chapter 13.

11.2 SPECIFICATIONS

A firm understanding of some of the specifications used to describe operational amplifiers is necessary to determine if an amplifier will be satisfactory in an intended application. Unfortunately, completely specifying a

---

[1] A few of these references are: Philbrick Researches, Inc., *Applications Manual of Computing Amplifiers*. G. A. Korn and T. M. Korn, *Electronic Analog and Hybrid Computers*, 2nd Edition, McGraw-Hill, New York, 1972. J. G. Graeme, G. E. Tobey, and L. P. Huelsman (Editors), *Operational Amplifiers, Design and Applications*, McGraw-Hill, New York, 1971. Analog Devices, Inc., *Product Guide*, 1973. National Semiconductor Corporation, *Linear Applications Handbook*, 1972.

complex circuit is a virtually impossible task. The problem is compounded by the fact that not all manufacturers specify the same quantities, and not all are equally conservative with their definitions of "typical," "maximum," and "minimum." As a result, the question of greatest interest to the designer (will it work in my circuit?) is often unanswered.

11.2.1 Definitions

Some of the more common specifications and their generally accepted definitions are listed below. Since there are a number of available operational amplifiers that are not intended for differential operation (e.g., amplifiers that use feedforward compensation are normally single-input amplifiers), the differences in specifications between differential- and single-input amplifiers are indicated when applicable.

Input offset voltage. The voltage that must be applied between inputs of a differential-input amplifier, or between the input and ground of a single-input amplifier, to make the output voltage zero. This quantity may be specified over a given temperature range, or its incremental change (drift) as a function of temperature, time, supply voltage, or some other parameter may be given.

Input bias current. The current required at the input of a single-input amplifier, or the average of the two input currents for a differential-input amplifier.

Input offset current. The difference between the two input currents of a differential-input amplifier. Both offset and bias current are defined for zero output voltage, but in practice the dependence of these quantities on output voltage level is minimal. The dependence of these quantities on temperature or other operating conditions is often specified.

Common-mode rejection ratio. The ratio of differential gain to common-mode gain.

Supply-voltage sensitivity. The change in input offset voltage per unit change in power-supply voltage. The reciprocal of this quantity is called the supply-voltage rejection ratio.

Input common-mode range. The common-mode input signal range for which a differential amplifier remains linear.

Input differential range. The maximum differential signal that can be applied without destroying the amplifier.

Output voltage range. The maximum output signal that can be obtained without significant distortion. This quantity is usually specified for a given load resistance.

Input resistance. Incremental quantities are normally specified for both differential (between inputs) and common-mode (either input to ground) signals.

Output resistance. Incremental quantity measured without feedback unless otherwise specified.

Voltage gain or open-loop gain. The ratio of the change in amplifier output voltage to its change in input voltage when the amplifier is in its linear region and when the input signal varies extremely slowly. This quantity is frequently specified for a large change in output voltage level.

Slew rate. The maximum time rate of change of output voltage. This quantity depends on compensation for an externally compensated amplifier. Alternatively, the maximum frequency at which an undistorted given amplitude sinusoidal output can be obtained may be specified.

Bandwidth specifications. The most complete specification is a Bode plot, but unfortunately one is not always given. Other frequently specified quantities include unity-gain frequency, rise time for a step input or half-power frequency for a given feedback connection. Most confusing is a gain-bandwidth product specification, which may be the unity-gain frequency, or may be the product of closed-loop voltage gain and half-power bandwidth in some feedback connection.

Even when operational-amplifier specifications are supplied honestly and in reasonable detail, the prediction of the performance of an amplifier in a particular connection can be an involved process. As an example of this type of calculation, consider the relatively simple problem of finding the output voltage of the noninverting amplifier connection shown in Fig. 11.1 when $v_I = 0$. The voltage offset of the amplifier is equal to E_O. It is assumed that the low-frequency voltage gain of the amplifier a_0 is very large so that the voltage between the input terminals of the amplifier is nearly equal to E_O. (Recall that with E_O applied to the input terminals, $v_0 = 0$. If a_0 is very large, any other d-c output voltage within the linear region of the

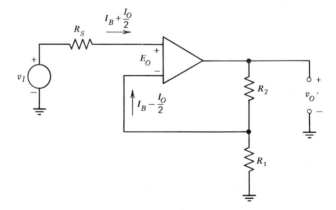

Figure 11.1 Noninverting amplifier.

amplifier can be obtained with a voltage of approximately E_O applied between inputs.) The currents at the amplifier input terminals expressed in terms of the bias current I_B and the offset current I_O are also shown in this figure.[2]

The voltage and current values shown in the figure imply that, for $v_I = 0$,

$$-R_S \left(I_B + \frac{I_O}{2} \right) - v_O \left(\frac{R_1}{R_1 + R_2} \right) + \left(I_B - \frac{I_O}{2} \right) \frac{R_1 R_2}{R_1 + R_2} = E_O \quad (11.1)$$

Solving for v_O yields

$$v_O = -\frac{(R_2 + R_1)}{R_1} E_O - \left(\frac{R_1 + R_2}{R_1} \right) R_S \left(I_B + \frac{I_O}{2} \right)$$
$$+ R_2 \left(I_B - \frac{I_O}{2} \right) \quad (11.2)$$

This equation shows that the output voltage attributable to amplifier input current can be reduced by scaling resistance levels, but that the error resulting from voltage offset is irreducible since the ratio $(R_1 + R_2)/R_1$ presumably must be selected on the basis of the required ideal closed-loop gain. Equation 11.2 also demonstrates the well-known result that balancing the resistances connected to the two inputs eliminates offsets attributable to input bias current, since with $R_1 R_2/(R_1 + R_2) = R_S$, the output voltage is independent of I_B.

As another example of the use of amplifier specifications, consider a device with an offset E_O, a d-c gain of a_0, and a maximum output voltage V_{OM}. The maximum differential input voltage required in order to obtain any static output voltage within the dynamic range of the amplifier is then[3]

$$V_{IM} = E_O + \frac{V_{OM}}{a_0} \quad (11.3)$$

This equation shows that values of a_0 in excess of V_{OM}/E_O reduce the amplifier input voltage (which must be low for the closed-loop gain to approximate its ideal value) only slightly. We conclude that if $a_0 > V_{OM}/E_O$, further operational-amplifier design efforts are better devoted to lowering offset rather than to increasing a_0.

---

[2] The specification of the input currents in terms of bias and offset current does not, of course, indicate which input-terminal current is larger for a particular amplifier. It has been arbitrarily assumed in Fig. 11.1 that the offset current adds to the bias current at the noninverting input and subtracts from it at the inverting input.

[3] The quantities in the equation are assumed to be maximum magnitudes. The possibility of cancellation because of algebraic signs exists for only one polarity of output voltage, and is thus ignored when calculating the maximum magnitude of the input voltage.

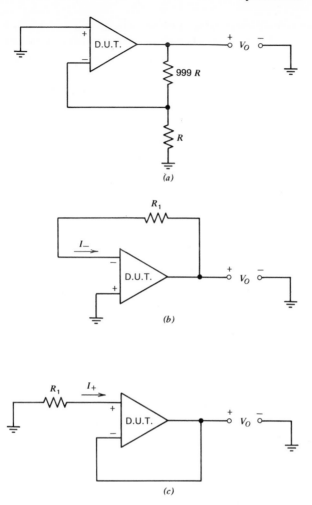

Figure 11.2 Circuits used to determine offset voltage and input currents. D.U.T. = Device Under Test. (*a*) Measurement of offset voltage. (*b*) Measurement of I_-. (*c*) Measurement of I_+.

11.2.2 Parameter Measurement

One way to bypass the conspiracy of silence that often surrounds amplifier specifications is to measure the parameters that are important in a particular application. Measurement allows the user to determine for himself how a particular manufacturer defines "typical," "maximum," and "minimum," and also permits him to grade circuits so that superior units can be used in the more demanding applications.

The d-c characteristics exclusive of open-loop gain are relatively straight-forward to measure. Circuits that can be used to measure the input offset voltage E_O and the input currents at the two input terminals, I_+ and I_-, are shown in Fig. 11.2. In the circuit of Fig. 11.2a, assume that resistor values are chosen so that $|E_O| \gg |I_-R|$. The quiescent voltages are

$$(-10^{-3}V_O + E_O)a_0 = V_O \tag{11.4}$$

for an appropriately chosen reference polarity for E_O. Solving this equation for V_O yields

$$V_O = \frac{a_0 E_O}{1 + 10^{-3}a_0} \simeq 10^3 E_O \tag{11.5}$$

Thus we see that this circuit uses the amplifier to raise its own offset voltage to an easily measured level.

If the resistor R_1 in Figs. 11.2b and 11.2c is chosen so that both $|I_-R_1|$ and $|I_+R_1| \gg |E_O|$, the output voltages are

$$V_O = I_-R_1 \tag{11.6}$$

and

$$V_O = -I_+R_1 \tag{11.7}$$

respectively. The measurement of I_- and I_+ allows direct calculation of offset and bias currents, since we recall from earlier definitions that the bias current is equal to the average of I_+ and I_-, while the offset current is equal to the magnitude of the difference between these two quantities.

A test box that includes a socket for the device under test and incorporates mode switching to select among the tests is easily constructed. Results can be displayed on an inexpensive D'Arsonval meter movement, since resistor values can be chosen to yield output voltages on the order of one volt. The low-pass characteristics of the meter movement provides a degree of noise rejection that improves the accuracy of the measurements. If further noise filtering is required, moderate-value capacitors may be used in parallel with resistors 999R and R_1 in Fig. 11.2.

The offset measurement circuit shown in Fig. 11.2a requires large loop-transmission magnitude for proper operation. If there is the possibility that the low-frequency loop transmission of a particular amplifier is too small, the alternative circuit shown in Fig. 11.3 can be used to measure offset. The second amplifier provides very large d-c gain with the result that the voltage out of the amplifier under test is negligible. At moderate frequencies, the second amplifier functions as a unity-gain inverter so that loop stability is not compromised by the integrator characteristics that result if the feedback resistor R is eliminated. Lowering the value of this

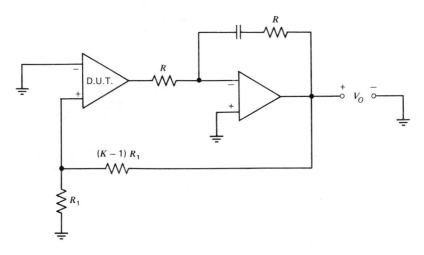

Figure 11.3   Offset-measurement circuit with increased loop transmission.

feedback resistor relative to the input resistor of the second amplifier may improve stability, particularly when the $(K - 1)R_1$ resistor is bypassed for noise reduction. Since this connection keeps the output voltage from the amplifier under test near ground, V_O will be (in the absence of input-current effects) simply equal to KE_O.

The supply-voltage rejection ratio of an amplifier can be measured with the same circuitry used to measure offset if provision is included to vary voltages applied to the amplifier. The supply-voltage rejection ratio is de-fined as the ratio of a change in supply voltage to the resulting change in input offset voltage.

The technique of including a second amplifier to increase loop transmis-sion simplifies the measurement of common-mode rejection ratio (see Fig. 11.4). If the differential gain a_0 of the amplifier is large compared to its common-mode gain a_{cm}, we can write

$$0 = a_{cm}v_{CM} + \frac{a_0 v_A}{K} \tag{11.8}$$

Thus

$$\left| \frac{a_0}{a_{cm}} \right| = \text{CMRR} = \left| \frac{K v_{CM}}{v_A} \right| \tag{11.9}$$

Since the voltage v_A also includes a component proportional to the offset voltage of the amplifier, it may be necessary to use incremental measure-ments to determine accurately rejection ratio.

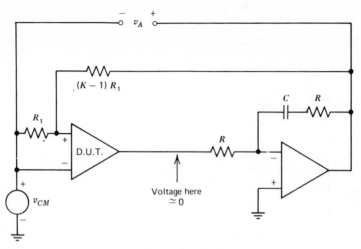

Figure 11.4 Circuit for measuring common-mode rejection ratio.

The slew rate of an amplifier may be dependent on the time rate of change of input-signal common-mode level and will be a function of compensation for an amplifier with selectable compensation. Standardized results that allow intercomparison of various amplifiers can be obtained by connecting the amplifier as a unity-gain inverter and applying a step input that sweeps the amplifier output through most of its dynamic range. Diode clamping at the inverting input of the amplifier can be used to prevent large common-mode signals. (See Section 13.3.7 for a representative circuit.) Alternatively, the maximum slew rate in a specific connection of interest can often be determined by applying a large enough input signal to force the maximum time rate of change of amplifier output voltage.

The open-loop transfer function of an amplifier is considerably more challenging to measure. Consider, for example, the problem of determining d-c gain. We might naively assume that the amplifier could be operated open loop (after all, we are measuring open-loop gain), biased in its linear region by applying an appropriate input quiescent level, and gain determined by adding an incremental step at the input and measuring the change in output level. The hazards of this approach are legion. The output signal is normally corrupted by noise and drift so that changes are difficult to determine accurately. We may also find that the amplifier exhibits bistable behavior, and that it is not possible to find a value for input voltage that forces the amplifier output into its linear region. This phenomenon results from positive thermal feedback in an integrated circuit, and can also occur in discrete designs because of self-heating, feedback through shared bias networks or power supplies, or for other reasons. The positive feedback that leads to this behavior is swamped by negative feedback applied around the amplifier in normal applications, and thus does not disturb performance in the usual connections.

After some frustration it is usually concluded that better results may be obtained by operating the amplifier in a closed-loop connection. Signal amplitudes can be adjusted for the largest output that insures linear performance at some frequency, and the corresponding input signal measured. The magnitude and angle of the transfer function can be obtained if the input signal can be accurately determined. Unfortunately, the signal at the amplifier input is usually noisy, particularly at frequencies where the open-loop gain magnitude is large. A wave analyzer or an amplifier followed by a phase-sensitive demodulator driven at the input frequency may be necessary for accurate measurements. This technique can even be used to determine a_0 if the amplifier is compensated so that the first pole in its open-loop transfer function is located within the frequency range of the detector.

There are also indirect methods that can be used to approximate the open-loop transfer function of the amplifier. The small-signal closed-loop frequency or transient response can be measured for a number of different values of frequency-independent feedback f_0. A Nichols chart or the curves of Fig. 4.26 may then be used to determine important characteristics of $a(j\omega)$ at frequencies near that for which $|a(j\omega)f_0| = 1$. Since various values of f_0 are used, $a(j\omega)$ can be determined at several different frequencies. This type of measurement often yields sufficient information for use in stability calculations.

A third possibility is to test the amplifier in a connection that provides a multiple of the signal at the amplifier input, in much the same way as does the circuit suggested earlier for offset measurement. Figure 11.5 shows one

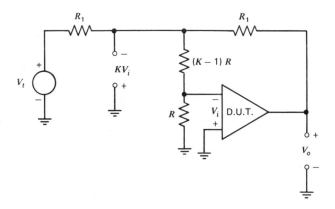

Figure 11.5 Connection for measuring the open-loop transfer function.

possibility. The signal at the junction of the two resistors labeled R_1 is K times as large as the input signal applied to the amplifier, and can be compared with either V_o or V_t (which is equal to $-V_o$ when the loop transmission is large) in order to determine the open-loop transfer function of the amplifier. Note that if the signal at the junction of the two equal-value resistors is compared with V_o, this method does not depend on large loop transmission.

While this method does scale the input signal, it does not provide filtering, with the result that some additional signal processing may be necessary to improve signal-to-noise ratio.

11.3   GENERAL PRECAUTIONS

The operational amplifier is a complex circuit that is used in a wide variety of connections. Frequently encountered problems include amplifier destruction because of excessive voltages or power dissipation and oscillation. The precautions necessary to avoid these hazards depend on the specific operational amplifier being used. We generally find, for example, that discrete-component amplifiers are more tolerant of abuse than are integrated-circuit units, since more sophisticated protective features are frequently included in discrete designs.

This section indicates some of the more common problem areas and suggests techniques that can be used to avoid them.

11.3.1   Destructive Processes

Excessive power-supply voltages are a frequent cause of amplifier damage. Isolation from uncertain supply-voltage levels via a resistor-Zener diode combination, as shown in Fig. 11.6, is one way to eliminate this hazard. The Zener diodes also conduct in the forward direction in the event of supply-voltage reversal. A better solution in systems that include a large number of operational amplifiers is to make sure that the power supplies include "crowbar"-type protection that limits voltages to safe levels in the event of power-supply failure.

Excessive differential input voltage applied to an operational amplifier may damage the input-transistor pair. If input voltages in excess of about 0.6 volt are applied to a normal differential-amplifier connection, the base-to-emitter junction of one of the members of the pair will be reverse biased. Further increases in applied differential voltage eventually result in reverse breakdown.

The base-to-emitter junction can be burned out if sufficient power is applied to it. A more subtle problem, however, is that base-to-emitter re-

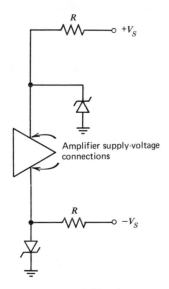

Figure 11.6 Supply-voltage limiting.

verse breakdown, even at low power levels, often irreversibly lowers the current gain of bipolar transistors. Thus the input current of an amplifier can be permanently increased by this mechanism. The potential for low power level base-to-emitter breakdown exists in many connections. Consider, for example, the usual integrator connection. If amplifier power is shut off with the feedback capacitor charged, the differential input voltage limit may be exceeded.

One practical solution is to include a pair of clamp diodes between the input terminals of the amplifier. Since the voltage between these terminals is nearly zero under normal operating conditions, the diodes have no effect until excessive voltage levels are reached.

Excessive power dissipation can result in some designs if the output terminal is shorted to ground. This possibility exists primarily in early integrated-circuit designs that do not include current-limiting circuits. More modern integrated-circuit designs, as well as most discrete circuits, are protected for output-to-ground shorts at normal supply-voltage levels and room-temperature operation. Some of these amplifiers are not protected for output shorts to either supply voltage, or output-to-ground shorts at elevated temperatures.

The compensation or balance terminals of operational amplifiers are frequently connected to critical low-level points, and any signal applied to these terminals invites disaster. The author is particularly adept at demonstrating this mode of destruction by shorting adjacent pins to each other or a pin to ground with an oscilloscope probe.

11.3.2   Oscillation

One of the most frequent complaints about operational amplifiers is that they oscillate in connections that the user feels should be stable. This phenomenon usually reflects a problem with the user rather than with the amplifier, and most of these instabilities can be corrected by proper design practice.

One frequent reason for oscillation is that dynamics associated with the load applied to the amplifier or the feedback network connected around it combine with the open-loop transfer function of the amplifier to produce feedback instabilities. The material presented in the chapters on feedback provides the general guidelines to eliminate these types of oscillations. Specific examples are given in Chapter 13.

Another common cause of oscillation is excessive power-connection impedance. This problem is particularly severe with high-frequency amplifiers because of the inductance of the leads that couple the power supply to the amplifier. In order to minimize difficulties, it is essential to properly decouple or bypass all power-supply leads to amplifiers without internal decoupling networks. Good design practice includes using a fairly large value ($> 1 \ \mu F$) solid-tantalum electrolytic capacitor from the positive and the negative power supply to ground on each circuit board. Individual amplifiers should have ceramic capacitors ($0.01 \ \mu F$ to $0.1 \ \mu F$) connected directly from their supply terminals to a common ground point. The single ground connection between the two decoupling capacitors should also serve as the tie point for the input-signal common, if possible. Lead length on both the supply voltage and the ground side of these capacitors is critical since series inductance negates their value. Ground planes may be mandatory in high-frequency circuits for acceptably low ground-lead inductance. If low supply currents are anticipated, crosstalk between amplifiers can be reduced by including small ($\simeq 22 \ \Omega$) series resistors in each decoupling network, as shown in Fig. 11.7.

In addition to reducing supply-line impedance, decoupling networks often lower the amplitude of any supply-voltage transients. Such transients are particularly troublesome with amplifiers that use capacitive minor-loop feedback for compensation, since the feedback element can couple transients applied to a supply-voltage terminal directly to the amplifier output.

The open-loop transfer function of many operational amplifiers is dependent on the impedance connected to the noninverting input of the amplifier. In particular, if a large resistor is connected in series with this terminal

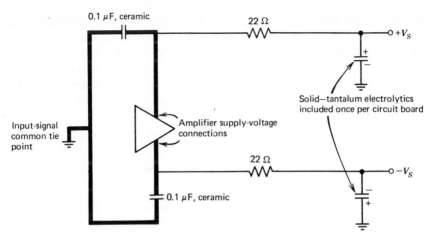

Figure 11.7 Power-supply decoupling network. *Note.* Heavy leads must be short.

(possibly to balance resistances seen at both inputs, thus reducing effects of bias currents), the bandwidth of the amplifier may deteriorate, leading to oscillation. In these cases a capacitor should be used to shunt the non-inverting input of the amplifier to the common input-signal and power-supply-decoupling ground point.

The input capacitance of an operational amplifier may combine with the feedback network to introduce a pole that compromises stability. Figure 11.8 is used to illustrate this problem. It is assumed that the input conductance of the amplifier is negligibly small, and that its input capacitance is modeled by the capacitor C_i shown in Fig. 11.8. If the capacitors shown with dotted connections in this figure are not present, the loop transmis-

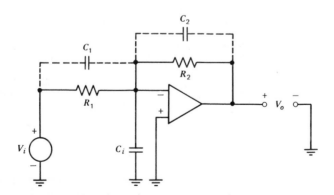

Figure 11.8 Effect of input capacitance.

sion includes a term $1/[(R_1 \| R_2)C_i s + 1]$. If capacitor C_2 is included and values are chosen so that $R_2 C_2 = R_1 C_i$, the transfer function of the feed-back network from the amplifier output to its inverting input becomes frequency independent. In practice, it is not necessary to match time constants precisely. A minor mismatch introduces a closely spaced pole-zero doublet, which normally has little effect on stability, into the loop-transmission expression.

A possible difficulty is that the inclusion of capacitor C_2 changes the ideal closed-loop gain of the amplifier to

$$\frac{V_o(s)}{V_i(s)} = - \frac{R_2}{R_1(R_2 C_2 s + 1)} \qquad (11.10)$$

An alternative is to include both capacitors C_1 and C_2. If $R_1 C_1 = R_2 C_2$ and $C_1 + C_2 \gg C_i$, the ideal closed-loop gain maintains its original value

$$\frac{V_o}{V_i} = - \frac{R_2}{R_1} \qquad (11.11)$$

while the feedback-network transfer function from the output to the inverting input of the operational amplifier becomes essentially frequency independent.

11.3.3   Grounding Problems

Improper grounding is a frequent cause of poor amplifier performance. While a detailed study of this subject is beyond the scope of this book, some discussion is in order.

One frequent grounding problem stems from voltage drops in ground lines as a consequence of current flow through these lines. Figure 11.9 illustrates an obviously poor configuration. Here both a signal source and a power supply are connected to a single ground point. However, the current through a load is returned to the low side of the power supply via a wire that also sets the potential at the noninverting input of the operational amplifier. If this current creates a potential V_g at the noninverting input with respect to system ground, the amplifier output voltage with respect to system ground will be

$$V_o = - \frac{R_2}{R_1} V_i + \frac{R_1 + R_2}{R_1} V_g \qquad (11.12)$$

The error term involving current flow through the ground return line can be substantial, since narrow printed-circuit conductors and connector pins can have considerable resistance and inductance.

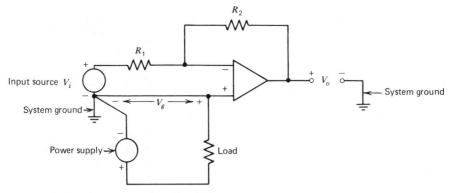

Figure 11.9 An example of poor grounding technique.

While the obvious topology illustrated in Fig. 11.9 is relatively easy to avoid, somewhat better disguised variations occur with disturbing frequency. The solution is to design the system with two different ground networks. One of these networks, called *signal ground*, serves as the return for critical points such as signal sources, feedback networks, and precision voltage references. Every attempt is made to keep both a-c and d-c currents in this network small so that it is essentially an equipotential network. High currents from noncritical loads (an excellent example is the logic often found in complex systems that include both analog and digital components) have their own ground-return network called *power ground*. These two grounds connect at *only one* point, which is also the low side of all system power supplies.

11.3.4 Selection of Passive Components

The passive components used in conjunction with operational amplifiers must be selected with care to obtain satisfactory performance.

Metal-film or carbon-film resistors with tolerances of 1% are inexpensive and readily available. These resistors can be obtained with temperature coefficients as low as 25 parts per million per degree Centigrade, and have fair long-term stability. They are acceptable in less demanding applications. We should note, however, that if 1% resistors are used, loop transmissions in excess of 100 are wasted in many connections.

Wire-wound resistors may be used where accuracy, stability, and temperature coefficient are of primary concern, since units are available that can maintain values to within 0.01% or better with time and over moderate temperature excursions. Disadvantages of these resistors include relatively large size and poor dynamic characteristics because of their distributed nature. It is important to realize that the excellent temperature stability of these resistors can easily be negated if they are combined with components such as potentiometers for trimming. The accepted procedure when adjustments are required is to use shunt or series connections of selectable, stable resistors to closely approximate the required value. A potentiometer with a total range of a fraction of a percent of the desired value can then be used to complete the trim. The temperature coefficient of the relatively less stable element has little effect since the potentiometer resistance is a very small fraction of the total.

At least one manufacturer offers precision relatively thick metal-film resistors with tolerances to 0.005 %. While the long-term stability and temperature coefficient of these units is not as good as that of the best wire wounds, their small size and excellent dynamic characteristics recommend them in many demanding applications.

The selection of acceptable capacitors is even more difficult. In addition to tolerance and stability problems, capacitors exhibit the phenomenon of *dielectric absorption*. One manifestation of this effect is that capacitors tend to "remember" and creep back toward the prior voltage if open circuited following a step voltage change. The time required to complete this transient ranges from milliseconds to thousands of seconds, while its magnitude can range from a fraction of a percent to as much as 25 % of the original change, depending primarily on the capacitor dielectric material. Dielectric absorption deteriorates the performance of any circuit using capacitors, with sample and holds (where step voltage changes are routine) and integrators being particularly vulnerable.

Teflon and polystyrene are the best dielectric materials from the point of view of dielectric absorption. The dielectric-absorption coefficient (roughly equal to the fractional recovery following a step voltage change) can be less than 0.1 % for capacitors properly constructed from these materials. They also have very high resistivity so that self-time constants (the product of capacitance and the shunt resistance that results from dielectric or case resistivity) in excess of 10^6 seconds are possible. While the temperature coefficient for either of these materials is normally the order of 100 parts per million per degree Centigrade, special processing or combinations of capacitors with two types of dielectrics can lower temperature coefficient to a few parts per million per degree Centigrade. The primary disadvantages are relatively large size and high cost (particularly for teflon) and a maximum temperature of 85° C for polystyrene.

Mica or glass capacitors often provide acceptable characteristics for lower-value units. Polycarbonate has considerably better volumetric efficiency than either teflon or polystyrene, and is used for moderate- and large-value capacitors. Dielectric absorption is somewhat poorer, but still acceptable in many applications. Mylar-dielectric capacitors are inexpensive and have an absorption coefficient of approximately 1%. These units are often used in noncritical applications.

Ceramic capacitors, particularly those constructed using high-dielectric-constant materials, have a particularly unfortunate combination of characteristics for most operational-amplifier circuits, and should generally be avoided except as decoupling components or in other locations where dielectric absorption is unimportant.

11.4 REPRESENTATIVE LINEAR CONNECTIONS

The objective of many operational-amplifier connections is to provide a linear gain or transfer function between circuit input and output signals. This section augments the collection of linear applications we have seen in preceding sections. As mentioned earlier, our objective in discussing these circuits is not to form a circuits handbook, but rather to encourage the creativity so essential to useful imaginative designs.

The connections presented in this and subsequent sections do not include the minor details that are normally strongly dependent on the specifics of a particular application and the operational amplifier used, and that would obscure more important and universal features. For example, no attempt is made to balance the resistances facing both input terminals, although we have seen that such balancing reduces errors related to amplifier input currents. We tacitly assume that the amplifier with feedback provides its ideal closed-loop gain unless specifically mentioned otherwise. Similarly, stability is assumed. The methods used to guarantee the latter assumption are the topic of Chapter 13.

11.4.1 Differential Amplifiers

We have seen numerous examples of both inverting and noninverting amplifier connections. Figure 11.10 shows a topology that combines the features of both of these connections. The ideal input-output relationship is easily determined by superposition. If V_b is zero,

$$V_o = -\frac{Z_2}{Z_1} V_a \tag{11.13}$$

If V_a is zero, the circuit is a noninverting amplifier preceded by an attenuator, and

$$V_o = \left(\frac{Z_4}{Z_3 + Z_4}\right)\left(\frac{Z_1 + Z_2}{Z_1}\right) V_b \tag{11.14}$$

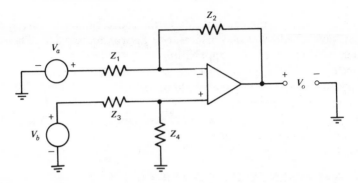

Figure 11.10 Differential connection.

Linearity insures that in general

$$V_o = \left(\frac{Z_4}{Z_3 + Z_4}\right)\left(\frac{Z_1 + Z_2}{Z_1}\right) V_b - \frac{Z_2}{Z_1} V_a \qquad (11.15)$$

If values are selected so that $Z_4/Z_3 = Z_2/Z_1$.

$$V_o = \frac{Z_2}{Z_1}(V_b - V_a) \qquad (11.16)$$

This connection is frequently used with four resistors to form a differential amplifier. Adjustment of any of the four resistors can be used to zero common-mode gain. Other possibilities involve combining two capacitors for Z_2 and Z_4 with two resistors for Z_1 and Z_3. If the time constants of the two combinations are equal, a differential or a noninverting integrator results.

It is important to note that the input current at the V_a terminal of the differential connection is dependent on both input voltages, while the current at the V_b terminal is dependent only on voltage V_b. This nonsymmetrical loading can cause errors in some applications. Two noninverting unity-gain amplifiers can be used as buffers to raise input impedance to very high levels if required.

If the design objective is a high-input-impedance differential amplifier with high common-mode rejection ratio, the connection shown in Fig. 11.11 can be used. Consider a common-mode input signal with $V_a = V_b = V_i$. In this case the two left-hand amplifiers combine to keep the voltage across R_2 zero. Thus for a common-mode input, the intermediate voltages V_c and V_d are related to inputs as

$$V_c = V_d = V_a \qquad V_a = V_b \qquad (11.17)$$

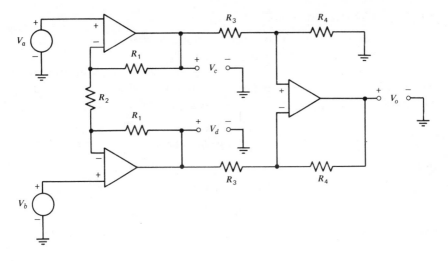

Figure 11.11 Buffered differential amplifier with high common-mode rejection ratio.

Alternatively, consider a pure differential input signal with $V_i/2 = V_a = -V_b$. In this case the midpoint of resistor R_2 is an incrementally grounded point, and each of the left-hand amplifiers functions as a noninverting amplifier with a gain of $(2R_1 + R_2)/R_2$. Linearity insures that the differential gain of the left-hand pair of amplifiers must be independent of common-mode level. Thus

$$\frac{V_c - V_d}{V_a - V_b} = \frac{2R_1 + R_2}{R_2} \qquad (11.18)$$

The right-hand amplifier has a gain of zero for the common-mode component of V_c and V_d, and a gain of R_4/R_3 for the differential component of these intermediate signals. Combining expressions shows that V_o is independent of the common-mode component of V_a and V_b, and is related to these signals as

$$V_o = \left(\frac{2R_1 + R_2}{R_2}\right)\frac{R_4}{R_3}(V_a - V_b) \qquad (11.19)$$

In addition to the high input impedance provided by the left-hand amplifiers, the differential gain of this pair makes the common-mode rejection of the overall amplifier less sensitive to ratio mismatches of the output-amplifier resistor networks.

11.4.2   A Double Integrator

We have seen that either inverting or noninverting integration can be accomplished with an operational amplifier. Figure 11.12 shows a connection that provides a second-order integration with a single operational amplifier. The circuit is analyzed by the virtual-ground method. Assuming that the inverting input of the amplifier is at ground potential

$$I_i(s) = \frac{V_i(s)}{2R(RCs + 1)} \tag{11.20}$$

and

$$I_f(s) = \frac{RC^2s^2V_o(s)}{2(RCs + 1)} \tag{11.21}$$

The negligible input current of the amplifier forces $I_f = -I_i$. Combining Eqns. 11.20 and 11.21 via this constraint shows that

$$\frac{V_o(s)}{V_i(s)} = -\frac{1}{(RCs)^2} \tag{11.22}$$

11.4.3   Current Sources

The operational amplifier can be used as a current source in a number of different ways. Figure 11.13 shows two simple configurations. In part *a* of this figure, the load serves as the feedback impedance of an inverting-connected operational amplifier. The virtual-ground method shows that the current through the load must be equal to the current through resistor *R*. In part *b*, the operational amplifier forces the voltage across *R* to be equal to the input voltage. Since the current required at the inverting input terminal of the amplifier is negligible, the load current is equal to the current through resistor *R*.

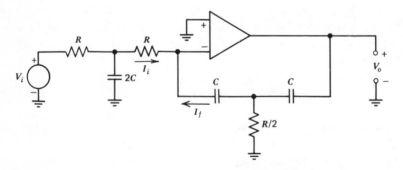

Figure 11.12   Double integrator.

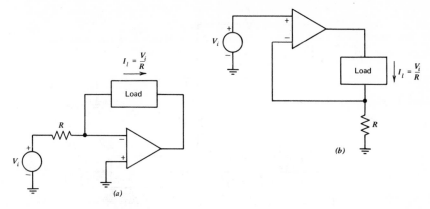

Figure 11.13 Current sources. (*a*) Inverting connection. (*b*) Follower connection.

Both of the current-source connections described above require that the load be floating. The configuration shown in Fig. 11.14 relaxes this requirement. Here the operational amplifier constrains the source current of a field-effect transistor. Provided that operating levels are such that the FET gate is reverse biased, the source and drain currents of this device are identical. Thus the operational amplifier controls the load current indirectly.

The relative operating levels of the circuit shown in Fig. 11.14 must be constrained to keep the FET in its forward operating region with its gate

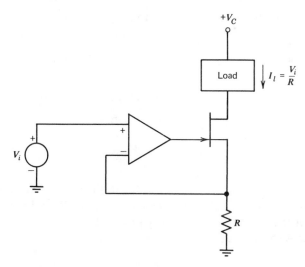

Figure 11.14 Current source using a field-effect transistor.

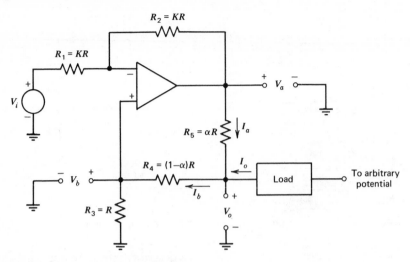

Figure 11.15   Howland current source.

reverse biased for satisfactory performance. The Howland current source shown in Fig. 11.15 allows further freedom in the choice of operating levels.

The analysis of this circuit is simplified by noting that the operational amplifier relates V_a to V_i and V_b as

$$V_a = -V_i + 2V_b \tag{11.23}$$

The circuit topology implies the relationships

$$I_o = I_b - I_a \tag{11.24}$$

$$I_a = \frac{V_a - V_o}{\alpha R} \tag{11.25}$$

$$I_b = \frac{V_o - V_b}{(1 - \alpha)R} \tag{11.26}$$

and

$$V_b = \frac{V_o}{2 - \alpha} \tag{11.27}$$

The transfer relationships of interest for this circuit are the input voltage to short-circuit output current transconductance I_o/V_i and the output conductance of the circuit I_o/V_o. Solving Eqns. 11.23 through 11.27 for these conductances shows that

$$\left. \frac{I_o}{V_i} \right|_{V_o = 0} = \frac{1}{\alpha R} \tag{11.28}$$

and

$$\frac{I_o}{V_o}\bigg|_{V_i = 0} = 0 \qquad (11.29)$$

Since the output current is independent of output voltage, we can model the circuit as a current source with a magnitude dependent on input voltage.

While the output resistance of this current source is independent of the quantity α, this parameter does affect scale factor. Smaller values of αR also allow a greater maximum output current for a given output voltage saturation level from the operational amplifier. There is a tradeoff involved in the selection of α, however, since smaller values for this parameter result in higher error currents for a given offset voltage referred to the input of the amplifier (see Problem P11.11).

There is further freedom in the selection of relative resistor ratios, since an extension of the above analysis shows that the output resistance is infinite provided $R_2/R_1 = (R_4 + R_5)/R_3$.

It is interesting to note that the success of this current source actually depends on *positive* feedback. Consider a voltage V_o applied to the output terminal of the circuit. The current that flows through resistor R_4 is exactly balanced by current supplied from the output of the operational amplifier via resistor R_5. The voltage at the output of the operational amplifier is the same polarity as V_o and has a larger magnitude than this variable.

We should further note that the resistor R_3 does not have to be connected to ground, but can also function as an input terminal. In this configuration the output current is proportional to the difference between the voltages applied to the two inputs.

11.4.4 Circuits which Provide a Controlled Driving-Point Impedance

We have seen examples of circuits designed to produce very high or very low input or output impedances. It is also possible to use operational amplifiers to produce precisely controlled output or driving-point impedances. Consider the circuit shown in Fig. 11.16. The operational amplifier is configured to provide a noninverting gain of two. As a result of this gain, the impedance connected between the amplifier output and its noninverting input has a voltage V_i across it with a polarity as shown in Fig. 11.16. Since there is negligible current required at the inverting input of the amplifier, the input current required from the source is

$$I_i = -I_a = -\frac{V_i}{Z} \qquad (11.30)$$

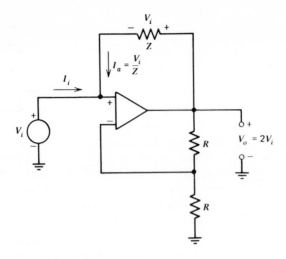

Figure 11.16 Negative impedance converter.

Solving Eqn. 11.30 for the input impedance of the circuit yields

$$\frac{V_i}{I_i} = -Z \tag{11.31}$$

Equation 11.31 shows that this circuit has sufficient positive feedback to produce negative input impedances.

The gyrator shown in Fig. 11.17 is another example of a circuit that provides a controlled driving-point impedance. The circuit relationships include

$$I_i = I_a + I_b \tag{11.32}$$

$$I_a = \frac{V_i - V_a}{R_1} = \frac{-V_i}{R_1} \tag{11.33}$$

$$I_b = \frac{V_i - V_b}{R_1} \tag{11.34}$$

$$V_b = -\frac{V_i Z}{R_2} \tag{11.35}$$

Combining Eqns. 11.32 through 11.35 and solving for the driving-point impedance shows that

$$\frac{V_i}{I_i} = \frac{R_1 R_2}{Z} \tag{11.36}$$

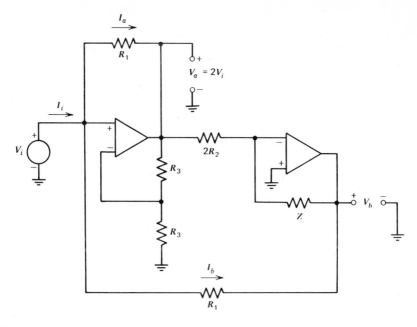

Figure 11.17   Gyrator.

We see that the gyrator provides a driving-point impedance that is reciprocally related to another circuit impedance. Applications include the synthesis of elements that function as inductors using only capacitors, resistors, and operational amplifiers. For example, if we choose impedance Z to be a 1-μF capacitor and $R_1 = R_2 = 1$ kΩ, the driving-point impedance of the circuit shown in Fig. 11.17 is s, equivalent to that of a 1-henry inductor.

11.5   NONLINEAR CONNECTIONS

The topologies presented in Section 11.4 were intended to provide linear gains, transfer functions, or impedances. While practical realizations of these circuits may include nonlinear elements, the feedback is arranged to minimize the effects of such nonlinearities. In many other cases feedback implemented by means of operational amplifiers is used to augment, control, or idealize the characteristics of nonlinear elements. Examples of these types of applications are presented in this section.

11.5.1   Precision Rectifiers

Many circuit connections use diodes to rectify signals. However, the forward voltage drop associated with a diode limits its ability to rectify low-

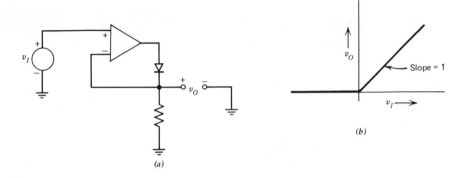

Figure 11.18   Precision rectifier. (*a*) Circuit. (*b*) Transfer characteristics.

level signals. The combination of a diode with an operational amplifier (Fig. 11.18) results in a circuit with a much lower threshold. Operation depends on the fact that the diode-amplifier combination can only pull the output voltage positive, so that negative input voltages result in zero output. With a positive input voltage, a negligibly small differential signal (equal to the threshold voltage of the diode divided by the open-loop gain of the amplifier) is amplified to provide sufficient amplifier output voltage to overcome the diode threshold, with the result that

$$v_O = v_I \qquad v_I > 0 \tag{11.37a}$$

$$v_O = 0 \qquad v_I < 0 \tag{11.37b}$$

Many variations of this precision rectifier or "superdiode" exist. For example, the circuit shown in Fig. 11.19 rectifies and provides a current-source drive for a floating load such as a D'Arsonval meter movement. Figure 11.20 illustrates another rectifier circuit. With v_I negative, voltage v_A is zero, and $v_O = -v_I$ because of the inversion provided by the right-hand amplifier. The transistor provides a feedback path for the first amplifier so that it remains in its linear region for negative inputs. Operation in the linear region keeps the inverting input of the first amplifier at ground potential, thereby preventing the input signal from driving voltage v_A via direct resistive feedthrough. Maintaining linear-region operation also eliminates the long amplifier recovery times that frequently accompany overload and saturation. While a diode could be used in place of the transistor, the transistor provides a convenient method for driving further amplifying circuits, which indicate input-signal polarity if this function is required. For positive input voltages, voltage $v_A = -v_I$, so that the resistor with value

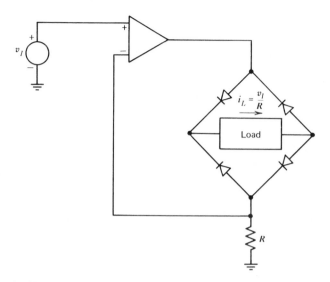

Figure 11.19   Full-wave precision rectifier for floating load.

$R/2$ also applies current to the input of the second amplifier, with the result that

$$v_O = -(v_I - 2v_I) = v_I \qquad v_I > 0 \qquad\qquad (11.38a)$$

$$v_O = -v_I \qquad\qquad\qquad v_I < 0 \qquad\qquad (11.38b)$$

11.5.2   A Peak Detector

The peak-detector circuit shown in Fig. 11.21 illustrates a further elaboration on the general theme of minimizing the effects of voltage drops in various elements by including these drops inside a feedback loop. If the output voltage is more positive than the input voltage, the output of the operational amplifier will be saturated in the negative direction. (Some form of clamping may be included to speed recovery from this state.) Under these conditions, the capacitor current consists only of diode and FET-gate leakage currents; thus the capacitor voltage changes very slowly. As a matter of practical concern, the circuit will function properly only if current levels are such that the capacitor voltage drifts negatively in this state. Otherwise, the connection will eventually saturate at its maximum positive output level.

If v_I becomes greater than v_O, the capacitor is charged rapidly from the output of the operational amplifier via the diode until equality is reestablished. Note that the capacitor voltage is not forced to be equal to v_I, but

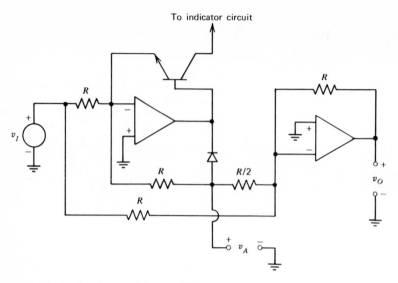

Figure 11.20 Full-wave precision rectifier.

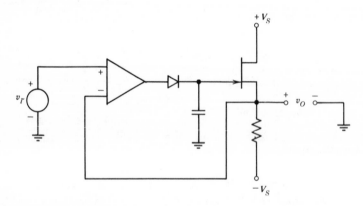

Figure 11.21 Peak detector.

rather to be equal to a voltage that, combined with the FET gate-to-source voltage, forces equality between v_O and v_I. In this way the output voltage "remembers" the most positive value of the input signal.

11.5.3 Generation of Piecewise-Linear Transfer Characteristics

Diodes can be combined with operational amplifiers to realize signal-shaping circuits other than rectifiers. Figure 11.22 shows a circuit that provides a compressive or limiting-type nonlinear transfer relationship. For

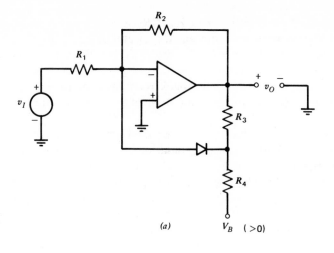

(a)

V_B (>0)

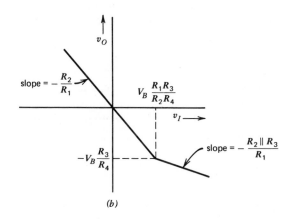

(b)

Figure 11.22   Limiter. (*a*) Circuit. (*b*) Transfer characteristics.

input voltages more negative than $V_B(R_1R_3/R_2R_4)$ the diode is an open circuit, and the incremental gain of the circuit is $-R_2/R_1$. When $v_I = V_B(R_1R_3/R_2R_4)$, the diode is on the threshold of conduction. Assuming a "perfect" diode (zero threshold voltage and zero on resistance in the forward direction), the effective feedback resistance for further increases in input voltage is $R_2 \parallel R_3$, and the magnitude of the incremental gain decreases to $-(R_2 \parallel R_3)/R_1$.

The operation of the limiter was described assuming perfect diode characteristics. If the performance degradation that results from actual diode characteristics is intolerable, a "superdiode" connection can be used as shown in Fig. 11.23. The lower operational amplifier cannot affect circuit operation for the positive values of v_A that correspond to input voltages more negative than $V_B(R_1R_3/R_2R_4)$ because the diode in series with its output is reverse biased. However, the lower amplifier can supply as much current as is required to keep the voltage at the junction of R_3 and R_4 from becoming negative, and thus this circuit provides hard limiting with the incremental gain dropping to zero for input voltages more positive than the threshold level. If softer limiting is required, a resistor can be included at the indicated point.

It is clear that additional resistor networks and diodes (or superdiodes) can be added to increase the number of break points in the transfer characteristics. However, the topology of Fig. 11.22 or Fig. 11.23 precludes increasing the magnitude of the incremental gain as input-voltage magnitude increases. Shifting the diode-resistor network to the amplifier input circuit (Fig. 11.24) is one way that expansive-type nonlinearities can be realized.

11.5.4 Log and Antilog Circuits

The exponential current-voltage characteristics of diodes or transistors can be exploited to realize circuits with exponential or logarithmic characteristics. Figure 11.25 illustrates a very simple circuit that provides a logarithmic relationship between output voltage and input current. Under normal operating conditions, the operational amplifier keeps the collector-to-base voltage of the transistor at zero. As a result, collector-to-base junction leakage currents are eliminated as are base-width modulation effects, and many types of transistors will accurately follow the relationship

$$i_C \simeq I_S e^{q v_{BE}/kT} \tag{11.39}$$

over a range of operating current levels that extends from picoamperes to a fraction of a milliamp. Deviation from purely exponential behavior oc-

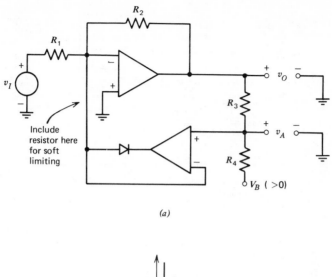

(a)

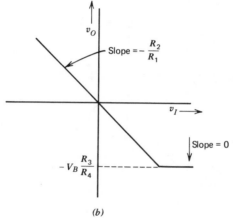

(b)

Figure 11.23 Limiter incorporating a super diode. (a) Circuit. (b) Transfer characteristics.

463

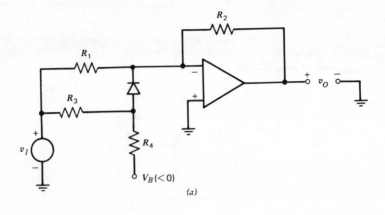

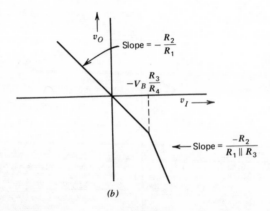

Figure 11.24 Expander. (*a*) Circuit. (*b*) Transfer characteristics.

464

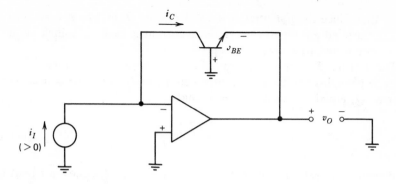

Figure 11.25 Log circuit.

curs at current levels comparable to I_S and at current levels where ohmic resistances become significant.[4]

For this circuit topology, $v_{BE} = -v_O$, and feedback keeps $i_C = i_I$. Substituting these constraints into Eqn. 11.39 shows that

$$i_I = I_S e^{-qv_O/kT} \tag{11.40}$$

or, if we solve for v_O,

$$v_O = -\frac{kT}{q} \ln \frac{i_I}{I_S} \tag{11.41}$$

Of course, the current applied to this circuit can be derived from an available input voltage via a resistor connected to the inverting input terminal of the operational amplifier. In this case, the voltage offset of the operational amplifier contributes an error term that normally limits dynamic range to three or four orders of magnitude. If the input signal is available as a current, as it is for some sensors such as ionization gauges, much wider dynamic range is possible for sufficiently low amplifier bias current.

One shortcoming of this simple circuit is that the quantity I_S is highly temperature dependent (see Section 7.2). The circuit shown in Fig. 11.26 offers improved performance with temperature. Feedback through the right-hand operational amplifier keeps the collector current of Q_2 equal to the reference current I_R; thus

$$v_{BE2} = \frac{kT}{q} \ln \frac{I_R}{I_{S2}} \tag{11.42}$$

[4] Theoretically, a diode could be used as a feedback element as indicated in Section 1.2.3 to obtain logarithmic closed-loop characteristics. In practice, the transistor connection illustrated here is preferable, since transistors generally display the desired characteristics over a far larger dynamic range than do diodes.

Note that, since the potential at the collector of Q_2 is held at zero volts by the operational amplifier, the reference current is easily obtained via a resistor connected to a positive supply voltage.

The left-hand operational amplifier adjusts the base voltage of Q_2, thereby changing the base-to-emitter voltage of Q_1 until the collector current of Q_1 equals i_I, with the result that

$$v_{BE1} = \frac{kT}{q} \ln \frac{i_I}{I_{S1}} \tag{11.43}$$

If values are selected so that the base current of Q_2 does not load the base-circuit attenuator, the voltage relationship is

$$v_{BE1} = v_{BE2} - \frac{1}{16.7} v_O \tag{11.44}$$

Combining Eqns. 11.42 through 11.44 and solving for v_O yields

$$v_O = -16.7 \frac{kT}{q} \left[\ln \frac{i_I}{I_{S1}} - \ln \frac{I_R}{I_{S2}} \right] = -16.7 \frac{kT}{q} \ln \left[\frac{i_I}{i_R} \frac{I_{S2}}{I_{S1}} \right] \tag{11.45}$$

If transistors Q_1 and Q_2 have well-matched values of I_S, Eqn. 11.45 becomes

$$v_O = -16.7 \frac{kT}{q} \ln \left[\frac{i_I}{i_R} \right] \tag{11.46}$$

The resistive-divider attenuation ratio of 16.7 is used so that at room temperature, Eqn. 11.46 reduces to

$$v_O = -1 \text{ volt } \log_{10} \left[\frac{i_I}{i_R} \right] \tag{11.47}$$

While the use of matched transistors as shown in Fig. 11.26 does eliminate the dependence of the output on I_S, Eqn. 11.46 shows that the scale factor of the circuit is proportional to absolute temperature. One common solution is to compensate by using a resistor with a value inversely proportional to absolute temperature as the smaller of the two resistors in the voltage divider.

The antilog circuit shown in Fig. 11.27 results from rearranging components. The reader should verify that, at room temperature and with matched transistors, the input-output relationship for this circuit is

$$v_O = R_1 I_R \times 10^{-(v_I/1 \text{ volt})} \tag{11.48}$$

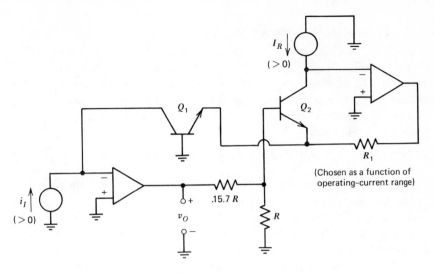

Figure 11.26 Improved log circuit.

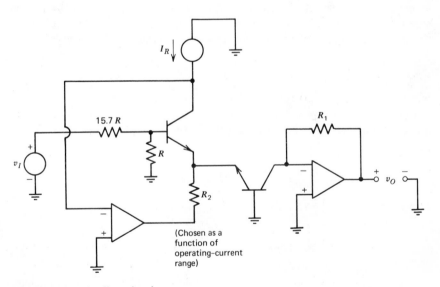

Figure 11.27 Antilog circuit.

467

11.5.5   Analog Multiplication

There are a number of configurations that perform analog multiplication, that is, provide an output voltage proportional to the product of two input voltages. For example, one or more log circuits can be combined with an antilog circuit to realize multipliers, dividers, or circuits that raise a voltage to a power. Another technique known as quarter-square multiplication exploits the relationship

$$(v_X + v_Y)^2 - (v_X - v_Y)^2 = 4v_X v_Y \tag{11.49}$$

The quadratic transfer characteristics can be approximated with piecewise-linear diode-operational amplifier connections.

A method known as transconductance multiplication is the basis for several available discrete and integrated-circuit analog multipliers because it is capable of moderate accuracy and requires relatively few components. A simplified transconductance multiplier (limited to two-quadrant operation because the voltage v_Y cannot be negative) is shown in Fig. 11.28.

If it is assumed the v_X attenuator is not loaded by the input current of transistor Q_1 and that the differential input voltage applied to the pair is small enough so that linear-region relationships are valid for the transistors, the difference between the two collector currents is

$$i_{C1} - i_{C2} = \alpha v_X g_m \tag{11.50}$$

where g_m is the (equal) transconductance of either transistor.

For small-signal operation, the quantity g_m is related to quiescent operating current, which is in turn determined by the input variable v_Y. Thus,

$$g_m = \frac{i_Y q}{2kT} = \frac{K v_Y q}{2kT} \tag{11.51}$$

Substituting Eqn. 11.51 into Eqn. 11.50 shows that

$$i_{C1} - i_{C2} = \frac{\alpha K q}{2kT} v_X v_Y \tag{11.52}$$

The reader should convince himself that the differentially connected operational amplifier provides an output voltage equal to R_2 times the difference between the two collector currents. Substituting this relationship into Eqn. 11.52 yields

$$v_O = \frac{\alpha K R_2 q}{2kT} v_X v_Y \tag{11.53}$$

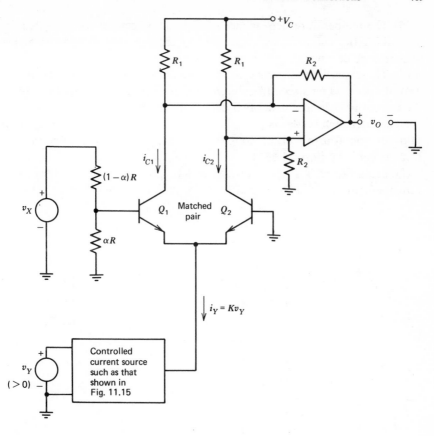

Figure 11.28 Two-quadrant transconductance multiplier.

There are a number of design constraints necessary for satisfactory operation or introduced for convenience, including the following.

(a) The current i_Y is normally limited to a fraction of a milliamp so that performance is not degraded by ohmic transistor resistance.

(b) The attenuation ratio α must be chosen to limit the input voltage applied to the transistor pair to a low level. Detailed calculations show that the inaccuracy attributable to the exponential transistor characteristics can be limited to less than 1% of maximum output if the maximum magnitude of the voltage into the differential pair is kept below approximately 8 mV.

(c) Because of the limited signal levels applied to the differential pair, its drift has a significant effect on overall performance. The circuit can be balanced by adjusting the ratio of the two resistors labeled R_1 in Fig. 11.28.

(d) The temperature dependence of Eqn. 11.53 can be compensated for by making the voltage-attenuator ratio or the current-source scale factor temperature dependent.

(e) The restriction of single-polarity values for the v_Y input can be removed by including a second differential pair of transistors, and by making the operating currents of the two pairs vary differentially as a function of v_Y. The interested reader is invited to show that the input-output relationship for the four-quadrant transconductance multiplier shown in Fig. 11.29 is given by Eqn. 11.53.

(f) Scale factor is frequently adjusted to give $v_O = v_X v_Y / 10$ volts, a value compatible with the signal levels common to many analog systems.

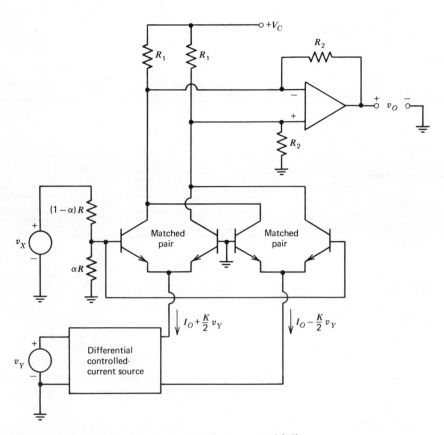

Figure 11.29   Four-quadrant transconductance multiplier.

In general, achieving highly accurate performance from a transconductance multiplier involves a rather complex series of adjustments to null various sources of error. This process is simplified somewhat by an innovation developed by Gilbert[5] which uses compensating diodes to eliminate scale-factor temperature dependence and to increase the signal levels that may be applied to the differential pairs. While there are problems that must be overcome, the technique is good enough so that several manufacturers offer inexpensive transconductance multipliers with errors from all sources of less than 1% of maximum output.

11.6 APPLICATIONS INVOLVING ANALOG-SIGNAL SWITCHING

Systems that combine operational amplifiers with analog switches add a powerful dimension to the data-processing capability of the amplifiers alone. The switches are often used to control analog operations with digital command signals, and the resultant *hybrid* (analog-digital) circuits such as analog-to-digital converters are used in a myriad of applications. While a detailed discussion of these advanced techniques is beyond the scope of this book, several simple examples of connections including analog switching are presented in this section.

Either junction-gate or MOS field-effect transistors are frequently used for low-level signal switching. One advantage of a field-effect transistor as an analog switch is that it has no inherent offset voltage. The drain-to-source characteristics of a FET in the on state are linear and resistive for small channel currents, and the drain-to-source voltage is zero for zero channel current. A second advantage is that the channel leakage current of a pinched-off FET is generally under 1 nA at room temperature. This level is insignificant in many operational-amplifier connections.

There are several integrated circuits available that combine FET's with drive circuitry to interface the switch to digital-signal levels. Alternatively, discrete-component circuits can be designed to take advantage of the lower on-state resistances generally available from discrete field-effect transistors.

A second possibility is to use a bipolar transistor as a switch. The current handling capacity of bipolar devices is generally higher than that of FET's. However, there is a collector-to-base offset voltage that can be as

[5] B. Gilbert, "A D.C.-500 MHz Amplifier/Multiplier Principle," Digest of Technical Papers, 1968 Solid-State Circuits Conference, Philadelphia, Pa.

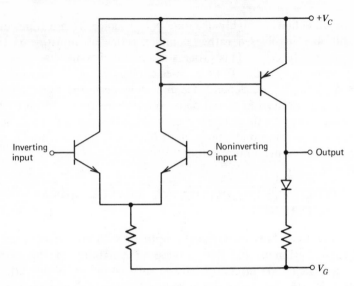

Figure 11.30   Gated operational amplifier.

high as several hundred millivolts.[6] Some high-current switching techniques arrange the feedback to eliminate offset-voltage effects.

A third type of switch combines the switching and amplification functions in a single circuit. Figure 11.30 shows a possible connection. With V_G negative, the amplifier is an example of the simple two-stage topology described in Section 8.2.3. If voltage V_G is switched to a positive potential, all three transistors and the diode become reverse biased, and thus both inputs and the output are open circuits. The gating feature can be retained in designs that expand the simple configuration shown in Fig. 11.30 into a complete operational amplifier. Several integrated-circuit examples of this type of design exist (see Section 10.4.2).

---

[6] One way to reduce the offset voltage of a bipolar transistor is to use it in an inverted or reverse mode with the roles of the emitter and collector interchanged, and offset voltages of a fraction of a millivolt are possible in this connection. The reason for the lower offset in the inverted mode is that the collector-to-emitter voltage of a saturated transistor is, in the absence of ohmic drops,

$$V_{offset} = \frac{kT}{q} \ln \frac{1}{\alpha}$$

The reverse common-base current gain α_R is used to determine forward-region offset, while the forward gain α_F is used to determine inverted offset voltage. Since α_F is generally close to one, inverted offset voltages can be quite small. Unfortunately, current gain and breakdown voltages are usually limited in the inverted connection. Consequently, as FET characteristics have improved, these devices have largely replaced inverted bipolar transistors as low-level switches.

One frequent use for analog switching is to multiplex a number of signals. The required circuit can be realized by using field-effect transistors to switch the signal applied to the input of a noninverting buffer amplifier. Another topology (see Fig. 11.31) results in an inverting multiplexer. The advantage of this configuration is that the drive circuit can be simpler than is the case with the noninverting connection. Recall that for a junction FET, it is necessary to make the gate potential approximately equal to the channel potential to turn on the transistor. If the noninverting connection is used, the channel of the on FET will be at the potential of the selected input. Furthermore, one end of the channel of all other switches will also be at the potential of the selected input. These uncertain levels complicate the drive-circuit requirements.

In the inverting topology, the channel of the on FET will be close to ground, and the diodes shown in Fig. 11.31 insure that the drain of the off FET will not be significantly more negative than ground. Thus a switch is turned on by grounding its gate, and turned off by making its gate more negative than the pinchoff voltage. An example of a common-base level shifter that converts T^2L logic signals to the required gate-drive levels is described in Section 12.3.3.

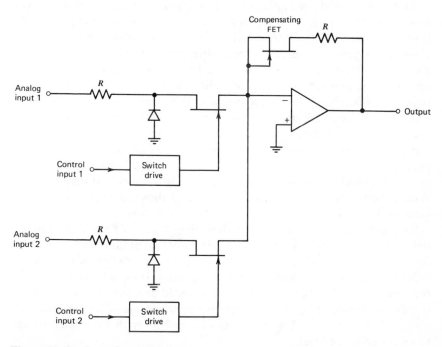

Figure 11.31   Inverting multiplexer.

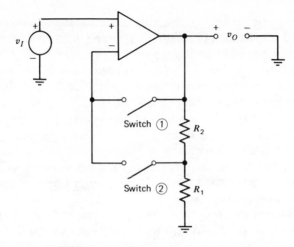

Figure 11.32   Gain-range amplifier.

The compensating FET is selected to have an on resistance matched to that of the switches. This device keeps the gain of the multiplexer equal to -1 as on resistance changes with temperature.

There are a variety of applications that require an amplifier with a select-able closed-loop gain. One topology for this type of gain-range amplifier is shown in Fig. 11.32. With switch ① closed and switch ② open, the ideal closed-loop gain is one, while reversing the state of the two switches changes the ideal gain to $(R_1 + R_2)/R_1$. The on resistance of the switches is relatively unimportant because only the low input current of the operational amplifier flows through a switch in this connection.

A related circuit function is that of an amplifier that provides a select-able gain of plus or minus one. One use for this kind of circuit is in square-wave modulators or demodulators. Figure 11.33 illustrates a possible connection. Assume initially the switch ② is not included in the circuit. With switch ① closed, the amplifier provides an ideal closed-loop gain of -1. With switch ① open, the voltage $v_A = v_I$, and thus the circuit provides an ideal gain of $+1$.

Switch ② may be included to reduce the effects of switch on-state resistance. Assume, for example, that design considerations dictate a value for R_1 equal to 10^3 times the on-state resistance of a switch. If only switch ① is used, a closed-loop gain error of 0.2% results from this resistance with the switch closed. If both switches are included and closed, the voltage v_A is reduced by a factor of 2.5×10^5 relative to v_I because of the resulting two stages of attenuation. This attenuation lowers the error from feedthrough to an insignificant level.

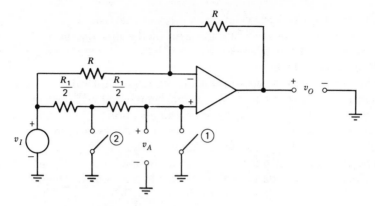

Figure 11.33   Amplifier that provides gain of ±1.

There are a number of topologies that combine operational amplifiers with switches to form a sample-and-hold circuit. Figure 11.34 shows one possibility. When the FET conducts, the loop drives the voltage v_O toward the value of v_I. The complementary emitter-follower pair amplifies the limited current available from the operational amplifier and FET combination so that large currents can be supplied to the capacitor to charge it rapidly. The resistive path between bases and emitters of the follower pair

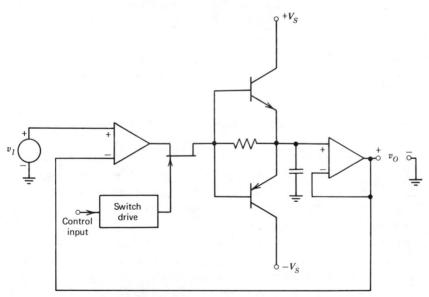

Figure 11.34   Sample-and-hold circuit.

eliminates the deadzone, which would result near equilibrium if the transistors alone were used. While the gain of the first operational amplifier insures that such a deadzone would not influence static characteristics, it could deteriorate stability.

When the switch opens, current into the capacitor is limited to buffer-amplifier input current and switch and emitter-follower leakage current. The base-to-emitter resistor prevents amplification of leakage currents in this state. Since the total capacitor current in the hold mode can be kept small, the held voltage maintains the desired value for prolonged periods.

Note that a field-effect transistor could be used as a buffer as was done in the peak detector described in Section 11.5.2 since the high open-loop gain of the first amplifier would drive the capacitor voltage to the value necessary to make $v_O = v_I$. However, the output resistance is higher in the hold mode if the FET buffer is used, since feedback is not available to reduce output impedance in the hold mode.

PROBLEMS

P11.1
The following results are obtained for measurements made using the circuit shown in Fig. 11.35a.

1. With switch ① open and switch ② closed, $V_O = 12$ mV.
2. With switch ① closed and switch ② closed, $V_O = 32$ mV.
3. With switch ① closed and switch ② open, $V_O = 10$ mV.

Determine values for the three bias generators shown in Fig. 11.35b. In this representation, the external generators model all bias voltage and current effects so that the input currents and differential input voltage at the terminals of the amplifier shown in the model are zero.

The amplifier is connected as shown in Fig. 11.35c. Express v_O in terms of v_I and the amplifier parameters shown in Fig. 11.35b.

P11.2
The circuit shown in Fig. 11.2a is used to measure the input offset voltage of an operational amplifier with a d-c open-loop voltage gain of 10^4. What error does limited loop transmission introduce into the offset measurement for these parameter values?

P11.3
A certain operational amplifier is specified to have a maximum input offset voltage magnitude of 5 mV. The amplifier is connected as a unity-gain inverter using two 2-MΩ resistors. The noninverting input is connected directly to ground. Measurements reveal that the output voltage is

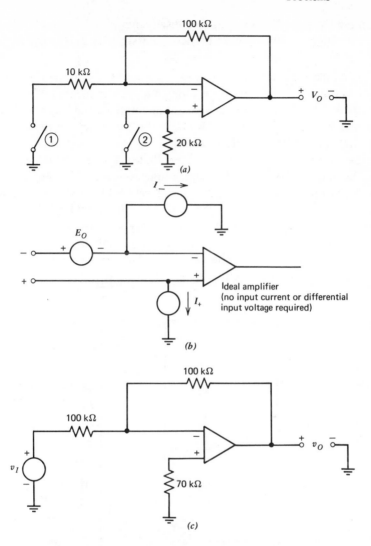

Figure 11.35 (*a*) Test circuit. (*b*) Model. (*c*) Amplifier connection.

+50 mV with zero input voltage in this connection. The amplifier in question has provision for reducing the input offset voltage at one temperature to zero by use of an appropriately connected external potentiometer that effectively changes the magnitude of current sources that load the amplifier input-stage transistors. It is found that by use of an extreme setting of the balance pot it is possible to make the output voltage of the inverter zero for zero input voltage. Discuss possible disadvantages of this method of adjustment. Suggest alternatives likely to yield superior performance.

P11.4

A simplified schematic for an integrated-circuit operational amplifier is shown in Fig. 11.36. Careful open-loop gain measurements indicate a gain of 300,000 at 1 kHz for the uncompensated amplifier and that the first pole in the amplifier transfer function is above this frequency. In the absence of load, the heating attributable to transistor Q_3 and its current-source load raise the temperature of Q_2 $0.1°$ C above that of Q_1 under static conditions with the output at its negative saturation level of -13 volts. Similarly, with the output at its positive saturation level ($+13$ volts) the temperature of transistor Q_1 is eventually raised $0.1°$ C above that of Q_2. Plot the v_O versus v_I characteristics that result for very slow variations in v_I. Now assume that the chip locations of transistors Q_1 and Q_2 are interchanged. Again plot the v_O versus v_I characteristics. Discuss how these results can complicate measurements of low-frequency open-loop gain.

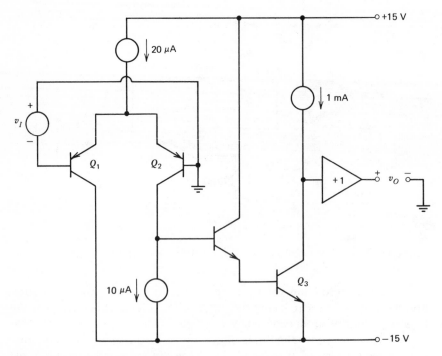

Figure 11.36 Operational amplifier.

P11.5

Integrated-circuit operational amplifiers that use an input stage similar to that of the LM101A (see Section 10.4.1) generally have a high maximum differential input voltage rating. Explain why differential input voltages of approximately 30 volts are possible with this stage compared with the 6-volt maximum level typically specified for a conventional differential amplifier.

P11.6

A low input current operational amplifier has an open-loop transfer function

$$a(s) = \frac{10^6}{(s + 1)(10^{-5}s + 1)}$$

This amplifier is connected to monitor the output current from an ionization gauge. The resultant circuit can be modeled as shown in Fig. 11.37. The capacitance shown at the input of the amplifier includes, in addition to the capacitance of the amplifier itself, the capacitance of the gauge and of the shielded cable used to connect the gauge to the amplifier. Investigate the stability of this circuit. Suggest a method for improving stability.

P11.7

An operational amplifier with high d-c open loop gain and 100-mA output current capacity is connected as shown in Fig. 11.38. Low-frequency measurements indicate an incremental gain $v_o/v_i = 1100$. Explain.

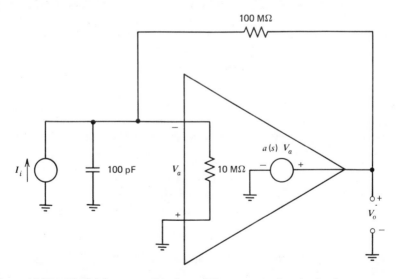

Figure 11.37 Model for operational amplifier connected to ionization gauge.

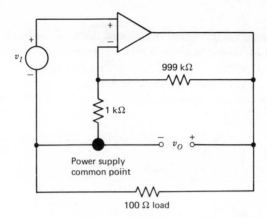

Figure 11.38 Noninverting amplifier connection.

P11.8

Measurements reveal that the dielectric absorption associated with a certain 1-μF capacitor can be modeled as shown in Fig. 11.39. Design a circuit that combines this capacitor with an ideal operational amplifier and any necessary passive components such that the closed-loop transfer function is $-1/s$.

P11.9

A differential connection as shown in Fig. 11.10 is constructed with $Z_1 = Z_3 = 1$ kΩ and $Z_2 = Z_4 = 10$ kΩ. The operational amplifier has very high d-c open-loop gain and a common-mode rejection ratio of 10^4. Assuming all other operational-amplifier characteristics are ideal, what output voltage results with both inputs equal to one volt? Suggest a modification that raises the common-mode rejection ratio for the connection.

P11.10

An operational amplifier with a d-c open-loop gain of 10^5 is connected as a current source with the topology shown in Fig. 11.14. The resistor

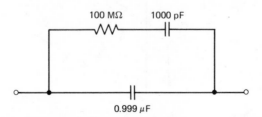

Figure 11.39 Capacitor with dielectric absorption.

value is $R = 10$ kΩ. With an input voltage of $+5$ volts, FET parameters are $y_{fs} = 1$ mmho and $y_{os} = 5$ μmho. (See Fig. 8.19 for a definition of terms.) What is the incremental output resistance of this connection?

P11.11

A Howland current source is constructed as shown in Fig. 11.40. Determine the current I_o as a function of V_a, V_b, V_o, and α. Assume that the offset voltage referred to the input of the amplifier is 5 mV and that the operational amplifier saturates at an output voltage level of ± 10 volts. Select the parameter α to maximize the output current available at zero output voltage subject to the constraint that $|i_o| < 5$ μA with $v_A = v_B = 0$.

P11.12

Design a circuit using no inductors that provides a driving-point impedance $Z = -1$ k$\Omega + 10^{-2}s$.

P11.13

A nonlinear lag network is required to compensate a servomechanism. (See Section 6.3.5 for a discussion of this type of network.) The network should have a transfer function

$$\frac{V_o(s)}{V_i(s)} = \frac{0.02s + 1}{s + 1}$$

for small input-signal levels. When the magnitude of the voltage across the capacitor exceeds 0.1 volt, the capacitor voltage should be clamped to pre-

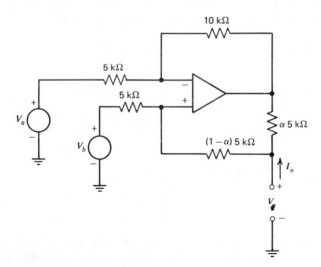

Figure 11.40 Differential current source.

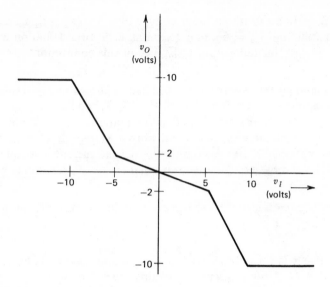

Figure 11.41 Nonlinear transfer characteristics.

vent further increases. Thus the large-signal transfer characteristics will approach $v_O/v_I \simeq 0.02$, independent of frequency.

Design the required network using a capacitor no larger than 5 μF. Provide buffering so that a power amplifier with 1-kΩ input resistance does not load the network appreciably. The capacitor-voltage limiting level for your design should be relatively temperature independent.

P11.14

Design a circuit that provides the transfer characteristics shown in Fig. 11.41. Use a configuration that makes the breakpoint locations well defined and relatively temperature independent. Select resistor values so that op-erational-amplifier input bias currents of 100 nA do not significantly affect performance and so that the loads applied to the outputs of the amplifiers used are less the 1 mA for any $|v_I| < 15$ volts.

P11.15

Design a circuit that provides an output

$$v_O = \frac{\sqrt{v_X v_Y{}^3}}{10 \text{ volts}}$$

You may assume that both v_X and v_Y are limited to a range of 0 to -10 volts. Assume that any operational amplifiers used can provide undistorted outputs of ± 10 volts. You should design your circuit so that various voltage levels are close to maximum values for maximum input signal levels in order to improve dynamic range. Comment on the temperature stability of your design.

P11.16

A sample-and-hold circuit is built using the topology shown in Fig. 11.34. The open-loop transfer function of the first operational amplifier is

$$a(s) = \frac{10^5}{(0.01s + 1)(5 \times 10^{-8}s + 1)^2}$$

and an LM110 amplifier with a closed-loop bandwidth in excess of 20 MHz is used as the output buffer. The sum of the FET on resistance and the resistor shunting the current-booster transistors is 1 kΩ, and the capacitor value is 1 μF. Investigate the stability of this system under small-signal conditions of operation. Suggest a circuit modification that can be used to improve stability. Comment on the effectiveness of your method under large-signal conditions (with the booster transistors conducting) as well as for linear-region operation.

CHAPTER XII

ADVANCED APPLICATIONS

12.1 SINUSOIDAL OSCILLATORS

One of the major hazards involved in the application of operational amplifiers is that the user often finds that they oscillate in connections he wishes were stable. An objective of this book is to provide guidance to help circumvent this common pitfall. There are, however, many applications that require a periodic waveform with a controlled frequency, waveshape, and amplitude, and operational amplifiers are frequently used to generate these signals.

If a sinusoidal output is required, the conditions that must be satisfied to generate this waveform can be determined from the linear feedback theory presented in earlier chapters.

12.1.1 The Wien-Bridge Oscillator

The Wien-bridge configuration (Fig. 12.1) is one way to implement a sinusoidal oscillator. The transfer function of the network that connects the output of the amplifier to its noninverting input is (in the absence of loading)

$$\frac{V_a(s)}{V_o(s)} = \frac{RCs}{R^2C^2s^2 + 3RCs + 1} \tag{12.1}$$

The operational amplifier is connected for a noninverting gain of 3. Combining this gain with Eqn. 12.1 yields for a loop transmission in this *positive*-feedback system

$$L(s) = \frac{3RCs}{R^2C^2s^2 + 3RCs + 1} \tag{12.2}$$

The characteristic equation

$$1 - L(s) = 1 - \frac{3RCs}{R^2C^2s^2 + 3RCs + 1} = \frac{R^2C^2s^2 + 1}{R^2C^2s^2 + 3RCs + 1} \tag{12.3}$$

has imaginary zeros at $s = \pm(j/RC)$, and thus the system can sustain constant-amplitude sinusoidal oscillations at a frequency $\omega = 1/RC$.

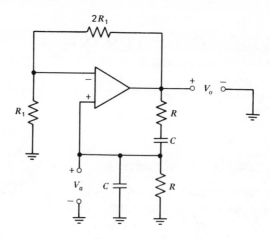

Figure 12.1   Wien-bridge oscillator.

12.1.2   Quadrature Oscillators

The quadrature oscillator (Fig. 12.2) combines an inverting and a non-inverting integrator to provide two sinusoids time phase shifted by 90° with respect to each other. The loop transmission for this connection is

$$L(s) = \left[-\frac{1}{R_1 C_1 s} \right] \left[\frac{R_3 C_3 s + 1}{(R_2 C_2 s + 1) R_3 C_3 s} \right] \qquad (12.4)$$

In this expression, the first bracketed term is the closed-loop transfer function of the left-hand operational amplifier (the inverting integrator),

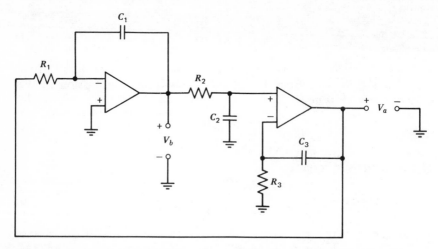

Figure 12.2   Quadrature oscillator.

while the second bracketed expression is the closed-loop transfer function of the right-hand operational amplifier. By proper selection of component values, the right-hand amplifier functions as a noninverting integrator. In fact, the discussion of this general connection in Section 11.4.1 shows that only the noninverting input of a differential connection is used as a signal input in this application.

If all three times constants are made equal so that $R_1C_1 = R_2C_2 = R_3C_3 = RC$, Eqn. 12.4 reduces to

$$L(s) = -\frac{1}{R^2C^2s^2} \tag{12.5}$$

The corresponding characteristic equation for this negative-feedback system is

$$1 - L(s) = 1 + \frac{1}{R^2C^2s^2} = \frac{R^2C^2s^2 + 1}{R^2C^2s^2} \tag{12.6}$$

Again, the imaginary zeros of Eqn. 12.6 indicate the potential for constant-amplitude sinusoidal oscillation. Note that, since there is an integration between V_a and V_b, these two signals will be phase shifted in time by 90° with respect to each other.

A similar type of oscillator (without an available quadrature output) can be constructed using a single amplifier configured as a double integrator (Fig. 11.12) with its output connected back to its input.

12.1.3 Amplitude Stabilization by Means of Limiting

There is a fundamental paradox that complicates the design of sinusoidal oscillators. A necessary and sufficient condition for the generation of constant-amplitude sinusoidal signals is that a pair of closed-loop poles of a feedback system lie on the imaginary axis and that no closed-loop poles are in the right half of the s plane. However, with this condition exactly satisfied (an impossibility in any but a purely mathematical system), the amplitude of the system output is determined by initial conditions. In any physical system, minor departure from ideal pole location results in an oscillation with an exponentially growing or decaying amplitude.

It is necessary to include some mechanism in the oscillator to stabilize its output amplitude at the desired level. One possibility is to design the oscillator so that its dominant pole pair lies slightly to the right of the imaginary axis for small signal levels, and then use a nonlinearity to limit amplitude to a controlled level. This approach was illustrated in Section 6.3.3 as an example of describing-function analysis and is reviewed briefly here.

Consider the Wien-bridge oscillator shown in Fig. 12.1. If the ratio of the resistors connecting the output of the amplifier to its inverting input is changed, it is possible to change the gain of the amplifier from 3 to $3(1 + \Delta)$. As a result, Eqn. 12.3 becomes

$$1 - L(s) = 1 - \frac{3(1 + \Delta)}{R^2C^2s^2 + 3RCs + 1} = \frac{R^2C^2s^2 - 3\Delta RCs + 1}{R^2C^2s^2 + 3RCs + 1} \quad (12.7)$$

The zeros of the characteristic equation (which are identically the closed-loop pole locations) become second order with $\omega_n = 1/RC$ and $\zeta = -(3/2)\Delta$. In practice, Δ is chosen to be large enough so that the closed-loop poles remain in the right-half plane for all anticipated parameter variations. For example, component-value tolerances or dielectric absorption associated with the capacitors alter the closed-loop pole locations.

Limiting can then be used to lower the value of Δ (in a describing-function sense) so that the output amplitude is controlled. Figure 12.3 shows one possible circuit where a value of $\Delta = 0.01$ is used. The oscillation frequency is 10^4 rad/sec or approximately 1.6 kHz. Output amplitude is (allowing for the diode forward voltage) approximately 20 V peak-to-peak. The symmetrical limiting is used since it does not add a d-c component or even harmonics to the output signal if the diodes are matched.

12.1.4 Amplitude Control by Parameter Variation

The use of a limiter to change a loop parameter in a describing-function sense after a signal amplitude has reached a specified value is one way to stabilize the output amplitude of an oscillator. This approach can result in significant harmonic distortion of the output signal, particularly when the oscillator is designed to function in spite of relatively large variations in element values. An alternative approach, which often results in significantly lower harmonic distortion, is to use an auxillary feedback loop to adjust some parameter value in such a way as to place the closed-loop poles precisely on the imaginary axis, precluding further changes in the amplitude of the oscillation, once the desired level has been reached. This technique is frequently referred to as automatic gain control, although in practice some quantity other than gain may be varied.

As an example of this type of amplitude stabilization, let us consider the effect on performance of varying resistor R_3 in the quadrature oscillator (Fig. 12.2). We assume that $C_1 = C_2 = C_3$, and that $R_1 = R_2 = R$, while $R_3 = (1 + \Delta)R$. In this case the loop transmission of the system (see Eqn. 12.4) is

$$L(s) = -\frac{(1 + \Delta)RCs + 1}{R^2C^2s^2(1 + \Delta)(RCs + 1)} \quad (12.8)$$

with a corresponding characteristic equation

$$1 - L(s) = \frac{R^3C^3(1 + \Delta)s^3 + R^2C^2(1 + \Delta)s^2 + RC(1 + \Delta)s + 1}{R^2C^2s^2(1 + \Delta)(RCs + 1)} \quad (12.9)$$

If we assume a small value for Δ, the zeros of the characteristic equation can be readily determined, since

$$R^3C^3(1 + \Delta)s^3 + R^2C^2(1 + \Delta)s^2 + RC(1 + \Delta)s + 1$$

$$\simeq \left[RC\left(1 + \frac{\Delta}{2}\right)s + 1 \right]\left[R^2C^2\left(1 + \frac{\Delta}{2}\right)s^2 + RC\frac{\Delta}{2}s + 1 \right]$$

$$|\Delta| \ll 1 \quad (12.10)$$

The performance of the oscillator is, of course, dominated by the complex-conjugate root pair indicated in Eqn. 12.10, and this pair has a natural frequency $\omega_n \simeq 1/RC$ and a damping ratio $\zeta \simeq \Delta/4$. The important feature is that the closed-loop poles can be made to lie in either the left half or the right half of the s plane according to the sign of Δ.

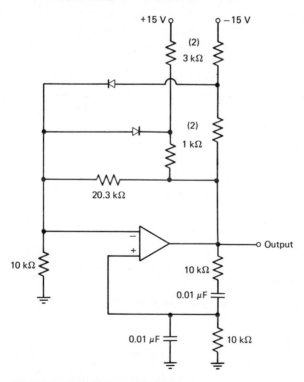

Figure 12.3 Wien-bridge oscillator with limiting.

The design of the amplitude-control loop for a quadrature oscillator provides an interesting and instructive example of the way that the feedback techniques developed in Chapters 2 to 6 can be applied to a moderately complex circuit, and for this reason we shall investigate the problem in some detail. The difficulties are concentrated primarily in the modeling phase of the analytical effort.

Our intent is to focus on amplitude control, and this control is to be accomplished by moving the closed-loop poles of the oscillator to the left- or the right-half plane according to whether the actual output amplitude is too large or too small, respectively. We assume that the signal $v_A(t)$ (see Fig. 12.2) has the form

$$v_A(t) = e_A(t) \sin \omega t \tag{12.11}$$

This representation, which models the signal as a constant-frequency sinusoid with a variable envelope $e_A(t)$, is not exact, because the instantaneous frequency of the sinusoidal component of v_A is a function of Δ. However, if the amplitude-control loop has a very low crossover frequency compared to the frequency of oscillation so that magnitude changes are relatively slow, we can consider the amplitude e_A alone and ignore the sinusoidal portion of the expression. In this case the exact frequency of the sinusoid is unimportant.

In order to find the dependence of v_A on the control parameter Δ, assume that the circuit is oscillating with $\Delta = 0$ so that the closed-loop poles of the oscillator are precisely on the imaginary axis. With this constraint the envelope is constant with some operating point value E_A so that

$$v_A(t) = E_A \sin \omega t \tag{12.12}$$

where $\omega = 1/RC$. If Δ undergoes an incremental step change to a new value Δ_1 at time $t = 0$, the oscillator poles move into the left-half plane (for positive Δ_1), and

$$v_A(t) \simeq E_A\, e^{-\zeta \omega_n t} \sin \omega_n t \tag{12.13}$$

Inserting values for ζ and ω_n from Eqn. 12.10 into Eqn. 12.13 yields

$$v_A(t) \simeq E_A\, e^{-(\Delta_1 t/4RC)} \sin \frac{t}{RC} \tag{12.14}$$

The envelope for this signal is

$$e_A(t) = E_A e^{-(\Delta_1 t/4RC)} = E_A \left[1 - \frac{\Delta_1 t}{4RC} + \frac{1}{2}\left(\frac{\Delta_1 t}{4RC}\right)^2 - \cdots + \right] \tag{12.15}$$

If $\Delta_1 t/4RC$ is small (a condition insured by a sufficiently small value of Δ_1), we can separate $e_A(t)$ into operating-point and incremental components as

$$e_A(t) = E_A + e_a(t) \simeq E_A - \frac{E_A \Delta_1}{4RC} t \qquad (12.16)$$

Thus a positive incremental step change in Δ leads to an incremental envelope change that is a linearly decreasing function of time. This condition implies that the linearized transfer function that relates envelope amplitude to Δ is

$$\frac{E_a(s)}{\Delta(s)} = -\frac{E_A}{4RCs} \qquad (12.17)$$

This linearized analysis confirms the feeling that control of the value of Δ is in fact a reasonable way to stabilize the amplitude of the oscillation, since the incremental change in the envelope of the oscillation is proportional to the time integral of Δ.

Further design of the amplitude-control loop depends on the actual topology of the system. Figure 12.4 shows one possible implementation in mixed circuit and functional block-diagram form. The envelope of the signal to be controlled is determined by an amplitude-measuring circuit. This circuit may be a simple diode-resistor-capacitor peak detector in cases where high precision is not required, or it may be an active "super-diode" type of connection (an example is given in Section 12.5.1) in more demanding applications. In either case, the design of this circuit is not particularly difficult and will not be discussed here. The envelope of the signal is compared with a reference value, and the resulting error signal passes through a linear controller with a transfer function $a(s)$. The output of the controller is used to drive a field-effect transistor that functions as a variable resistor whose value determines Δ.

The FET connection incorporates local compensation to linearize its characteristics as shown in the following development. If a junction FET is biased into conduction with a small voltage applied across its channel, and its gate reverse biased with respect to its channel, the drain current is approximately related to terminal voltages as

$$i_D = K\left[(v_{GS} + V_P)v_{DS} - \frac{v_{DS}^2}{2}\right] \qquad (12.18)$$

where K is a constant dependent on transistor construction, and V_P is the magnitude of the gate-to-source pinch-off voltage.

The dependence of i_D on the square of the drain-to-source voltage is undesirable, since this term represents a nonlinearity in the channel resist-

492

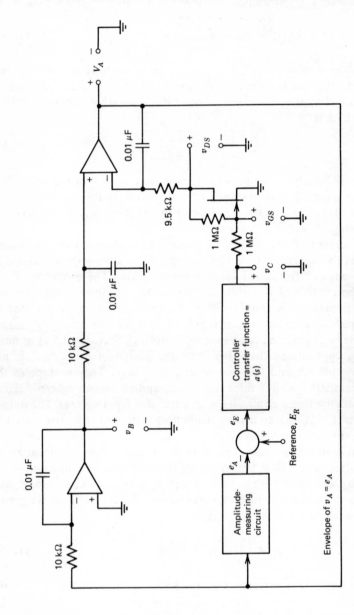

Figure 12.4 Quadrature oscillator with amplitude stabilization.

ance of the device, and this nonlinearity will introduce harmonic distortion into the oscillator output. The nonlinearity can be eliminated by adding half of the drain-to-source voltage to the gate-to-source voltage via resistors as shown in Fig. 12.4. The resistors are large enough so that they do not significantly shunt the drain-to-source resistance of the FET under normal operating conditions. With the topology shown,

$$v_{GS} = \tfrac{1}{2}\,(v_C + v_{DS}) \tag{12.19}$$

Substituting Eqn. 12.19 into Eqn. 12.18 shows that

$$i_D = K\left[\left(\frac{v_C}{2} + \frac{v_{DS}}{2} + V_P\right) v_{DS} - \frac{v_{DS}^2}{2}\right] = K\left(\frac{v_C}{2} + V_P\right) v_{DS} \tag{12.20}$$

or

$$R_{DS} = \frac{\partial v_{DS}}{\partial i_D} = \frac{1}{K[(v_C/2) + V_P]} \tag{12.21}$$

This equation indicates that the incremental resistance of the FET is independent of drain-to-source voltage when the network is included.

For purposes of design, we assume that the FET is characterized by $V_P = 4$ volts and $K = 10^{-3}$ mho per volt. Recall that stable-amplitude oscillations require that all three R-C time constants be identical; thus the operating point value of R_{DS} is 500 ohms. Equation 12.21 combined with FET parameters indicates that this value results with an operating-point value for the control voltage of -4 volts. The incremental change in R_{DS} as a function of the control voltage at this operating point, obtained by differentiating Eqn. 12.21 with respect to v_C,

$$\left.\frac{\partial R_{DS}}{\partial v_C}\right|_{v_C \,=\, -4\text{ V}} = -125 \ \Omega/\text{V} \tag{12.22}$$

Earlier modeling was done in terms of Δ, the fractional deviation of the resistance R_3 in Fig. 12.2 from its nominal value. This resistor consists of the FET plus a 9.5 kΩ resistor in the actual implementation. The incremental dependence of Δ on the control voltage is determined by dividing Eqn. 12.22 by the anticipated operating-point value of the total resistance, 10 kΩ. Thus

$$\left.\frac{\partial \Delta}{\partial v_C}\right|_{v_C \,=\, -4\text{ V}} = -0.0125 \ \text{V}^{-1} \tag{12.23}$$

The relationships summarized in Eqns. 12.17 and 12.23 combined with the system topology and an assumed operating-point value for the envelope $E_A = 10$ volts lead to the linearized block diagram for the amplitude-

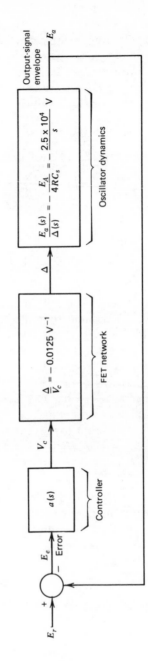

Figure 12.5 Linearized block diagram for amplitude-control loop.

control loop shown in Fig. 12.5. The negative of the loop transmission for this system is

$$\frac{E_a(s)}{E_e(s)} = a(s) \times \frac{312.5}{s} \qquad (12.24)$$

A number of factors govern the choice of $a(s)$ for this application including:

(a) The actual FET gate-to-source voltage required under quiescent conditions is strongly dependent on FET parameters and the exact values of the other components used in the circuit. The easiest way to insure that the difference between the envelope and the reference is constant in spite of these variable parameters is to include an integration in $a(s)$ since this integration forces the operating-point value of the error to zero.

(b) The analysis is predicated on a much lower crossover frequency for the amplitude-control loop than the frequency of oscillation, 10^4 radians per second. However, a very low frequency control loop accentuates the effect on amplitude of rapid changes in quantities like the supply voltages. A somewhat arbitrary compromise is to choose a crossover frequency of 100 radians per second.

(c) Since the analysis is based on a hierarchy of approximations, the system should be designed to have a very conservative phase margin.

(d) The controller transfer function should include low-pass filtering. The detector signal that indicates the envelope amplitude invariably includes components at the oscillation frequency or its harmonics. If these components are not filtered so that they are at an insignificant level when applied to the FET gate, the resultant channel-resistance modulation introduces distortion into the oscillator output signal.

A controller transfer function that incorporates these features is

$$a(s) = \frac{3.2(0.1s + 1)}{s(10^{-3}s + 1)^2} \qquad (12.25)$$

The negative of the loop transmission with this value for $a(s)$ is

$$\frac{E_a(s)}{E_e(s)} = \frac{10^3(0.1s + 1)}{s^2(10^{-3}s + 1)^2} \qquad (12.26)$$

The system crossover frequency is 100 radians per second, and phase margin exceeds 70° with this value for $a(s)$.

A possible circuit that provides the negative of the desired $a(s)$ is shown in Fig. 12.6. In many cases of practical interest, this inversion can be cancelled by some rearrangement of the amplitude-measuring circuit. The second required filter pole is obtained with a passive network. The filter

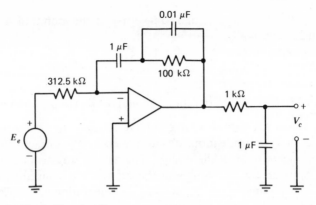

Figure 12.6 Controller circuit.

network impedance level is low enough so that the network is not disturbed by the 2-MΩ load connected to it.

The reference level required to establish oscillator amplitude can be applied to the controller by adding another input resistor to the operational amplifier. It may also be possible to realize part of the amplitude-measuring circuitry with this amplifier. An example of this type of function combination is given in Section 12.5.1.

Two practical considerations involved in the design of this oscillator deserve special mention. First, the signal v_B normally has lower harmonic distortion than does v_A since the integration of the first amplifier filters any harmonics that may be introduced by the FET. Second, it is possible to vary the reference amplitude for this circuit and thus modulate the amplitude of the oscillator output. However, the control bandwidth in this mode will be relatively small, and performance will change as a function of quiescent envelope amplitude since the loop-transmission magnitude is dependent on operating levels.

The performance of an oscillator of this type can be quite impressive. Amplitude control to within 1 mV peak-to-peak is possible if "superdiodes" are used in the envelope detector. Harmonic distortion of the output signal can be kept a factor of 10^4 or more below the fundamental component.

12.2 NONLINEAR OSCILLATORS

The discussion of oscillators up to this point has focused on the design of circuits that provide sinusoidal output signals. The basic approach is to use a linear, second-order feedback loop to generate the sinusoid, and then incorporate some mechanism to control amplitude.

Operational amplifiers are also frequently used in nonlinear oscillator circuits that intentionally produce nonsinusoidal output signals. The analysis of these types of oscillators is complicated by the fact that transform methods normally cannot be used. One frequently used technique for evaluating the performance of these types of oscillators is to determine the output and internal signals directly via time-domain calculations.

12.2.1 A Square- and Triangle-Wave Generator

A function generator that produces square and triangle waves as its outputs was used as an example of describing-function analysis in Section 6.3.3. This topology combines an integrator with a Schmitt-trigger circuit. The Schmitt trigger can be realized by applying *positive* feedback around an operational amplifier, as shown in Fig. 12.7.[1] Consider operation with v_I a large positive voltage. In this case the amplifier will be saturated with a positive output voltage.

It is assumed that the output-voltage magnitude is limited to a maximum value of V_M. This limiting can be accomplished in several ways. If relatively crude level control is sufficient, the saturation levels may be determined simply by power-supply voltages and internal amplifier voltage drops. Somewhat better control is possible if an amplifier such as the LM101A (see Section 10.4.1) is used. The output level of this circuit can be limited by connecting diode clamps to a compensation terminal. A third possibility is to follow the operational amplifier shown with a precision limiter similar to those described in Section 11.5.3, and to apply positive feedback around the entire connection. This approach has the further advantage that the output element is operating with local negative feedback and thus has very low output resistance.

In order to force the circuit to change state, the input voltage is lowered. When the input level reaches approximately $-(R_1/R_2)V_M$, the noninverting input of the amplifier is close to ground potential and the device enters its linear operating region. The massive positive feedback that results with the amplifier active sweeps its output negative until a level of $-V_M$ is reached. Further negative changes in input voltage do not affect the output. If the input voltage is raised, the amplifier enters its active region at an

---

[1] In many practical circuits, a comparator rather than an operational amplifier is used to implement a Schmitt trigger. A comparator, like an operational amplifier, is a high-gain, direct-coupled amplifier. However, since it is not intended for use in negative-feedback connections, the frequency-response compromises that must be made to insure the stability of an operational amplifier need not be included in the comparator design. Consequently, the response time of a Schmitt trigger realized via a comparator can be significantly faster than that obtained using an operational amplifier.

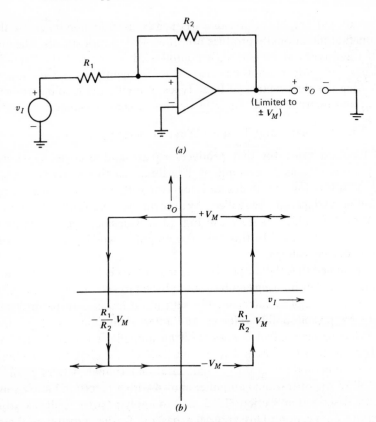

(a)

(b)

Figure 12.7 Schmitt trigger. (a) Circuit. (b) Characteristics.

input level of $+(R_1/R_2)V_M$, and is then driven to positive saturation. These transition points combine to give the characteristics shown in Fig. 12.7b.

A possible oscillator connection using this type of Schmitt trigger is shown in Fig. 12.8. With the modulating voltage $v_C = 0$, signal waveforms are as shown in part b of this figure. The period of oscillation is determined by noting that the magnitude of the slope of the triangle wave is always $10/RC$, and that the total change in the voltate level of v_A is 40 volts for one complete cycle. Therefore

$$\tau = \frac{40}{10/RC} = 4RC \tag{12.27}$$

The corresponding frequency of oscillation is

$$f = \frac{1}{\tau} = \frac{1}{4RC} \tag{12.28}$$

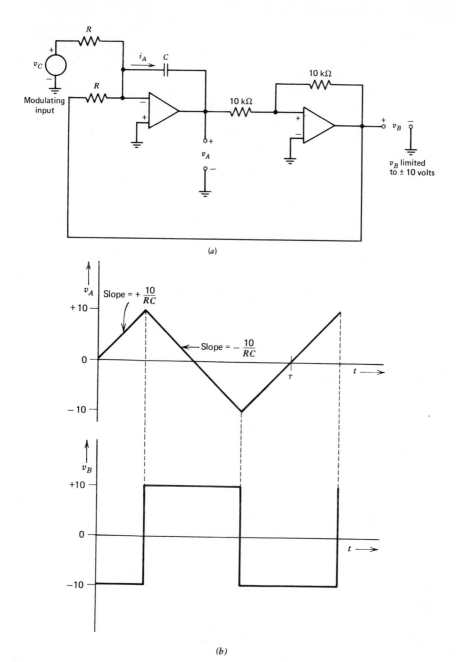

Figure 12.8 Nonlinear oscillator. (*a*) Circuit. (*b*) Waveforms with $v_C = 0$. (*c*) Waveforms with $|v_C| < 10$ volts.

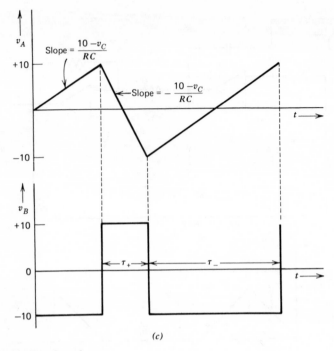

(c)

Figure 12.8—Continued

In commercial versions of this circuit, decade frequency switching is frequently accomplished by changing capacitors, while variation of the value of resistor R provides vernier control in any one decade.

12.2.2 Duty-Cycle Modulation

The current that charges the capacitor can be modulated by means of an applied voltage v_C, with this current given by

$$i_A = \frac{v_C + v_B}{R} \qquad (12.29)$$

A positive value for v_C increases capacitor charging current when v_B is positive and decreases this current when v_B is negative. The net result is to duty-cycle modulate the signal v_B as shown in Fig. 12.8c. The fraction of the time this signal stays positive is

$$\frac{\tau_+}{\tau_+ + \tau_-} = \frac{20RC/(10 + v_C)}{20RC/(10 + v_C) + 20RC/(10 - v_C)} = \frac{1}{2}\left(1 - \frac{v_C}{10}\right) \quad (12.30)$$

This duty-cycle modulator has a number of interesting features that make it useful in a variety of applications. Equation 12.30 shows that the duty cycle is linearly proportional to v_C and changes from one to zero as v_C changes from -10 volts to $+10$ volts. However, maximum capacitor charging current is limited to twice its value with zero v_C, so that the time spent in the shorter of the two periods is never less than half its quiescent value. The frequency of operation is a nonlinear function of v_C and is given by

$$f = \frac{1}{\tau_+ + \tau_-} = \frac{1}{20RC/(10 + v_C) + 20RC/(10 - v_C)} = \frac{100 - v_C^2}{400RC} \quad (12.31)$$

This equation shows that the frequency is lowered by any nonzero value of v_C.

Applications include the control of switching power amplifiers and the realization of the type of analog multiplier shown in Fig. 12.9. In this circuit, the duty-cycle modulator controls the state of a switch that is frequently realized with field-effect transistors. The circuit is arranged so that the switch arm is connected to a voltage $+v_Y$ for a fraction of the time $\frac{1}{2}[1 + (v_X/V_R)]$, and to a voltage $-v_Y$ for the remainder of the time, a fraction equal to $\frac{1}{2}[1 - (v_X/V_R)]$. (Alternative implementations use current rather than voltage switching to increase switching speed.) The output filter (usually a multiple-order active filter rather than the simple network shown) averages the switch voltage v_S, so that

$$v_O = \overline{v_S} = +v_Y\left[\frac{1}{2}\left(1 + \frac{v_X}{V_R}\right)\right] - v_Y\left[\frac{1}{2}\left(1 - \frac{v_X}{V_R}\right)\right] = \frac{v_X v_Y}{V_R} \quad (12.32)$$

where the over bar indicates time averaging. Note that the voltage V_R (which is equal to the maximum magnitude of the signal out of the Schmitt

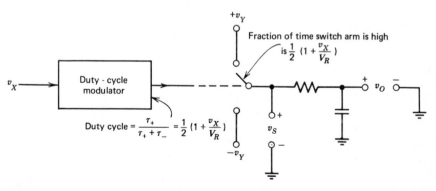

Figure 12.9 Time-division multiplier.

trigger) can be varied to mechanize division. A technique for varying the signal from the Schmitt trigger is described below.

Versions of this type of multiplier that limit errors to 0.05 % of maximum output have been designed.

12.2.3 Frequency Modulation

Another variation of the basic nonlinear oscillator shown in Fig. 12.10 results in an oscillator with a voltage-controlled operating frequency. Here the Schmitt trigger determines the state of a switch that allows a variable-level voltage to be applied to the integrator. If the Schmitt trigger switches at input-signal levels of $\pm V_T$ the total excursion of the signal v_A will be $4 V_T$ volts per cycle. The slope of signal v_A has a magnitude of v_F/RC volts per second, and thus the frequency of oscillation is

$$f = \frac{v_F/RC}{4V_T} = \frac{v_F}{4V_T RC} \tag{12.33}$$

12.2.4 A Single-Amplifier Nonlinear Oscillator

The operational amplifier used as an integrator in the nonlinear oscillator described above can be replaced with a passive resistor-capacitor network a shown in Fig. 12.11, resulting in a configuration first reported by Bose.[2] The Schmitt trigger functions in an inverting mode in this connection so that a sufficiently positive level for v_A saturates the amplifier output at $-V_M$. Switching points occur at $v_A = \pm V_M R_1/(R_1 + R_2)$. If the dotted modulating resistor is omitted, the waveforms are as shown in Fig. 12.11c. The capacitor voltage is a sequence of exponential segments rather than a true triangular wave. The duty cycle of the signal can be modulated by including the dotted resistor shown in Fig. 12.11a. If the width of the hysterisis region is made very small by choosing $R_1 \ll R_2$, the current into the capacitor becomes nearly constant in each state, since the circuit keeps the capacitor voltage close to zero. In this case, the duty cycle of the voltage v_O is linearly related to control voltage v_C.

12.3 ANALOG COMPUTATION

It was mentioned in Chapter 1 that operational amplifiers were initially used primarily for analog computation. The objective in analog computation is to build an electrical network, using operational amplifiers and

---

[2] A. G. Bose, "A Two-State Modulation System," 1963 Wescon Convention Record, Part 6, Paper 7.1.

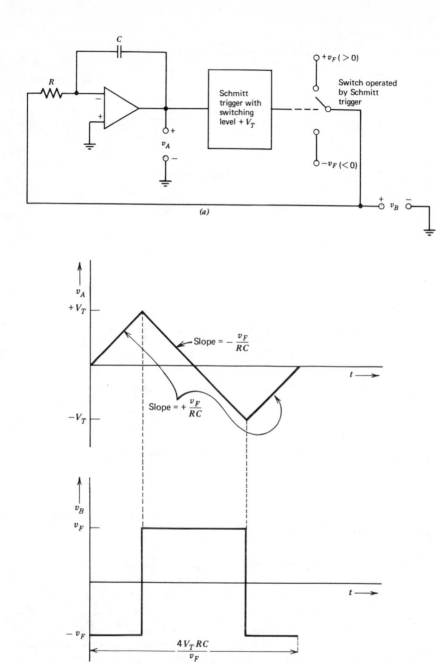

Figure 12.10 Voltage-controlled oscillator. (*a*) Circuit. (*b*) Waveforms.

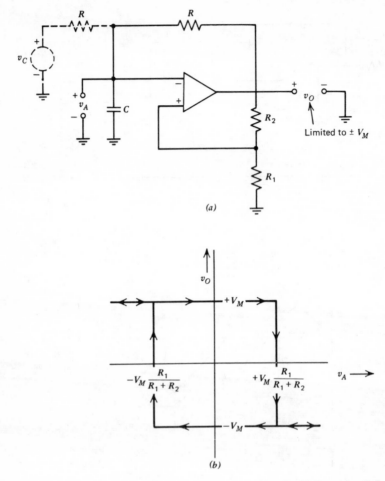

Figure 12.11 One amplifier nonlinear oscillator. (*a*) Circuit. (*b*) Inverting Schmitt-trigger characteristics. (*c*) Waveforms.

associated components, that obeys the same differential equation as does the system under study. The answers obtained consist of the responses of the electrical analog to particular inputs and initial conditions.

Analog computers are available from several manufacturers. These machines incorporate, in addition to the necessary hardware, a considerable human-engineering effort. Summing amplifiers and integrators included in these machines are normally constructed with fixed scale factors so that external components need not be used. For example, several inputs with gains of -1 and -10 are typically provided for each summing amplifier.

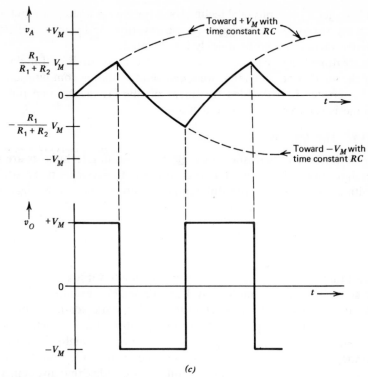

Figure 12.11—Continued

Potentiometers are also included, and these devices are combined with fixed-gain amplifiers to provide arbitrary gain levels. Thus a gain of -3.12 might be realized by preceding a gain of -10 amplifier with a potentiometer set for an attenuation of 0.312. Nonlinear elements such as function generators and multipliers are frequently included. The inputs and outputs of the various elements are usually connected to jacks of some type. The interconnections necessary to simulate a particular system are then made with patchcords that connect the various jacks. In many cases, the programming (inserting the patchcords to establish the proper connection pattern) is done on a board physically removed from the computer while other users, with their own boards, solve their problems. The board makes the required connections when it is inserted into a mating plate located on the machine.

While the accuracy of solutions obtained via analog computation is limited by component tolerances, it normally far exceeds the accuracy required for the simulation of physical systems, which are themselves constructed with imprecise components. A further consideration is that it is frequently possible to get a good physical feeling for a system via analog computation, since many variables are available for observation, and since the effects of parameter variations can be quickly investigated.

Our treatment here can only cover the barest essentials and highlight a few of the ancillary circuits that were evolved for analog computation. The reader interested in a detailed treatment of this fascinating and powerful technique is referred to Korn and Korn.[3]

12.3.1 The Approach

Our objective here is to show how electronic-analog techniques are used to simulate differential equations that describe the systems to be studied. We initially assume that the differential equation under investigation is linear and has the general form

$$a_n \frac{d^n x}{dt^n} + a_{n-1} \frac{d^{n-1}x}{dt^{n-1}} + \cdots + a_1 \frac{dx}{dt} + a_0 x = f(t) \qquad (12.34)$$

It is certainly not necessary that the independent variable of the system under study be time as implied by Eqn. 12.34. For example, if we were investigating the deflection of a bridge under static load, we might be interested in vertical displacements from equilibrium as a function of distance from one end of the bridge. However, since our analog will use time as its independent variable, we substitute time for the independent variable if necessary in the original equation. Similarly, we realize that any dependent variables in our analog will have to be voltages, regardless of the variables they actually represent in the system under study.

Equation 12.34 is rewritten so that the highest derivative of x is expressed in terms of the other variables in the form

$$\frac{d^n x}{dt^n} = - \frac{a_{n-1}}{a_n} \frac{d^{n-1}x}{dt^{n-1}} - \cdots - \frac{a_1}{a_n} \frac{dx}{dt} - \frac{a_0 x}{a_n} + \frac{1}{a_n} f(t) \qquad (12.35)$$

Equation 12.35 can be represented as the block diagram shown in Fig. 12.12. In this representation, the variable $d^n x/dt^n$ appears as the output of a summation point. Inputs to the summation point are scaled multiples of the driving function and the lower-order derivatives of x. The lower-order derivatives are obtained by successive integrations of $d^n x/dt^n$, with a total of n integrations required to complete the block diagram.

Note that the only elements included in the block diagram are a multiple-input summation point, inverters to precede some inputs on the summer,

[3] G. A. Korn and T. M. Korn, *Electronic Analog and Hybrid Computers*, 2nd Edition, McGraw-Hill, New York, 1972.

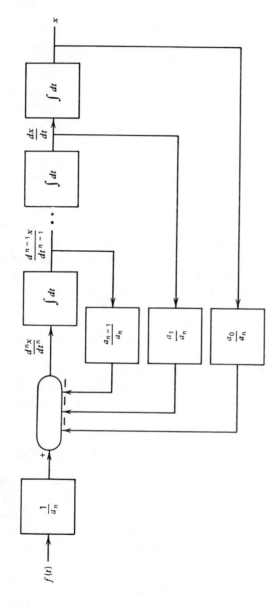

Figure 12.12 Block diagram of Eqn. 12.35.

gain blocks, and integrators. Since each of these elements can be readily constructed using operational amplifiers and passive components, the block diagram can be implemented using these devices. When the analog realization is excited with a voltage equal to $f(t)$, voltages equal in value to x and its derivatives will be available as the outputs of the integrators.

As an example of this process, consider the differential equation

$$\frac{d^4x}{dt^4} + 2.61 \frac{d^3x}{dt^3} + 3.42 \frac{d^2x}{dt^2} + 2.61 \frac{dx}{dt} + x = f(t) \qquad (12.36)$$

(We recall from Section 3.3.2 that this equation represents a fourth-order Butterworth filter.) Solving for d^4x/dt^4 yields

$$\frac{d^4x}{dt^4} = -2.61 \frac{d^3x}{dt^3} - 3.42 \frac{d^2x}{dt^2} - 2.61 \frac{dx}{dt} - x + f(t) \qquad (12.37)$$

One possible simulation of this equation is shown in Fig. 12.13. The voltages expected at the output of various amplifiers are indicated by writing the value of the variable the voltage represents at appropriate nodes. Note that in contrast to traditional analog-computer methods, gains are established by selecting impedances[4] used around operational amplifiers rather than by combining potentiometers with fixed-gain amplifiers and integrators. Also, functions have been combined in order to reduce the number of amplifiers required. The use of inverting connections only is traditional in analog computation, and reflects that fact that an operational-amplifier design technique frequently used to improve d-c performance results in an amplifier that can only be used in inverting connections. (See Section 12.3.3.) It may, of course, be possible to use noninverting integrators or summing amplifiers (realized with resistive summing at the input to a noninverting-amplifier connection) if general-purpose operational amplifiers are used for this simulation.

The four integrators appear along the top of the diagram. Since it is assumed that there is no need to have a voltage representing d^4x/dt^4 available, the summing operation is included in the first integrator connection. The output of this integrator is $-(d^3x/dt^3)$ when the indicated current is equal to $(10^{-6}\text{ A}) d^4x/dt^4$. Since inverting integrators are used, the signs associated with successive derivatives alternate. The scaling and inversions required by the coefficients of x and its second derivative are obtained with the bottom amplifier.

----

[4] The relative impedance levels shown in Fig. 12.13 are high if general-purpose operational amplifiers such as the LM101A are used. Since only ratios are important in establishing the transfer function, all impedance levels can be scaled to reduce errors that result from amplifier input currents.

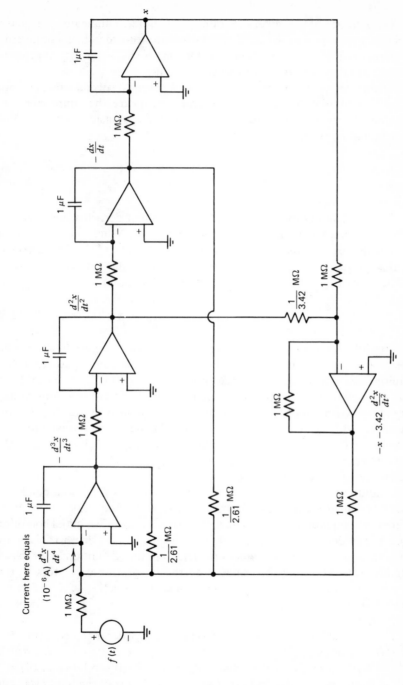

Figure 12.13 Simulation of fourth-order Butterworth equation.

509

The number of amplifiers required in Fig. 12.13 indicates the general rule. If this topology is used, simulating an nth-order linear differential equation requires n integrators and one amplifier that inverts appropriate signals as necessary to complete feedback paths.

Analog-computing techniques can also be used to solve a variety of non-linear differential equations by including hardware that implements the nonlinearity in the simulation. As an example, consider Van der Pol's differential equation

$$\frac{d^2x}{dt^2} + \mu(x^2 - 1)\frac{dx}{dt} + x = 0 \qquad (12.38)$$

where μ is a positive constant.

For small values of x, the coefficient of the first derivative term is negative, and increasing-amplitude oscillations result. When the amplitude of the oscillation becomes large enough, the coefficient of the first derivative will be positive over part of the cycle, and a limit cycle can result. Equation 12.38 is rewritten in a form convenient for simulation as

$$\frac{d^2x}{dt^2} = -\mu x^2\frac{dx}{dt} + \mu\frac{dx}{dt} - x \qquad (12.39)$$

Multipliers are required to generate x^2 and form the $x^2(dx/dt)$ product necessary for the simulation of Eqn. 12.39. Two techniques for analog multiplication were described in Sections 11.5.5 and 12.2.2. Practical multipliers based on these methods are often designed to have an output voltage equal to the product of the two input voltages divided by 10 volts for compatibility with the dynamic range of most solid-state operational amplifiers. Figure 12.14 shows a possible simulation of Eqn. 12.39 assuming that multipliers with this scale factor are used.

Van der Pol's equation is an example of an undriven differential equation, and excitation is by initial conditions only. While initial conditions were not mentioned in our earlier discussion of the simulation of linear differential equations, we recognize that we must specify n initial conditions in order to determine the complete (homogeneous plus driven) solution of an nth-order differential equation. These initial conditions can be set simply by establishing the voltages on the integrating capacitors at time $t = 0$, since these voltages are proportional to the values of x and its first $n - 1$ derivatives. A circuit for setting initial conditions is described in Section 12.3.3.

The value of x as a function of time for Van der Pol's equation with $\mu = 0.25$ is shown in Fig. 12.15. The initial conditions used for parts a and b of this figure are $x(0) = 0.5$, $(dx/dt)(0) = 0$ and $x(0) = 3$, $(dx/dt)(0) = 0$, respectively. We see that in both cases the amplitude of the limit cycle con-

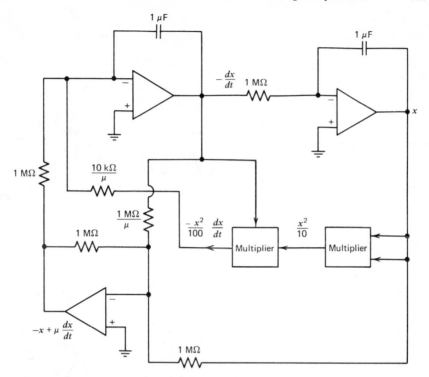

Figure 12.14 Simulation of Van der Pol's equation.

verges to a peak-to-peak value of approximately 4. Part c of this figure is a plot of dx/dt versus $x(t)$. This representation, in which time is a parameter along the curve, is called a *phase-plane* plot. The responses for both values of initial conditions are included. The convergence to equal-amplitude limit-cycles for both sets of initial conditions is evident in this figure.

The formal procedure described here is certainly not the only one which results in a correct analog representation of a problem. While it does lead to a compact realization, other realizations may maintain better correspondence with the physical system that is being modeled. One popular alternative technique involves simply drawing a block diagram for the system under study, and then implementing the block diagram on a block-by-block basis without ever writing down the complete system differential equation. While this approach often requires more hardware to complete the simulation, it is convenient in that voltages proportional to the actual variables of interest in the problem under study are avaliable. Furthermore, it is generally possible using this alternative to associate scale factors with the parameters of physical elements in the simulated systems on a one-to-one basis.

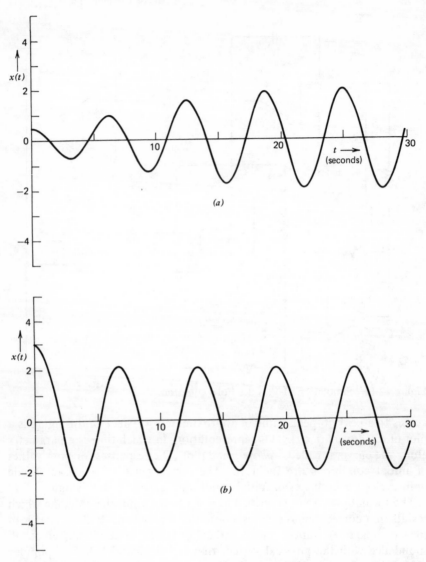

Figure 12.15 Solution to $d^2x/dt^2 + 0.25(x^2 - 1)(dx/dt) + x = 0$. (a) With initial conditions $x(0) = 0.5$, $(dx/dt)(0) = 0$. (b) With initial conditions $x(0) = 3$, $(dx/dt)(0) = 0$, (c) Parts a and b repeated in phase-plane form.

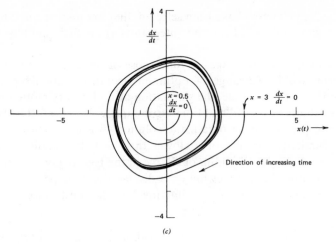

(c)

Figure 12.15—Continued

12.3.2 Amplitude and Time Scaling

Practical considerations constrain the amplitude and frequency range of the signals that arise in analog computation. We normally prefer maximum signal levels that are comfortably below amplifier saturation levels, but well above noise and offset uncertainties. Similarly, very low-frequency signals are difficult to integrate accurately, while the limited gain of an operational amplifier at high frequencies compromises accuracy in this frequency range. *Amplitude scaling* and *time scaling* are used to standardize signals to convenient amplitude levels and spectral content.

Amplitude scaling involves little more than some additional bookkeeping effort. Since we are using voltages for all of the dependent variables in our simulation, there must be a dimensioned scale factor that relates the machine variables to the problem variables when the problem variables are quantities other than voltages. For example, if x is a displacement in meters and some voltage in a simulation represents this variable on a 1 meter = 1 volt basis, the machine variable should really be labeled (1 volt/meter)x rather than simply x as is frequently done. We should realize that the number associated with the scale factor can readily be selected to be other than unity. Thus we might use $10x$ as the label for some voltage, or, preferably (10 volts/meter)x. If this voltage were 7 volts, the corresponding displacement would be $x = $ (7 volts) (1 meter/10 volts) = 0.7 meter. The appropriate values for scale factors can only be determined with a knowledge of approximate problem-variable levels, since the corresponding machine variables should have peak values slightly below the saturation level. Once scale factors have been selected, they are implemented by modifying the gains of amplifiers and integrators from their initially selected values.

Time scaling has advantages beyond those of centering signal-frequency components within the range of optimum operational-amplifier performance. Consider, for example, the simulation of a planetary motion problem that may require years of "real time" to complete. Using a faster "machine time" scale permits us to obtain the solution in a more reasonable time interval. Similarly, the use of a slower than real time scaling procedure allows us to display the buildup of charge in the base region of a transistor at a rate comfortable for viewing on a display oscilloscope.

The technique used for time scaling involves the substitution

$$t = \sigma\tau \qquad (12.40)$$

where τ is machine time and is equal to real time divided by a scale factor σ. A value of σ greater than one implies that the machine solution is *faster* than the actual solution so that one second of real time is represented by a shorter period τ of machine time.

This process is illustrated using the form for a differential equation given in Eqn. 12.34 and repeated here for convenience.

$$a_n \frac{d^n x}{dt^n} + a_{n-1} \frac{d^{n-1} x}{dt^{n-1}} + \cdots + a_1 \frac{dx}{dt} + a_0 x = f(t) \qquad (12.34)$$

In order to apply the substitution of Eqn. 12.40, we change $f(t)$ to $f(\sigma\tau)$ and change $d^m x/dt^m$ to $(1/\sigma^m)(d^m x/d\tau^m)$. Thus the time-scaled version of Eqn. 12.34 is

$$\frac{a_n}{\sigma^n} \frac{d^n x}{d\tau^n} + \frac{a_{n-1}}{\sigma^{n-1}} \frac{d^{n-1} x}{d\tau^{n-1}} + \cdots + \frac{a_1}{\sigma} \frac{dx}{d\tau} + a_0 x = f(\sigma\tau) \qquad (12.41)$$

The equation when simulated will have a solution identical in form to that of Eqn. 12.34, but will run a factor of σ faster than the original equation.

A second way to implement time scaling is to realize that the dynamics of the simulation are implemented by means of integrations, and that changing the scale factor of every integrator in the simulation by some factor must change the time scale of the simulation by precisely the same factor. Thus problems can be time scaled by first simulating the problem for a real-time solution and then dividing the value of every capacitor by a factor of σ. Alternatively, every resistor used to implement all integrators can be reduced in value by a factor of σ, or the scale-factor change can be apportioned between resistors and capacitors. The net result of any of these modifications will be to make the problem on the machine run a factor of σ faster than the real-time solution. It is, of course, still necessary to increase the speed of driving functions applied to the system by a factor of σ if these signals are derived from sources that are not implemented using scaled integrators.

The coefficients of the original differential equation often can be used to determine the time scale appropriate to a particular problem. If the roots of the characteristic equation have approximately equal magnitudes, the natural frequencies of the undriven solution will be the order of

$$\omega = \left(\frac{a_0}{a_n}\right)^{1/n} \tag{12.42}$$

Conversely, if the system is dominated by one pole, the characteristic frequency is the order of

$$\omega = \frac{a_0}{a_1} \tag{12.43}$$

The characteristic frequencies given by Eqn. 12.42 or 12.43 can be changed to values convenient for display and compatible with operational-amplifier performance by appropriate selection of σ.

The element values that occur in a problem simulation often provide clear indications of the need to modify amplitude or time scales. If, for example, we find that high gain is required at the input of every amplifier being supplied with some particular signal, the scale factor of that signal is probably too small relative to other amplitude scale factors used. Similarly, if one input resistor to a summing amplifier or an integrator is much larger than all other input resistors associated with the amplifier, the implication is that the term applied to the input in question contributes little to the output of the summer or integrator. In the case of time-scale selection, an inappropriate choice is usually reflected by unreasonable resistor values, capacitor values, or both associated with integrators.

The Van der Pol equation simulated earlier (Eqn 12.38) is used as a simple example of time and amplitude scaling. For the range of initial conditions used previously and with $\mu = 0.25$, the maximum magnitudes of x and dx/dt are approximately 3 and 3 sec^{-1}, respectively, while the maximum magnitude of d^2x/dt^2 is slightly greater than 3 sec^{-2}. Accordingly, if 10-volt maximum amplifier outputs are assumed, scale factors of 3 volts per unit for x and dx/dt, combined with a scale factor of 2 volts per unit for d^2x/dt^2 are reasonable. If Eqn. 12.39 is rewritten using these scale factors, we obtain

$$2\frac{d^2x}{dt^2} = -\frac{2}{27}\mu(3x)^2\left(3\frac{dx}{dt}\right) + \frac{2}{3}\mu\left(3\frac{dx}{dt}\right) - \frac{2}{3}(3x) \tag{12.44}$$

The simulation diagram, again assuming that multipliers with outputs equal to the product of the inputs divided by 10 are used, is shown in Fig. 12.16. It has also been assumed in forming this diagram that a voltage proportional to d^2x/dt^2 is required. Note that the input signals applied to the first amplifier are negatives of the right-hand side of Eqn. 12.44 because of the inversion associated with this amplifier. The transfer function of the first integrator is $-(3/2s)$ so that it provides an output of $-3(dx/dt)$ when driven with $2(d^2x/dt^2)$. Alternate scaling may be advantageous if different values of μ are used to keep the maximum magnitudes of the voltages proportional to dx/dt and d^2x/dt^2 at optimum levels.

If a value of $RC = 1$ second is used, the solution will run at real time, and the oscillation frequency will be about one radian per second. Changing this product will time scale the solution. For example, the use of $RC = 1$ ms results in limit-cycle oscillation at approximately 1000 radians per second.

12.3.3 Ancillary Circuits

There are several interesting circuit configurations that are frequently employed in analog computation and that also can be used in other more general applications.

One of these topologies is the three-mode integrator. We have seen that it is necessary to apply initial conditions to integrators in order to obtain complete (homogeneous plus driven) solutions for simulated differential

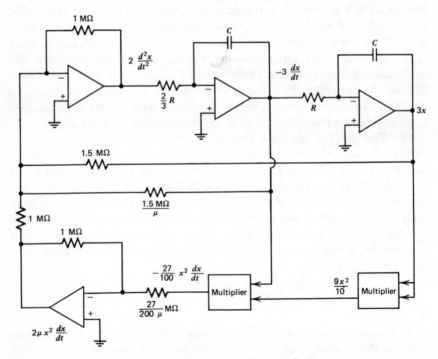

Figure 12.16   Scaled simulation of Van der Pol's equation.

equations. Another useful computing mode results if all integrators are simultaneously switched to a state where their outputs become time invariant and thus hold the values that were present at the switching time. The values of problem variables at the switching time can then be determined accurately with a digital voltmeter.

The three-mode integrator shown in Fig. 12.17 permits application of initial conditions and allows holding an output voltage in addition to functioning as an integrator. The reset (or initial condition), operate, and hold modes are selected by appropriate choice of switch positions. With switch ① open and switch ② closed, the amplifier closed-loop transfer function is

$$\frac{V_o(s)}{V_a(s)} = -\frac{1}{R_2Cs + 1} \tag{12.45}$$

If v_A is time invariant in this mode, the capacitor will charge so that the output voltage eventually becomes the negative of v_A. The capacitor voltage can then provide initial conditions for subsequent operations.

If switch ① is closed and switch ② is open, the amplifier integrates v_B in the usual fashion.

With both switches open, capacitor current is limited to operational-amplifier input current and capacitor self-leakage; thus capacitor voltage is ideally time invariant.

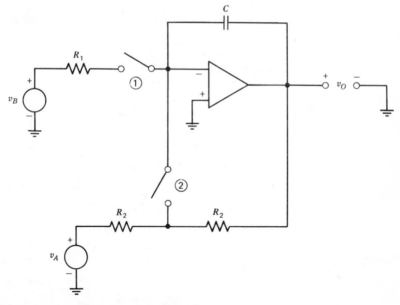

Figure 12.17   Three-mode integrator.

The required reset time of the connection shown in Fig. 12.17 can be quite long if reasonable values are used for the resistors labeled R_2. The use of a second operational amplifier connected as a voltage follower and supplying a low-resistance drive for the inverting input of the integrator can substantially shorten reset times. A practical three-mode integrator circuit that incorporates this feature is shown in Fig. 12.18.

The bipolar-transistor drivers are compatible with T^2L logic signals, and drive the gate potential of field-effect-transistor switches to ground on inputs that exceed two diode forward voltages. With a high level for the "operate" signal and the "reset" signal at ground, Q_1 is on and Q_2 is off. This combination puts the circuit in the normal integrating mode. FET Q_1 has a drain-to-source on resistance of approximately 25 ohms, and this value is compensated for by reducing the integrating-resistor size by a

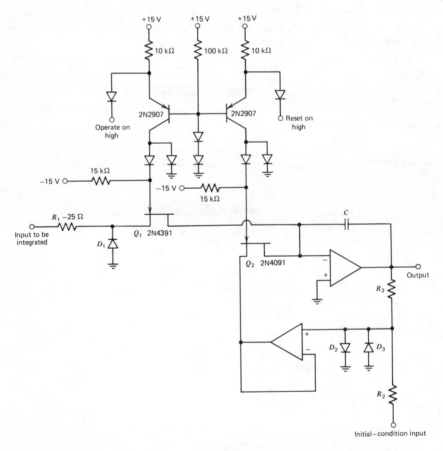

Figure 12.18   Circuit for three-mode integrator.

corresponding amount. Diode D_1 does not conduct significant current in this state. Diodes D_2 and D_3 keep the output of the follower within approximately 0.6 volt of ground. One benefit of this clamping is that the source of Q_2 cannot become negative enough to initiate conduction with its gate at -15 volts, since the maximum pinchoff voltage of the 2N4391 is 10 volts. Clamping the follower input level also keeps its signal levels near those anticipated during reset thus avoiding long slewing periods when the circuit is switched to apply initial conditions.

With the gate of Q_1 at -15 volts (corresponding to a low level on the "operate" control line), diode D_1 prevents source potentials that would initiate conduction of transistor Q_1. If Q_2 is on, the output voltage is driven toward the negative of the initial-condition input-signal level. The details of the transient for a large error depend on diode, FET, and amplifier characteristics. As the error signal becomes smaller, the reset loop enters its linear operating region. The reader should convince himself that the linear-region transmission of the reset loop (assuming ideal operational amplifiers) is $-1/2r_{ds}Cs$, where r_{ds} is the incremental drain-to-source on resistance of the FET. Thus the low FET resistance, rather than R_2, determines linear-region dynamics.

The hold mode results with both the "operate" and the "reset" signals at ground so that both FET's are off. In this state the current supplied to the capacitor is determined by FET leakage and amplifier input current.

One application for this type of circuit in addition to its use in analog computation is as a sample-and-hold circuit. In this case the operate switch is not needed, and the circuit is switched from sampling the negative of an input voltage to hold with Q_2.

Sinusoidal signals are frequently used as test inputs in analog-computer simulations. A quadrature oscillator that includes limiting and that is easily assembled using components available on most analog computers is shown in Fig. 12.19. The diagram implies a simulated differential equation, prior to limiting, of

$$ -R^2C^2 \frac{d^2v_O}{dt^2} = -\frac{RC}{K}\frac{dv_O}{dt} + v_O \tag{12.46} $$

We recognize this equation as a linear, second-order differential equation with $\omega_n = 1/RC$ and $\zeta = -1/2K$. The value of K is chosen small enough to guarantee oscillation with anticipated capacitor losses and amplifier imperfections, thus insuring that signal amplitudes will be determined primarily by the diode-resistor networks shown.

A precisely known voltage reference is required in many simulations to apply constant input signals, provide initial-condition voltages, function as a bias level for nonlinearities, or for other purposes. Voltage references are also used regularly in a host of applications unrelated to analog simulation.

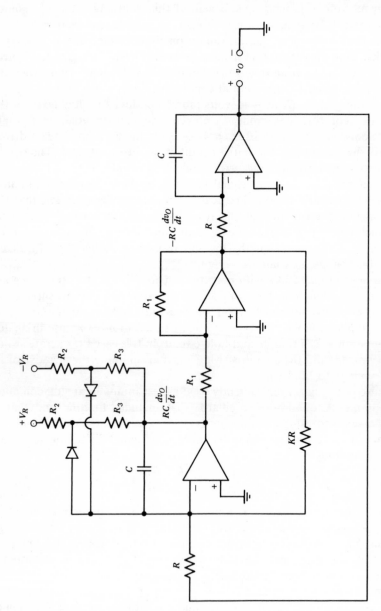

Figure 12.19 Quadrature oscillator with limiting.

The circuit shown in Fig. 12.20 is a simple yet highly stable voltage reference. The operational amplifier is connected for a noninverting gain of slightly more than 1.5 so that a 10-volt output results with 6.4 volts applied to the noninverting amplifier input.

With the topology as shown, the voltage across the resistor connected from the amplifier output to its noninverting input is constrained by the amplifier closed-loop gain to be $0.562\,V_Z$ where V_Z is the forward voltage of the Zener diode. The current through this resister is the bias current applied to the Zener diode. Zener-diode current is thus established by the stable value of the Zener voltage itself. The Zener output resistance does not deteriorate voltage regulation since the diode is operated at constant current in this connection. The filter following the Zener diode helps to attentuate noise fluctuations in its output voltage.

An emitter follower is included inside the operational-amplifier loop to increase output current capacity (current limiting circuitry as discussed in Section 8.4 is often a worthwhile precaution) and to lower output impedance, particularly at higher frequencies. While the low-frequency output impedance of the circuit would be small even without the follower because of feedback, this impedance would increase to the amplifier open-loop output impedance at frequencies above crossover. The emitter follower reduces open-loop output impedance to improve performance when pulsed or high-frequency load-current changes are anticipated. A shunt capacitor at the output may also be used to lower high-frequency output impedance. (See Section 5.2.2.)

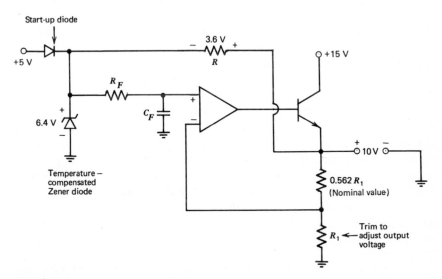

Figure 12.20   Voltage reference.

The bootstrapping used to excite the Zener diode is of course a form of positive feedback and would deteriorate performance if the magnitude of this feedback approached unity. The low-frequency transmission of the positive feedback loop is

$$L = 1.562 \frac{r_d}{R + r_d} \tag{12.47}$$

where r_d is the incremental resistance of the Zener diode. This expression is evaluated using parameters for a 1N829A, a temperature-compensated Zener diode. The diode is designed for an operating current of 7.5 mA, and thus R will be approximately 500 Ω. The incremental resistance of the diode is specified as a maximum of 10 Ω. Thus the loop transmission is, from Eqn. 12.47, 0.03. This small amount of positive feedback does not significantly affect performance.

The positive feedback can result in the circuit operating with the diode in its forward-conducting state rather than its normal reverse-breakdown mode. This state, which leads to a negative output of approximately one volt, can be eliminated with the start-up diode shown. The start-up diode insures that the Zener diode is forced into its reverse region, but does not contribute to Zener current under normal operating conditions.

The expected operational-amplifier imperfections have relatively little effect on the overall performance of the reference circuit. A value of 30,000 for supply-voltage rejection ratio (typical for integrated-circuit amplifiers) causes a change in output voltage of approximately 50 μV per volt of supply change. (This 33 μV/V sensitivity is amplified by the closed-loop gain of 1.5.) The typical input-voltage drift for many inexpensive operational amplifiers is the order of 5 μV per degree Centigrade. This figure is not significant compared to the temperature coefficient of 5 parts per million per degree Centigrade or approximately 32 μV per degree Centigrade of a high-quality Zener diode such as the 1N829A.

The designers of the large analog computers that evolved during the period from the early 1950s to the mid-1960s often devoted almost fanatical effort to achieving high static accuracy in their computing elements. Toward this end, operational amplifiers were surrounded with high-precision wire-wound resistors and capacitors that could be accurately trimmed to desired values. These passive components were often placed in temperature-stable ovens to eliminate variations with ambient temperature.

The low-frequency errors (particularly input voltage offset) characteristic of vacuum-tube operational amplifiers were largely eliminated by means of an imaginative technique known as *chopper stabilization*.[5] This method

[5] E. A. Goldberg, "Stabilization of Wide-Band Direct Current Amplifiers for Zero and Gain," *RCA Review*, Vol. II, No. 2, June 1950, pp. 296–300.

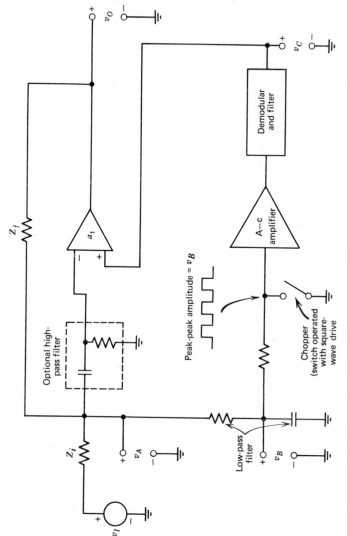

Figure 12.21 Chopper-stabilized amplifier.

is still incorporated into some modern operational-amplifier designs, and it provides a way of reducing the voltage drift and input current of an amplifier to vanishingly small levels. The usual implementation of this technique can be viewed as an extreme example of feedforward (see Section 8.2.2) and thus results in an amplifier that can only be used in inverting connections.

Figure 12.21 illustrates the concept. Assume that the optional network is eliminated so that the junction of Z_f and Z_i is connected directly to the inverting input of the top amplifier. The resulting connection clearly functions as an inverting amplifier if the voltage v_C is zero. Observe that one necessary condition for the amplifier closed-loop gain to be equal to its ideal value is that $v_A = 0$. The objective of chopper stabilization is to reduce v_A to nearly zero by applying an appropriate signal to the noninverting input of the top amplifier.

The d-c component of the voltage v_A is determined with a low-pass filter, and this component (v_B) is "chopped" (converted to a square wave with peak-to-peak amplitude v_B) using a periodically operated switch. (Early designs used vibrating-reed mechanical switches, while more modern units often use periodically illuminated photoresistors or field-effect transistors as the switch.) The chopped a-c signal can be amplified without offset by an a-c amplifier and demodulated to produce a signal v_C proportional to v_B. If the gain of the a-c amplifier is high, the low-frequency gain $v_C/v_A = a_{02}$ will be high. If a_{02} is negative, the signal applied to the positive gain input of the top amplifier will be of the correct polarity to drive v_A toward zero. Arbitrarily small d-c components of v_A can theoretically be obtained by having a sufficiently high magnitude for a_{02}, although in practice achievable offsets are limited by errors such as thermally induced voltages in the switch itself. The low-pass filter is necessary to prevent sampling errors that arise if signals in excess of half the chopping frequency are applied to the chopper.

An alternative way to view the operation of a chopper-stabilized amplifier is to notice that high-frequency signals pass directly through the top amplifier, while components below the cutoff frequency of the low-pass filter are amplified by both the bottom amplifier and the top amplifier in cascade. (It is interesting to observe that low-frequency open-loop gain magnitudes in excess of 10^9 have been achieved in this way.) It is therefore not necessary to apply low-frequency signals directly to the top amplifier, and a high-pass filter (shown as the optional network) can be included in series with the inverting input of the top amplifier. As a result, both voltage offset and input current to the operational amplifier can be reduced by chopper stabilization, yielding an amplifier with virtually ideal low-frequency characteristics.

Several manufacturers offer packages that combine discrete-component choppers with integrated-circuit amplifiers. More recently, integrated-

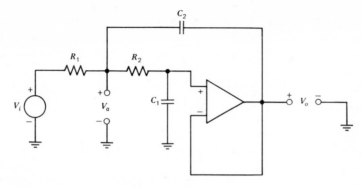

Figure 12.22   Second-order low-pass active filter.

circuit manufacturers have been able to fabricate complete chopper-stabilized amplifiers either in monolothic form or by combining several monolithic chips to form a hybrid circuit. These circuits incorporate topological improvements that permit true differential operation. The large capacitors required are connected externally to the package. Drifts of a fraction of a microvolt per degree Centigrade, coupled with input currents in the picoampere range, are available at surprisingly low cost.

12.4   ACTIVE FILTERS

There are numerous applications that require the realization of a particular transfer function. One of the many limitations of the design of filter networks using only passive components is that inductors are required to obtain complex pole locations. This restriction is removed if active elements are included in the designs, and the resultant *active filters* permit the realization of complex poles using only resistors and capacitors in addition to the active elements. Further advantages of active-filter synthesis include the possibility of a wide range of relative input and output impedances, and the use of smaller, less expensive reactive components than is normally possible with passive designs.

There is a fair amount of present research devoted toward improving techniques for active-filter synthesis, and the probability is that better designs, particularly with respect to *sensitivity* (the dependence of the transfer function on variations in parameter values), will evolve. This section describes two presently popular topologies that can be used to realize active filters.

12.4.1   The Sallen and Key Circuit[6]

Figure 12.22 shows an active-filter circuit that uses a unity-gain-connected operational amplifier. Node equations for the circuit are easily

---

[6] R. P. Sallen and E. L. Key, "A Practical Method of Designing RC Active Filters," Institute of Radio Engineers, *Transactions on Circuit Theory*, March, 1955, pp. 74–85.

written by noting that the voltage at the noninverting input of the amplifier is equal to the output voltage and are

$$G_1 V_i = (G_1 + G_2 + C_2 s) V_a - (G_2 + C_2 s) V_o \qquad (12.48)$$

$$0 = -G_2 V_a + (G_2 + C_1 s) V_o$$

Solving for the transfer function yields

$$\frac{V_o(s)}{V_i(s)} = \frac{1}{R_1 R_2 C_1 C_2 s^2 + (R_1 + R_2) C_1 s + 1} \qquad (12.49)$$

This equation represents a second-order transfer function with standard-form parameters

$$\omega_n = \frac{1}{\sqrt{R_1 R_2 C_1 C_2}} \qquad (12.50)$$

and

$$\zeta = \frac{R_1 + R_2}{2\sqrt{R_1 R_2}} \sqrt{\frac{C_1}{C_2}} \qquad (12.51)$$

Since only two quantities are required to characterize the second-order filter, the four degrees of freedom represented by the four passive-component values are redundant. Part of this redundancy is frequently eliminated by choosing $R_1 = R_2 = R$. In this case, the standard-form parameters become

$$\omega_n = \frac{1}{R\sqrt{C_1 C_2}} \qquad (12.52)$$

and

$$\zeta = \sqrt{\frac{C_1}{C_2}} \qquad (12.53)$$

The addition of another section to the second-order low-pass active filter as shown in Fig. 12.23 allows the synthesis of a third-order transfer function with a single amplifier. If equal-value resistors are used as shown, the transfer function is

$$\frac{V_o(s)}{V_i(s)} = \frac{1}{C_1 C_2 C_3 R^3 s^3 + 2(C_1 C_3 + C_2 C_3) R^2 s^2 + (C_1 + 3C_3) R s + 1} \qquad (12.54)$$

An nth-order low-pass filter is often designed by combining $n/2$ second-order sections in the case of n even, or one third-order section with $n/2 - 3/2$ second-order sections when n is odd. Tables[7] that simplify

[7] Farouk Al-Nasser, "Tables Speed Design of Low-Pass Active Filters," *EDN*, March 15, 1971, pp. 23–32.

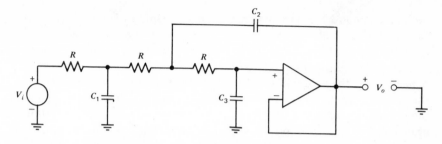

Figure 12.23   Third-order low-pass active filter.

element-value selection are available for filters up to the tenth order with a number of different pole patterns.

Interchanging resistors and capacitors as shown in Fig. 12.24 changes the second-order low-pass filter to a high-pass filter. The transfer function for this configuration is

$$\frac{V_o(s)}{V_i(s)} = \frac{R_1 R_2 C_1 C_2 s^2}{R_1 R_2 C_1 C_2 s^2 + R_2(C_1 + C_2)s + 1} \tag{12.55}$$

If, in a development analogous to that used for the low-pass filter, we choose $C_1 = C_2 = C$, Eqn. 12.55 reduces to

$$\frac{V_o(s)}{V_i(s)} = \frac{s^2/\omega_n^2}{(s^2/\omega_n^2) + (2\zeta s/\omega_n) + 1} \tag{12.56}$$

where

$$\omega_n = \frac{1}{C\sqrt{R_1 R_2}}$$

and

$$\zeta = \sqrt{\frac{R_2}{R_1}}$$

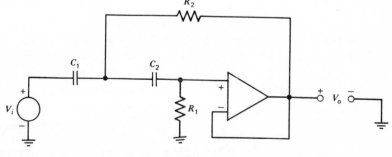

Figure 12.24   Second-order high-pass active filter.

The Sallen and Key circuit can be designed with an amplifier gain other than unity (see Problem P12.8). This modification allows greater flexibility, since the low- or high-frequency gain of the circuit can be made other than one. However, the damping ratio of transfer functions realized in this way is dependent on the values of resistors that set the closed-loop amplifier gain; thus poles may be somewhat less reliably located. A further advantage of the unity-gain version is that it may be constructed using the LM110 integrated circuit (see Section 10.4.4). The bandwidth of this amplifier far exceeds that of most general-purpose integrated-circuit units, and corner frequencies in the low megahertz range can be obtained using it.

12.4.2 A General Synthesis Procedure

The Sallen and Key configuration, together with many other active-filter topologies, allows complete freedom in the choice of pole location, but does not permit arbitrary placement of transfer-function zeros. The application of the analog-computation concepts described in Section 12.3.1 allows the synthesis of any realizable transfer function that is expressable as a ratio of polynomials in s, provided that the number of poles is equal to or greater than the number of zeros in the transfer function.

Consider the transfer function

$$\frac{V_o(s)}{V_i(s)} = \frac{b_n s^n + b_{n-1} s^{n-1} + \cdots + b_1 s + b_0}{a_n s^n + a_{n-1} s^{n-1} + \cdots + a_1 s + a_0} \tag{12.57}$$

The first step is to introduce an intermediate variable $V_a(s)$ such that $V_a(s)/V_i(s)$ contains only the poles of the transfer function, or

$$\frac{V_a(s)}{V_i(s)} = \frac{1}{a_n s^n + a_{n-1} s^{n-1} + \cdots + a_1 s + a_0} \tag{12.58}$$

Proceeding in a way exactly parallel to the time-domain development of Section 12.3.1, we write

$$s^n V_a(s) = -\frac{a_{n-1}}{a_n} s^{n-1} V_a(s) - \cdots - \frac{a_1}{a_n} s V_a(s) - \frac{a_0}{a_n} V_a(s) + \frac{V_i(s)}{a_n} \tag{12.59}$$

The block-diagram representation of Eqn. 12.59 is shown in Fig. 12.25. This block diagram can be readily implemented using summers and integrators. In order to complete the synthesis of our transfer function (Eqn. 12.57) we recognize that

$$V_o(s) = V_a(s)(b_n s^n + b_{n-1} s^{n-1} + \cdots + b_1 s + b_0) \tag{12.60}$$

The essential feature of Eqn. 12.60 is that it indicates $V_o(s)$ is a linear combination of $V_a(s)$ and its first n derivatives. Since all of the necessary vari-

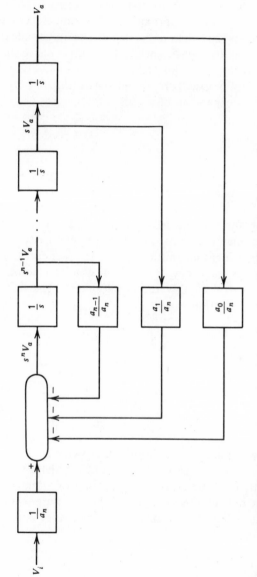

Figure 12.25 Block diagram representation of transfer function that contains only poles.

529

ables appear in the block diagram, $V_o(s)$ can be generated by simply scaling and summing these variables, without the need for differentiation.

This synthesis procedure is illustrated for an approximation to a pure time delay known as the Padé approximate. The time delay has a transfer function $e^{-s\tau}$, where τ is the length of the delay. The magnitude of this transfer function is one at all frequencies, while its negative phase shift is linearly proportional to frequency. The time delay has an essential singularity at the origin, and thus cannot be exactly represented as a ratio of polynomials in s.

The Taylor's series expansion of $e^{-s\tau}$ is

$$e^{-s\tau} = 1 - s\tau + \frac{s^2\tau^2}{2!} - \cdots + \cdots + (-1)^m \frac{s^m\tau^m}{m!} + \cdots \quad (12.61)$$

The Padé approximates locate an equal number of poles and zeros so as to agree with the maximum possible number of terms of the Taylor's series expansion. This approximation always leads to an *all-pass network* that has right-half-plane zeros and left-half-plane poles located symmetrically with respect to the imaginary axis. This type of singularity pattern results in a frequency-independent magnitude for the transfer function.

Since we can always frequency or time scale at a later point, we consider a unit time delay e^{-s} to simplify the development. The first-order Padé approximate to this function is

$$P_1(s) = \frac{1 - (s/2)}{1 + (s/2)} = 1 - s + \frac{s^2}{2} - \frac{s^3}{4} + \cdots - \cdots + \quad (12.62)$$

The expansion for e^{-s} is

$$e^{-s} = 1 - s + \frac{s^2}{2} - \frac{s^3}{6} + \frac{s^4}{24} - \frac{s^5}{120} + \frac{s^6}{720} - \cdots + \quad (12.63)$$

The first-order approximation matches the first two coefficients of s of the complete expansion, and is in reasonable agreement with the third coefficient. This match is all that can be expected, since only two degrees of freedom (the location of the pole and the location of the zero) are available for the first-order approximation. The second-order Padé approximate to a one-second time delay is

$$P_2(s) = \frac{1 - (s/2) + (s^2/12)}{1 + (s/2) + (s^2/12)}$$

$$= 1 - s + \frac{s^2}{2} - \frac{s^3}{6} + \frac{s^4}{24} - \frac{s^5}{144} + \cdots - \cdots + \quad (12.64)$$

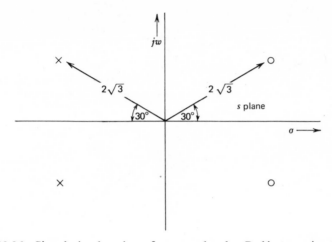

Figure 12.26 Singularity locations for second-order Padé approximate to one-second time delay.

As expected, the first four time-delay coefficients of s are matched by the approximation. The s-plane plot for $P_2(s)$ is shown in Fig. 12.26. Simple vector manipulations confirm the fact that the magnitude of this function is one at all frequencies.

The phase shift of the approximating function is (from Eqn. 12.64)

$$\angle P_2(j\omega) = 2 \angle \left[1 - \frac{j\omega}{2} + \frac{(j\omega)^2}{12} \right] = -2 \tan^{-1} \left\{ \frac{\omega}{2[1 - (\omega^2/12)]} \right\} \quad (12.65)$$

This function is compared with an angle of $-57.3°\omega$ (the value for a one-second time delay) in Fig. 12.27. We note excellent agreement to frequencies of approximately 2 radians per second implying that the approximation represents the actual function well for sinusoidal excitation to this frequency, with increasing discrepancy at higher frequencies. The error reflects the fact that the maximum negative phase shift of the Padé approximate is 360°, while the time delay provides unlimited negative phase shift at sufficiently high frequency.

Synthesis is initiated by defining an intermediate variable $V_a(s)$ in accordance with Eqns. 12.58 and 12.59, or

$$\frac{V_a(s)}{V_i(s)} = \frac{1}{(s^2/12) + (s/2) + 1} \quad (12.66)$$

and

$$s^2 V_a(s) = -6s V_a(s) - 12 V_a(s) + 12 V_i(s) \quad (12.67)$$

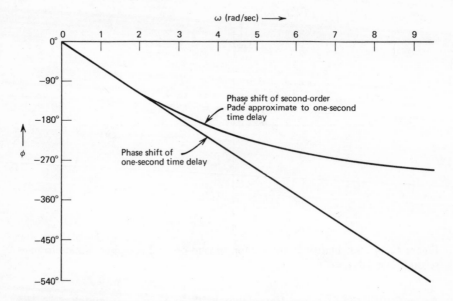

Figure 12.27 Comparison of time delay and Padé approximate phase characteristics.

The output voltage is

$$V_o(s) = \frac{s^2}{12} V_a(s) - \frac{s}{2} V_a(s) + V_a(s) \qquad (12.68)$$

The operational-amplifier synthesis shown in Fig. 12.28 provides the required transfer function if $RC = 1$ second. The reader should convince himself that the liberties taken with inversions and various resistor values do in fact lead to the desired relationship.

Anticipated amplitudes depend on the input-signal level and its spectral content. For example, if a step is applied to the input of the circuit, the magnitude of the signal out of the first amplifier must initially be 12 times as large as the step amplitude, since the outputs of the integrators cannot change instantaneously to subtract from the input-signal level. Note, however, that the input-to-output transfer function of the circuit remains the same for any values of $R_1 = R_2$. If, for example, 10-V step changes are expected at the input, selection of $R_1 = R_2 = 120$ kΩ will limit the signal level at the output of the first amplifier to 10 volts while maintaining the correct input-to-output gain.

The circuit shown in Fig. 12.28 was constructed using $R = 100$ kΩ and $C = 0.01$ μF, values resulting in an approximation to a 1-ms time delay. This choice of time scale is convenient for oscilloscope presentation. The

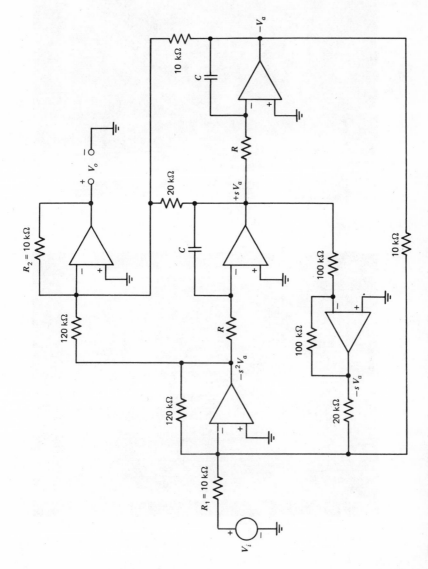

Figure 12.28 Synthesis of second-order Padé approximate to a time delay.

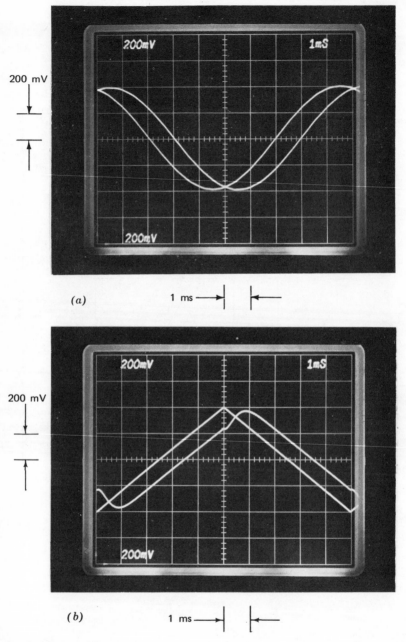

Figure 12.29 Input and output signals for second-order Padé approximate to a 1-ms time delay. (*a*) Sine-wave excitation. (*b*) Triangular-wave excitation. (*c*) Square-wave excitation.

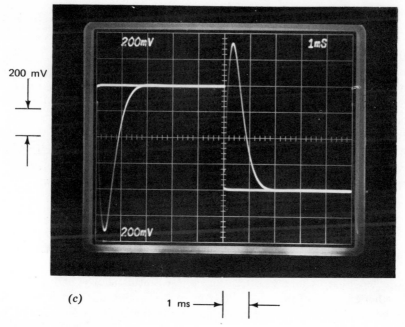

(c)

1 ms

Figure 12.29—Continued

input and output signals for 100-Hz sine-wave excitation are shown in Fig. 12.29a. The time delay between these two signals is 1 ms to within instrumentation tolerances. This performance reflects the prediction of Fig. 12.27, since good agreement to 2000 rad/sec or 300 Hz is anticipated for the approximation to a 1-ms delay.

Input and output signals for 100-Hz triangular-wave excitation are compared in Fig. 12.29b. The triangular wave contains only odd harmonics, and these harmonics fall off as the square of their frequency. Thus the amplitude of the third harmonic of the triangular wave is approximately 11% of the amplitude of the fundamental, the amplitude of the fifth harmonic is 4% of the fundamental, while higher harmonics are further attentuated. We notice that the circuit does very well in approximating a 1-ms time delay most of the time. The aberration that results immediately following a change in slope reflects the inability of the circuit to provide proper phase shift to the higher-frequency components.

The performance of the circuit when excited with an 100-Hz square wave is shown in Fig. 12.29c. The relatively poorer behavior in the vicinity of a transition in this case results from the higher harmonic content of the square wave. (Recall that the square wave contains odd harmonics that fall off only as the first power of the frequency.)

12.5 FURTHER EXAMPLES

It was mentioned in the introduction to Chapter 11 that the objective of the application portion of this book was to illustrate concepts for design rather than to provide specific, detailed examples in the usually futile hope that the reader could apply them directly to his own problems. Successful design almost always involves combining bits and pieces, a concept here, a topology there, to ultimately arrive at the optimum solution. In this section we will see how some of the ideas introduced earlier are combined into relatively more sophisticated configurations. The three examples that are presented are all "real world" in that they reflect actual requirements that the author has encountered recently in his own work.

12.5.1 A Frequency-Independent Phase Shifter

There are a number of operational-amplifier connections, such as the approximation to a time delay described in the previous section, that have a transfer-function magnitude independent of frequency combined with specified phase characteristics. The phase shifter shown in Fig. 12.30 is another example of this type of circuit. We recognize this circuit as a differential-amplifier connection, and thus realize that its transfer function is

$$\frac{V_o(s)}{V_i(s)} = \left(\frac{2RCs}{RCs + 1} - 1 \right) = \frac{RCs - 1}{RCs + 1} \tag{12.69}$$

This transfer function (which is the negative of a first-order Padé approximate to a time delay of $2RC$ seconds) produces a phase shift that varies from $-180°$ at low frequencies to $0°$ at high frequencies. If a potentiometer or a field-effect transistor is used for R, the phase shift can be manually or electronically varied.

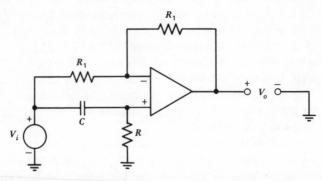

Figure 12.30 Adjustable phase shifter.

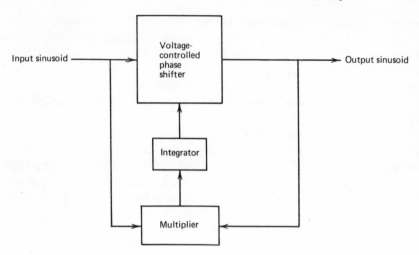

Figure 12.31   Constant phase shifter using a phase detector.

One technique for converting resolver[8] signals to digital form requires that a fixed 90° phase shift be applied to a sinusoidal signal with no change in its amplitude. The frequency of the signal to be phase shifted may change by a few percent. Unfortunately, there are no finite-polynomial linear transfer functions that combine frequency-independent magnitude characteristics with a constant 90° phase shift. While approximating functions do exist over restricted frequency ranges, the arc-minute phase-shift constancy required in this application precluded the use of such functions. We note that since a very specific class of input signals (single-frequency sinusoids) is to be applied to the phase shifter, linearity may not be a necessary constraint. Nonlinear circuits, in spite of our inability to analyze them systematically, often have very interesting properties.

Consider the configuration shown diagrammatically in Fig. 12.31 as a possible solution to our problem. In this circuit, an all-pass phase shifter with a voltage-variable amount of phase shift is the central element. The circuit shown in Fig. 12.30 with a field-effect transistor used for the resistor R can perform this function. The multiplier is used as a phase detector. If the magnitude of the phase shift between the input and output signals is less than 90°, the average value of the multiplier output will be positive, while if this magnitude is between 90° and 180°, the average multiplier output signal will be negative. The integrator, which provides the control

---

[8] A resolver is basically a transformer with a primary-to-secondary coupling that can be varied by mechanically changing the relative alignment of these windings. This device is used as a rugged and highly accurate mechanical-angle transducer.

voltage for the FET in the phase shifter, filters the second harmonic that results from the multiplication and supplies the loop gain necessary to keep the average value of the multiplier output at zero, thus forcing a 90° phase shift between input and output signals. Although the circuit described above can result in moderate accuracy, a detailed investigation indicated that meeting the required specifications probably was not practical with this topology.

It is worth noting that while the basic approach described above was not used in this case, it is a valuable technique that has a number of interesting and useful variations. For example, the phase shift of a second-order high- or low-pass active filter is ±90° when excited at its corner frequency. Tracking filters can be realized by replacing the fixed resistors in an active filter with voltage-controlled resistors and using a phase comparison to locate the corner frequency of the filter at its excitation frequency.

In some applications, other types of phase detectors are used. One possibility involves high-gain limiters that produce square waves with zero crossing synchronized to those of the sine waves of interest. The duty cycle of an exclusive OR gate operating on the square waves indicates the relative phase of the original signals.

The previous circuit combined an all-pass network that provides a transfer-function magnitude that is independent of frequency with feedback which forces 90° of phase shift at the operating frequency. An alternative approach is to combine a network that provides 90° of phase shift at all frequencies (an integrator) with feedback that forces its gain magnitude to be one at the operating frequency.

The circuit that evolved to implement the above concept is shown in only slightly simplified form in Fig. 12.32. The signal integrator provides the required 90° of phase shift. Its scale factor is adjusted by means of the field-effect transistor so that a gain magnitude of one is provided at frequencies close to the nominal operating value of 400 Hz. Half of the drain-to-source voltage of the field-effect transistor is applied to its gate to linearize the drain-to-source resistance as described in Section 12.1.4. The unity-gain buffer amplifier prevents current flowing through the FET-gate network from being integrated. The capacitor in series with the signal-integrator input resistor and the resistor shunting the integrating capacitor are required to keep this integrator from saturating as a consequence of input voltage offset and bias current. While they change the ideal phase shift by a total of approximately eight arc minutes, this value is trimmed out along with other phase-shift errors with a network (not shown) following the integrator.

The two full-wave precision-rectifier connections combine with the loop-gain integrator to provide an average current into the capacitor of this

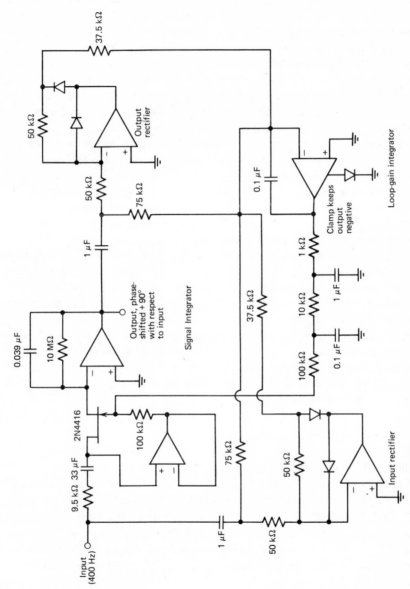

Figure 12.32 Precision phase shifter with amplitude control.

integrator that is proportional to the difference between the magnitudes of the input and output signals. If, for example, the output-signal magnitude exceeds the input-signal magnitude, the voltage out of the loop-gain integrator is driven negative. This action increases the incremental resistance of the FET, thus decreasing the signal-integrator scale factor and lowering the magnitude of the output signal. The inputs to the precision rectifiers are a-c coupled so that d-c components of these signals do not influence the rectifier output signals. A two-pole low-pass filter follows the loop-gain integrator to further filter harmonics that would degrade signal-integrator performance.

The maximum positive output level of the loop-gain integrator is clamped via an internal node to a maximum output level of zero volts in order to eliminate a latch-up mode. If this voltage became positive, the FET would conduct gate current, and this current could cause the signal-integrator output to saturate. As a result, the a-c component of the signal-integrator output would be eliminated, and the loop, in an attempt to restore equilibrium, would drive the output of the loop-gain integrator further positive. The diode clamp prevents initiation of this unfortunate chain of events.

The circuit shown in Fig. 12.32 has been built and tested at operating frequencies between 395 and 405 Hz over the temperature range of 0° to 50° Centigrade. (The feedback also eliminates the effects of signal-integrator component-value changes with temperature.) The input- and output-signal amplitudes remain equal within 1 mV peak-to-peak at any input-signal level up to 20 volts peak-to-peak. The phase shift of the circuit with a 20-volt peak-to-peak input remains constant within one arc minute. While the actual phase shift is not precisely 90°, the constant component of the phase error can be trimmed out as described earlier.

12.5.2   A Sine-Wave Shaper

We have discussed certain aspects of a function-generator circuit that combines an integrator and a Schmitt trigger to produce square and triangle waves in Sections 6.3.3 and 12.2.1. Commercial versions of this circuit usually also provide a sine-wave output that is synthesized by the seemingly improbable method of shaping the triangle wave with a piecewise-linear network. This technique is practical because of the ease of generating variable-frequency triangular waves, and because the use of relatively few segments in the shaping network gives surprisingly good sine-wave fidelity.

Part of the design problem is to determine how the characteristics of the shaping network should be chosen to best approximate a sine wave. The parameters that define the network are shown in Fig. 12.33. A total of n break points are located over the input-variable range of 0° to 90°. The

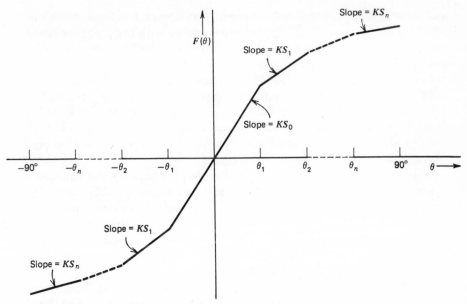

Figure 12.33 Piecewise-linear network characteristics.

slope of the input-output transfer relationship is KS_m between $\theta = \theta_m$ and $\theta = \theta_{m+1}$. The multiplying constant K reflects the fact that only relative slopes are important, since a multiplicative change in all slopes changes only the magnitude of the input-output transfer characteristics. The symmetry of the transfer characteristics about the origin insures that the output signal will have no d-c component and will contain no even harmonics when a zero-average-value triangular signal is used as the input.

The network specification involves the choice of n values of θ (the breakpoint locations) and $n + 1$ relative slopes. It can be shown that if the θ's are selected such that

$$\theta_m = \frac{m\ 180°}{2n + 1} \qquad 0 \leq m \leq n \qquad (12.70)$$

and slopes selected as

$$S_m = \sin \theta_{m+1} - \sin \theta_m \qquad 0 \leq m < n \qquad (12.71a)$$

$$S_m = 0 \qquad m = n \qquad (12.71b)$$

the first n odd harmonics will be eliminated from the output signal.

The decision to use four break points in the realization of the sine shaper was based on two considerations. With this number of break points, output distortion resulting from imprecise break-point locations and slope values is comparable to the distortion associated with the piecewise-linear approximation unless expensive components are used to establish these parameters. Furthermore, an inexpensive integrated-circuit five-diode array is available. This matched-diode array can be used for the four break points, with the fifth diode providing temperature compensation as described in material to follow. Equations 12.70 and 12.71 evaluated for $n = 4$ suggest break points located at input-variable values of 20°, 40°, 60°, and 80°, with relative segment slopes (normalized to a minimum nonzero slope of one) of 2.879, 2.532, 1.879, 1, and 0, respectively.

With the transfer characteristics of the shaping network determined, it is necessary to design the circuit that synthesizes the required function. The discussion of Section 11.5.3 mentioned the use of superdiode connections to improve the sharpness of break points compared to that which can be achieved with diodes alone. This technique was not used for the sine shaper, since the rounding associated with the normal diode forward characteristics actually improves the quality of the fit to the sine curve.

The compressive type nonlinearities described in Section 11.5.3 were realized using diodes to increase the feedback around an operational amplifier, thus reducing its incremental closed-loop gain when a break-point level was exceeded. An alternative is to use diodes to decrease the drive signal applied to the amplifier to lower incremental gain. This approach simplifies temperature compensation. The topology used is shown in Fig. 12.34.

The input-signal level of 20 volts peak-to-peak corresponds to the input variable range of ±90° shown in Fig. 12.33. Thus the break-point locations of ±20°, ±40°, ±60°, and ±80° correspond to input-voltage levels of ±2.22 volts, ±4.44 volts, ±6.67 volts, and ±8.89 volts, respectively. Resistor values are determined as follows. It is initially assumed that the diodes are ideal, in that they have a threshold voltage of zero volts, zero resistance in the forward direction, and zero conductance in the reverse direction. Assume that the R_1-R_2 path is to provide the break points at input voltages of ±8.89 volts. Since the inverting input of the operational amplifier is at ground potential, the resistor ratio necessary to make the voltage at the midpoint of these two resistors ±1.5 volts with ±8.89 volts at the input is

$$\frac{R_2}{R_1 + R_2} = \frac{1.5}{8.89} = 0.1687 \qquad (12.72)$$

The ratios of resistor pairs R_3-R_4, R_5-R_6, and R_7-R_8 are chosen in a similar way to locate the remaining break points.

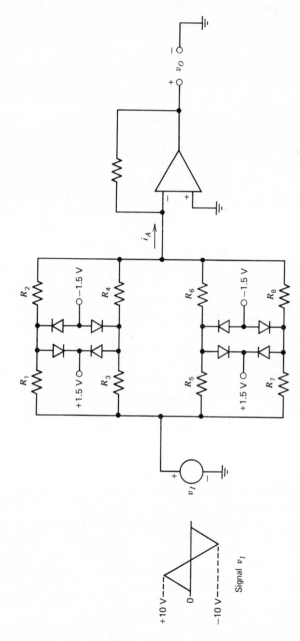

Figure 12.34 Simplified sine-wave shaper.

The relative conductances of the resistive paths between the triangular-wave signal source and the inverting input of the operational amplifier are constrained by the relative slopes of the desired transfer characteristics as follows. The closed-loop incremental gain of the connection is proportional to the incremental transfer conductance from the signal source to the current i_A defined in Fig. 12.34. With the ratio of the two resistors in each path chosen in accordance with relationships like Eqn. 12.72, the incremental transfer conductance is zero (for ideal diodes) when the input-signal magnitude exceeds 8.89 volts, increases to $1/(R_1 + R_2)$ for input-signal magnitudes between 6.67 and 8.89 volts,. increases further to $[1/(R_1 + R_2)] + [1/(R_3 + R_4)]$ for input-signal magnitudes between 4.44 and 6.67 volts, etc. If we define $1/(R_1 + R_2) = G$, realizing the correct relative slope for input-signal magnitudes between 4.44 and 6.67 volts requires

$$\cdot \quad \frac{1}{R_1 + R_2} + \frac{1}{R_3 + R_4} = 1.879G \qquad (12.73)$$

The satisfaction of Eqn. 12.73 makes the slope in this input signal range 1.879 times as large as the slope for input signals between 6.67 and 8.89 volts. Corresponding relationships couple other resistor-pair values to the R_1-R_2 pair.

The sets of equations that parallel Eqns. 12.72 and 12.73, together with the selection of any one resistor value, determine resistors R_1 through R_8. The general resistance level set by choosing the one free resistor value is selected based on loading considerations and to insure that stray capacitance does not deteriorate dynamic performance.

The circuit used for the sine shaper (Fig. 12.35) uses the standard 1%-tolerance resistor values that best approximate calculated values. The five diodes labeled A and those labeled B are from two CA3039 integrated-circuit diode arrays. One member of each array modifies the bias voltages to account for the diode threshold voltages and to provide temperature compensation. The compensating diodes are operated at a current level of approximately half the maximum operating current level of the shaping diodes. While this type of compensation clearly has no effect on the conductance characteristics of the shaping diodes, the exponential diode characteristics actually improve the performance of the circuit as described earlier.

Since this circuit is intended to operate to 1 MHz (a high-speed integrated-circuit operational amplifier with a discrete-component buffer to increase output-current capacity is used), capacitors are necessary at the output of the reference-voltage amplifiers to lower their output impedance at the switching frequency of the diodes. The 1.5-V levels are derived from the

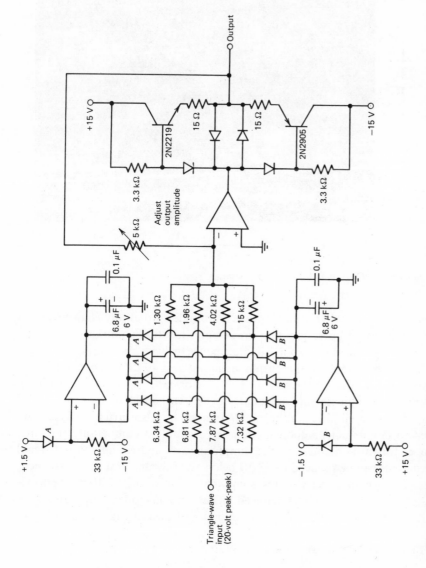

Figure 12.35 Sine-wave shaper.

545

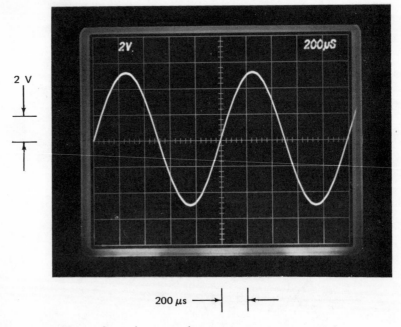

Figure 12.36 Output from sine-wave shaper.

voltages that establish triangle-wave amplitude so that any changes in this amplitude cause corresponding break-point location changes.

The circuit produces approximate sine waves with the amplitude of any individual harmonic in the output signal at least 40 dB (a voltage ratio of 100:1) below the fundamental. This performance is obtained with no trimming. If adjustments are made to null the offset of the operational .amplifier, and empirical adjustments (guided by a spectrum analyzer) are used to counteract component-value errors and to compensate for finite diode forward resistance, the amplitude of individual output-signal harmonics can be reduced to 55 dB below the fundamental at low frequencies. Performance deteriorates somewhat at frequencies above approximately 10 kHz because of reduced signal-amplifier open-loop gain.

A 1-kHz output signal from the circuit is shown in Fig. 12.36.

12.5.3 A Nonlinear Three-Port Network

The realization of a device analog that may be of value in teaching the dynamic behavior of bipolar transistors requires a three-port network

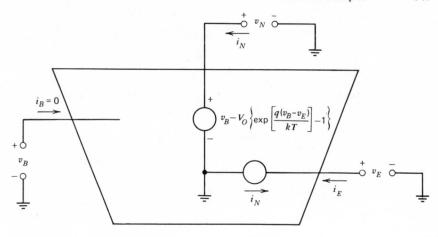

Figure 12.37 Three-port network.

defined by Fig. 12.37. The synthesis of this network is initiated by first designing a circuit that provides the relationship

$$v_N = v_B - V_O \left\{ \exp\left[\frac{q(v_B - v_E)}{kT} \right] - 1 \right\} \qquad (12.74)$$

The parameter V_O, as we might expect, is related to the quantity I_S for the transistor being simulated, and consequently a corresponding temperature dependence is desirable.

There are a number of ways to simulate Eqn. 12.74. One topology that is adaptable to further requirements is shown in Fig. 12.38. Since an eventual constraint is that the current at the v_B input be zero, a buffer amplifier is used at this terminal. The second amplifier is differentially connected with an output voltage.

$$v_A = 2v_E - v_B \qquad (12.75)$$

The third amplifier is also connected as a differential amplifier, so that

$$v_N = v_E - (i_T + i_A)R \qquad (12.76)$$

Since feedback keeps the inverting input terminal of the third amplifier at potential v_E,

$$i_A = \frac{v_A - v_E}{R} = \frac{v_E - v_B}{R} \qquad (12.77)$$

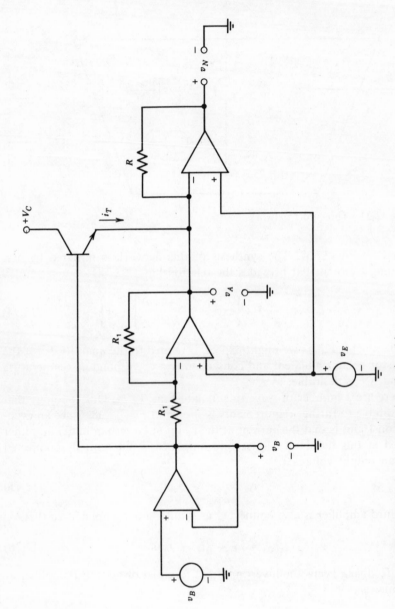

Figure 12.38 Synthesis of exponential relationship.

If we assume the usual transistor characteristics,

$$i_T \simeq I_S \left\{ \exp\left[\frac{q(v_B - v_E)}{kT} \right] - 1 \right\} \tag{12.78}$$

Substituting Eqns. 12.77 and 12.78 into Eqn. 12.76 yields the form required by Eqn. 12.75:

$$v_N = v_B - RI_S \left\{ \exp\left[\frac{q(v_B - v_E)}{kT} \right] - 1 \right\} \tag{12.79}$$

In order to complete the synthesis, it is necessary to sample the current flowing at terminal N and make the current flowing at terminal E the negative of this current. A modification of the Howland current source (see Section 11.4.3) can be used. The basic circuit with differential inputs is shown in Fig. 12.39a. (The reason for the seemingly strange input-voltage connection and the split resistor will become apparent momentarily.) The current i_O for these parameter values is

$$i_O = \frac{2(v_A - v_C)}{R} = \frac{2(v_A - v_A - v_I)}{R} = \frac{-2v_I}{R} \tag{12.80}$$

In Fig. 12.39b, the voltage source v_I and half of the split resistor are replaced with a Norton-equivalent circuit. For equivalence, it is necessary to make $i_I = 2v_I/R$. Expressing Eqn. 12.80 in terms of i_I shows

$$i_O = -i_I \tag{12.81}$$

The topology of Fig. 12.39b shows that the i_T current source can be returned to ground rather than to voltage source v_A. This modification is shown in Fig. 12.39c, the current-controlled current source necessary in our present application. Note that the output is independent of v_A, the common-mode input voltage applied to the current source.

The circuits of Figs. 12.38 and 12.39c are combined to form the three-port network as shown in Fig. 12.40. In this circuit, the feedback for the voltage v_N is taken from the output side of the current-sampling resistor so that voltage drops in this resistor do not influence v_N. It is necessary to buffer the 100-kΩ feedback resistor with a unity-gain follower to insure that current through this resistor does not flow through the current-sampling resistor and thus alter i_E.

The trim potentiometer allows precise matching of resistor ratios to make current i_E independent of common-mode voltage levels at various points in the current source and thus dependent only on i_N. In this application, it was not necessary to have exactly unity gain between i_N and $-i_E$, so no trim is included for this ratio. The general magnitude of the resistors

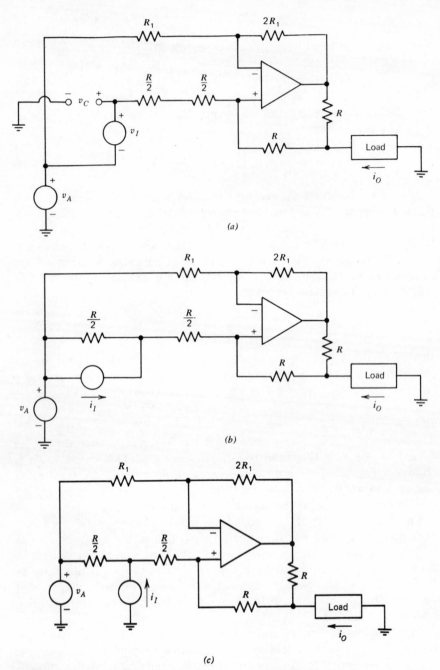

Figure 12.39 Current-controlled current source. (*a*) Basic Howland current source. (*b*) Current source following Norton substitution. (*c*) Final configuration.

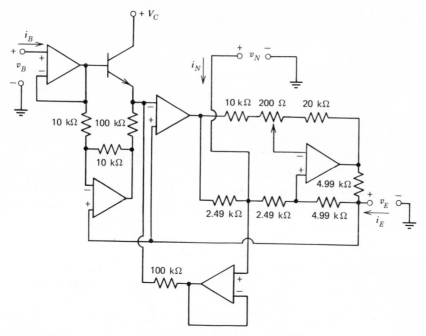

Figure 12.40 Complete nonlinear three-port network.

in the current source is chosen for compatibility with required current levels and amplifier characteristics and is not important for purposes of this discussion.

PROBLEMS

P12.1

Consider a Wien-Bridge oscillator as shown in Fig. 12.1. Show that if the output signal is of the general form $v_o = E \sin [(t/RC) + \theta]$ where θ is a constant, the signals applied to the two inputs of the operational amplifier are virtually identical, a necessary condition for satisfactory performance. Note that if the inverting and noninverting inputs are interchanged and it is assumed that the output has the form indicated above, the signals at the two inputs will also be identical. However, this modified topology will not function as an oscillator. Explain.

P12.2

A Wien-Bridge oscillator is constructed using the basic topology shown in Fig. 12.1. Because of component tolerances, the time constants of the series and parallel arms of the frequency-dependent feedback network differ by 5%. How must component values in the frequency-independent feedback path be related to guarantee oscillation?

P12.3
Use a describing-function approach to analyze the circuit shown in Fig. 12.3, assuming that the operational amplifier is ideal and that the diodes have zero conductance until a forward voltage of 0.6 volt is reached and zero resistance in the forward-conducting state. In particular, determine the magnitude of the signal applied to the noninverting input of the amplifier and the third-harmonic distortion present at the amplifier output.

P12.4
A sinusoidal oscillator is constructed by connecting the output of a double integrator (see Fig. 11.12) to its input. Show that amplitude can be controlled by varying the magnitude of the $(R/2)$-valued resistor shown in this figure. Design a complete circuit that can produce a 20-V peak-to-peak output signal at 1 kHz. Use a FET with parameters given in Section 12.1.4 for the control element. Analyze your amplitude-control loop to show that it has acceptable stability and a crossover frequency compatible with the 1-kHz frequency of oscillation. If you have confidence in your design, build it. The 2N4416 field-effect transistor is reasonably well characterized by the parameters referred to above.

P12.5
The discussions of Sections 12.2.2 and 12.2.3 suggest operating electronic switches connected to symmetrical, variable voltages from the output of a

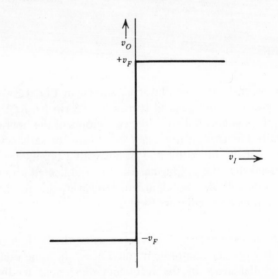

Figure 12.41 Infinite-gain limiter.

Schmitt trigger for two different applications. An alternative to the use of switches is to use a circuit that has the transfer characteristic shown in Fig. 12.41 for the necessary shaping function. (In this diagram, the voltage v_F is a positive variable.) Design a circuit that uses operational amplifiers to synthesize this transfer characteristic. Your output levels should be insensitive to temperature variations.

P12.6

A magnetic-suspension system was described in Section 6.2.3. Develop an electronic analog simulation of this system that permits determination of the transients that result from disturbing forces applied to the ball. Assume that, in addition to operational amplifiers and appropriate passive components, multipliers with a scale factor $v_O = v_X v_Y / 10$ volts are available. A way to perform the division required in this simulation using a multiplier and an operational amplifier is outlined in Section 6.2.2.

You may leave the various element values in the simulation defined in terms of system parameters, without developing final amplitude-scaled values.

P12.7

A circuit intended for use as a precision voltage reference for an analog-to-digital converter is shown in Fig. 12.42. The circuit uses a fraction of the Zener-diode voltage as its output. While this method involving resistive attenuation results in relatively high output resistance compared with using the voltage at the output of the amplifier as the reference, the output voltage becomes essentially independent of operational-amplifier offset voltage.

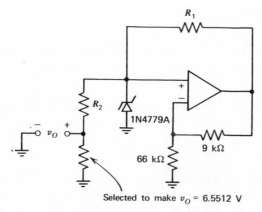

Figure 12.42   Voltage reference.

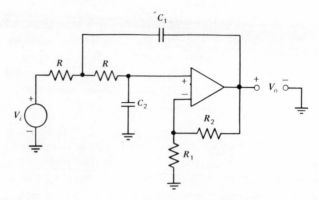

Figure 12.43 Low-pass Sallen and Key circuit with voltage gain.

The specified breakdown voltage of the 1N4779A is 8.5 volts $\pm 5\%$. The indicated resistor is selected during testing to obtain the required output voltage independent of the actual value of the Zener-diode voltage.

The breakdown voltage range and the temperature coefficient of the device are guaranteed at an operating current of 0.500 mA. By proper choice of R_1 and R_2, it is possible to make the current through the Zener diode independent of the actual Zener voltage after the single indicated selection has been completed. Such a choice is advantageous since it simplifies circuit calibration as opposed to methods that require two or more interdependent adjustments to set output voltage and Zener-diode operating current. Find values for R_1 and R_2 that result in this simplification. (Please excuse the somewhat unwieldy numbers involved in this problem, but it is drawn directly from an existing application.)

P12.8

A Sallen and Key low-pass circuit with an amplifier closed-loop voltage gain greater than unity is shown in Fig. 12.43. Determine the transfer function $V_o(s)/V_i(s)$ for this circuit. Compare the sensitivity of this circuit to component variations with that of the unity-gain version.

P12.9

One way to analyze the Sallen and Key circuit shown in Fig. 12.43 is to recognize the configuration as a *positive-feedback* circuit. If the loop is broken at the noninverting input to the operational amplifier, analysis techniques based on loop-transmission properties can be used.

(a) Indicate the loop-transmission singularity pattern that results when the loop is broken at the point mentioned above. It is not necessary to determine singularity locations exactly in terms of element values.

(b) Show how the closed-loop poles of the system move as a function of the closed-loop gain of the operational amplifier by using root-locus methods that have been appropriately modified for positive feedback systems.

P12.10

Design a sixth-order Butterworth filter with a 1 kHz corner frequency by cascading three unity-gain Sallen and Key circuits.

P12.11

The fifth-order Padé approximate to a one-second time delay is

$$P_5(s) = \frac{1 - 0.5s + 0.111s^2 - 1.39 \times 10^{-2}s^3 + 9.92 \times 10^{-4}s^4 - 3.31 \times 10^{-5}s^5}{1 + 0.5s + 0.111s^2 - 1.39 \times 10^{-2}s^3 + 9.92 \times 10^{-4}s^4 + 3.31 \times 10^{-5}s^5}$$

Design an active filter that synthesizes this transfer function.

P12.12

Develop a linearized block-diagram for the system shown in Fig. 12.32, assuming that the FET is characterized by the parameters given in Section 12.1.4. Show that the loop crossover frequency is low compared to 400 Hz for any input-voltage level up to 20 volts peak-to-peak. Estimate the time required for the system to restore equilibrium following an incremental perturbation (initiated, for example, by a change in input frequency) when the input-signal amplitude is 100 mV peak-to-peak. Note that the system is not significantly disturbed by a change in input amplitude when operating under equilibrium conditions, and that therefore this relatively long settling time does not deteriorate performance.

CHAPTER XIII

COMPENSATION REVISITED

13.1 INTRODUCTION

Proper compensation is essential for achieving optimum performance from virtually any sophisticated feedback system. Objectives extend far beyond simply guaranteeing acceptable stability. If stability is our only concern, the relatively unimaginative approaches of lowering loop-transmission magnitude or creating a sufficiently low-frequency dominant pole usually suffice for systems that do not have right-half-plane poles in their loop transmissions. More creative compensation is required when high desensitivity over an extended bandwidth, wideband frequency response, ideal closed-loop transfer functions with high-pass characteristics, or operation with uncertain loop parameters is essential. The type of compensation used can also influence quantities such as noise, drift, and the class of signals for which the system remains linear.

A detailed general discussion has already been presented in Chapter 5. In this chapter we become more specific and look at the techniques that are most appropriate in the usual operational-amplifier connections. It is assumed that the precautions suggested in Section 11.3.2 have been observed so that parasitic effects resulting from causes such as inadequate power-supply decoupling or feedback-network loading at the input of the amplifier do not degrade performance.

It is cautioned at the outset that there is no guarantee that particular specifications can be met, even with the best possible compensation. For example, earlier developments have shown how characteristics such as the phase shift from a pure time delay or a large number of high-frequency poles set a very real limit to the maximum crossover frequency of an amplifier-feedback network combination. Somewhat more disturbing is the reality that there is usually no way of telling when the best compensation for a particular application has been realized, so there is no clear indication when the trial-and-error process normally used to determine compensation should be terminated.

The attempt in this chapter is to introduce the types of compensation that are most likely to succeed in a variety of applications, as well as to indicate some of the hazards associated with various compensating techniques. The suggested techniques for minor-loop compensation are illustrated with experimental results.

13.2 COMPENSATION WHEN THE OPERATIONAL-AMPLIFIER TRANSFER FUNCTION IS FIXED

Many available operational amplifiers have open-loop transfer functions that cannot be altered by the user. This inflexibility is the general rule in the case of discrete-component amplifiers, and many integrated-circuit designs also include internal (and thus fixed) compensating networks. If the manufacturers' choice of open-loop transfer function is acceptable in the intended application, these amplifiers are straightforward to use. Conversely, if loop dynamics must be modified for acceptable performance, the choices available to the designer are relatively limited. This section indicates some of the possibilities.

13.2.1 Input Compensation

Input compensation consists of shunting a passive network between the input terminals of an operational amplifier so that the characteristics of the added network, often combined with the properties of the feedback network, alter the loop transmission of the system advantageously. This form of compensation does not change the ideal closed-loop transfer function of the amplifier-feedback network combination. We have already seen an example of this technique in the discussion of lag compensation using the topology shown in Fig. 5.13. That particular example used a noninverting amplifier connection, but similar results can be obtained for an inverting amplifier connection by shunting an impedance from the inverting input terminal to ground.

Figure 13.1 illustrates the topology for lag compensating the inverting connection. The loop transmission for this system (assuming that loading at the input and the output of the amplifier is insignificant) is

$$L(s) = - \frac{a(s)R_1}{(R_1 + R_2)} \frac{(RCs + 1)}{[(R_1 \parallel R_2 + R)Cs + 1]} \tag{13.1}$$

The dynamics of this loop transmission include a lag transfer function with a pole located at $s = -[1/(R_1 \parallel R_2 + R)C]$ and a zero located at $s = -1/RC$.

The example of lead compensation using the topology shown in Fig. 5.11 obtained the lead transfer function by paralleling one of the feedback-network resistors with a capacitor. A potential difficulty with this approach

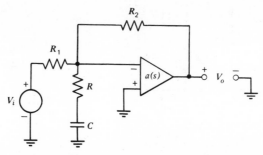

Figure 13.1 Lag compensation for the inverting amplifier connection.

is that the ideal closed-loop transfer function is changed. An alternative is illustrated in Fig. 13.2. Since component values are selected so that $R_1C_1 = R_2C_2$, the ideal closed-loop transfer function is

$$\frac{V_o(s)}{V_i(s)} = -\frac{R_2/(R_2C_2s + 1)}{R_1/(R_1C_1s + 1)} = -\frac{R_2}{R_1} \qquad (13.2)$$

The loop transmission for this connection in the absence of loading and following some algebraic manipulation is

$$L(s) = -\frac{a(s)R_1R}{(R_1R_2 + R_1R + R_2R)} \frac{[(R_1C_1)s + 1]}{[(R_1 \parallel R_2 \parallel R)(C_1 + C_2)s + 1]} \qquad (13.3)$$

A disadvantage of this method is that it lowers d-c loop-transmission magnitude compared with the topology that shunts R_2 only with a capacitor. The additional attenuation that this method introduces beyond that provided by the R_1-R_2 network is equal to the ratio of the two break frequencies of the lead transfer function.

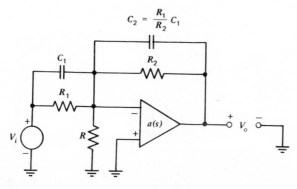

Figure 13.2 Lead compensation for the inverting amplifier connection.

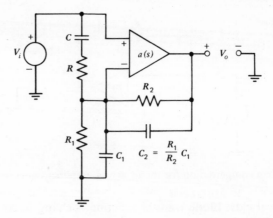

Figure 13.3 Lead and lag compensation for the noninverting amplifier connection.

This basic approach can also be used to combine lead and lag transfer functions in one loop transmission. Figure 13.3 illustrates one possibility for a noninverting connection. The equality of time constants in the feedback network insures that the ideal gain for this connection is

$$\frac{V_o(s)}{V_i(s)} = \frac{R_1 + R_2}{R_1} \tag{13.4}$$

Some algebraic reduction indicates that the loop transmission (assuming negligible loading) is

$$L(s) = -\frac{a(s)R_1}{(R_1 + R_2)} \frac{(RCs + 1)(R_1 C_1 s + 1)}{\{RR_1 CC_1 s^2 + [(R_1 \| R_2 + R)C + R_1 C_1]s + 1\}} \tag{13.5}$$

The constraints among coefficients in the transfer function related to the feedback and shunt networks guarantee that this expression can be factored into a lead and a lag transfer function, and that the ratios of the singularity locations will be identical for the lead and the lag functions.

The way that topologies of the type described above are used depends on the dynamics of the amplifier to be compensated and the load connected to it. For example, the HA2525 is a monolithic operational amplifier (made by a process more involved than the six-mask epitaxial process) that combines a unity-gain frequency of 20 MHz with a slew rate of 120 volts per microsecond. The dynamics of this amplifier are such that stability is guaranteed only for loop transmissions that combine the amplifier open-loop transfer function with an attenuation of three or more. Figure 13.4 shows how a stable, unity-gain follower can be constructed using this amplifier. Component values are selected so that the zero of the lag network is lo-

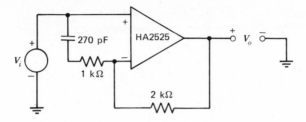

Figure 13.4   Unity-gain follower with input compensation.

cated approximately one decade below the compensated loop-transmission crossover frequency. Of course, the capacitor could be replaced by the short circuit, thereby lowering loop-transmission magnitude at all frequencies. However, advantages of the lag network shown include greater desensitivity at intermediate and low frequencies and lower output offset for a given offset referred to the input of the amplifier.

There are many variations on the basic theme of compensating with a network shunted across the input terminals of an operational amplifier. For example, many amplifiers with fixed transfer functions are designed to be stable with direct feedback provided that the unloaded open-loop transfer function of the amplifier is not altered by loading. However, a load capacitor can combine with the open-loop output resistance of the amplifier to create a pole that compromises stability. Performance can often be improved in these cases by using lead input compensation to offset the effects of the second pole in the vicinity of the crossover frequency or by using lag input compensation to force crossover below the frequency where the pole associated with the load becomes important.

In other connections, an additional pole that deteriorates stability results from the feedback network. As an example, consider the differentiator shown in Fig. 13.5a. The ideal closed-loop transfer function for this connection is

$$\frac{V_o(s)}{V_i(s)} = -s \qquad (13.6)$$

It should be noted at the outset that this connection is not recommended since, in addition to its problems with stability, the differentiator is an inherently noisy circuit. The reason is that differentiation accentuates the input noise of the amplifier because the ideal gain of a differentiator is a linearly increasing function of frequency.

Many amplifiers are compensated to have an approximately single-pole open-loop transfer function, since this type of transfer function results in excellent stability provided that the load element and the feedback network

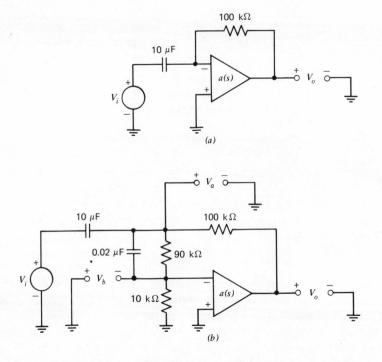

Figure 13.5   Differentiator. (*a*) Uncompensated. (*b*) With input lead compensation.

do not introduce additional poles. Accordingly, we assume that for the amplifier shown in Fig. 13.5*a*

$$a(s) = \frac{10^5}{0.01s + 1} \tag{13.7}$$

If loading is negligible, the feedback network and the amplifier open-loop transfer function combine to produce the loop transmission

$$L(s) = - \frac{10^5}{(0.01s + 1)(s + 1)} \tag{13.8}$$

The unity-gain frequency of this function is 3.16×10^3 radians per second and its phase margin is less than 2°.

   Stability is improved considerably if the network shown in Fig. 13.5*b* is added at the input of the amplifier. In the vicinity of the crossover frequency, the impedance of the 10-μF capacitor (which is approximately equal to the Thevenin-equivalent impedance facing the compensating network) is much lower than that of the network itself. Accordingly the transfer function from $V_o(s)$ to $V_a(s)$ is not influenced by the input network at these frequencies. The lead transfer function that relates $V_b(s)$ to $V_a(s)$ combines with other elements in the loop to yield a loop transmission in the vicinity of the crossover frequency

$$L(s) \simeq - \frac{10^6(1.8 \times 10^{-3}s + 1)}{s^2(1.8 \times 10^{-4}s + 1)} \tag{13.9}$$

(Note that this expression has been simplified by recognizing that at frequencies well above 100 radians per second, the two low-frequency poles can be considered located at the origin with appropriate modification of the scale factor.) The unity-gain frequency of Eqn. 13.9 is 1.8×10^3 radians per second, or approximately the geometric mean of the singularities of the lead network. Thus (from Eqn. 5.6) the phase margin of this system is $55°$.

One problem with the circuit shown in Fig. 13.5b is that its output voltage offset is 20 times larger than the offset referred to the input of the amplifier. The output offset can be made equal to amplifier input offset by including a capacitor in series with the 10-kΩ resistor, thereby introducing both lead and lag transfer functions with the input network. If the added capacitor has negligibly low impedance at the crossover frequency, phase margin is not changed by this modification. In order to prevent conditional stability this capacitor should be made large enough so that the phase shift of the negative of the loop transmission does not exceed $-180°$ at low frequencies.

13.2.2 Other Methods

The preceding section focused on the use of a shunt network at the input of the operational amplifier to modify the loop transmission. In certain other cases, the feedback network can be changed to improve stability without significantly altering the ideal closed-loop transfer function. As an example, consider the circuit shown in Fig. 13.6. The ideal closed-loop transfer function for this circuit is

$$\frac{V_o(s)}{V_i(s)} = - \frac{s}{3 \times 10^{-4}s + 1} \tag{13.10}$$

and thus it functions as a differentiator at frequencies well below 3.3×10^3 radians per second. The transfer function from the output of the operational amplifier to its inverting input via the feedback network includes a zero because of the 30-Ω resistor. The resulting loop transmission is

$$L(s) = - \frac{a(s)(3 \times 10^{-4}s + 1)}{s + 1} \tag{13.11}$$

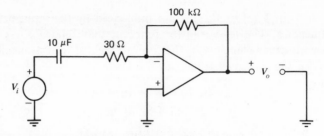

Figure 13.6 Differentiator with feedback-network compensation.

If it is assumed that the amplifier open-loop transfer function is the same as in the previous differentiator example $[a(s) = 10^5/(0.01s + 1)]$, Eqn. 13.11 becomes

$$L(s) \simeq - \frac{10^7(3 \times 10^{-4}s + 1)}{s^2} \qquad (13.12)$$

in the vicinity of the unity-gain frequency. The unity-gain frequency for Eqn. 13.12 is approximately 4×10^3 radians per second and the phase margin is 50°. Thus the differentiator connection of Fig. 13.6 combines stability comparable to that of the earlier example with a higher crossover frequency. While the ideal closed-loop gain includes a pole, the pole location is above the crossover frequency of the previous connection. Since the actual closed-loop gain of any feedback system departs substantially from its ideal value at the crossover frequency, this approach can yield performance superior to that of the circuit shown in Fig. 13.5.

In the examples involving differentiation, loop stability was compromised by a pole introduced by the feedback network. Another possibility is that a capacitive load adds a pole to the loop-transmission expression. Consider the capacitively loaded inverter shown in Fig. 13.7. The additional pole results because the amplifier has nonzero output resistance (see the amplifier model of Fig. 13.7b). If the resistor value R is much larger than R_o and loading at the input of the amplifier is negligible, the loop transmission is

$$L(s) = - \frac{a(s)}{2(R_o C_L s + 1)} \qquad (13.13)$$

The feedback-path connection can be modified as shown in Fig. 13.8a to improve stability. It is assumed that parameter values are selected so that $R \gg R_o + R_C$ and that the impedance of capacitor C_F is much larger in

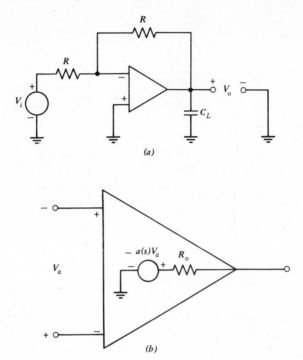

Figure 13.7 Capacitively loaded inverter. (*a*) Circuit. (*b*) Amplifier model.

magnitude than R_o at all frequencies of interest. With these assumptions, the equations for the circuit are

$$V_a = \frac{(V_i + V_o)}{2[(RC_F/2)s + 1]} + \frac{V_b(RC_F/2)s}{(RC_F/2)s + 1} \tag{13.14a}$$

$$V_b = \frac{a(s)V_a(R_CC_{LS} + 1)}{(R_o + R_C)C_{LS} + 1} \tag{13.14b}$$

$$V_o = -\frac{a(s)V_a}{(R_o + R_C)C_{LS} + 1} \tag{13.14c}$$

These equations lead to the block diagram shown in Fig. 13.8*b*.

Two important features are evident from the block diagram or from physical arguments based on the circuit configuration. First, since the transfer functions of blocks ① and ② are identical and since the outputs of both of these blocks are summed with the same sign to obtain V_a, the ideal output is the negative of the input at frequencies where the signal propagated

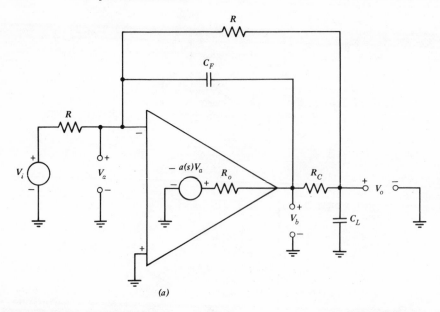

Figure 13.8 Feedback-network compensation for capacitively loaded inverter. (*a*) Circuit. (*b*) Block diagram.

through path 1 is insignificant. We can argue the same result physically, since Fig. 13.8*a* indicates an ideal transfer function $V_o = -V_i$ if feedback through C_F is negligible.

The second conclusion involves the stability of the system. If the loop is broken at the indicated point, the loop transmission is

$$L(s) = -a(s) \underbrace{\frac{[(RC_F/2)s]}{[(RC_F/2)s + 1]} \frac{(R_C C_L s + 1)}{[(R_o + R_C)C_L s + 1]}}_{\text{term from feedback via path 1}}$$

$$\underbrace{- \frac{a(s)}{2[(R_o + R_C)C_L s + 1][(RC_F/2)s + 1]}}_{\text{term from feedback via path 2}} \qquad (13.15)$$

At sufficiently high frequencies, Eqn. 13.15 reduces to

$$L(s) \simeq - \frac{a(s)R_C}{R_o + R_C} \qquad (13.16)$$

because path transmission of path 1 reaches a constant value, while that of path 2 is progressively attenuated with frequency. If parameters are

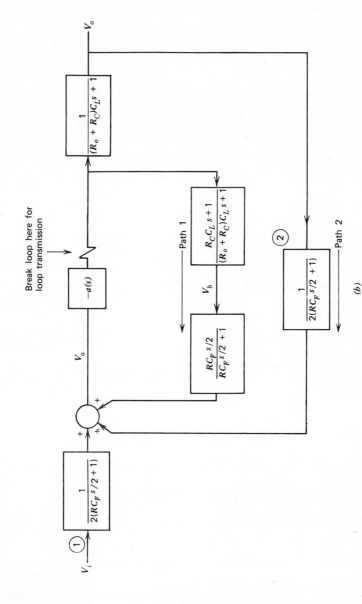

Figure 13.8—Continued

chosen so that crossover occurs where the approximation of Eqn. 13.16 is valid, system stability is essentially unaffected by the load capacitor. The same result can be obtained directly from the circuit of Fig. 13.8a. If parameters are chosen so that at the crossover frequency ω_c

$$\frac{1}{C_L \omega_c} \ll R_o + R_C \quad \text{and} \quad R_o \ll \frac{1}{C_F \omega_c} \ll \frac{R}{2}$$

the feedback path around the amplifier is frequency independent at crossover.

Previous examples have indicated how dynamics related to the feedback network or the load can deteriorate stability. Stability may also suffer if an active element that provides voltage gain greater than one is used in the feedback path, since this type of feedback element will result in a loop-transmission crossover frequency in excess of the unity-gain frequency of the operational amplifier itself. Additional negative phase shift then occurs at crossover because the higher-frequency poles of the amplifier open-loop transfer function are significant at the higher crossover frequency.

The simple log circuit described in Section 11.5.4 demonstrates this type of difficulty. The basic circuit is illustrated in Fig. 13.9a. We recall that the ideal input-output transfer relationship for this circuit is

$$v_O = -\frac{kT}{q} \ln \frac{v_I}{RI_s} \tag{13.17}$$

where the quantity I_s characterizes the transistor.

A linearized block diagram for the connection, formed assuming that loading at the input and the output of the amplifier is negligible, is shown in Fig. 13.9b. Since the quiescent collector current of the transistor is related to the operating-point value of the input voltage V_I by $I_F = V_I/R$, the transistor transconductance is

$$g_m = \frac{qV_I}{kTR} \tag{13.18}$$

Consequently, the loop transmission for the linearized system is

$$L(s) = -a(s)\frac{qV_I}{kT} \tag{13.19}$$

If it is assumed that the maximum operating-point value of the input signal is 10 volts, the feedback network can provide a voltage gain of as much as 400. It is clear that few amplifiers with fixed open-loop transfer functions will be compensated conservatively enough to remain stable with this increase in loop-transmission magnitude.

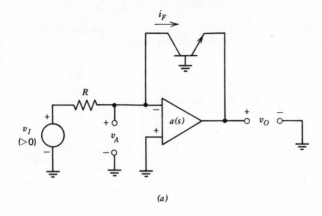

(a)

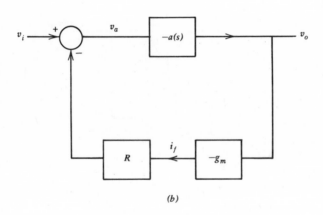

(b)

Figure 13.9 Evaluation of log-circuit stability. (a) Circuit. (b) Linearized block diagram.

A method that can be used to reduce the high voltage gain of the feed-back network is illustrated in Fig. 13.10. By including a transistor emitter resistor equal in value to the input resistor, we find that the maximum voltage gain of the resultant common-base amplifier is reduced to one. Note that the feedback effectively constrains the voltage at the emitter of the transistor to be logarithmically related to input voltage independent of load current, and thus the ideal transfer relationship of the circuit is inde-pendent of load. Also note that at least in the absence of significant output-current drain, the maximum voltage across the emitter resistor is equal to

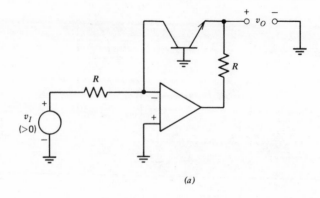

(a)

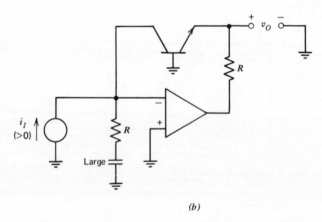

(b)

Figure 13.10   Reduction of loop-transmission magnitude for log circuit. (a) Use of emitter resistor. (b) Emitter resistor combined with input lag compensation.

the maximum input voltage so that output-voltage range is often unchanged by this modification.

It was mentioned in Section 11.5.4 that the log circuit can exhibit very wide dynamic range when the input signal is supplied from a current source, since the voltage offset of the amplifier does not give rise to an error current in this case. Figure 13.10b shows how input lag compensation can be combined with emitter degeneration to constrain loop-transmission magnitude without deteriorating dynamic range for current-source inputs.

13.3 COMPENSATION BY CHANGING THE AMPLIFIER TRANSFER FUNCTION

If the open-loop transfer function of an operational amplifier is fixed, this constraint, combined with the requirement of achieving a specified ideal closed-loop transfer function, severely restricts the types of modifications that can be made to the loop transmission of connections using the amplifier. Significantly greater flexibility is generally possible if the open-loop transfer function of the amplifier can be modified. There are a number of available integrated-circuit operational amplifiers that allow this type of control. Conversely, very few discrete-component designs are intended to be compensated by the user. The difference may be historical in origin, in that early integrated-circuit amplifiers used shunt impedances at various nodes for compensation (see Section 8.2.2) and the large capacitors required could not be included on the chip. Internal compensation became practical as the two-stage design using minor-loop feedback for compensation evolved, since much smaller capacitors are used to compensate these amplifiers. Fortunately, the integrated-circuit manufacturers choose to continue to design some externally compensated amplifiers after the technology necessary for internal compensation evolved.

In this section, some of the useful open-loop amplifier transfer functions that can be obtained by proper external compensation are described, and a number of different possibilities are analytically and experimentally evaluated for one particular integrated-circuit amplifier.

13.3.1 General Considerations

An evident question concerning externally compensated amplifiers is why they should be used given that internally compensated units are available. The answer hinges on the wide spectrum of applications of the operational amplifier. Since this circuit is intended for use in a multitude of feedback applications, it is necessary to choose its open-loop transfer function to insure stability in a variety of connections when this quantity is fixed.

The compromise most frequently used is to make the open-loop transfer function of the amplifier dominated by one pole. The location of this pole is chosen such that the amplifier unity-gain frequency occurs below frequencies where other singularities in the amplifier transfer function contribute excessive phase shift.

This type of compensation guarantees stability if direct, frequency-independent feedback is applied around the amplifier. However, it is overly conservative if considerable resistive attenuation is provided in the feedback path. In these cases, the crossover frequency of the amplifier with feedback drops, and the bandwidth of the resultant circuit is low. Conversely, if the feedback network or loading adds one or more intermediate-frequency poles to the loop transmission, or if the feedback path provides voltage gain, stability suffers.

However, if compensation can be intelligently selected as a function of the specific application, the ultimate performance possible from a given amplifier can be achieved in all applications. Furthermore, the compensation terminals make available additional internal circuit nodes, and at times it is possible to exploit this availability in ways that even the manufacturer has not considered. The creative designer working with linear integrated circuits soon learns to give up such degrees of freedom only grudgingly.

In spite of the clear advantages of user-compensated designs, internally compensated amplifiers outsell externally compensated units. Here are some of the reasons offered by the buyer for this contradictory preference.

(a) The manufacturers' compensation is optimum in my circuit. (This is true in about 1% of all applications.)

(b) It's cheaper to use an internally compensated amplifier since components and labor associated with compensation are eliminated. (Several manufacturers offer otherwise identical circuits in both internally and externally compensated versions. For example, the LM107 series of operational amplifiers is identical to the LM101A family with the single exception that a 30-pF compensating capacitor is included on the LM107 chip. The current prices of this and other pairs are usually identical. However, this excuse has been used for a long time. As recently as 1970, unit prices ranged from $0.75 to $5.00 *more* for the compensated designs depending on temperature range. A lot of 30-pF capacitors can be bought for $5.00.)

(c) Operational amplifiers can be destroyed from the compensating terminals. (Operational amplifiers can be destroyed from any terminals.)

(d) The compensating terminals are susceptible to noise pickup since they connect to low-signal-level nodes. (This reason is occasionally valid. For example, high-speed logic can interact with an adjacent operational amplifier through the compensating terminals, although inadequate power-supply bypassing is a far more frequent cause of such coupling.)

(e) Etc.

After sufficient exposure to this type of rationalization, it is difficult to escape the conclusion that the main reason for the popularity of internally compensated amplifiers is the inability of many users to either determine appropriate open-loop transfer functions for various applications or to implement these transfer functions once they have been determined. A primary objective of this book is to eliminate these barriers to the use of externally compensated operational amplifiers.

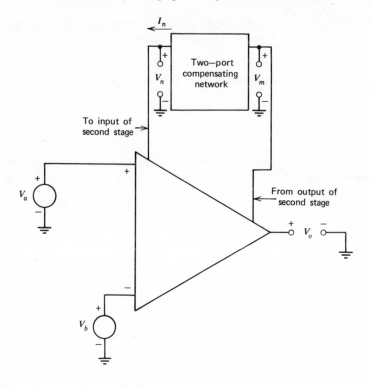

Figure 13.11   Operational amplifier compensated with a two-port network.

Any detailed and specific discussion of amplifier compensating methods must be linked to the design of the amplifier. It is assumed for the remainder of this chapter that the amplifier to be compensated is a two-stage design that uses minor-loop feedback for compensation. This assumption is realistic, since many modern amplifiers share the two-stage topology, and since it is anticipated that new designs will continue this trend for at least the near future.

It should be mentioned that the types of open-loop transfer functions suggested for particular applications can often be obtained with other than two-stage amplifier designs, although the method used to realize the desired transfer function may be different from that described in the material to follow.

Figure 13.11 illustrates the topology for an amplifier compensated with a two-port network. This basic configuration has been described earlier in Sections 5.3 and 9.2.3. While the exact details depend on specifics of the amplifier involved, the important general conclusions introduced in the earlier material include the following:

(a) The unloaded open-loop transfer function of a two-stage amplifier compensated this way (assuming that the minor loop is stable) is

$$a(s) = \frac{V_o(s)}{V_a(s) - V_b(s)} \simeq \frac{K}{Y_c(s)} \qquad (13.20)$$

over a wide range of frequencies.

The quantity K is related to the transconductance of the input-stage transistors, while $Y_c(s)$ is the short-circuit transfer admittance of the network.

$$Y_c(s) = \frac{I_n(s)}{V_m(s)}\bigg|_{V_n} = 0 \qquad (13.21)$$

This result can be justified by physical reasoning if we remember that at frequencies where the transmission of the minor loop formed by the second stage and the compensating network is large, the input to the second stage is effectively a virtual incremental ground. Furthermore, the current required at the input to the second stage is usually very small. Thus it can be shown by an argument similar to that used to determine the ideal closed-loop gain of an operational amplifier that incremental changes in current from the input stage must be balanced by equal currents into the compensating network. System parameters are normally selected so that major-loop crossover occurs at frequencies where the approximation of Eqn. 13.20 is valid and, therefore, this approximation can often be used for stability calculations.

(b) The d-c open-loop gain of the amplifier is normally independent of compensation. Accordingly, at low frequencies, the approximation of Eqn. 13.20 is replaced by the constant value a_0. The approximation fails because the usual compensating networks include a d-c zero in their transfer admittances, and at low frequencies this zero decreases the magnitude of the minor-loop transmission below one.

(c) The approximation fails at high frequencies for two reasons. The minor-loop transmission magnitude becomes less than one, and thus the network no longer influences the transfer function of the second stage. This transfer function normally has at least two poles at high frequencies, reflecting capacitive loading at the input and output of the second stage. There may be further departure from the approximation because of singularities associated with the input stage and the buffer stage that follows the second stage. These singularities cannot be controlled by the minor loop since they are not included in it.

(d) The open-loop transfer function of the compensated amplifier can be estimated by plotting both the magnitude of the approximation (Eqn. 13.20) and of the uncompensated amplifier transfer function on common log-magnitude versus log-frequency coordinates. If the dominant amplifier dynamics are associated with the second stage, the compensated open-loop transfer-function magnitude is approximately equal to the lower of the two curves at all frequencies. The uncompensated amplifier transfer function must reflect loading by the compensating network at the input and the output of the second-stage. In practice, designers usually determine experimentally the frequency range over which the approximation of Eqn. 13.20 is valid for the amplifier and compensating networks of interest.

Several different types of compensation are described in the following sections. These compensation techniques are illustrated using an LM301A operational amplifier. This inexpensive, popular amplifier is the commercial-temperature-range version of the LM101A amplifier described in Section 10.4.1. Recall from that discussion that the quantity K is nominally 2×10^{-4} mho for this amplifier, and that its specified d-c open-loop gain is typically 160,000. Phase shift from elements outside the minor loop (primarily the lateral-PNP transistors in the input stage) becomes significant at 1 MHz, and feedback connections that result in a crossover frequency in excess of approximately 2 MHz generally are unstable.

A number of oscilloscope photographs that illustrate various aspects of amplifier performance are included in the following material. A single LM301A was used in all of the test circuits. Thus relative performance reflects differences in compensation, loading, and feedback, but not in the uncompensated properties of the amplifier itself. (The fact that this amplifier survived the abuse it received by being transferred from one test circuit to another and during testing is a tribute to the durability of modern integrated-circuit operational amplifiers.)

13.3.2 One-Pole Compensation

The most common type of compensation for two-stage amplifiers involves the use of a single capacitor between the compensating terminals. Since the short-circuit transfer admittance of this "network" is $C_c s$ where C_c is the value of the compensating capacitor, Eqn. 13.20 predicts

$$a(s) \simeq \frac{K}{C_c s} \qquad (13.22)$$

The approximation of Eqn. 13.22 is plotted in Fig. 13.12 along with a representative uncompensated amplifier transfer function. As explained in

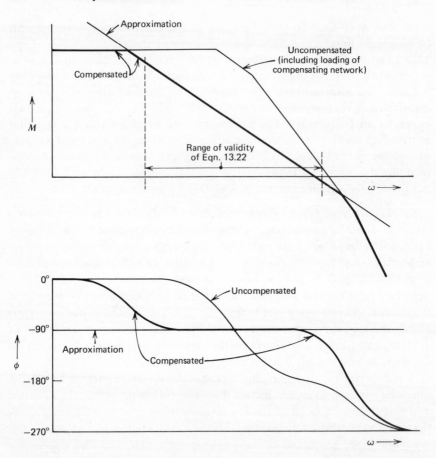

Figure 13.12   Open-loop transfer function for one-pole compensation.

the previous section, the compensated open-loop transfer function is very nearly equal to the lower of the two curves at all frequencies.

The important feature of Fig. 13.12, which indicates the general-purpose nature of this type of one-pole compensation, is that there is a wide range of frequencies where the magnitude of $a(j\omega)$ is inversely proportional to frequency and where the angle of this open-loop transfer function is approximately $-90°$. Accordingly, the amplifier exhibits essentially identical stability (but a variable speed of response) for many different values of frequency-independent feedback connected around it.

Two further characteristics of the open-loop transfer function of the amplifier are also evident from Fig. 13.12. First, the approximation of

Eqn. 13.22 can be extended to zero frequency if the d-c open-loop gain of the amplifier is known, since the geometry of Fig. 13.12 shows that

$$a(s) \simeq \frac{a_0}{(a_0 C_c / K)s + 1} \qquad (13.23)$$

at low and intermediate frequencies. Second, if the unity-gain frequency of the amplifier is low enough so that higher-order singularities are unimportant, this frequency is inversely related to C_c and is

$$\omega_u = \frac{K}{C_c} \qquad (13.24)$$

Stability calculations for feedback connections that use this type of amplifier are simplified if we recognize that provided the crossover frequency of the combination lies in the indicated region, these calculations can be based on the approximation of Eqn. 13.22.

Several popular internally compensated amplifiers such as the LM107 and the μA741 combine nominal values of K of 2×10^{-4} mho with 30-pF capacitors for C_c. The resultant unity-gain frequency is 6.7×10^6 radians per second or approximately 1 MHz. This value insures stability for any resistive feedback networks connected around the amplifier, since, with this type of feedback, crossover always occurs at frequencies where the loop transmission is dominated by one pole.

The approximate open-loop transfer function for either of these internally compensated amplifiers is $a(s) = 6.7 \times 10^6/s$. This transfer function, which is identical to that obtained from an LM101A compensated with a 30-pF capacitor, may be optimum in applications that satisfy the following conditions:

(a) The feedback-network transfer function from the amplifier output to its inverting input has a magnitude of one at the amplifier unity-gain frequency.

(b) Any dynamics associated with the feedback network and output loading contribute less than 30° of phase shift to the loop transmission at the crossover frequency.

(c) Moderately well-damped transient response is required.

(d) Input signals are relatively noise free.

If one or more of the above conditions are not satisfied, performance can often be improved by using an externally compensated amplifier that allows flexibility in the choice of compensating-capacitor value. Consider,

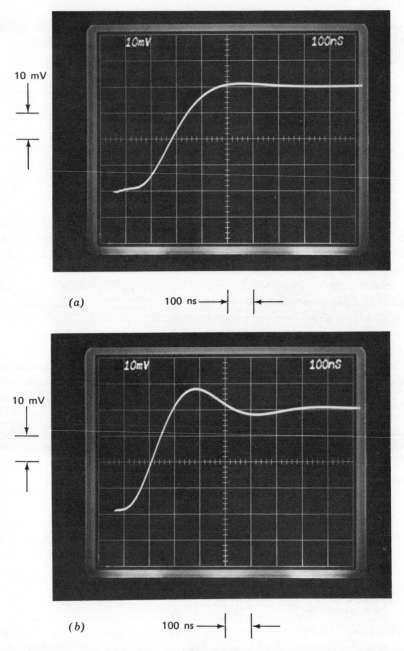

(a)

100 ns ⟶

(b)

100 ns ⟶

Figure 13.13 Step response of unity-gain follower as a function of compensating-capacitor value. (Input-step amplitude is 40 mV.) (a) $C_c = 30$ pF. (b) $C_c = 18$ pF. (c) $C_c = 68$ pF.

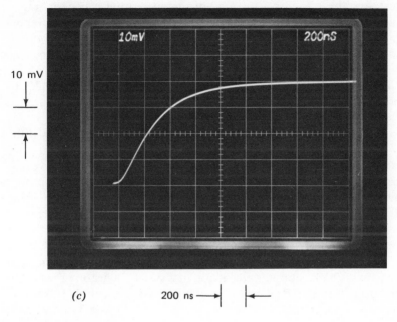

10 mV

(c)   200 ns →| |←

Figure 13.13—Continued

for example, a feedback connection that combines $a(s)$ as approximated by Eqn. 13.22 with frequency-independent feedback f_0. The closed-loop transfer function for this combination is

$$A(s) = \frac{a(s)}{1 + a(s)f_0} = \frac{1}{f_0}\left[\frac{1}{(C_c/Kf_0)s + 1}\right] \tag{13.25}$$

The closed-loop corner frequency (in radians per second) is

$$\omega_h = \frac{Kf_0}{C_c} \tag{13.26}$$

This equation shows that the bandwidth can be maintained at the maximum value consistent with satisfactory stability (recall the phase shift of terms ignored in the approximation of Eqn. 13.22) if C_c is changed with f_0 to keep the ratio of these two quantities constant. Alternatively, the closed-loop bandwidth can be lowered to provide improved filtering for noisy input signals by increasing the size of the compensating capacitor. A similar increase in capacitor size can also force crossover at lower frequencies to keep poles associated with the load or a frequency-dependent feedback network from deteriorating stability.

Figure 13.13 shows the small-signal step responses for the LM301A test amplifier connected as a unity-gain follower ($f_0 = 1$). Part a of this figure illustrates the response with a 30-pF compensating capacitor, the value used in similar, internally compensated designs. This transient response is quite well damped, with a 10 to 90% rise time of 220 ns, implying a closed-loop bandwidth (from Eqn. 3.57) of approximately 10^7 radians per second or 1.6 MHz.[1]

The response with a 18-pF compensating capacitor (Fig. 13.13b) trades considerably greater overshoot for improved rise time. Comparing this response with the second-order system responses (Fig. 3.8) shows that the closed-loop transient is similar to that of a second-order system with $\zeta = 0.47$ and $\omega_n = 13.5 \times 10^6$ radians per second. Since the amplifier open-loop transfer function satisfies the conditions used to develop the curves of Fig. 4.26, we can use these curves to approximate loop-transmission properties. Figure 4.26a estimates a phase margin of 50° and a crossover frequency of 11×10^6 radians per second. Since the value of f is one in this connection, these quantities correspond to compensated open-loop parameters of the amplifier itself.

Figure 13.13c illustrates the step response with a 68-pF compensating capacitor. The response is essentially first order, indicating that crossover now occurs at a frequency where only the dominant pole introduced by compensation is important. Equation 13.25 predicts an exponential time constant

$$\tau = \frac{C_c}{Kf_0} \tag{13.27}$$

under these conditions. The zero to 63% rise time shown in Fig. 13.12c is approximately 300 ns. Solving Eqn. 13.27 for K using known parameter values yields

$$K = \frac{C_c}{\tau f_0} = \frac{68 \text{ pF}}{300 \text{ ns}} = 2.3 \times 10^{-4} \text{ mho} \tag{13.28}$$

We notice that this value for K is slightly higher than the nominal value of 2×10^{-4} mho, reflecting (in addition to possible experimental errors) a somewhat higher than nominal first-stage quiescent current for this particular amplifier.[2] Variations of as much as 50% from the nominal value

---

[1] If the amplifier open-loop transfer function were exactly first order, the closed-loop half-power frequency in this connection would be identically equal to the unity-gain frequency of the amplifier itself. However, the phase shift of higher-frequency singularities ignored in the one-pole approximation introduces closed-loop peaking that extends the closed-loop bandwidth.

[2] The quiescent first-stage current of this amplifier can be measured directly by connecting an ammeter from terminals 1 and 5 to the negative supply (see Fig. 10.19). The estimated value of K is in excellent agreement with the measured total (the sum of both sides) quiescent current of 24 μA for the test amplifier.

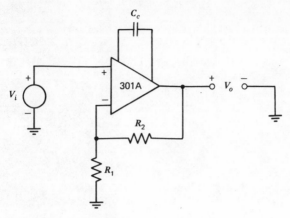

Figure 13.14   Noninverting amplifier.

for K are not unusual as a consequence of uncertainties in the integrated-circuit process.

The amplifier was next connected in the noninverting configuration shown in Fig. 13.14. The value of R_1 was kept less than or equal to 1 kΩ in all connections to minimize the loading effects of amplifier input capacitance. Figure 13.15a shows the step response for a gain-of-ten connection ($R_1 = 1$ kΩ, $R_2 = 9$ kΩ) with $C_c = 30$ pF. The 10 to 90% rise time has increased significantly compared with the unity-gain case using identical compensation. This change is expected because of the change in f_0 (see Eqn. 13.25).

Figure 13.15b is the step response when the capacitor value is lowered to get an overshoot approximately equal to that shown in Fig. 13.13a. While this change does not return rise time to exactly the same value displayed in Fig. 13.13a, the speed is dramatically improved compared to the transient shown in Fig. 13.15a. (Note the difference in time scales.)

Our approximate relationships predict that the effects of changing f_0 from 1 to 0.1 could be completely offset by lowering the compensating capacitor from 30 pF to 3 pF. The actual capacitor value required to obtain the response shown in Fig. 13.15b was approximately 4.5 pF. At least two effects contribute to the discrepancy. First, the approximation ignores higher-frequency open-loop poles, which must be a factor if there is any overshoot in the step response. Second, there is actually some *positive* minor-loop capacitive feedback in the amplifier. The schematic diagram for the LM101A (Fig. 10.19) shows that the amplifier input stage is loaded with a current repeater. The usual minor-loop compensation is connected to the output side of this current repeater. However, the input side of the

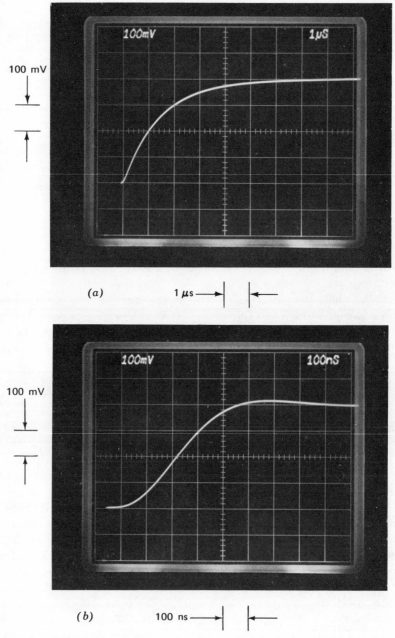

(a)   1 μs →| |←

(b)   100 ns →| |←

Figure 13.15 Step responses of gain-of-ten noninverting amplifier. (Input-step amplitude is 40 mV.) (a) $C_c = 30$ pF. (b) $C_c = 4.5$ pF.

582

repeater is also brought out on a pin to be used for balancing the amplifier. Any capacitance between a part of the circuit following the high-gain stage and the input side of the current repeater provides positive minor-loop feedback because of the inversion of the current repeater. An excellent stray-capacitance path exists between the amplifier output (pin 6) and the balance terminal connected to the input side of the current repeater (pin 5).[3] Part of the normal compensating capacitance is "lost" cancelling this positive feedback capacitance.

The important conclusion to be drawn from Fig. 13.15 is that, by properly selecting the compensating-capacitor value, the rise time and bandwidth of the gain-of-ten amplifier can be improved by approximately a factor of 10 compared to the value that would be obtained from an amplifier with fixed compensation. Furthermore, reasonable stability can be retained with the faster performance.

Figures 13.16 and 13.17 continue this theme for gain-of-100 ($R_1 = 100 \, \Omega$, $R_2 = 10 \, k\Omega$) and gain-of-1000 ($R_1 = 10 \, \Omega$, $R_2 = 10 \, k\Omega$) connections, respectively. The rise time for 30-pF compensation is linearly related to gain, and has a 10 to 90% value of approximately 350 microseconds in Fig. 13.17a, implying a closed-loop bandwidth of 1 kHz for an internally compensated amplifier in this gain-of-1000 connection. Compensating-capacitor values of 1 pF for the gain-of-100 amplifier and just a pinch (obtained with two short, parallel wires spaced for the desired transient response) for the gain-of-1000 connection result in overshoot comparable to that of the unity-gain follower compensated with 30 pF. The rise time does increase slightly at higher gains reflecting the fact that the uncompensated amplifier high-frequency open-loop gain is limited. However, a rise time of approximately 2 μs is obtained in the gain-of-1000 connection. The corresponding closed-loop bandwidth of 175 kHz represents a nearly 200:1 improvement compared with the value expected from an internally compensated general-purpose amplifier.

It is interesting to note that the closed-loop bandwidth obtained by properly compensating the inexpensive LM301A in the gain-of-1000 connection compares favorably with that possible from the best available discrete-component, fixed-compensation operational amplifiers. Unity-gain frequencies for wideband discrete units seldom exceed 100 MHz; consequently these single-pole amplifiers have closed-loop bandwidths of 100 kHz or less in the gain-of-1000 connection. The bandwidth advantage com-

---

[3] A wise precaution that reduces this effect is to clip off pin 5 close to the can when the amplifier is used in connections that do not require balancing. This modification was not made to the demonstration amplifier in order to retain maximum flexibility. Even with pin 5 cut close to the can, there is some header capacitance between it and pin 6.

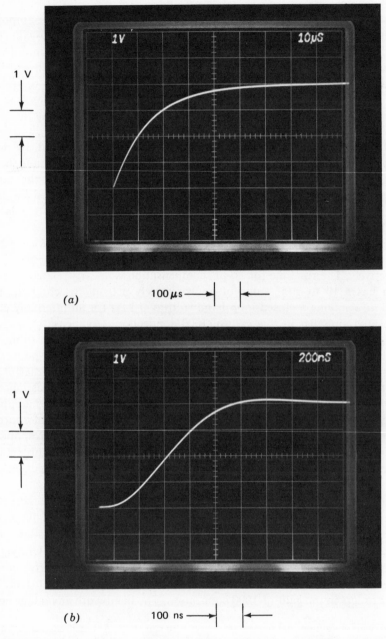

Figure 13.16 Step responses of gain-of-100 noninverting amplifier. (Input-step amplitude is 40 mV.) (*a*) $C_c = 30$ pF. (*b*) $C_c = 1$ pF.

584

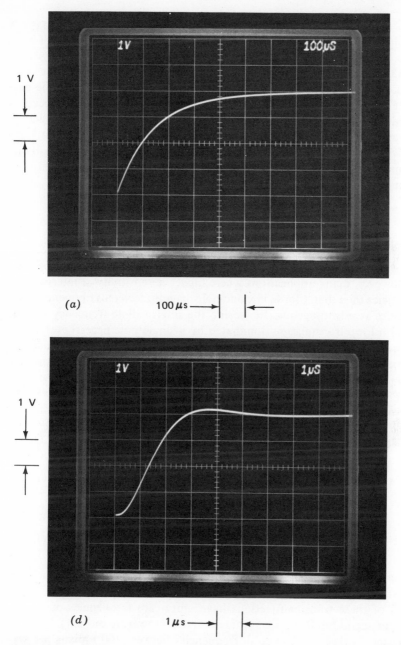

Figure 13.17 Step responses of gain-of-1000 noninverting amplifier. (Input-step amplitude is 4 mV.) (a) $C_c = 30$ pF. (b) Very small C_c.

pared with wideband internally compensated integrated-circuit amplifiers such as the LM118 is even more impressive.

We should realize that obtaining performance such as that shown in Fig. 13.17b requires careful adjustment of the compensating-capacitor value because the optimum value is dependent on the characteristics of the particular amplifier used and on stray capacitance. Although this process is difficult in a high-volume production situation, it is possible, and, when all costs are considered, may still be the least expensive way to obtain a high-gain wide-bandwidth circuit. Furthermore, compensation becomes routine if some decrease in bandwidth below the maximum possible value is acceptable.

13.3.3   Two-pole Compensation

The one-pole compensation described above is a conservative, general-purpose compensation that is widely used in a variety of applications. There are, however, many applications where higher desensitivity at intermediate frequencies than that afforded by one-pole magnitude versus frequency characteristics is advantageous. Increasing the intermediate-frequency magnitude of a loop transmission dominated by a single pole necessitates a corresponding increase in crossover frequency. This approach is precluded in systems where irreducible phase shift constrains the maximum crossover frequency for stable operation.

The only way to improve intermediate-frequency desensitivity without increasing the crossover frequency is to use a higher-order loop-transmission rolloff at frequencies below crossover. For example, consider the two amplifier open-loop transfer functions

$$a(s) = \frac{10^5}{10^{-2}s + 1} \tag{13.29}$$

and

$$a'(s) = \frac{10^5(10^{-6}s + 1)}{(10^{-4}s + 1)^2} \tag{13.30}$$

The magnitude versus frequency characteristics of these two transfer functions are compared in Fig. 13.18.

Both of these transfer functions have unity-gain frequencies of 10^7 radians per second and d-c magnitudes of 10^5. However, the magnitude of $a'(j\omega)$ exceeds that of $a(j\omega)$ at all frequencies between 100 radians per second and 10^6 radians per second. The advantage reaches a factor of 100 at 10^4 radians per second.

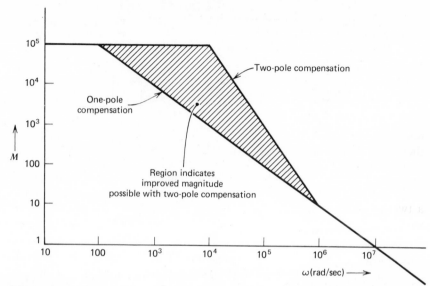

Figure 13.18 Comparison of magnitudes of one- and two-pole open-loop transfer functions.

The same advantage can be demonstrated using error coefficients. If amplifiers with open-loop transfer functions given by Eqns. 13.29 and 13.30 are connected as unity-gain followers, the respective closed-loop gains are

$$A(s) = \frac{a(s)}{1 + a(s)} = \frac{10^5}{10^{-2}s + 1 + 10^5} \qquad (13.31)$$

and

$$A'(s) = \frac{a'(s)}{1 + a'(s)} = \frac{10^5(10^{-6}s + 1)}{(10^{-4}s + 1)^2 + 10^5(10^{-6}s + 1)} \qquad (13.32)$$

The corresponding error series are

$$1 - A(s) \simeq \frac{10^{-2}s + 1}{10^{-2}s + 10^5} = 10^{-5} + 10^{-7}s - \cdots + \qquad (13.33)$$

and

$$1 - A'(s) \simeq \frac{10^{-8}s^2 + 2 \times 10^{-4}s + 1}{10^{-8}s^2 + 0.1s + 10^5} = 10^{-5} + 2 \times 10^{-9}s + \cdots + \qquad (13.34)$$

Identifying error coefficients shows that while these two systems have identical values for e_0, the error coefficient e_1 is a factor of 50 times smaller for the system with the two-pole rolloff. Thus, dramatically smaller errors

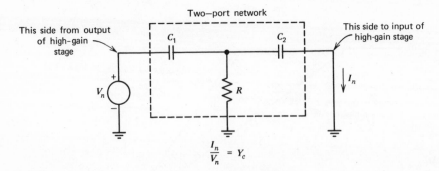

Two—port network

This side from output of high-gain stage

This side to input of high-gain stage

C_1 C_2

R

$+$

V_n

I_n

$$\frac{I_n}{V_n} = Y_c$$

Figure 13.19   Network for two-pole compensation.

result with the two-pole system for input signals that cause the e_1 term of the error series to dominate.

It is necessary to use a true two-port network to implement this compensation, since the required s^2 dependence of Y_c cannot be obtained with a two-terminal network. The short-circuit transfer admittance of the network shown in Fig. 13.19,

$$\frac{I_n(s)}{V_n(s)} = \frac{RC_1C_2s^2}{R(C_1 + C_2)s + 1} \tag{13.35}$$

has the required form. The approximate open-loop transfer function with this type of compensating network is (from Eqn. 13.20)

$$a(s) \simeq \frac{K}{Y_c} \simeq \frac{K'(\tau s + 1)}{s^2} \tag{13.36}$$

where $\tau = R(C_1 + C_2)$ and $K' = K/RC_1C_2$.

An estimation of the complete open-loop transfer function based on Eqn. 13.36 and a representative uncompensated transfer function are shown in Fig. 13.20. We note that while this type of transfer function can yield significantly improved desensitivity and error-coefficient magnitude compared to a one-pole transfer function, it is not a general-purpose compensation. The zero location and constant K' must be carefully chosen as a function of the attenuation provided by the feedback network in a particular application in order to obtain satisfactory phase margin. While lowering the frequency of the zero results in a wider frequency range of acceptable phase margin, it also reduces desensitivity, and in the limit leads to a one-pole transfer function. This type of open-loop transfer function is also intolerant of an additional pole introduced in the feedback network or by capacitive loading. If the additional pole is located at an intermediate frequency below the zero location, instability results. Another problem is that

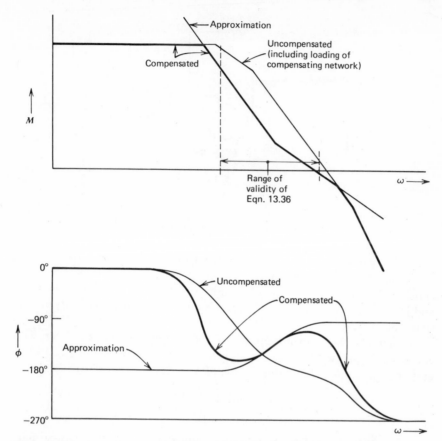

Figure 13.20 Open-loop transfer function for two-pole compensation.

there is a wide range of frequencies where the phase shift of the transfer function is close to $-180°$. While this transfer function is not conditionally stable by the definition given in Section 6.3.4, the small phase margin that results when the crossover frequency is lowered (in a describing-function sense) by saturation leads to marginal performance following overload.

In spite of its limitations, two-pole compensation is a powerful technique for applications where signal levels and the dynamics of additional elements in the loop are well known. This type of compensation is demonstrated using the unity-gain inverter shown in Fig. 13.21. The relatively low feedback-network resistors are chosen to reduce the effects of capacitance at the inverting input of the amplifier. This precaution is particularly important since the voltage at this input terminal is displayed in several of the oscilloscope photographs to follow. An LM310 voltage follower (see Section

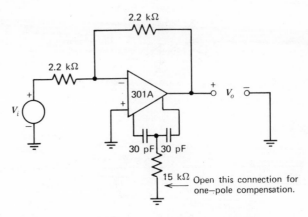

Figure 13.21 Unity-gain inverter with two-pole compensation.

10.4.4) was used to isolate this node from the relatively high oscilloscope input capacitance for these tests. The input capacitance of the LM310 is considerably lower than that of a unity-gain passive oscilloscope probe, and its bandwidth exceeds that necessary to maintain the fidelity of the signal of interest.

The approximate open-loop transfer function for the amplifier with compensating-network values as shown in Fig. 13.21 is (from Eqn. 13.36 using the previously determined value of $K = 2.3 \times 10^{-4}$ mho)

$$a(s) \simeq \frac{1.7 \times 10^{13}(9 \times 10^{-7}s + 1)}{s^2} \tag{13.37}$$

Since the value of f_0 for the unity-gain inverter is $1/2$, the approximate loop transmission for this system is

$$L(s) \simeq - \frac{0.85 \times 10^{13}(9 \times 10^{-7}s + 1)}{s^2} \tag{13.38}$$

The crossover frequency predicted by Eqn. 13.38 is 7.7×10^6 radians per second, and the zero is located at 1.1×10^6 radians per second, or a factor of 7 below the crossover frequency. Consequently, the phase margin of this system is $\tan^{-1}(1/7) = 8°$ less than that of a unity-gain inverter using one-pole compensation adjusted for the same crossover frequency.

Figure 13.22 compares the step responses of the inverter-connected LM301A with one- and two-pole compensation. Part *a* of this figure was obtained with the lower end of the 15-kΩ resistor removed from ground as indicated in Fig. 13.21. In this case the compensating element is equivalent to a single 15-pF capacitor. Note that Eqn. 13.22 combined with the value $f_0 = 1/2$, which applies to the unity-gain inverter, predicts a 7.7×10^6-radian-per-second crossover frequency for 15-pF compensation. The same result can be obtained by realizing that at frequencies beyond the zero location the parallel impedance of the capacitors in the two-pole compensating network must be smaller than that of the resistor, and thus removing the resistor does not alter the amplifier open-loop transfer function substantially in the vicinity of crossover.

The response shown in Fig. 13.22a is quite similar to that shown previously in Fig. 13.13a. Recall that Fig. 13.13a was obtained with a unity-gain follower ($f_0 = 1$) and $C_c = 30$ pF. As anticipated, lowering f_0 and C_c by the same factor results in comparable performance for single-pole systems.

There is a small amount of initial undershoot evident in the transient of Fig. 13.22a. This undershoot results from the input step being fed directly to the output through the two series-connected resistors. This fed-forward signal can drive the output negative initially because of the nonzero output impedance and response time of the amplifier. The magnitude of the initial undershoot would shrink if larger-value resistors were used around the amplifier.

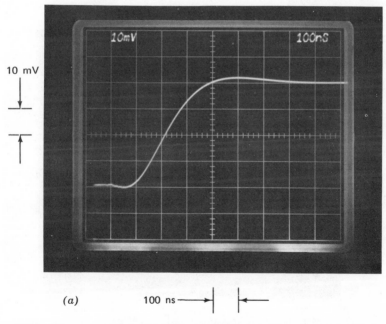

(a)   100 ns

Figure 13.22   Step responses of unity-gain inverter. (Input-step amplitude is −40 mV.) (a) One-pole compensation. (b) Two-pole compensation. (c) Repeat of part b with slower sweep speed.

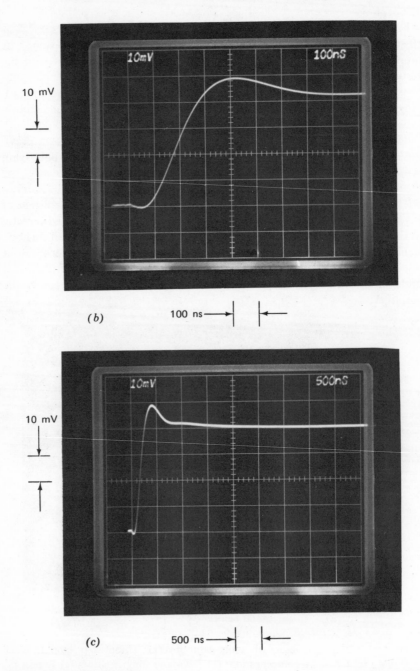

(b)

100 ns

(c)

500 ns

Figure 13.22—Continued

592

The step response of Fig. 13.22b results with the 15-kΩ resistor connected to ground and is the response for two-pole compensation. Three effects combine to speed the rise time and increase the overshoot of this response compared to the single-pole case. First, the phase margin is approximately 8° less for the two-pole system. Second, the T network used for two-pole compensation loads the output of the second stage of the amplifier to a greater extent than does the single capacitor used for one-pole compensation, although this effect is small for the element values used in the present example. The additional loading shifts the high-frequency poles associated with limited minor-loop transmission toward lower frequencies. Third, a closed-loop zero that results with two-pole compensation also influences system response.

The root-locus diagram shown in Fig. 13.23 clarifies the third reason. (Note that this diagram is not based on the approximation of Eqn. 13.38, but rather on a more complete loop transmission assuming a representative amplifier using these compensating-network values.) With the value of $a_0 f_0$ used to obtain Fig. 13.22b, one closed-loop pole is quite close to the zero at

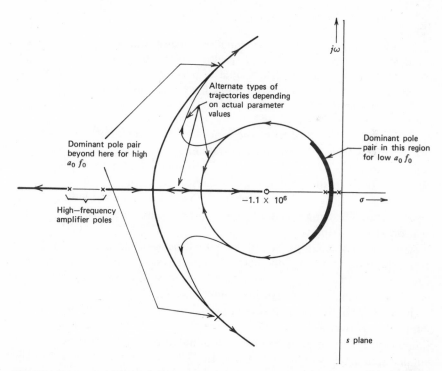

Figure 13.23   Root-locus diagram for inverter with two-pole compensation.

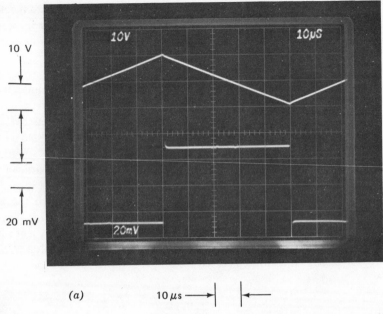

10 V

20 mV

(a)   10 μs ⟶ | |← ───

Figure 13.24 Unity-gain inverting-amplifier response with triangle-wave input. (Input amplitude is 20 volts peak-to-peak.) (*a*) One-pole compensation—upper trace: output; lower trace: inverting input of operational amplifier. (*b*) Two-pole compensation—upper trace: output; lower trace: inverting input of operational amplifier. (*c*) Repeat of lower trace, part *b*, with faster sweep speed.

$-1.1 \times 10^6 \text{ sec}^{-1}$ regardless of the exact details of the diagram. Since the zero is in the forward path of the system, it appears in the closed-loop transfer function. The resultant closed-loop doublet adds a positive, long-duration "tail" to the response as explained in Section 5.2.6. The tail is clearly evident in Fig. 13.22*c*, a repeat of part *b* photographed with a slower sweep speed. The time constant of the tail is consistent with the doublet location at approximately -10^6 sec^{-1}.

We recall that this type of tail is characteristic of lag-compensated systems. The loop transmission of the two-pole system combines a long $1/s^2$ region with a zero below the crossover frequency. This same basic type of loop transmission results with lag compensation.

The root-locus diagram also shows that satisfactory damping ratio is obtained only over a relative small range of $a_0 f_0$. As $a_0 f_0$ falls below the optimum range, system performance is dominated by a low-frequency poorly damped pole pair as indicated in Fig. 13.23. As $a_0 f_0$ is increased above the optimum range, a higher-frequency poorly damped pole pair dominates

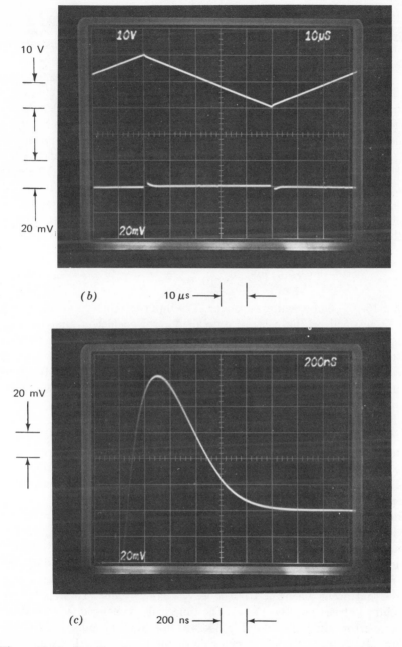

10 V

20 mV

(b)   10 μs

20 mV

(c)   200 ns

Figure 13.24—Continued

performance since the real-axis pole closest to the origin is very nearly cancelled by the zero.

The error-reducing potential of two-pole compensation is illustrated in Fig. 13.24. The most important quantity included in these photographs is the signal at the inverting input of the operational amplifier. The topology used (Fig. 13.21) shows that the signal at this terminal is (in the absence of loading) half the error between the actual and the ideal amplifier output. Part *a* of this figure indicates performance with single-pole compensation achieved via a 15-pF capacitor. The upper trace indicates the amplifier output when the signal applied from the source is a 20-volt peak-to-peak, 10-kHz triangle wave. This signal is, to within the resolution of the measurement, the negative of the signal applied by the source. The bottom trace is the signal at the inverting input terminal of the operational amplifier.

The approximate open-loop transfer function from the inverting input to the output of the test amplifier is

$$-a(s) = -\frac{2.3 \times 10^{-3}}{1.5 \times 10^{-11}s} = -\frac{1.5 \times 10^{7}}{s} \tag{13.39}$$

with a 15-pF compensating capacitor. The input-terminal signal illustrated can be justified on the basis of a detailed error-coefficient analysis using this value for $a(s)$. A simplified argument, which highlights the essential feature of the error coefficients for this type of compensation, is to recognize that Eqn. 13.39 implies that the operational amplifier itself functions as an integrator *on an open-loop basis*. Since the amplifier output signal is a triangle wave, the signal at the inverting input terminal (proportional to the derivative of the output signal) must be a square wave. The peak magnitude of the square wave at the input of the operational amplifier should be the magnitude of the slope of the output, 4×10^{5} volts per second, divided by the scale factor 1.5×10^{7} volts per second per volt from Eqn. 13.39, or approximately 27 mV. This value is confirmed by the bottom trace in Fig. 13.24*a* to within experimental errors.

Part *b* of Fig. 13.24 compares the output signal and the signal applied to the inverting input terminal of the operational amplifier with the two-pole compensation described earlier. A substantial reduction in the amplifier input signal, and thus in the error between the actual and ideal output, is clearly evident with this type of compensation. There are small-area error pulses that occur when the triangle wave changes slope. These pulses are difficult to observe in Fig. 13.24*b*. The time scale is changed to present one of these error pulses clearly in Fig. 13.24*c*. Note that this pulse is effectively an impulse compared to the time scale of the output signal. As might be anticipated, when compensation that makes the amplifier behave like a double integrator is used, the signal at the amplifier input is approximately the second derivative of its output, or a train of alternating-polarity impulses.

Eqn. 13.37 shows that

$$a(s) \simeq \frac{1.7 \times 10^{13}}{s^2} \tag{13.40}$$

at frequencies below approximately 10^6 radians per second for the two-pole compensation used. A graphically estimated value for the area of the impulse shown in Fig. 13.24c is 5×10^{-8} volt-seconds. Multiplying this area by the scale factor 1.7×10^{13} volts per second squared per volt from Eqn. 13.40 predicts a change in slope of 8.5×10^5 volts per second at each break of the triangle wave. This value is in good agreement with the actual slope change of 8×10^5 volts per second.

We should emphasize that the comparisons between one- and two-pole compensation presented here were made using one-pole compensation tailored to the attenuation of the feedback network. Had the standard 30-pF compensating-capacitor value been used, the error of the one-pole-compensated configuration would have been even larger.

13.3.4   Compensation That Includes a Zero

We have seen a number of applications where the feedback network or capacitive loading at the output of the operational amplifier introduces a pole into the loop transmission. This pole, combined with the single dominant pole often obtained via minor-loop compensation, will deteriorate stability.

Figure 13.25 shows how capacitive loading decreases the stability of the LM301A when single-pole compensation is used. The amplifier was connected as a unity-gain follower and compensated with a single 30-pF capacitor to obtain these responses. The load-capacitor values used were 0.01 μF and 0.1 μF for parts a and b, respectively.

These transient responses can be used to estimate the open-loop output resistance of the operational amplifier. We know that the open-loop transfer function for this amplifier compensated with a 30-pF capacitor is

$$a(s) \simeq \frac{7.7 \times 10^6}{s} \tag{13.41}$$

in the absence of loading. This transfer function is also the negative of the unloaded loop transmission for the follower connection. When capacitive loading is included, the loop transmission changes to

$$L(s) \simeq -\frac{7.7 \times 10^6}{s(R_o C_L s + 1)} \tag{13.42}$$

where R_o is the open-loop output resistance of the amplifier and C_L is the value of the load capacitor.

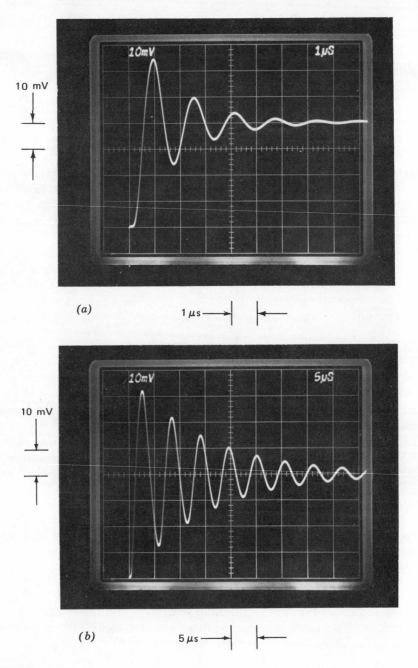

Figure 13.25 Step response of capacitively loaded unity-gain follower with one-pole compensation. (Input-step amplitude is 40 mV.) (a) 0.01-μF load capacitor. (b) 0.1-μF load capacitor.

598

The ringing frequency shown in Fig. 13.25b is approximately 1.1×10^6 radians per second. Since this response is poorly damped, the ringing frequency must closely approximate the crossover frequency. Furthermore, the poor damping also indicates that crossover occurs well above the break frequency of the second pole in Eqn. 13.42.

These relationships, combined with the known value for C_L, allow Eqn. 13.42 to be solved for R_o, with the result

$$R_o \simeq \frac{7.7 \times 10^6}{\omega_c^2 C_L} = 65 \ \Omega \qquad (13.43)$$

One simple way to improve stability is to include a zero in the unloaded open-loop transfer function of the amplifier to partially offset the negative phase shift of the additional pole in the vicinity of crossover. If a series resistor-capacitor network with component values R_c and C_c is used for compensation, the short-circuit transfer admittance of the network is

$$Y_c = \frac{C_c s}{R_c C_c s + 1} \qquad (13.44)$$

The approximate value for the corresponding unloaded open-loop transfer function of the amplifier is

$$a(s) \simeq \frac{K(R_c C_c s + 1)}{C_c s} \qquad (13.45)$$

An estimation of the complete, unloaded open-loop transfer with this type of compensation, based on Eqn. 13.45 and representative uncompensated amplifier characteristics, is shown in Fig. 13.26. Note that in this case the slope of the approximating function is zero when it intersects the uncompensated transfer function. The geometry involved shows that the approximation fails at lower frequencies than was the case with other types of compensation.

The approximate loop transmission for the unity-gain follower with capacitive loading and this type of compensation is

$$L(s) \simeq - \frac{K(R_c C_c s + 1)}{C_c s (R_o C_L s + 1)} \qquad (13.46)$$

Appropriate R_c and C_c values for the LM301A loaded with a 0.1-μF capacitor are 33 kΩ and 30 pF, respectively. Substituting these and other previously determined values into Eqn. 13.46 yields

$$L(s) \simeq - \frac{7.7 \times 10^6 (10^{-6} s + 1)}{s(6.5 \times 10^{-6} s + 1)} \qquad (13.47)$$

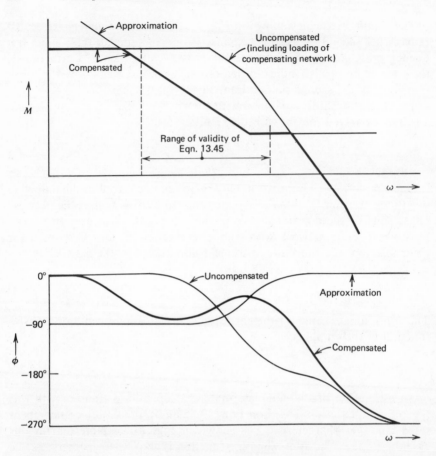

Figure 13.26 Unloaded open-loop transfer function for compensation that includes a zero.

The crossover frequency for Eqn. 13.47 is 1.4×10^6 radians per second and the phase margin is approximately 55°, although higher-frequency poles ignored in the approximation will result in a lower phase margin for the actual system.

The step response of the test amplifier connected this way is shown in Fig. 13.27a. Although the basic structure of the transient response is far superior to that shown in Fig. 13.25b, there is a small-amplitude high-frequency ringing superimposed on the main transient. This component, at a frequency well above the major-loop crossover frequency, reflects potential minor-loop instability.

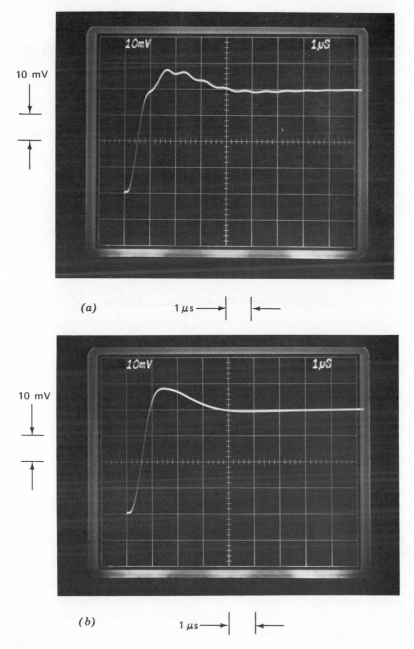

(a)

1 μs →| |←

(b)

1 μs →| |←

Figure 13.27 Step response of unity-gain inverter loaded with 0.1 μF capacitor and compensated with a zero. (Input-step amplitude is 40 mV for parts *a* and *b*, 2 mV for part *c*.) (*a*) With series resistor-capacitor compensation. (*b*) With compensating network of Fig. 13.29. (*c*) Smaller amplitude input signal.

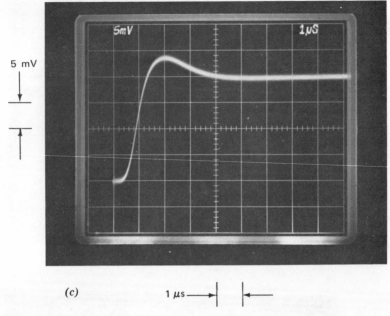

(c)

1 µs ——→

Figure 13.27—Continued

Figure 13.28 illustrates the mechanism responsible for the instability. This diagram combines an idealized model for the second stage of a two-stage amplifier with a compensating network. (The discussion of Section 9.2.3 justifies this general model for a high-gain stage.) When the crossover frequency of the loop formed by the compensating network is much higher than $1/R_iC_i$, the second stage input looks capacitive at crossover. If the compensating-network transfer admittance is capacitive in the vicinity of crossover, the phase margin of the inner loop approaches 90°. Alternatively, if the compensating network is resistive, the input capacitance introduces a second pole into the inner-loop transmission and the phase margin of this loop drops.

The solution is to add a small capacitor to the compensating network as indicated in Fig. 13.29. The additional element insures that the network transfer admittance is capacitive at the minor-loop crossover frequency, thus improving stability. The approximate loop transmission of the major loop is changed from that given in Eqn. 13.47 to

$$L(s) \simeq - \frac{7.7 \times 10^6(10^{-6}s + 1)}{s(6.5 \times 10^{-6}s + 1)(10^{-7}s + 1)} \tag{13.48}$$

The effect on the major loop is to introduce a pole at a frequency approximately a factor of 7 above crossover, thereby reducing phase margin by 8°.

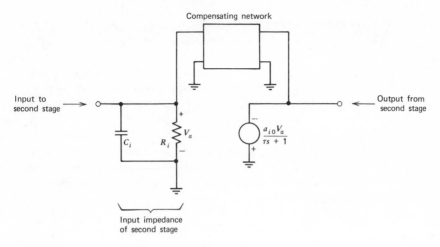

Figure 13.28   Model for second stage of a two-stage amplifier.

The step response shown in Fig. 13.27*b* results with this modification. An interesting feature of this transient is that it also has a tail with a duration that seems inconsistent with the speed of the initial rise. While there is a zero included in the closed-loop transfer function of this connection since the zero in Eqn. 13.48 occurs in the forward path, the zero is close to the crossover frequency of the major loop. Consequently, any tail that resulted from a doublet formed by a closed-loop pole combining with this zero would have a decay time consistent with the crossover frequency of the major loop. In fact, the duration of the tail evident in Fig. 13.27*b* is reasonable in view of the 1.4×10^6 radian-per-second crossover frequency of the major loop. The inconsistency stems from an *initial rise that is too fast*.

The key to explaining this phenomenon is to note that the output-signal slope reaches a maximum value of approximately 6×10^4 volts per second, implying a 6-mA charging current into the 0.1-μF capacitor. This current level is substantially above the quiescent current of the output stage of the LM301A, and results in a lowered output resistance from the active emitter follower during the rapid transition. Consequently, the pole asso-

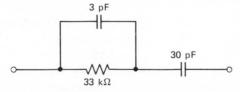

Figure 13.29   Compensating network used to obtain transients shown in Figs. 13.27*b* and 13.27*c*.

ciated with capacitive loading moves toward higher frequencies during the initial high-current transient, and the speed of response of the system improves in this portion.

Figure 13.27c verifies this reasoning by illustrating the response of the capacitively loaded follower to a 2-mV input signal step. A gain-of-10 amplifier (realized with another appropriately compensated LM301A) amplified the output signal to permit display at the 5 mV-per-division level indicated in the photograph. While this transient is considerably more noisy (reflecting the lower-amplitude signals), the relative speed of various portions of the transient is more nearly that expected of a linear system.

The fractional change in output resistance with output current level is probably less than 25 % for this amplifier because the dominant component of output resistance is the value at the high-resistance node divided by the current gain of the buffer amplifier. Consequently, the differences between Figs. 13.27b and 13.27c are minor. It should also be noted that the estimated value for R_o (Eqn. 13.43) is probably slightly low because of this effect.

Many applications, such as sample-and-hold circuits or voltage regulators, apply capacitive loading to an operational amplifier. Other connections, such as a differentiator, add a pole to the loop transmission because of the transfer function of the feedback network. The method of adding a zero to single-pole compensation can improve performance substantially in these types of applications.

The comparison between Figs. 13.25b and 13.27b shows how changing from 30-pF compensation to compensation that includes a zero can greatly improve stability and can reduce settling time by more than a factor of 10 for a capacitively loaded voltage follower.

It should be emphasized that this type of compensation is not suggested for general-purpose use, since the compensating-network element values must be carefully chosen as a function of loop-parameter values for acceptable stability. If, for example, the pole that introduced the need for this type of compensation is eliminated or moved to a higher frequency, the crossover frequency increases and instability may result.

13.3.5 Slow-Rolloff Compensation

The discussion of the last section showed how compensation can be designed to introduce a zero into the compensated open-loop transfer function of an operational amplifier. The zero can be used to offset the effects of a pole associated with other elements in the loop. Since the zero location is selected as a function of other loop parameters, this type of compensation is effectively specifically tailored for one fixed feedback network and load.

There are applications where the transfer functions of certain elements in an operational-amplifier loop vary as a function of operating conditions or as the components surrounding the amplifier change. The change in amplifier open-loop output resistance described in connection with Figs. 13.27b and 13.27c is one example of this type of parameter variation.

Operational amplifiers that are used (often with the addition of high-current output stages) to supply regulated voltages are another example. The total capacitance connected to the output of a supply is often dominated by the decoupling capacitors included with the circuits it powers. The output resistance of the power stage may also be dependent on load current, and these two effects can combine to produce a major uncertainty in the location of the pole associated with capacitive loading. One approach to stabilizing this type of regulator was described in Section 5.2.2.

A third example of a variable-parameter loop involves the use of an operational amplifier, an incandescent lamp, and a photoresistor in a feedback loop intended to control the intensity of the lamp. In this case, the dynamics of both the lamp and the photoresistor as well as the low-frequency "gain" of the combination depend on light level.

The stabilization of variable-parameter systems is often difficult and compromises, particularly with respect to settling time and desensitivity, are frequently necessary. This section describes one approach to the stabilization of such systems and indicates the effect of the necessary compromises on performance.

Consider a variable-parameter system that has a loop transmission

$$L(s) = -a(s)\left(\frac{k}{\tau s + 1}\right) \tag{13.49}$$

where k and τ represent the uncertain values associated with elements external to the operational amplifier. It is assumed that these parameters can have any positive values. If the amplifier open-loop transfer function is selected such that

$$a(s) = \frac{K'}{\sqrt{s}} \tag{13.50}$$

the phase margin of Eqn. 13.49 will be at least 45° for any values of k and τ, since the phase shift of the function $1/\sqrt{s}$ is $-45°$ at all frequencies.

In order to obtain the open-loop transfer function indicated by Eqn. 13.50 from a two-stage amplifier, it is necessary to use a network that has a short-circuit transfer admittance proportional to $\sqrt{s}$. While the required

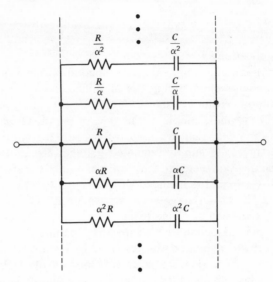

Figure 13.30 Network used to approximate an admittance proportional to $\sqrt{s}$.

admittance cannot be realized with a lumped, finite network, it can be approximated by the ladder structure shown in Fig. 13.30. The driving-point admittance of this network (which is, of course, equal to its short-circuit transfer admittance) is

$$Y_c(s) = +\cdots+ \frac{(C/\alpha^2)s}{(RC/\alpha^4)s + 1} + \frac{(C/\alpha)s}{(RC/\alpha^2)s + 1} + \frac{Cs}{RCs + 1}$$

$$+ \frac{\alpha Cs}{\alpha^2 RCs + 1} + \frac{\alpha^2 Cs}{\alpha^4 RCs + 1} + \cdots + \quad (13.51)$$

The poles of Eqn. 13.51 are located at

$$\vdots$$

$$p_{n+2} = -\frac{\alpha^4}{RC}$$

$$p_{n+1} = -\frac{\alpha^2}{RC}$$

$$p_n = -\frac{1}{RC}$$

$$p_{n-1} = -\frac{1}{\alpha^2 RC}$$

$$p_{n-2} = -\frac{1}{\alpha^4 RC} \tag{13.52}$$

$$\vdots$$

while its zeros are located at

$$\vdots$$

$$z_{n+2} = -\frac{\alpha^3}{RC}$$

$$z_{n+1} = -\frac{\alpha}{RC}$$

$$z_n = -\frac{1}{\alpha RC}$$

$$z_{n-1} = -\frac{1}{\alpha^3 RC}$$

$$z_{n-2} = -\frac{1}{\alpha^5 RC} \tag{13.53}$$

$$\vdots$$

This admittance function has poles and zeros that alternate along the negative real axis, with the ratio of the locations of any two adjacent singularities a constant equal to α. On the average, the magnitude of this function will increase proportionally to the square root of frequency since on an asymptotic log-magnitude versus log-frequency plot it alternates equal-duration regions with slopes of zero and one.

If this network is used to compensate a two-stage amplifier, the amplifier open-loop transfer function

$$a(s) \simeq \frac{K}{Y_c(s)} \tag{13.54}$$

will approximate the relationship given in Eqn. 13.50. If the magnitude of uncompensated amplifier open-loop transfer function is adequately high, the range of frequencies over which the approximation is valid can be made arbitrarily wide by using a sufficiently large number of sections in the ladder network. Note that it is also possible to make the compensated transfer function be proportional to $1/s^r$, where r is between zero and one, by appropriate selection of relative pole-zero spacing in the compensating network.

Since the usual objective of this type of compensation is to maintain satisfactory phase margin in systems with uncertain parameter values, guidelines for selecting the frequency ratio between adjacent singularities α are best determined by noting how the phase of the actual transfer function is influenced by this quantity. If the poles and zeros are closely spaced, the phase shift of the compensated open-loop transfer function will be approximately $-45°$ over the effective frequency range of the network. As α is increased, the magnitude of the phase ripple with frequency, which is symmetrical with respect to $-45°$, increases. The maximum negative phase shift of $a(j\omega)$ (see Eqns. 13.51 and 13.54) is plotted as a function of α in Fig. 13.31.

This plot shows that reasonably large values of α can be used without the maximum negative phase shift becoming too large. If, for example, a spacing between adjacent singularities of a factor of 10 in frequency is

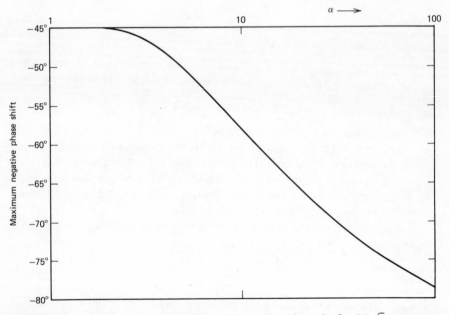

Figure 13.31 Maximum negative phase shift as a function of α for $1/\sqrt{s}$ compensation.

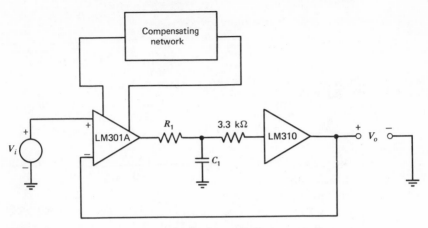

Figure 13.32   Circuit used to evaluate slow-rolloff compensation.

used, the maximum negative phase shift is $-58°$. Since the phase ripple is symmetrical with respect to $-45°$, the phase shift varies from $-32°$ to $-58°$ as a function of frequency for this value of α. If an amplifier compensated using a network with $\alpha = 10$ is combined in a loop with an element that produces an additional $-90°$ of phase shift at crossover, the system phase margin will vary from 58° to 32°.

The performance of a $1/\sqrt{s}$ system is compared with that of alternatively compensated systems using the connection shown in Fig. 13.32. Providing that the open-loop output resistance of the operational amplifier is much lower than R_1, the R_1-C_1 network adds a pole with a well-determined location to the loop transmission of the system. The LM310 unity-gain follower is used to avoid loading the network. The 3.3-kΩ resistor included in series with the LM310 input is recommended by the manufacturer to improve the stability of this circuit. The bandwidth of this follower is high enough to have a negligible effect on loop dynamics.

The circuit shown in Fig. 13.32 has a forward-path transfer function equal to $a(s)/(RCs + 1)$ and a feedback transfer function of one.

Three different types of compensation were evaluated with this connection. One type was single-pole compensation using a 220-pF capacitor. The approximate open-loop transfer function of the LM301A is

$$a(s) \simeq \frac{10^6}{s} \qquad (13.55)$$

with this compensation. The corresponding loop transmission is

$$L(s) = -\frac{10^6}{s(R_1C_1s + 1)} \qquad (13.56)$$

The closed-loop transfer function is

$$\frac{V_o(s)}{V_i(s)} = A(s) = \frac{1}{10^{-6}R_1C_1s^2 + 10^{-6}s + 1} \tag{13.57}$$

Second-order parameters for Eqn. 13.57 are

$$\omega_n = \frac{10^3}{\sqrt{R_1C_1}} \quad \text{and} \quad \zeta = \frac{5 \times 10^{-4}}{\sqrt{R_1C_1}}$$

As expected, increasing the R_1-C_1 time constant lowers both the natural frequency and the damping ratio of the system.

The second compensating network was an 11 "rung" ladder network of the type shown in Fig. 13.30. The sequence of resistor-capacitor values used for the rungs was 330 Ω–10 pF, 1 kΩ–33 pF, 3.3 kΩ–100 pF $\cdots$ 10 MΩ–0.33 μF, 33 MΩ–1 μF.

This network combines with operational-amplifier parameters to yield an approximate open-loop transfer function

$$a'(s) \simeq \frac{10^3}{\sqrt{s}} \tag{13.58}$$

over a frequency range that extends from below 0.1 radian per second to above 10^6 radians per second. The value of α for the approximation is $\sqrt{10}$. The curve of Fig. 13.31 shows that the maximum negative phase shift of the open-loop transfer function at intermediate frequencies should be $-46.5°$, corresponding to a peak-to-peak phase ripple of 3°.

The approximate loop transmission that results with this compensation is

$$L'(s) = -\frac{10^3}{\sqrt{s}(R_1C_1s + 1)} \tag{13.59}$$

The third compensation used the two-rung slow-rolloff network shown in Fig. 13.33. The resultant amplifier open-loop transfer function is

$$a''(s) \simeq \frac{10^5(10^{-4}s + 1)}{s(10^{-5}s + 1)} \tag{13.60}$$

This transfer function is a very crude approximation to a $1/\sqrt{s}$ rolloff that combines a basic $1/s$ rolloff with a decade-wide zero-slope region realized by placing a zero two decades and a pole one decade below the unity-gain frequency. Alternatively, the open-loop transfer function can be viewed as the result of adding a lead network located well below the unity-gain frequency to a single-pole transfer function.

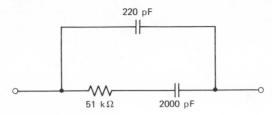

Figure 13.33   Slow-rolloff network.

The loop transmission in this case is

$$L''(s) = -\frac{10^5(10^{-4}s + 1)}{s(10^{-5}s + 1)(R_1C_1s + 1)} \qquad (13.61)$$

Bode plots for the three compensated open-loop transfer functions of Eqns. 13.55, 13.58, and 13.60 are shown in Fig. 13.34. Note that parameters are selected so that unity-gain frequencies are identical for the three transfer functions.

The step responses for the test system with $R_1C_1 = 0$ are compared for the three types of compensation in Fig. 13.35. Part a shows the step response for one-pole compensation. The expected exponential response with a 1-μs 0 to 63% rise time is evident.

Part b shows the response with the $1/\sqrt{s}$ compensation. An interesting feature of this response is that while it actually starts out faster than that of the previous system with the same crossover frequency (compare, for example, the times required to reach 25% of final value), it settles much more slowly. Note that the transient shown in part b has only reached 75% of final value after 4.5 μs (the input-step amplitude is 40 mV for both parts a and b), while the system using one-pole compensation has settled to within 2% of final value by this time. Part c is a repeat of part b with a slower sweep speed. Note that even after 180 μs, the transient has reached only 95% of final value. This type of very slow creep toward final value is characteristic of many types of distributed systems. Long transmission lines, for example, often exhibit step responses similar in form to that illustrated.

Parts d and e of Fig. 13.35 show the response for the system using slow-rolloff compensation at two different time scales. The transient consists of a 1-μs time constant exponential rise to 90% of final value, followed by a 100-μs time constant rise to final value. The reader should use Eqn. 13.61 to convince himself that the long tail is anticipated in view of the location of the closed-loop pole-zero doublet that results in this case. Note that even with this tail, settling to a small fraction of final value is substantially shorter than for the $1/\sqrt{s}$ system.

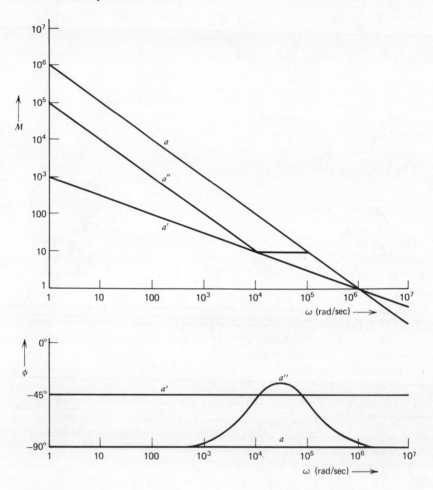

Figure 13.34 Comparison of approximate open-loop transfer functions for three types of compensation.

Figure 13.36 indicates responses for $R_1C_1 = 1$ μs for the three different types of compensation. This R_1-C_1 product adds a pole slightly above the resultant crossover frequency of the loop for all of the compensations. The phase margin of the system with one pole compensation is about 50°, with the resulting moderate damping shown in part a. The phase margin for $1/\sqrt{s}$ compensation exceeds 90° in this case, and the main effect of the extra pole is to make the initial portion of the response (see part b) look somewhat more exponential. The very slow tail is not altered substantially. The step response of the system with slow-rolloff compensation (Fig. 13.36c)

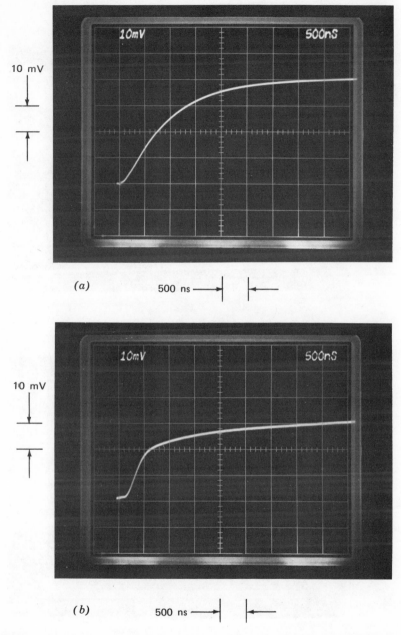

(a)

500 ns ⟶

(b)

500 ns ⟶

Figure 13.35 Comparison of step responses as a function of compensation with $R_1C_1 = 0$. (Input-step amplitude is 40 mV.) (a) One-pole compensation. (b) $1/\sqrt{s}$ compensation. (c) Repeat of part b with slower sweep speed. (d) Slow-rolloff compensation. (e) Repeat of part d with slower sweep speed.

613

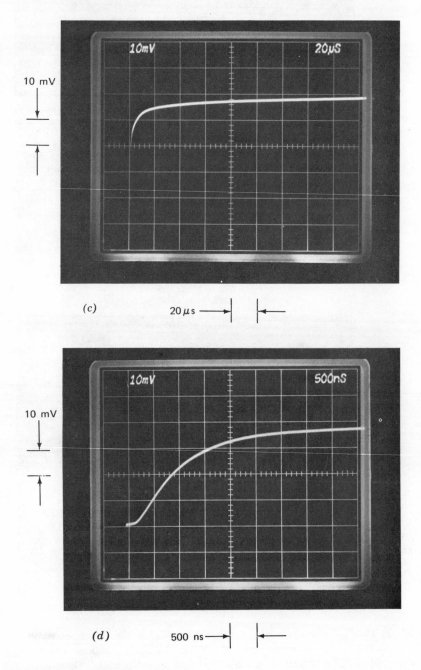

(c)

10 mV

20 µs

(d)

10 mV

500 ns

Figure 13.35—Continued

614

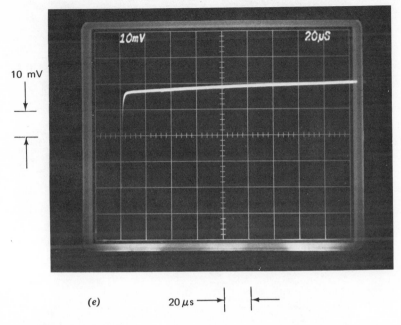

(e)   20 μs →| |←

Figure 13.35—Continued

has slightly less peak overshoot (measured from the final value shown in the figure) than does the system shown in part *a*. The difference reflects the 5° phase-margin advantage of the slow-rolloff system. The tail is unaltered by the additional pole.

The experimentally measured step response of the system with one-pole compensation is shown in Fig. 13.37 for a number of values of the R_1-C_1 time constant. The deterioration of stability and settling time that results as R_1C_1 is increased is clearly evident in this sequence. The value for natural frequency predicted by Eqn. 13.57 can be verified to within experimental tolerances. However, the actual system is actually somewhat *better* damped than the analysis indicates, particularly in the relatively lower-damped cases. The unit-step response for a second-order system is

$$v_o(t) = \left[1 - \frac{1}{\sqrt{1 - \zeta^2}} e^{-\zeta \omega_n t} \sin\left(\sqrt{1 - \zeta^2}\ \omega_n t + \Phi \right) \right] \qquad (13.62)$$

where

$$\Phi = \tan^{-1}\left[\frac{\sqrt{1 - \zeta^2}}{\zeta} \right]$$

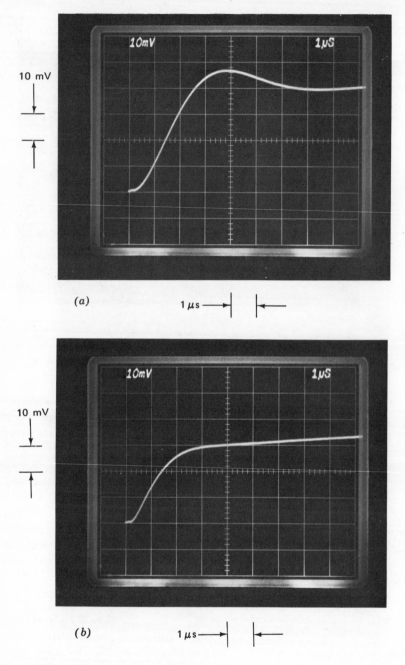

Figure 13.36 Comparison of step responses as a function of compensation for $R_1C_1 = 1$ μs. (Input-step amplitude is 40 mV.) (a) One-pole compensation. (b) $1/\sqrt{s}$ compensation. (c) Slow-rolloff compensation.

616

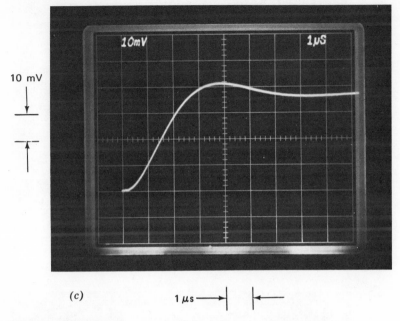

10 mV

(c)   1 μs ⟶

Figure 13.36—Continued

This relationship shows that the exponential time constant of the envelope of the transient should have a value of $1/\zeta\omega_n$, or, from Eqn. 13.57, $2R_1C_1$. Thus, for example, the transient illustrated in Fig. 13.37e, which has analytically determined values of $\omega_n = 3.1 \times 10^3$ radians per second and $\zeta = 1.6 \times 10^{-3}$, should have a decay time approximately five times longer than that actually measured.

The reason for this discrepancy is as follows. An extension of the curves shown in Fig. 4.26 estimates that a damping ratio of 1.6×10^{-3} corresponds to a phase margin of 0.184°. Accordingly, very small changes in the angle of the loop transmission at the crossover frequency can change damping ratio by a substantial factor.

There are at least three effects, which (in apparent violation of Murphy's laws) combine to improve phase margin in the actual system. First, the compensated amplifier open-loop pole is not actually at the origin, and thus contributes less than 90° of negative phase shift to the loop transmission at crossover. Second, any series resistance associated with the connections made to the capacitor adds a zero to the loop transmission that contributes positive phase shift at crossover. Third, the losses associated with dielectric absorption or dissipation factor of the capacitor also improve the phase margin of the system.

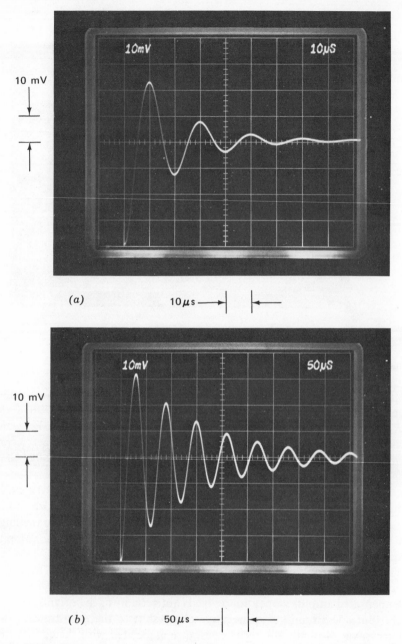

(a)   10 μs ——| |←——

(b)   50 μs ——| |←——

Figure 13.37 Step response of system with one-pole compensation as a function of R_1C_1. (Input-step amplitude is 40 mV.) (a) $R_1C_1 = 10$ μs. (b) $R_1C_1 = 100$ μs. (c) $R_1C_1 = 1$ ms. (d) $R_1C_1 = 10$ ms. (e) $R_1C_1 = 100$ ms. (f) $R_1C_1 = 100$ ms with polystyrene capacitor.

618

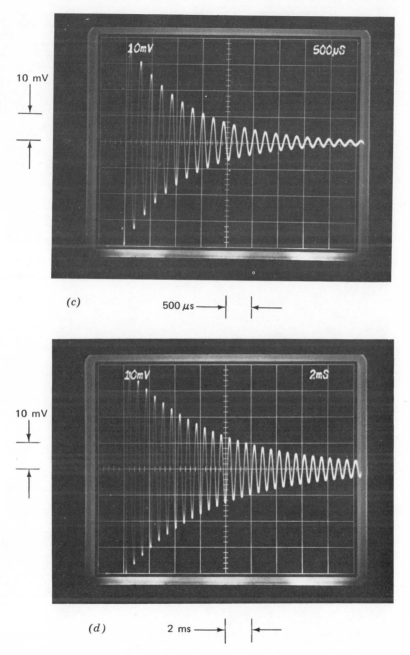

(c)   500 μs →| |←

(d)   2 ms →| |←

Figure 13.37—Continued

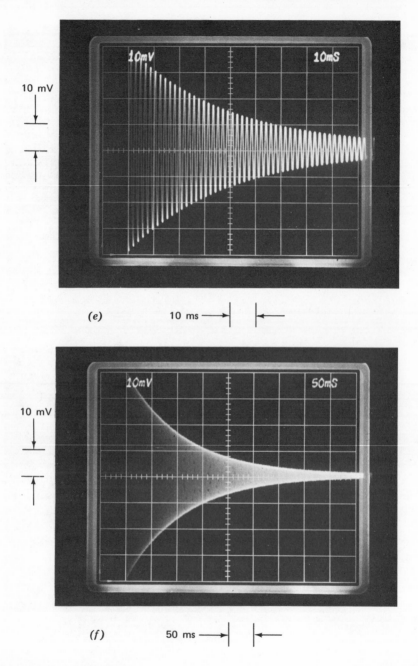

10 mV

(e)   10 ms →| |←

10 mV

(f)   50 ms →| |←

Figure 13.37—Continued

620

The importance of the third effect can be seen by comparing parts e and f of Fig. 13.37. For part e (and all preceding photographs) a ceramic capacitor was used. The transient indicated in part f, with a decay time approximately three times that of part e and within 60% of the analytically predicted value, results when a low-loss polystyrene capacitor is used in place of the ceramic unit. This comparison demonstrates the need to use low-loss capacitors in lightly damped systems such as oscillators.

It should also be noted that, in addition to very low damping, this type of connection can lead to inordinately high signal levels with the possibility of saturation at certain points inside the loop. Since the frequency of the ringing at the system output is higher than the cutoff frequency of the R_1-C_1 low-pass network, the signal out of the LM301A will be larger than the system output signal during the oscillatory period. In fact, the peak signal level at the output of the LM301A exceeded 20 volts peak-to-peak during the transients shown in Figs. 13.37e and 13.37f. Longer R_1-C_1 time constants would have resulted in saturation with the 40-mV step input.

Figure 13.38 shows the responses of the system compensated with a $1/\sqrt{s}$ amplifier rolloff as a function of the R_1-C_1 product. Several analytically predictable features of this system are demonstrated by these responses. Since the magnitude of the loop transmission falls as $1/\omega^{3/2}$ at frequencies above the pole of the R_1-C_1 network and as $1/\omega^{1/2}$ below the pole location, the loop crossover frequency decreases as $1/R_1C_1^{2/3}$. A factor of 10 increase in the R_1-C_1 product lowers crossover by a factor of $10^{2/3} = 4.64$, while a three-decade change in this product changes crossover by two decades. Comparing, for example, parts c and f or d and g of Fig. 13.38 shows that while the general shapes of these responses are similar, the speeds differ by a factor of 100, reflecting the change in crossover frequency that occurs for a factor of 1000 change in the R_1-C_1 product. Part h is somewhat faster than predicted by the above relationship because, with an R_1-C_1 value of 100 seconds, the corresponding pole lies at frequencies below the $1/\sqrt{s}$ region of the compensation. (Recall that the longest time constant in the compensating network is 33 seconds.) The $1/\sqrt{s}$ rolloff could be extended to lower frequencies by using more sections in the network, but very long time constants would be required. The amplifier d-c gain of approximately 10^5 would permit a $1/\sqrt{s}$ rolloff from 10^{-4} radian per second to unity gain at 10^6 radians per second.

The crossover frequencies for parts a, b, and c are located at factors of approximately 2.16, 4.64, and 10, respectively, above the break frequency of the R_1-C_1 network. Accordingly, the pole associated with the R_1-C_1 network produces somewhat less than $-90°$ of phase shift in these cases, with the result that the phase margin is above $45°$. This effect is negligible in

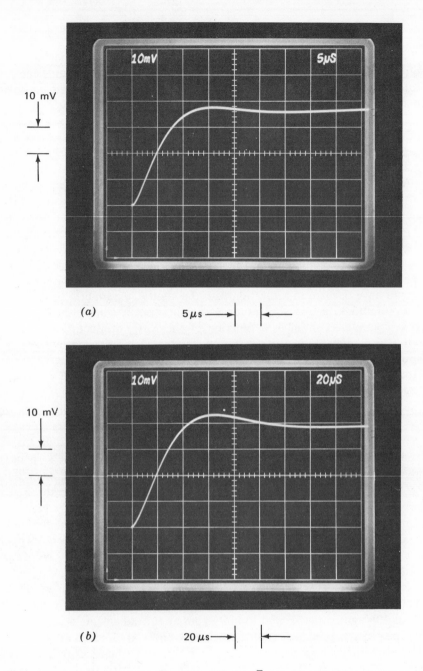

(a)

5 μs →| |←

(b)

20 μs →| |←

Figure 13.38 Step response of system with $1/\sqrt{s}$ compensation as a function of R_1C_1. (Input-step amplitude is 40 mV.) (a) $R_1C_1 = 10\mu s$. (b) $R_1C_1 = 100$ μs. (c) $R_1C_1 = 1$ ms. (d) $R_1C_1 = 10$ ms. (e) $R_1C_1 = 100$ ms. (f) $R_1C_1 = 1$ second.

622

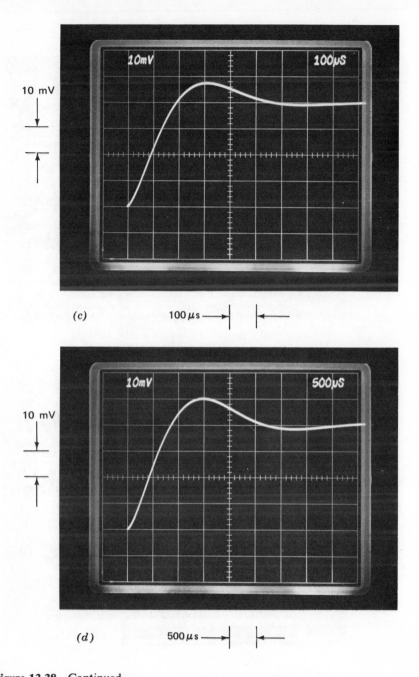

(c)

100 μs ⟶| |←

(d)

500 μs ⟶| |←

Figure 13.38—Continued
(g) $R_1C_1 = 10$ seconds. (h) $R_1C_1 = 100$ seconds. (i) Comparison of part d with second-order system.

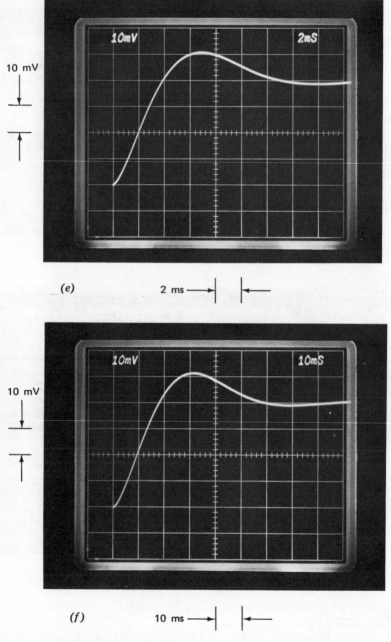

(e)

2 ms

(f)

10 ms

Figure 13.38—Continued

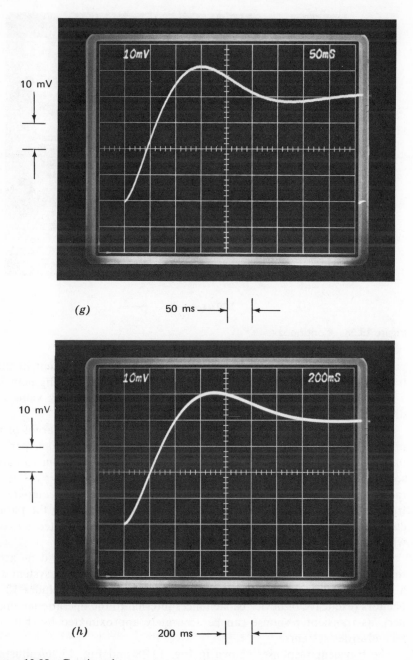

(g)   50 ms ➝ | |◄—

(h)   200 ms ➝ | |◄—

Figure 13.38—Continued

625

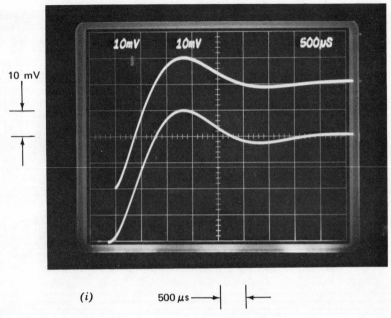

(i)   500 μs ⟶| |⟵

Figure 13.38—Continued

part *d* through *h*, and the slight differences in damping evident in these transients arise because of the phase ripple of the compensating network. The actual ripple is probably larger than the 3° peak-to-peak value predicted in Fig. 13.31 as a consequence of component tolerances.

The transient shown in Fig. 13.38*d* results from a phase margin of approximately 45° and a crossover frequency of 2.16×10^3 radians per second. The curves of Fig. 4.26 indicate that the appropriate approximating second-order system in this case is one with $\zeta = 0.42$ and $\omega_n = 2.5 \times 10^3$ radians per second. The two responses shown in Fig. 13.38*i* compare the transient shown in part *d* with a second-order response using the parameters developed with the aid of Fig. 4.26. The reader is invited to guess which transient is which.

The remarkable similarity of these two transients is a further demonstration of the validity of approximating the response of a complex system with a far simpler transient. Note that while the actual system includes 12 capacitors (exclusive of device capacitances internal to the operational amplifier), its transient response can be accurately approximated by that of a second-order system.

The transient responses shown in Fig. 13.38 and Fig. 13.36*b* illustrate how $1/\sqrt{s}$ compensation can maintain remarkably constant relative stability as a system pole location is varied over eight decades of frequency. Actually, even lower values for the R_1-C_1 break frequencies[3] yield comparable results, although difficulties associated with obtaining the very long time constants required and photographing the resulting slow transients prevented including additional responses.

Since this type of compensation eliminates the relatively high-frequency oscillations of the output signal, the signal levels at the output of the LM301A are considerably smaller than when one-pole compensation is used.

The step response of the system with slow-rolloff compensation is indicated in Fig. 13.39 for values of R_1C_1 from 10 μs to 100 ms. The important point illustrated in these photographs is that the response of the system remains moderately well damped for R_1-C_1 products as large as 1 ms, and that the damping is superior to that of the system using single-pole compensation for any R_1-C_1 value shown. The reason is explained with the aid of Fig. 13.40, which is a plot of phase margin as a function of $1/R_1C_1$ for this system. Note that the phase margin exceeds 30° for any value of R_1C_1 less than approximately 3 ms.

While the variation in phase margin with R_1C_1 is larger for this system than for the system with $1/\sqrt{s}$ compensation, this type of compensation can result in reasonable stability as the location of the variable pole changes from more than three decades below the unity-gain frequency of the amplifier upward. In exchange for the somewhat greater variation in phase margin as a function of the R_1-C_1 time constant and a more limited range of this product for acceptable stability, the complexity of the amplifier compensating network is reduced from 22 to three components.

Simple slow rolloff networks typified by that described above provide useful compensation in many practical variable-parameter systems because the range of parameter variation is seldom as great as that used to illustrate the performance of the system with $1/\sqrt{s}$ compensation. Furthermore, other effects may combine with slow-rolloff compensation to increase its effectiveness in actual systems. Consider the voltage regulator with an arbitrarily large capacitive load mentioned earlier as an example of a variable parameter system. The series resistance and dissipation characteristic of electrolytic capacitors add a zero to the pole associated with the capacitive load, and this zero can aid slow-rolloff compensation in stabilizing a regulator for a very wide range of load-capacitor values.

It is also evident that adding one or more rungs to the compensating network can increase the range of this type of compensation when required.

---

[3] While R_1-C_1 time constants in excess of about 10 seconds cause some deviation from $1/\sqrt{s}$ characteristics in the vicinity of the R_1-C_1 pole, the phase margin is determined by the network characteristics at the crossover frequency. The system phase margin will remain approximately 45° for R_1-C_1 time constants as large as 10^4 seconds.

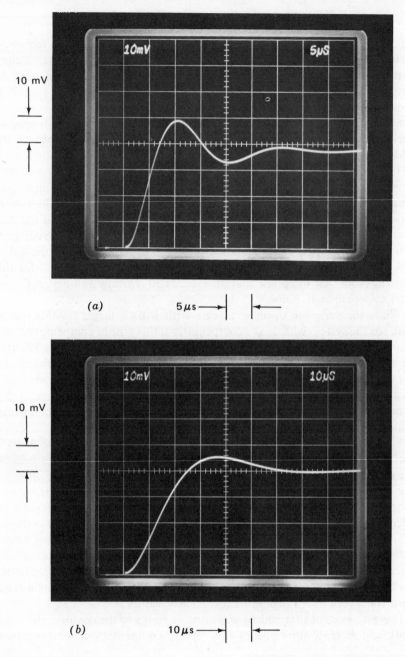

Figure 13.39 Step response for system with slow-rolloff compensation as a function of R_1C_1. (Input-step amplitude = 40 mV.) (a) $R_1C_1 = 10\ \mu s$. (b) $R_1C_1 = 100\ \mu s$. (c) $R_1C_1 = 1$ ms. (d) $R_1C_1 = 10$ ms. (e) $R_1C_1 = 100$ ms.

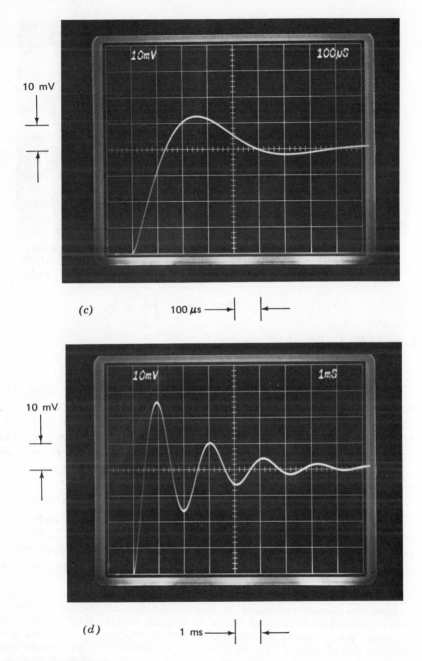

10 mV

(c)   100 µs ⟶| |⟵

10 mV

(d)   1 ms ⟶| |⟵

Figure 13.39—Continued

629

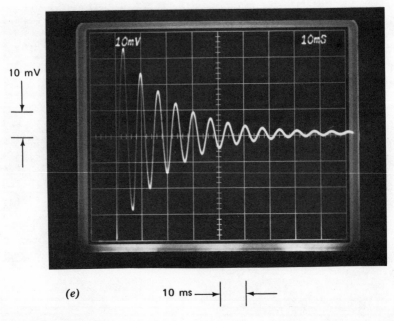

(e)   10 ms →

Figure 13.39—Continued

13.3.6 Feedforward Compensation

Feedforward compensation was described briefly in Section 8.2.2. This method, which differs in a fundamental way from minor-loop compensation, involves capacitively coupling the signal at the inverting input terminal of an operational amplifier to the input of the final voltage-gain stage. This final stage is assumed to provide an inversion. The objective is to eliminate the dynamics of all but the final stage from the amplifier open-loop transfer function in the vicinity of the unity-gain frequency.

This approach, which has been used since the days of vacuum-tube operational amplifiers, is not without its limitations. Since only signals at the inverting input terminal are coupled to the output stage, the feedforward amplifier has much lower bandwidth for signals applied to its noninverting input and generally cannot be effectively used in noninverting configurations.[4]

----

[4] There have been several attempts at designing amplifiers that use dual feed-forward paths to a differential output stage. One of the difficulties with this approach arises from mismatches in the feedforward paths. A common-mode input results in an output signal because of such mismatches. The time required for the error to settle out is related to the dynamics of the bypassed amplifier, and thus very long duration tails result when these amplifiers are used differentially.

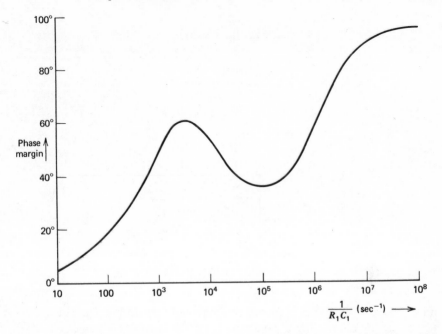

Figure 13.40 Phase margin as a function of $1/R_1C_1$ for $L''(s) = -[10^5(10^{-4}s + 1)/s(10^{-5}s + 1)(R_1C_1s + 1)]$.

Another difficulty is that the phase shift of a feedforward amplifier often approaches $-180°$ at frequencies well below its unity-gain frequency. Accordingly, large-signal performance may be poor because the amplifier is close to conditional stability. The excessive phase shift also makes feedforward amplifiers relatively intolerant of capacitive loading.

Feedforward is normally most useful for three or more stage amplifiers, and results in relatively little performance improvement for many two-stage designs because the first stage of these amplifiers is often faster than the rest of the amplifier. The LM101A[5] is an exception to this generality. Recall that the input stage of this amplifier includes lateral-PNP transistors. Because NPN transistors are used for voltage gain in the second stage, the first stage represents the bandwidth bottleneck for the entire amplifier.[6] Since the input to the second amplifier stage is available as a compensating ter-

[5] R. C. Dobkin, *Feedforward Compensation Speeds Op Amp*, Linear Brief 2, National Semiconductor Corporation.

[6] The unity-gain output stage of this amplifier also uses lateral-PNP transistors. However, the buffer stage has approximately unity small-signal voltage gain even at frequencies where the common-emitter current gain of the lateral PNP's in this stage is zero. Thus these transistors do not have the dominant effect on amplifier bandwidth that the input transistors do.

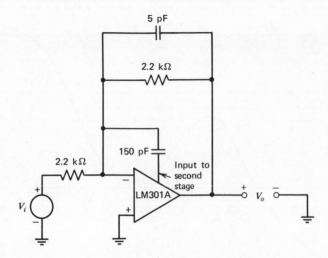

Figure 13.41 Unity-gain inverter with feedforward compensation.

minal, feedforward that bypasses the narrow-bandwidth stage can be implemented by connecting a capacitor from the inverting input terminal of the amplifier to this compensation terminal.

The LM301A was connected as shown in Fig. 13.41. Low-value resistors and the 5-pF capacitor are used to reduce the effects of amplifier input capacitance on loop transmission. The 150-pF feedforward capacitor is the value recommended by National Semiconductor Corporation, although other values may give better performance in some applications. This capacitor value can be selected to minimize the signal at the inverting input terminal (which is proportional to the error between actual and ideal output) if optimum performance is required.

The step response of this inverter is shown in Fig. 13.42. There is substantial overshoot evident in the figure, as well as some "teeth" on the rising portion of the waveform that are probably at least partially related to high-speed grounding problems in the test set up. The 10 to 90% rise time of the circuit is approximately 50 ns, or a factor of three faster than the fastest rise time obtained with minor-loop compensation (see Fig. 13.13*b*).

13.3.7 Compensation to Improve Large-Signal Performance

The discussion in earlier parts of this chapter has focused on how compensation influences the linear-region performance of an operational amplifier. The compensation used in a particular connection also has a profound effect on the large-signal performance of the amplifier, particularly the

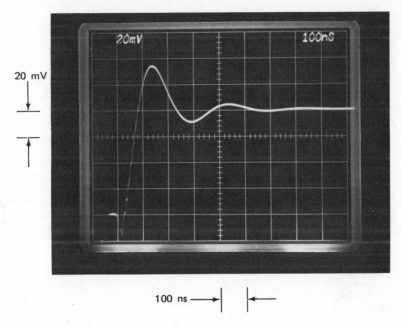

20 mV

100 ns

Figure 13.42 Step response of unity-gain inverter with feedforward compensation. (Input-step amplitude is −80 mV.)

slew rate or maximum time rate of change of output signal and how gracefully the amplifier recovers from overload.

The simplified two-stage amplifier representation shown in Fig. 13.43 illustrates how compensation can determine the linear operating region of the amplifier. This model, which includes a current repeater, can be slightly modified to represent many available two-stage integrated-circuit operational amplifiers such as the LM101A. Elimination of the current repeater, which does not alter the essential features of the following argument, results in a topology adaptable to the discrete designs that do not include these transistors.

The key to understanding the performance of this amplifier is to recognize that, with a properly designed second stage, the input current required by this stage is negligible when it is in its linear operating region. Accordingly,

$$i_N = i_{C4} - i_{C2} \tag{13.63}$$

This relationship, coupled with the fact that the incremental voltage change at the input of the second stage is also small under many operating con-

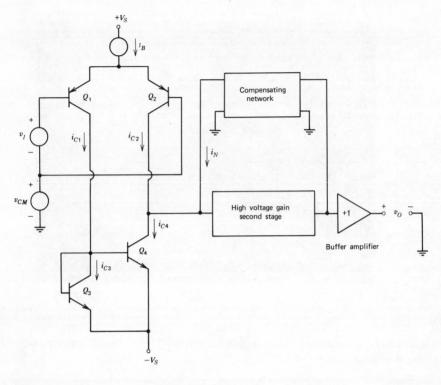

Figure 13.43 Operational-amplifier model.

ditions, is sufficient to determine the open-loop transfer function of the amplifier as a function of the compensating network.

Since the second stage normally operates at current levels large compared to those of the first stage, the limits of linear-region operation are usually determined by the first stage. Note that first-stage currents are related to the total quiescent bias current of this stage, I_B, and the differential input voltage. For the topology shown, the relationship between the relative input-stage collector currents and input voltage becomes highly nonlinear when the differential input voltage v_I exceeds approximately kT/q. At room temperature, for example, a $+25$-mV value for v_I raises the collector current of Q_2 a factor of 1.46 above its quiescent level, while input voltages of 60 mV and 120 mV increase i_{C2} by factors of 1.82 and 1.98, respectively, above the quiescent value.

When a differential input-voltage level in excess of 100 mV is applied to the amplifier, the magnitude of the current $i_{C4} - i_{C2}$ will have nearly its maximum value of I_B. Regardless of how much larger the differential input signal to the amplifier becomes, the current from the input stage and, thus, the current i_N, remains relatively constant.

Even if series emitter resistors are included (as they are in some amplifiers at the expense of drift referred to the input of the amplifier), or if more junctions are connected in the input-signal path as in the LM101A amplifier, the maximum magnitude of the current supplied by the first stage is bounded by its bias level. Since the value of i_N cannot exceed the current supplied by the first stage (at least if the second stage remains linear), the output voltage can not have characteristics that cause i_N to exceed a fixed limit. If, for example, a capacitor with a value C_c is used for compensation,

$$i_N \simeq C_c \dot{v}_O \tag{13.64}$$

where the dot indicates time differentiation. Thus, for the values shown in Fig. 13.43, the maximum magnitude of $\dot{v}_O$ is

$$|\dot{v}_O|_{\max} = \frac{I_B}{C_c} \tag{13.65}$$

One of the more restrictive design interrelationships for a two-stage amplifier is that with single-capacitor compensation and without emitter degeneration in the input stage, both the maximum time rate of change of output voltage and the unity-gain frequency of the amplifier are directly proportional to first-stage bias current. Hence increases in slew rate can only be obtained in conjunction with identical increases in unity-gain frequency. Since stability considerations generally bound the unity-gain frequency, the maximum slew rate is also bounded.

The large-signal performance of the LM301A operational amplifier used in all previous tests is demonstrated using the connection shown in Fig. 13.44. This connection is identical to the inverter used with feedforward compensation, except for the addition of Schottky diodes that function as an input clamp. If clamping were not used, the voltage at the inverting input of the amplifier would become approximately half the magnitude of a step input-voltage change immediately following the step because of direct resistive coupling. This type of transient would add currents to the output of the first stage because of signals fed through the collector-to-base junctions of the input transistors and because of transient changes in current-source levels. The Schottky diodes are used in preference to the usual silicon P-N junction diodes because they have superior dynamic characteristics and because their threshold voltage of approximately 0.3 volt is closer to the minimum value that guarantees complete input-stage current steering for the LM301A.

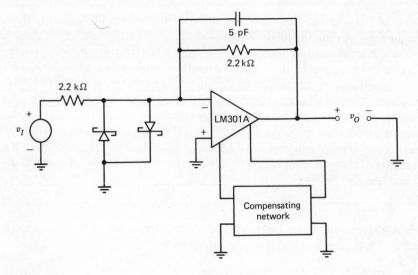

Figure 13.44   Inverter used to evaluate large-signal response.

The square-wave response of the circuit of Fig. 13.44 with a 30-pF compensating capacitor is shown in Fig. 13.45a. The positive and negative slew rates are equal and have a magnitude of approximately 0.85 volt per microsecond. Note that there is no discernible overshoot as the amplifier output voltage reaches final value, indicating that the amplifier with this compensation recovers quickly and cleanly from the overload associated with a 20-volt step input signal.

The transient of Fig. 13.45b is the response of a unity-gain voltage follower, compensated with a 30-pF capacitor, to the same input. In this case, the positive transition has a step change followed by a slope of approximately 0.8 volt per microsecond, while the negative-transition slew rate is somewhat slower. The lack of symmetry reflects additional first-stage currents related to rapidly changing common-mode signals. Figure 13.43 indicates that a common-mode input applied to this type of amplifier forces voltage changes across the collector-to-base capacitances of the input transistors and the current source. The situation for the LM301A is somewhat worse than that depicted in Fig. 13.43 because of the gain provided to bias current source variations by the lateral PNP's used in the input stage (see Section 10.4.1). The nonsymmetrical slewing that results when the amplifier is used differentially is the reason that the inverter connection was selected for the following demonstrations.

An earlier development showed that slew rate is related to input-stage bias current and compensating-capacitor size with single-pole compensa-

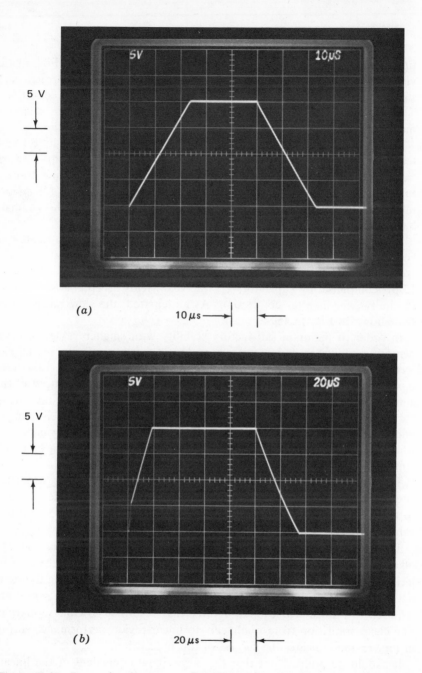

Figure 13.45 Large-signal response of LM301A with a 30-pF compensating capacitor. (Input square-wave amplitude is 20 volts peak-to-peak.) (a) Unity-gain inverter. (b) Unity-gain voltage follower.

637

tion. Solving Eqn. 13.65 for I_B using values associated with Fig. 13.45a yields a bias current of 25.5 μA, with half this current flowing through each side of the input-stage differential connection under quiescent conditions. The transconductance of the input-stage transistors, based on this estimated value of quiescent current, is approximately 5 $\times$ 10^{-4} mho.

Recall that the constant which relates the linear-region open-loop transfer function of the LM301A to the reciprocal of the compensating-network transfer admittance is one-half the transconductance of the input transistors. The value for $g_m/2$ of 2.3 $\times$ 10^{-4} mho determined in Eqn. 13.28 from linear-region measurements is in excellent agreement with the estimate based on slew rate.

Since slew rate with single-pole compensation is inversely related to compensating-capacitor size, one simple way to increase slew rate is to decrease this capacitor size. The transient shown in Fig. 13.46a results with a 15-pF compensating capacitor, a value that yields acceptable stability in the unity-gain inverter connection. As anticipated, the slew rate is twice that shown in Fig. 13.45a.

In order to maintain satisfactory stability with smaller values of compensating capacitor, it is necessary to lower the transmission of the elements surrounding the amplifier. The connection shown in Fig. 13.47 uses input lag compensation to increase the attenuation from the output of the amplifier to its inverting input to approximately a factor of 10 at intermediate and high frequencies. It was shown in Section 13.3.2 that well-damped linear-region performance results with a 4.5-pF compensating capacitor when the network surrounding the amplifier provides this degree of attenuation.

The response of Fig. 13.46b results with a 5-pF compensating capacitor and input lag compensation as shown in Fig. 13.47. The slew rate increases to the value of 5 volts per microsecond predicted by Eqn. 13.65 with this value for C_c.

The large capacitor is used in the lag network to move the two-pole roll-off region that results from lag compensation well below crossover. This location improves recovery from the overload that results during the slewing period because large gain changes (in a describing-function sense) are required to reduce crossover to a value that results in low phase margin. The clean transition from nonlinear- to linear-region performance shown in Fig. 13.46b indicates the success of this precaution.

It should be pointed out that the large-signal equivalent of the linear-region tail associated with lag compensation exists with this connection, although the scale factor used in Fig. 13.46b is not sensitive enough to display this effect. The voltage v_A reaches its clamped value of approximately

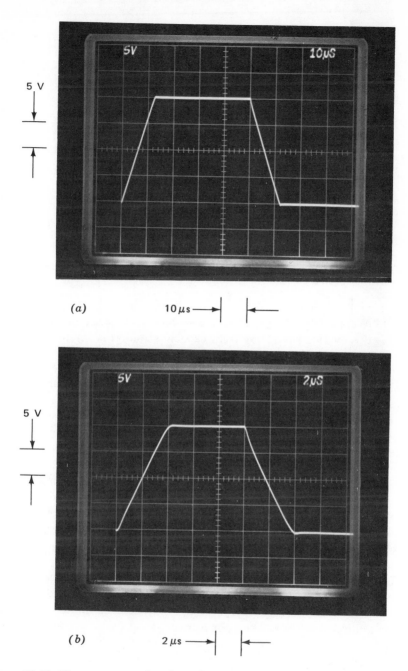

(a)

5 V

10 μs →| |←

(b)

5 V

2 μs →| |←

Figure 13.46 Slew rate as a function of compensating capacitor for unity-gain inverter. (Input square-wave amplitude is 20 volts peak-to-peak.) (a) $C_c = 15$ pF. (b) $C_c = 5$ pF and input lag compensation.

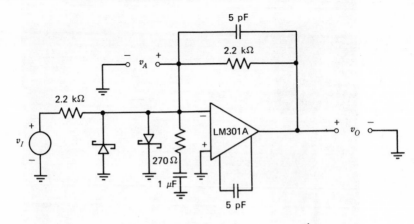

Figure 13.47   Unity-gain inverter with input lag compensation.

0.3 volt for the slewing period of 4 μs following a 20-volt transition at the input. Accordingly, the 1-μF capacitor charges to approximately 4 mV during this overloaded interval. The capacitor voltage is amplified by a factor of 2.2 kΩ/270 $\Omega \simeq 8$, with the result that the output voltage is in error by 32 mV immediately following the transition. The decay time associated with the error is 270 $\Omega \times 1 \mu$F = 270 μs. Note that increasing the lag-network capacitor value decreases the amplitude of the nonlinear tail but increases its duration.

We might suspect that two-pole compensation could improve the slew rate because the network topology shown in Fig. 13.19 has the property that the steady-state value of the current i_N is zero for any magnitude ramp of v_N. The output using a 30 pF–15 kΩ–30 pF two-pole compensating network (the values indicated earlier in Fig. 13.21) with the inverter connection is shown in Fig. 13.48a for a 20-volt peak-to-peak, 50 kHz sine-wave input signal. The maximum slew rate demonstrated in this photograph is approximately 3.1 volts per microsecond, a value approximately twice that obtained with a single 15-pF compensating capacitor.

Unfortunately, the large negative phase shift close to the crossover frequency that results from two-pole compensation proves disastrous when saturation occurs because the system approaches the conditions necessary for conditional stability. The poor recovery from the overload that results with large-signal square-wave excitation is illustrated in Fig. 13.48b. The collector-to-base junctions of the second-stage transistors are forward biased during part of the cycle because of the overshoot, and the resultant charge storage further delays recovery from overload.

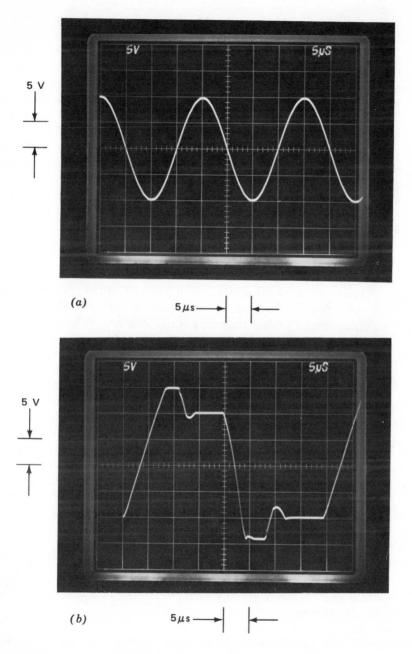

(a)

5μs ——→| |←——

(b)

5μs ——→| |←——

Figure 13.48 Large-signal performance of unity-gain inverter with two-pole compensation. (a) Sine-wave input. (b) Square-wave input.

It is of passing interest to note that the circuit using two-pole compensation exhibits a phenomenon called *jump resonance*. If the frequency of the 20-volt peak-to-peak sinusoid is raised slightly above the 50-kHz value used in Fig. 13.48*a*, the output signal becomes severely distorted. Further increases in excitation frequency result in an abrupt jump to a new mode of limiting with a recognizably different (though equally distorted) output signal. The process exhibits hysteresis, in that it is necessary to lower the excitation frequency measurably below the original jump value to reestablish the first type of nonlinear output signal. One of the few known virtues of jump resonance is that it can serve as the basis for very difficult academic problems in advanced describing-function analysis.[7]

Feedforward compensation results in a high value for slew rate because capacitive feedback around the second stage is eliminated. The response of the inverter with 150-pF feedforward compensation to a 20-volt peak-to-peak, 200-kHz triangle wave is shown in Fig. 13.49*a*. While some distortion is evident in this photograph, the amount is not excessive considering that a slew rate of 8 volts per microsecond is achieved. The square-wave response shown in Fig. 13.49*b* indicates problems similar to those associated with two-pole compensation. The response also indicates that the negative slew rate is substantially faster than the positive slew rate with this compensation. The reason for the nonsymmetry is that with feedforward compensation, the slew rate of the amplifier is limited by the current available to charge the node at the output of the second stage (the collector of Q_{10} in Fig. 10.19). The current to charge this node in a negative direction is derived from Q_{10}, and relatively large currents are possible from this device. Conversely, postive slew rate is established by the relatively lower bias currents available. The way to improve symmetry is to increase the collector bias current of Q_{10}. While this increase could be accomplished with an external current source applied via a compensation terminal, a simpler method is available because of the relationship between the voltage at the collector of Q_{10} and the output voltage of the amplifier. Level shifting in the buffer stage raises the output voltage one diode potential above the collector voltage of Q_{10} in the absence of load. If a resistor is connected between the amplifier output and the compensation terminal at the output of second stage, the resistor will act like a current source because the voltage across it is "bootstrapped" by the buffer amplifier. Furthermore, the level shift in the buffer is of the correct polarity to improve slew-rate symmetry when the resistor is used.

The square-wave response of Fig. 13.49*c* illustrates the performance of the inverter with feedforward compensation when a 1-kΩ resistor is con-

[7] G. J. Thaler and M. P. Pastel, *Analysis and Design of Nonlinear Feedback Control Systems*, McGraw-Hill, New York, 1962, pp. 221–225.

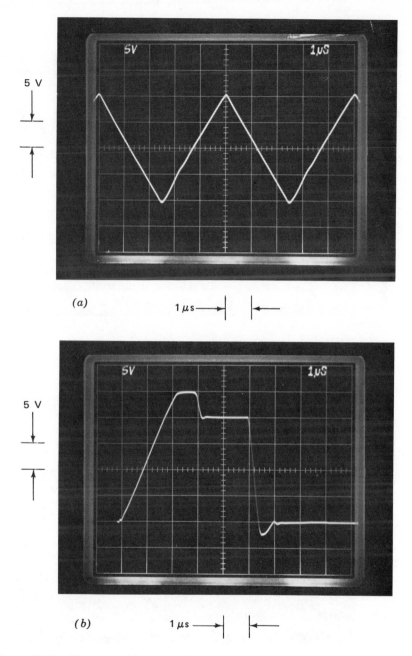

(a)

1 μs

(b)

1 μs

Figure 13.49 Slew rate of inverter with feed-forward compensation. (*a*) Triangle-wave input. (*b*) Square-wave input. (*c*) Square-wave input with increased second-stage bias current. (*d*) Triangle-wave input with increased second-stage current.

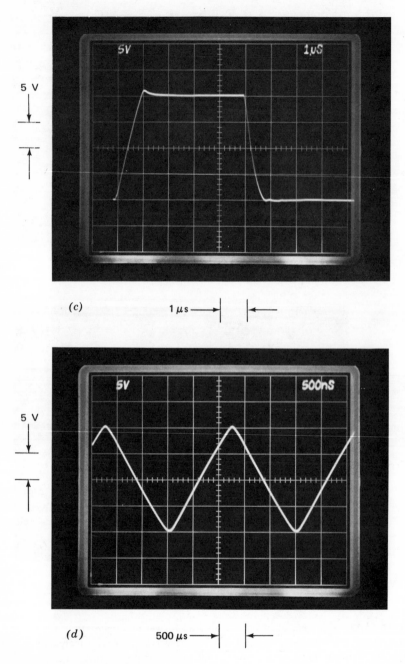

(c)   1 μs →| |←

(d)   500 μs →| |←

Figure 13.49—Continued

644

nected from the amplifier output to the collector side of the second stage. The positive-going slew rate is increased to approximately 20 volts per microsecond. Furthermore, the overload recovery characteristics improve, probably as a result of better second-stage dynamics at higher bias currents. The response to a 400-kHz triangle wave with feedforward compensation and increased bias current is illustrated in Fig. 13.49d. This signal is reasonably free of distortion and has a slew rate of 16 volts per microsecond.

While the method of combining feedforward compensation with increased operating levels is not necessarily recommended for routine use, it does illustrate the flexibility that often accompanies the availability of external compensating terminals. In this case, it is possible to raise the dynamic performance of an inexpensive, general-purpose integrated-circuit amplifier to levels usually associated with more specialized wideband units by means of appropriate connections to the compensating terminals.

13.3.8 Summary

The material presented earlier in this section has given some indication of the power and versatility associated with the use of minor-loop compensation for two-stage operational amplifiers. We should recognize that the relative merits of various forms of open-loop transfer functions remain the same regardless of details specific to a particular feedback system. For example, tachometric feedback is often used around a motor-amplifier combination to form a minor loop included as part of a servomechanism. This type of compensation is entirely analogous to using a minor-loop feedback capacitor for one-pole compensation. Similarly, if a tachometer is followed with a high-pass network, two-pole minor-loop compensation results. It should also be noted that in many cases transfer functions similar to those obtained with minor-loop compensation can be generated via forward-path compensation.

While the compensation networks have been illustrated in connections that use relatively simple major-loop feedback networks, this limitation is unnecessary. There are many sophisticated systems that use operational amplifiers to provide gain and to generate compensating transfer functions for other complex elements. The necessary transfer functions can often be realized either with major-loop feedback around the operational amplifier, or by compensating the amplifier to have an open-loop transfer function of the required form. The former approach results in somewhat more stable transfer functions since it is relatively less influenced by amplifier parameters, while the latter often requires fewer components, particularly when a differential-input connection is necessary.

It is emphasized that a fair amount of experience with a particular amplifier is required to obtain the maximum performance from it in demanding applications. Quantities such as the upper limit to crossover frequency for reliably stable operation and the uncompensated open-loop transfer function are best determined experimentally. Furthermore, many amplifiers have peculiarities that, once understood, can be exploited to enhance performance. The feedforward connection used with the LM101A is an example. Another example is that the performance of certain amplifiers is enhanced when the compensating network (or some portion of the compensation) is connected to the output of the complete amplifier rather than to the output of the high-gain stage because effects of loading by the network are reduced and because more of the amplifier is included inside the minor loop. The time a system designer spends understanding the subtleties of a particular amplifier is well rewarded in terms of the performance that he can obtain from the device.

Important features of the various types of compensation discussed in this section are summarized in Table 13.1. This table indicates the open-loop transfer functions obtained with the different compensations. The solid lines represent regions where the transfer function is controlled by the compensating network, while dotted lines are used when uncompensated amplifier characteristics dominate. The minor-loop feedback networks used to obtain the various transfer functions from two-stage amplifiers are also shown. Comments indicating relative advantages and disadvantages are included.

Table 13.1 Implementation and Effects of Various Types of Compensation

One Pole	
Transfer Function	Network

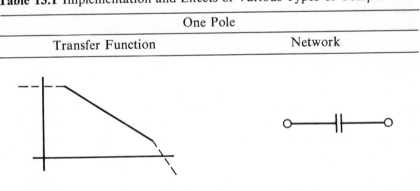

Conservative, general-purpose compensation for systems with frequency-independent feedback and loading. Changing capacitor value optimizes bandwidth as a function of attenuation provided by feedback network. Slew rate inversely proportional to capacitor size.

Table 13.1—(Continued)

Two Pole	
Transfer Function	Network

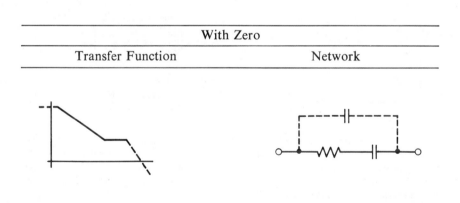

Improved desensitivity and lower error coefficients compared with one-pole systems with identical crossover frequencies. Loop parameters must be selected to insure that crossover occurs in the $1/s$ region of the characteristics for adequate stability. Instability generally results with capacitive loading or low-pass major-loop feedback networks. Poor recovery from overload.

With Zero	
Transfer Function	Network

Zero is used to offset effects of pole associated with load or feedback network, and must be located as a function of this pole. Major loop becomes unstable if pole is eliminated. A small-value capacitor, indicated with dotted lines, improves minor-loop stability.

Table 13.1—(Continued)

Slow Rolloff	
Transfer Function	Network

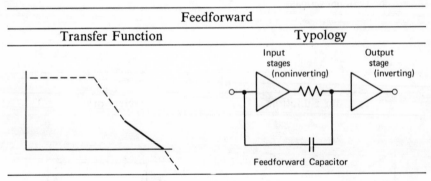

Useful for systems with an additional loop-transmission pole at an uncertain location. Adding more rungs to ladder network increases the range of frequencies over which the additional pole can be located and results in greater uniformity of phase-shift characteristics. Prolonged settling time compared to one-pole compensation when additional loop-transmission pole not present.

Feedforward	
Transfer Function	Typology

Input stages (noninverting)   Output stage (inverting)

Feedforward Capacitor

Highest bandwidth. Most useful when bandwidth of first stage or stages is less than that of rest of amplifier. Can result in substantial slew-rate improvement. Limits amplifier to use in inverting connections only. Values and results critically dependent on specific details of amplifier performance. Sensitive to capacitive loading or other sources of negative loop-transmission phase shift.

PROBLEMS

P13.1

An operational amplifier is available with a fixed, unloaded open-loop transfer function

$$a(s) \simeq \frac{10^5}{10^{-2}s + 1} \cdot$$

This amplifier is to be used as a unity-gain inverter. A load capacitor adds a pole at $s = -10^6$ sec^{-1} to the unloaded open-loop transfer function. Compensate this configuration with an input lead network so that its loop-transmission magnitude is inversely proportional to frequency from low frequencies to a factor of five beyond the crossover frequency. Choose element values to maximize crossover frequency subject to this constraint. You may assume high input impedance for the amplifier.

P13.2

Design an input lag network and an input lead-lag network to compensate the capacitively loaded inverter described in Problem P13.1. Maximize crossover frequency for your designs subject to the constraint that the loop transmission is inversely proportional to frequency over a frequency range that extends from a factor of five below to a factor of five above the crossover frequency.

P13.3

An operational amplifier is connected as shown in Fig. 13.50 in an attempt to obtain a closed-loop transfer function

$$\frac{V_o(s)}{V_i(s)} = -s(0.1s + 1)$$

Determine element values that yield an ideal closed-loop gain given by this expression.

Measurements indicate that the open-loop transfer function of the amplifier is approximately single pole and that the transfer-function magnitude is 10^4 at $\omega = 10^3$ radians per second. Needless to say, the configuration shown in Fig. 13.50 is hopelessly unstable with this amplifier.

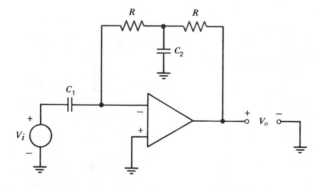

Figure 13.50   Double differentiator.

Find appropriate topological modifications that will stabilize the system (without changing the amplifier) and will result in a closed-loop transfer function that approximates the desired one at frequencies below 100 radians per second.

P13.4

A sample-and-hold circuit is constructed as shown in Fig. 13.51. The unloaded open-loop transfer function of the amplifier is

$$a(s) \simeq \frac{10^6}{(0.1s + 1)(10^{-7}s + 1)}$$

The sum of the open-loop output resistance of the amplifier and the on resistance of the switch is $100 \, \Omega$.

(a) With $R = 0$, is this circuit stable in the sample mode (switch closed)?

(b) Determine a value for R that results in approximately 45° of phase margin in the sample mode.

(c) Estimate the time required for $v_O(t)$ to reach 1 % of final value following initiation of sampling when the value of R determined in part b is used. You may assume that the capacitor is initially discharged, that v_I is time invariant, and that the circuit remains linear during the transient.

P13.5

An externally compensated operational amplifier that uses minor-loop feedback to generate an approximate open-loop transfer function

$$a(s) \simeq \frac{2 \times 10^{-4}}{Y_c(s)}$$

is available. The amplifier is connected as a unity-gain voltage follower. The spectral content of anticipated input signals is such that a closed-loop

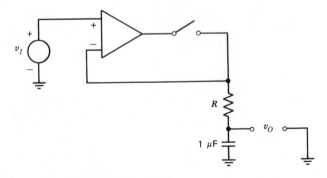

Figure 13.51   Sample-and-hold circuit.

bandwidth in excess of 10^6 radians per second degrades the noise performance of the connection. Determine a compensating element that will result
in a closed-loop transfer function

$$A(s) \simeq \frac{1}{10^{-6}s + 1}$$

for the voltage-follower connection.

P13.6

An operational amplifier of the type described in Problem P13.5 is connected in the log circuit shown in Fig. 13.9a. Experimental evaluation shows
that this connection will be acceptably stable if the loop crossover frequency
is limited to 1 MHz. Determine a compensating element that insures stability for any input-signal level between 0 and $+10$ volts. Estimate the
time required for the incremental output signal of the circuit to settle to
1% of final value when a small step change in input voltage is applied at an
operating point $V_I = 0.1$ volt.

P13.7

A two-stage operational amplifier has a d-c open-loop gain of 10^6 and
is acceptably stable in connections involving frequency-independent feedback provided that compensation is selected which limits the crossover
frequency to 1 MHz. This amplifier is used as a unity-gain inverter to
amplify 10-kHz sinusoids, and a major design objective is to have the
input and the output signals of the inverter exactly $180°$ out of phase.

Discuss the relative merits of one- and two-pole compensation in this
application. Also indicate the effect that the two types of compensation
have on the magnitude of the closed-loop transfer function at 10 kHz.

P13.8

The uncompensated, open-loop transfer function of a two-stage amplifier
is

$$a(s) = \frac{10^5}{(10^{-4}s + 1)(10^{-5}s + 1)(5 \times 10^{-8}s + 1)^2}$$

The two lowest-frequency poles result from dynamics that can be modified
by compensation, while the location of the higher-frequency pole pair is
independent of the compensation that is used.

The amplifier is compensated and connected for a noninverting gain of
10. You may assume that the compensation used does not cause significant
loading of the minor loop. This closed-loop connection is excited with an
input ramp having a slope of 10^4 volts per second. The differential input
signal applied to the amplifier is observed, and it is found that after a
starting transient, the steady-state value of the signal is 10 mV.

(a) Determine a single-pole approximation to the amplifier open-loop transfer function.
(b) Refine your estimate of part a, taking advantage of all the information you have available about the amplifier.
(c) Assuming that this amplifier described is an LM301A, what compensating element is used?
(d) Suggest alternate compensation that results in the same crossover frequency as obtained with the compensation described, a phase margin in excess of 60°, and essentially zero steady-state ramp error. Determine element values that implement the required compensation for an LM301A.

P13.9

The material discussed in connection with Fig. 13.24 indicated that the steady-state error of a closed-loop operational-amplifier connection in response to a ramp can be reduced to insignificant levels by using two-pole compensation. An extension of this line of reasoning implies that if three-pole compensation is used, the steady-state error will be nearly zero for parabolic excitation. Linear-system considerations show that stability is possible if two zeros are combined with a three-pole rolloff. For example, a loop transmission

$$L(s) = - \frac{10^{16}(10^{-5}s + 1)^2}{s^3}$$

has approximately 80° of phase margin at its crossover frequency.

Find a compensating-network topology that can be used in conjunction with minor-loop compensated amplifiers to provide this general type of open-loop transfer function. Discuss practical difficulties you anticipate with this form of transfer function.

P13.10

A two-stage operational amplifier that uses minor-loop compensation is loaded with a capacitor that adds a pole at $s = -10^6$ sec^{-1} to the unloaded open-loop transfer function of the amplifier. The desired open-loop transfer function including loading effects is

$$a(s) \simeq \frac{2 \times 10^{11}(5 \times 10^{-6}s + 1)}{s^2}$$

Find a compensating-network topology that can be used to effect this form of compensation. Determine appropriate element values assuming that the effective input-stage transconductance of the operational amplifier used is 2×10^{-4} mho.

P13.11

A two-stage operational amplifier is connected as an inverting differentiator with a feedback resistor of 100 kΩ and an input capacitor of 1 μF. What type of minor-loop compensating network should be used to stabilize this configuration? Determine element values that result in a predicted crossover frequency of 10^4 radians per second with a value of 2×10^{-4} mho for input-stage transconductance.

When this type of compensation is tried using an LM301A operational amplifier, minor loop stability is unacceptable, and it is necessary to shunt the compensation terminals with a 3-pF capacitor in addition to the network developed above for satisfactory performance. Describe the effect of this modification on closed-loop performance.

P13.12

A certain application necessitates an operational amplifier with an approximate open-loop transfer function

$$a(s) \simeq \frac{10^4}{s^{2/3}}$$

Find a compensating network that can be used in conjunction with an LM301A to approximate this transfer function. The phase shift of the approximating transfer function should be $-60° \pm 5°$ over a frequency range from 1 radian per second to 10^6 radians per second.

INDEX